YOU HAVE SAVED MONEY
PURCHASING THIS
USED BOOK
Spartan Bookstore

JAN 9 1973

SAN JOSE STATE COLLEGE
Official College Bookstore
PRICE $

SEASHORE ANIMALS
of the
PACIFIC COAST

Myrtle Elizabeth Johnson, Ph. D.
and
Harry James Snook, M.S.

DOVER PUBLICATIONS, INC.
NEW YORK

TO
WILLIAM EMERSON RITTER
Naturalist, Teacher, and Friend

Copyright © 1927, 1955 by Myrtle Elizabeth Johnson and Harry James Snook.
All rights reserved under Pan American and International Copyright Conventions.

Published in Canada by General Publishing Company Ltd., 30 Lesmill Road, Don Mills, Toronto, Ontario.
Published in the United Kingdom by Constable and Company, Ltd., 10 Orange Street, London WC 2.

This Dover edition, first published in 1967, is an unabridged and unaltered republication of the work originally published by the Macmillan Company in 1927. The publisher wishes to thank the Oberlin College Library for the use of their copy for reproduction.

Standard Book Number: 486-21819-8
Library of Congress Catalog Card Number: 67-19700

Manufactured in the United States of America
Dover Publications, Inc.
180 Varick Street
New York, N. Y. 10014

PREFACE

This book has grown out of the need felt among naturalists, teachers, and others interested in natural history, for a non-technical, illustrated account of the structure and habits of the common seashore animals of the west coast of the United States.

It is our experience that most people, although they avoid technical scientific literature, are greatly interested in knowing about living animals that they find, and are eager to learn their common names. Accordingly we have gathered material from many sources and have made illustrations of most of the species mentioned, in the hope that the reader, even though he may not be equipped with a zoological vocabulary, may be able to identify most of his collections by the help of the pictures and the brief text, and that he may find here information on habits and life-histories that will be of interest to him.

Such a book as this is, of necessity, largely a compilation of material from many sources. We are indebted to a greater or less extent to all the authors whose works are included in the bibliography. From the publications listed we have secured much data for descriptions of species, though owing to the necessity for brevity and for keeping the text as free as possible from technical wording it has often been impossible to make exact quotations and to note the source of our data. We wish especially to acknowledge the help of Dr. W. E. Ritter, Dr. W. K. Fisher, Dr. Harry B. Torrey, Dr. F. M. MacFarland, Dr. R. S. Bassler, Dr. Waldo L. Schmitt, Dr. Mary J. Rathbun, Mr. Clarence R. Shoemaker, Dr. C. O. Esterly, Dr. C. Essenberg, Dr. Fred Baker, Mr. F. W. Kelsey, Mrs. Kate Stephens, Mr. W. E. Allen, and Dr. Tage Skogsberg, all of whom have made helpful suggestions and have read portions of the manuscript.

To Dr. W. K. Fisher, Director of the Hopkins Marine Station and to Dr. T. C. Frye, Director, and Professor Trevor Kincaid of the Puget Sound Biological Station we express our thanks for the many courtesies extended to us. To Dr. W. E. Ritter and Dr. T. W. Vaughan, Directors of the Scripps Institution, and to Mr. W. C. Crandall, and Mr. P. S. Barnhart of the staff, we express our appreciation for courtesies extended throughout our prolonged stay at the institution.

We extend thanks to the many friends who helped us in our numerous collecting expeditions, in reading manuscript, and in numberless other ways and mention especially Miss Mabel Pierson, Miss Gertrude Pierson, Miss Olive Kelso, Miss Emma Bee Mundy, and Mrs. Snook.

Although most of our figures are original, we are indebted to a number of authors, and others whose names appear with the figures, for permission to use their illustrations.

We wish to add a word of appreciation for the work of the U. S. National Museum, the University of California Press, and the California Fish and Game Commission. The publications of these organizations contain much of the data we have presented here and the work they are continually carrying on is making valuable knowledge of the seashore forms available to students.

And finally, we express our great debt to Miss Ellen Browning Scripps, who, when our work was complete, advanced our share of the cost of publication.

TABLE OF CONTENTS

CHAPTER	PAGE
I THE COLLECTING GROUNDS AND HOW TO COLLECT	1
THE CHARACTER OF THE COAST LINE	1
FACTORS DETERMINING THE DISTRIBUTION OF MARINE ANIMALS	1
THE PUGET SOUND FAUNA	5
THE FAUNA OF THE CENTRAL COAST REGION	6
THE MARINE FAUNA SOUTH OF POINT CONCEPTION	8
THE PELAGIC FAUNA	8
COLLECTING EQUIPMENT	10
II CLASSIFICATION OF ANIMALS	12
THE NECESSITY FOR SCIENTIFIC NAMES	12
VERNACULAR NAMES	13
THE LINNAEAN SYSTEM OF CLASSIFICATION	14
THE BASIS OF CLASSIFICATION	16
A KEY TO THE ARTHROPODA	20
A KEY TO THE DECAPODA	21
A KEY TO THE ANIMAL PHYLA	22
III LIFE IN THE OCEAN	25
A COMPARISON OF SHORE CONDITIONS, THE OPEN OCEAN, AND THE GREAT DEPTHS OF THE OCEAN	25
GLOBIGERINA OOZE	26
TEMPERATURE AND FAUNA OF THE OCEAN DEPTHS	28
PLANKTON	31
"PHOSPHORESCENCE"	32
IV PHYLUM PORIFERA (SPONGES)	34
CLASS CALCAREA	39
CLASS HEXACTINELLIDA	39
CLASS DEMOSPONGIAE	39

CONTENTS

CHAPTER	PAGE

V PHYLUM COELENTERATA (HYDROIDS, JELLYFISH, SEA ANEMONES, CORALS, ETC.).................. 42
    CLASS HYDROZOA, HYDROMEDUSAE (THE HYDROIDS AND MEDUSAE)... 43
    CLASS SCYPHOZOA, SCYPHOMEDUSAE (JELLYFISH).......... 77
    CLASS ANTHOZOA OR ACTINOZOA (SEA ANEMONES AND CORALS)... 85
    CLASS CTENOPHORA................................... 109

VI THE LOWER WORMS................................. 113
    PHYLUM PLATYHELMINTHES (FLATWORMS)................ 114
    CLASS TURBELLARIA................................. 114
    CLASS TREMATODA................................... 119
    CLASS CESTODA..................................... 121
    CLASS NEMERTINEA.................................. 122
    PHYLUM NEMATHELMINTHES (ROUND WORMS)............. 126
    PHYLUM TROCHELMINTHES (WHEEL ANIMALS)............. 127
    SUBPHYLUM CHAETOGNATHA (ARROW WORMS)............. 128

VII PHYLUM MOLLUSCOIDEA (MOSS ANIMALS AND LAMP SHELLS).................................... 130
    CLASS BRYOZOA OR POLYZOA (THE MOSS ANIMALS)........ 130
    CLASS PHORONIDA................................... 151
    CLASS BRACHIOPODA (THE LAMP SHELLS)................ 152

VIII PHYLUM ANNULATA (THE SEGMENTED WORMS). 158
    CLASS CHAETOPODA.................................. 158
    CLASS GEPHYREA.................................... 173

IX PHYLUM ECHINODERMATA (STARFISH, BRITTLE STARS, SEA URCHINS, AND SEA CUCUMBERS).... 180
    CLASS ASTEROIDEA (STARFISH)........................ 181
    CLASS OPHIUROIDEA (BRITTLE STARS).................. 213
    CLASS CRINOIDEA................................... 225
    CLASS ECHINOIDEA (SEA URCHINS)..................... 227
    CLASS HOLOTHURIOIDEA (SEA CUCUMBERS).............. 238

CONTENTS

CHAPTER	PAGE
X PHYLUM ARTHROPODA	249
CLASS CRUSTACEA (LOBSTERS, SHRIMPS, CRABS, AND BARNACLES)	250
SUBCLASS ENTOMOSTRACA	252
SUBCLASS MALACOSTRACA	270
CLASS ARACHNOIDEA (SEA SPIDERS)	402
XI PHYLUM MOLLUSCA (CHITONS, SNAILS, CLAMS, AND DEVIL FISH)	411
CLASS PELECYPODA OR LAMELLIBRANCHIATA (CLAMS AND OYSTERS)	413
CLASS SCAPHOPODA (TOOTH SHELLS)	473
CLASS GASTROPODA (SNAILS AND SEA SLUGS)	475
CLASS AMPHINEURA (CHITONS)	557
CLASS CEPHALOPODA (SQUIDS, OCTOPODS, AND NAUTILUS)	567
XII PHYLUM CHORDATA	582
SUBPHYLUM ENTEROPNEUSTA (BALANOGLOSSIDA)	583
SUBPHYLUM TUNICATA (SALPA AND THE SEA SQUIRTS)	584
SUBPHYLUM LEPTOCARDIA (AMPHIOXUS)	599
APPENDIX: METHODS OF PRESERVING ANIMALS	601
DEVICES FOR KEEPING MARINE ANIMALS ALIVE IN THE LABORATORY	601
COLLECTING EQUIPMENT	602
PRESERVING FLUIDS SUITABLE FOR USE ON MARINE ANIMALS	603
METHODS OF DRYING SPECIMENS	603
BIBLIOGRAPHY	605
GLOSSARY	621
INDEX	633

LIST OF COLOR PLATES

[BETWEEN PAGES 116 AND 117]

FRONTISPIECE.
  FIG. 1.— A marine slug or mollusk, *Flabellina iodinea;* x 2. FIG. 2.—The striped shore crab, *Pachygrapsus crassipes;* x ⅔. FIG. 3.—A mollusk, *Galvina olivacea;* x 7. FIG. 4.—A mollusk, *Hermissenda crassicornis;* x 6 (small specimen). FIG. 5.—*Ciona intestinalis,* a cosmopolitan "sea squirt" or ascidian found on floats and piling in harbors; about ⅔ natural size.

PLATE I.
  FIG. 1.—*Velella lata,* from a color sketch of a small, living specimen; x 2. FIG. 2.— *Chrysaora gilberti,* a jellyfish of the California coast; x ½.

PLATE II.
  The jellyfish, *Pelagia* sp.; x ⅓. Although jellyfish like this are large and heavy, the body tissues contain less than five per cent of solid substance.

PLATE III.
  FIG. 1.—A burrowing anemone, *Edwardsiella californica.* Color drawing from life; x 4/3. FIG. 2.—An anemone that lives with its column buried in the sand and its tentacles and disk at the surface of the sand. Color sketch from living specimen, fully expanded, slightly enlarged. FIG. 3.—*Corynactis* sp., a sea anemone with capitate tentacles. One specimen expanded, the other contracted. About natural size. FIG. 4.—*Balanophyllia elegans,* a coral that is abundant alongshore at Pacific Grove. Drawn from a living specimen in an aquarium. About natural size. FIG. 5.—The striped sea anemone, *Sagartia* sp.; drawn from a living specimen fully expanded; x 10.

PLATE IV.
  FIG. 1.—The sea pansy, *Renilla amethystina,* a large colony with rachis and zoöids expanded, viewed from above. The peduncle, when buried in the sand, serves as an anchor. The zoöids extend above the surface of the sand when the tide is in and the colony is expanded. (About natural size.) FIG. 2.—*Cribrina xanthogrammica.* A small individual, about natural size. (From a color sketch by C. E. von Geldern.) FIG. 3.—A flatworm, *Planocera burchami;* x 3. FIG. 4.—A flatworm from Puget Sound; x 2 (a small specimen).

PLATE V. FLATWORMS.
  FIG. 1.—*Thysanozoon* sp. Drawn from a specimen taken at La Jolla, California; x 2. FIG. 2.—*Prosthiostomum* (?) a southern California species; x 3. FIG. 3.— *Eurylepta aurantiaca;* x 5. FIG. 4.—The striped flatworm. Drawn from a specimen found at La Jolla; x 2.

PLATE VI. NEMERTEAN WORMS.
  FIG. 1.—*Emplectonema gracile;* x 3. FIG. 2.—*Paranemertes peregrina;* x 1. FIG. 3.—*Amphiporus bimaculatus;* x 1⅓. FIG. 4.—*Lineus pictifrons;* x 2.

LIST OF COLOR PLATES

PLATE VII.

FIG. 1.—*Eudistylia polymorpha*. The gills of the worm project from the canals of a sponge in which the worm has made its tube; x ⅔. FIG. 2.—*Serpula columbiana*. The brightly colored gills project from the limy tube in which the worm lives; x 1⅓. FIG. 3.—*Cucumaria curata*. This sea cucumber broods the eggs under the ventral surface; x 1⅔. FIG. 4.—*Thyonepsolus nutrians*, a sea cucumber that carries the young upon its back. The structures shown on the back of this specimen are tube feet. (Small specimen, x 3⅓.) FIG. 5.—A sea slug, *Tritonia festiva*; x 2⅔.

PLATE VIII. NUDIBRANCH MOLLUSKS OR SEA SLUGS.

FIG. 1.—*Navanax inermis*, a sea slug often found in muddy bays as well as in rocky tide pools; x ⅔. FIG. 2.—*Archidoris montereyensis*; x 1⅓. FIG. 3.—*Anisodoris nobilis*; x ⅔ (small specimen). FIG. 4.—*Rostanga pulchra*. Most specimens show much more red than this specimen did; x 6. FIG. 5.—*Aldisa sanguinea*; x 2⅔.

PLATE IX. NUDIBRANCH MOLLUSKS OR SEA SLUGS.

FIG. 1.—*Cadlina marginata*; x 1⅓. FIG. 2.—*Cadlina flavomaculata*; x 6. FIG. 3.—*Laila cockerelli*; x 2. FIG. 4.—*Chromodoris californiensis*; x ⅔.

PLATE X. NUDIBRANCH MOLLUSKS OR SEA SLUGS.

FIG. 1.—*Triopha carpenteri*; x 4/3. FIG. 2.—*Triopha maculata*, a small specimen x 4. FIG. 3.—*Triopha maculata*, a large specimen showing bluish white spots not seen in young specimens; x 2. FIG. 4.—*Polycera atra*; x 3.

PLATE XI. NUDIBRANCH MOLLUSKS OR SEA SLUGS.

FIG. 1.—*Acanthodoris rhodoceras*; x 6. FIG. 2.—*Chromodoris macfarlandi*; x 1. FIG. 3.—*Ancula pacifica*; x 5. FIG. 4.—*Hopkinsia rosacea*; small specimen; x 2. FIG. 5.—*Doriopsis fulva*; x 4/3.

CHAPTER I

THE COLLECTING GROUNDS AND HOW TO COLLECT

The surf of the Pacific, on our western coast, beats mainly against a bold line of cliffs broken here and there by stretches of sandy beach. Low tide in some cases exposes a narrow sandy strand at the base of the rocky precipices, but in many instances the low water uncovers only more rocks and rugged reefs.

A glance at a map will show that from the Mexican border to Cape Flattery the coast is relatively straight, with but few bays or other indentations to break the force of the waves, while north of that point, from Puget Sound to Bering Sea, the coast line becomes very irregular and enormously lengthened by countless inlets and islands. In this region a great variety of conditions exists even in contiguous localities.

Marine animals are limited in their distribution by many factors. For example, they are affected by the temperature of the water, its salinity, the velocity of currents, the nature of the rock surface if they are fixed forms, or by its hardness if they are burrowing creatures. The presence or absence of the proper food greatly influences the range of a species as well as its abundance. One would not look for mud-eating species on a rock bound coast, but rather in sheltered lagoons where mud has a chance to accumulate. The way in which an animal escapes its enemies may determine the kind of place in which it will be found. The little flat crabs, *Petrolisthes*, which are just fitted to hide in cracks and crevices would not be at home on a sandy shore. If one

FIG. 1.—Mud flats of a shallow bay near San Diego, known as Mission or False Bay. The shore has a gradual slope so that there is a long stretch of mud exposed at low tide. On these mud flats, one finds *Corymorpha*, crabs, burrowing sea anemones, clams, worms, many species of snail-like mollusks, and the burrowing echinoderm, *Synapta*. Far up on the shore near high-tide line are the burrows of fiddler crabs.

FIG. 2.—A rocky shore a few miles north of Carmel. A spot like this is an excellent collecting ground at low tide: here one would expect to find sea urchins, abalones, sea stars, sea anemones, crabs, and many kinds of mollusks.

COLLECTING GROUNDS

wishes to find them, he has only to visit a shore where rocks abound, to turn these rocks over, and thus take the animals by surprise.

Sand is easily penetrated by burrowing creatures so that a sandy beach is usually much more thickly populated than it may appear to be. From South America to Oregon the wave-washed sand is the home of innumerable sand crabs and clams. High up on shore where seaweed accumulates, amphipods or beach hoppers are often to be seen jumping about or digging shallow burrows in the sand. These bunches of seaweed often yield interesting finds of tiny bryozoans, brittle stars, or more rarely, brilliant nudibranchs and sea anemones. Such a beach is also usually strewn with shells, worn smooth by wave action. Among them may be the hard parts of animals that have lived in the sand or on the rocks far out where the water is deep and have been washed ashore. At times, floating organisms are stranded in great numbers; jellyfish and the barrel-shaped, transparent *Salpa; Velella*, a fairy boat with a tiny triangular sail; or the purple sea-going snail *Janthina*, which sometimes is driven in in such numbers that the beach appears blue.

One who delights in a multitude of forms or who loves to feast his eyes upon the colors which nature has evolved in living organisms, will find the reefs and rocky beaches a more fertile field than the sandy patches of the shore. Here the crevices in the rocks and the mat of seaweed that covers much of the rocky surface afford shelter for a great diversity of forms while the hard rocks give firm places of attachment for the animals that lead a sedentary life. Moreover, the outgoing tide leaves pools in little rocky basins which become traps to catch the free-swimming kinds, such as the shrimps. These rock bound pools are wonderful natural aquaria where the animals can be seen in serene possession of their own familiar neighborhood and all that goes with it. Even sand-loving species may be present on such a beach in the sandy patches between the rocks.

FIG. 3.—The marine station of the University of Washington at Friday Harbor on San Juan Island in Puget Sound. This picture was taken in 1921. The station is now housed in new buildings.

COLLECTING GROUNDS

We may roughly divide our west coast into three regions, the northern, central, and southern, the northern region extending from Alaska to Oregon, the central from Oregon to Point Conception, and the southern from Point Conception to Mexico. Each of the regions is characterized in the main by certain animals. Indeed, if you should list the forms that you have found in the course of a day's collecting, a student of marine zoology could tell you just about where you did your collecting. Many species are found from Alaska to Mexico, close along shore, others are found on shore in Alaska but in deep water off our southern coast, while still others are strictly limited to the northern, central, or southern regions.

The Puget Sound region offers a wonderful collecting ground. The inlets and islands furnish, in turn, rocky, sandy, and muddy beaches. At Friday Harbor, on San Juan Island, where the University of Washington maintains a biological station and summer school, one sees an enormous variety of sea life. One of the most characteristic forms in the tide pools is the large red sea cucumber, *Cucumaria miniata*, which lies along the bottom of the pool with its branched tentacles extended so that they resemble a large starry flower. There are hosts of the spiny green sea urchins, *Strongylocentrotus dröbachiensis*, and the flower-like sea anemone, *Metridium dianthus*. Here and there, the serpulid worms thrust out little bouquets of bright-colored tentacles from the openings of their tubes, only to hastily withdraw them when our shadow falls across them. On the rocks are numerous mollusks such as *Thais*, or clinging still more closely to their support, are limpets and chitons in considerable variety. The large whelk *Argobuccinium* is often seen, perhaps in the act of depositing eggs which look like gelatinous kernels of corn standing on end in a spirally arranged group each firmly attached to the seaweed-covered rocks. Now and then a crab, probably *Cancer oregonensis*, ventures out from its hiding place. To see these more timid forms, one has but to turn over the rocks that lie in the pools at low tide. Great numbers of the flat crabs, *Petrolisthes*, are uncovered, probably also the orange-red ascidian, *Styel stimpsoni*, or the crab-like *Hapalogaster mertensii* which has an abdomen like a soft sack. As is the case in the more southern regions, the rocks are covered with sharp, jagged barnacles which may puncture the rubber boots of the unwary collector.

In the central region, the shore of Tomales Bay offers an interesting collecting ground. Low tide exposes a large area of mud flat which harbors a multitude of creatures. The large "rubber neck" clam or gaper, *Schizothaerus nuttallii*, is a most conspicuous inhabitant since it shoots jets of water a foot or more above the surface of mud whenever an intruder invades the flats. The clams live 2-4 feet below the surface, but thrust their long siphons up to the water above in order to obtain food and air. Smaller clams of several kinds are abundant and are preyed upon by the large pinkish starfishes, *Pisaster brevispinus*, which are often numerous in the shallow pools. The large snail, *Polinices lewisii*, is another predatory species and will be found plowing along through the mud. In some parts of the flat the clam digger encounters sand-encrusted, gelatinous worm tubes standing like miniature organ pipes, massed together just below the surface of the mud. A short journey across the sand spit from the bay brings one to the ocean shore where the life is much like that at Pacific Grove.

Monterey Bay and Pacific Grove have been called the conchologists' paradise. Other students beside conchologists may well consider the region a paradise for the variety and beauty of the shore life is remarkable. Stanford University maintains the Hopkins Marine Laboratory at Pacific Grove with college classes during the summer and a small corps of resident workers throughout the year.

On the rocks in this region one finds great numbers of bright colored starfish and in certain localities, the sea

FIG. 4.—The Hopkins Marine Station at Pacific Grove, maintained by Stanford University (Photograph by Dr. W. K. Fisher.)

urchins form a veritable carpet in the shallow tide pools. Abalones abound and crabs scuttle to find new hiding places whenever a rock is turned over. Many of the rocks uncovered at low tide seem to bear splashes of bright red paint. On close examination these patches of red color generally prove to be encrusting sponges and one often finds on the sponge, bright red nudibranchs or slug-like creatures which match the sponge perfectly and apparently find it a most satisfactory place in which to live. Bright colors are all around us, red sea anemones, some of them six inches or more across, and tiny orange-red corals. Nudibranch mollusks, slug-like and colored often in brilliant hues of red, purple, or yellow, abound on the seaweeds of the tide pools. The big twenty-rayed starfish, a soft and oozy-looking member of the family is frequently seen here.

One is often told that the sea life south of Point Conception is less abundant than it is farther north. This may be true of the more conspicuous forms that greet one as he walks along the shore, but a search under rocks, in tide pools, or on wharf piles always reveals small shore forms in great variety and numbers. The Scripps Institution of Oceanography, a department of the University of California, is situated in this southern region and near it one finds many interesting forms. The California spiny lobster, common in the markets, is caught along shore from Point Conception south into Mexican waters. The sea pansy, *Renilla*, found on the sandy mud flats is a beautiful creature and always provokes many questions as to "how it works" when it is shown to groups of students. Another southerner is *Corymorpha* whose delicate curling tentacles and graceful fairy form call forth admiration. One of our starfishes, *Linckia*, seems to be preyed upon often by its enemies. A perfect individual of this species is seldom found, the rays in most specimens all showing various stages of regeneration.

Jellyfish and "floating cartridge-belts" or chains of salpae are often noted by sea-going travelers. The jellyfish are

FIG. 5.—The Scripps Institution of Oceanography of the University of California, located near La Jolla, California. The larger building houses the library and the museum, the smaller one the laboratory rooms, and the cottages in the foreground are a few of the number built for the use of those who are doing work at the institution.

especially conspicuous at night since most of them are luminescent. The appearance of these shining argonauts is somewhat sporadic and they are borne along more or less by the tidal currents as well as by their own activities. Jellyfish of many kinds are especially abundant at the Friday Harbor station in Puget Sound. One can stand on the station float and see a never-ending stream of these dainty creatures go by, borne along by the tidal current.

A few suggestions as to collecting equipment; old clothes and rubber boots make a good combination though some prefer to don bathing suit and tennis shoes. Most women collectors discard skirts in favor of breeches. For storing the catch, a pail and glass jar or bottle are useful. The bottle will house the smaller, choice specimens that are apt to be lost or eaten if they are put into the pail with the common herd.

It is well if the collector has no desire to carry home much of his catch since the animals soon die and lose their attractive appearance when removed from the cool, well-aerated water of the seashore. One often finds more enjoyment in putting the creatures he has caught into the tide pools and studying their behavior there.

If the animals are taken away, they may survive a few days if only a small number of them occupy the container and if the sea water in the dish is renewed frequently or kept well aerated. In aerating large aquaria, the air is usually carried in by means of pumps or through a special arrangement of the water supply tubes, but it is easy to aerate water in a small aquarium by pouring it from one dish to another or by driving in air by means of a pipette. The water can be kept clear by filtering it through filter paper and washing the dishes frequently. In this way one can keep *Corymorpha*, for example, in good condition in a tumbler of sea water for a week or more. The aerating should be done at least once a day and the filtering and dish washing as often as necessary to keep the little aquarium clear, probably every other day being sufficient. It must be remembered, however, that it is almost impossible to keep some species in aquaria at all, while others, such as sea anemones and certain crabs can be kept indefinitely. Small individuals of most species live longer in captivity than large ones do. Without special equipment, it is hard to keep large starfish, devil fish, and sea urchins alive over a day but small ones are easily kept for two weeks or more.

If one is collecting on the rocks or from wharf piles, a heavy case knife for prying off mollusks is a necessity, for detaching abalones a small crowbar is essential, while on sandy or muddy beaches a long, narrow spade is useful.

The collector soon learns to read the signs along the shore. The tiny holes in the sand, mounds, or tracks all lead one to find new creatures together with interesting items as to their life-histories and habits. The rocks uncovered at low tide may seem to be devoid of active animal life but if you turn over some of them, numbers of animals will usually be found hiding there or crawling over the sheltered surface.

One is apt to find the greatest variety of material when there are extremely low or minus tides, but on many beaches one can see much even at a $+1.0$ foot tide. The tide tables are frequently given in the local papers of the coast towns or a list of the tides for the year may often be secured at stores selling nautical supplies or sporting goods. Collecting from floats has the great advantage of being much the same at any tide.

Finally, the collector must remember that the fauna is more or less changeable from year to year. Certain creatures can be counted upon to be present on a given beach, others may usually be present, while still others may be found there once and then be absent for months or years together. The uncertainty as to what they may find lends zest to the collectors as they sally forth in old clothes and rubber boots, armed with spades, pails, and a never failing enthusiasm and interest in the lowly inhabitants along shore.

CHAPTER II

CLASSIFICATION OF ANIMALS

People often ask, "Why do scientists use such formidable names for creatures that have names already?" In the first place the names are neither so long nor so difficult, on the average, as beginners are apt to imagine. Some, to be sure, are involved but most of them are little longer than the English names and they are pronounced according to the Latin rules for accent. Usually they appear difficult merely because they are strange.

Scientific names are not given to animals for the purpose of mystifying the public, as some seem to believe, but for the sake of accuracy. Between 500,000 and 600,000 kinds of animals are now known and large portions of the earth's surface are very inadequately explored by the zoologist. The use of the microscope is constantly enabling us to get acquainted with new species and many think that at least a million kinds of animals are now living on earth. But science is also interested in animals that have lived in the past and are now extinct and this means the addition of an enormous and constantly growing number of different forms. Obviously, there must be some means of listing and cataloging this immense body of knowledge. Imagine a library of half a million or more books of all kinds and descriptions without any scheme of classification. How long it would take to find the book you desire and how difficult to tell another person where to find it! The books in the library, moreover, would be assembled in one place while animals and plants are found scattered in every latitude, longitude, altitude, and depth, on land or sea.

CLASSIFICATION OF ANIMALS 13

For another reason the common or vernacular name does not suffice. We sometimes find the same name in use for two or more different animals. For example, our robin is the namesake of a European bird not closely related to it. In the western part of the United States, *gopher* refers to a small rodent, in Georgia the name is commonly used for a burrowing tortoise. In California, the word *lobster* is usually applied to a crustacean belonging to an entirely different family from the animal originally so named. Some call it the crayfish or crawfish, terms which pertain more particularly to fresh water crustaceans allied to the true lobster. In this case the use of *lobster* and *crayfish* expresses a popularly recognized relationship between the eastern and western animals but it is confusing, nevertheless. Consequently, in the attempt to be more definite, our species is called the spiny lobster, rock lobster, Pacific coast lobster, California lobster, salt water crayfish, Pacific coast crayfish, and doubtless other terms are used. Some of these apply with equal force to related species in other parts of the world. A very similar form exists in European waters. The French call it the *langouste*, the Germans name it *Languste*, the English know it as the rock lobster and the ancient Romans referred to it as *locusta*.

After carefully defining the peculiarities of the species and after a description of the common features of the spiny lobsters of the world, the form found on the California coast has been named *Panulirus interruptus*, *Panulirus* referring to all spiny lobsters without a median spine on the front margin of the carapace (the hard shell-like covering of the front part of the body), and *interruptus* to the particular kind. This name it retains in modern scientific literature whether written in English, French, Japanese, or any other language. In this way it is distinguished from the Hawaiian and Japanese *Panulirus japanocum* and the somewhat more distant European relative, *Palinurus vulgaris*, as well as from a host of similar forms in other parts of the world.

Many of the smaller or less common animals that are of great importance to mankind have no vernacular names. Comparatively few of the parasitic worms related to the hookworms and tapeworms have an English name, but their importance is so great that many volumes have been written about their structure, habits, and the harm they do, and much has been done to find out how they may be prevented from entering the body and how they may be destroyed after they have gained an entrance. It has been found that slight variations in structure may be accompanied by great differences in habits and in effect upon the host. Without a methodical scheme of classification it would be impossible for students to make clear what particular creature they are discussing when they announce their discoveries.

It should be kept in mind, however, that the technical nomenclature does not bar the use of the more familiar terms. If the animal to which reference is made has an English name it may be given with the scientific one and we have followed that practice throughout the following pages whenever we could find a common name in use.

There are two main types of classification which may be designated as the artificial and the natural. In the first, plants and animals are grouped according to superficial resemblances as an aid to memory. At present every effort is being made to group them according to their genetic relationships, *i.e.*, their actual kinship. While uniformity has not been completely obtained in this way and systematic zoologists are often at variance with respect to the grouping of species, much has been accomplished. The attempt is not to form an arbitrary system but to express the relationships found in nature. The present method, known as the binomial nomenclature, is based upon the work of the Swedish naturalist, Linné, better known by his Latin name, Linnaeus, and is a little older than the United States. His great work, the *Systema Naturae*, passed through many revisions until, in the tenth edition (1758), 4236 species were named and tersely described in Latin. Some of the animals found on the Pacific coast were known to him. Linnaeus did not use the term *family* but grouped species into classes and orders. Since his time, the general scheme of classification has been greatly expanded and of late years international rules have been adopted to govern the naming of animals and plants. The names of the genus and species are still employed to designate a particular kind and both are expressed. The name of the genus is a noun and refers to a group of similar forms, *e.g.*, *Panulirus*. It always begins with a capital letter. When used twice in the same paragraph in connection with the specific name it may be abbreviated, the initial letter standing for the whole, providing no confusion results. The specific name is an adjective or a modifying noun and is never capitalized even when derived from a proper noun, hence *Panulirus interruptus*, *P. argus*, and *Astropecten californicus*, *Sabellaria californica*, *Murex carpenteri*. The latter may be translated Carpenter's Murex. Owing to the wide distribution of zoologists and the number of publications in all languages, it sometimes happens that two or more workers name the same species. In that case it is the rule that the name first published with a description is adopted, the other remaining a synonym. Consequently, it is customary to give the source of authority in the matter and we have *Panulirus interruptus* (Randall), and *Murex carpenteri* Dall. The use of the parenthesis in the first case indicates that Randall was the first to describe the species but that he did not assign it to the genus in which it is now placed. It was *Palinurus interruptus* Randall. Another way of showing a change in the generic name is to give the authority followed in making the change, thus: *Ophiopholis aculeata* (Linnaeus) Gray. In many late works this rather cumbersome but lucid custom is followed. When the discovery of an early publication changes a long established name, as recently happened when an old work of Weber (1795) was found that antedated that of Fabricius on crustacea (1798), long accepted as having priority, the well known but incorrect name can be given parenthetically, *e.g.*, *Crago* (*Crangon*) *franciscorum* (Stimpson).

The basis of classification lies in similarities of structure more or less fundamental in significance. In the first place all animals as well as all plants are composed of units known

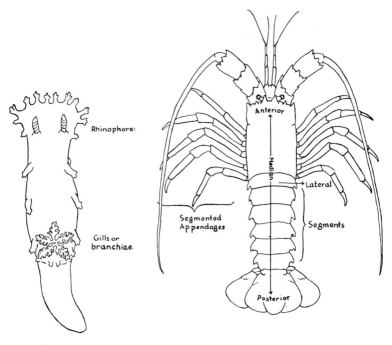

FIG. 6.—A nudibranch mollusk, showing unsegmented body.
FIG. 7.—Spiny lobster, showing segmented appendages and bilateral symmetry.

as cells. The most fundamental difference is between animals of one cell, *Protozoa*, and those composed of many interdependent cells, the *Metazoa*. The latter group is divided by a number of important characters into several divisions called phyla. The arrangement of the cells cannot be seen without the help of a powerful microscope, but is of large importance. Cells are usually arranged in the body

CLASSIFICATION OF ANIMALS

in definite layers, some animals being built up of two and others of three layers of cells. Again, the body may be segmented or non-segmented (figs. 6 and 7). The organs of the body may be arranged radially (fig. 8) around a central axis (radial symmetry) or the equivalent parts may lie bilaterally (figs. 6 and 7) on the two sides of a median line (bilateral symmetry). Some groups have a sack-like body with a single cavity; others have an inner tube for digestion with a fluid-filled space around it, the latter called the body cavity.

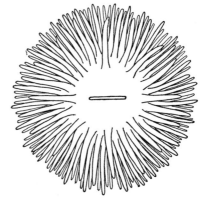

FIG. 8.—Sea anemone, viewed from above, showing radial symmetry.

Other differences are to be found in the form of the appendages, the presence or absence of a skeleton, the structure of the skeleton, whether an outer exoskeleton, as in the spiny lobster, or an internal endoskeleton, as in the chordates.

It is not difficult to understand the idea underlying the classification of animals and plants if one keeps in mind the methods by which other things are classified. Perhaps the best example for our purpose is the system used in addressing a letter. Let us begin as far away from home as possible. Suppose you were in Europe, Asia, or Australia, how would you communicate with a friend in the United States? There are many million people in the world but you wish your letter to be delivered to but one of them. How is this one to be picked out? It seems easy enough, having sealed your letter and placed a stamp on it you write something like this:

John Smith,
430 E. Vine St., Stockton,
California, U. S. A.

18 SEASHORE ANIMALS OF THE PACIFIC COAST

If you were very particular, the name of the county might be included between the names of the city and the state but probably you would consider this addition unnecessary. Let us turn this address about, writing out some of the abbreviations.

Continent	America (North)
Country	United States
State	California
County	San Joaquin
City	Stockton
Street	East Vine
Residence	No. 430
Family	Smith
Individual	John

It is evident that, for the most part, this is a grouping on the basis of geography. The mind is carried at each step from the large division to one of the subdivisions composing it, a process of sorting out, passing always from the more general to the more particular. The same mental processes based on structural affinities instead of geographical divisions are used to distinguish plants and animals and we may compare the way the post office catalogs John Smith with reference to geographic location with the scientific method of locating the spiny lobster. However, since the structural groups are not so familiar as the geographic ones it may be helpful to add a brief characterization of each.

Organic World
(Including all living things)

    Animal Kingdom
    (In distinction to plant kingdom)

        Subkingdom—*Metazoa*
        (Animals composed of organizations of cells, each kind with a special function)

CLASSIFICATION OF ANIMALS 19

Phylum—*Arthropoda*
(Metazoa with bilateral symmetry, bodies and appendages segmented)

  Class—*Crustacea*
  (Arthropods with gills and 2 pairs of antennae)

    Subclass—*Malacostraca*
    (Crustaceans with abdominal appendages)

      Order—*Decapoda*
      (Ten-footed Malacostraca)

        Suborder—*Reptantia*
        (Decapoda with bodies flattened or partly so)

          Tribe—*Palinura*
          (Reptantia with carapace fused at sides to epistome, 3rd legs like first)

            Family—*Palinuridae*
            (Palinura with cephalothorax sub-cylindrical none of legs chelate)

              Genus—*Panulirus*
              (Palinuridae without rostrum but with long antennae)

                Species—*interruptus*
                (A form of Panulirus having all of following features; transverse furrows on abdominal segments, 4 teeth on 1st antennal segment, carapace spinose anteriorly).

As in the case of the address quoted above, it is seldom necessary to include the names of all the groups. In practice, all but the genus and species are omitted but the text should tell whether reference is to an animal or plant and should indicate some of the major groups to which it belongs. Where greater definition is needed, the addition of superfamilies, subfamilies, subgenera, subspecies, and varieties makes the system more flexible.

Another way of expressing relationships is given below:

PHYLUM	CLASS	SUBCLASS	ORDERS
ARTHROPODA (Jointed-limbed animals)	CRUSTACEA	ENTOMOSTRACA	PHYLLOPODA* COPEPODA OSTRACODA CIRRIPEDIA
		MALACOSTRACA	AMPHIPODA ISOPODA SCHIZOPODA STOMATOPODA CUMACEA* DECAPODA
	ARACHNOIDEA (Spiders, mites, horse-shoe crabs)		XIPHOSURA* (Horse-shoe crabs) ARACHNIDA (Spiders, etc.)

[Some include the pycnogonids here]

| | TRACHEATA* (Insects, cenpedes) | | ONYCHOPHORA* (Peripatus) MYRIAPODA* (Centipedes, etc.) INSECTA* (Insects) |

---

*Outside the scope of this work.

CLASSIFICATION OF ANIMALS

To carry one order somewhat further:

ORDER — SUBORDER — TRIBE

DECAPODA (Decapods, or 10-footed crustacea)

- NATANTIA (Shrimps, prawns. Body compressed, etc.)
  - STENOPIDES* (Natantia with 3rd legs chelate and stouter than first). Shrimps.
  - PENEIDES* (Natantia with 3rd legs chelate, not stouter than first). Shrimps or prawns.
  - CARIDES (Natantia with 3rd legs not chelate) Shrimps.
- REPTANTIA (Crabs, lobsters, spiny lobsters, hermit crabs, etc. Body not compressed)
  - PALINURA (Reptantia with 3rd legs like first whether chelate or not, rostrum small or absent). Spiny lobsters.
  - ASTACURA* (Reptantia with 3rd legs like first, rostrum of good size). Lobsters and crayfish.
  - ANOMURA (Reptantia with 3rd legs unlike first, never chelate, carapace not fused with episome. Abdomen reduced.) Sand crabs, hermit crabs, etc.
  - BRACHYURA (Reptantia with 3rd legs unlike first, never chelate, carapace fused with episome. Abdomen not only reduced but folded under body.) True crabs.

To bring this system to completion through the families, subfamilies, genera and species would require an enormous expansion. Note that the characters distinguishing the groups are less fundamental as we pass from the larger to the smaller divisions. This affords the opportunity for keys.

In this book the use of keys has been reduced to a minimum because it would necessitate the employment of too many technical terms in order to be sufficiently concise, but the

*Not included in the scope of this work.

specialist would make little progress without them as they make for accuracy and brevity. Now that we have the fundamentals of classification before us, it would be well to learn how to use a key. There are usually two alternatives. Choose the one that fits the specimen and then pass on to the set of alternatives subordinate to it. The alternatives are commonly listed in such a way as to show their relative importance. In the key below, if A does not fit the case, pass to AA and then determine the applicability of a_1, or a_2, if a_1 applies choose between b_1 and b_2, etcetera. Where more than two items are given equal importance in the scheme, it will be shown by the similar letters that stand before them.

KEY YO THE PHYLA OF ANIMALS

(In the main, this is an adaptation of the familiar system of Parker and Haswell*)

A. Unicellular animals, either free-living or loosely associated in colonies..*Protozoa.*
AA. Multicellular animals with definite organs.
  a_1 Body radially symmetrical.
    b_1 Digestive cavity not distinct from body cavity.
      c_1 Tentacles absent, numerous pores lead to a central cavity, skeletal fibres or spicules present and frequently visible externally. Animals attached and often forming complex colonies in which the fundamental radial structure may be obscured.................*Porifera.*
      c_2 Tentacles present, surrounding a single opening (mouth) leading to a central cavity. Body soft or gelatinous, some secrete hard protective skeletons. Free-swimming or attached, often colonial.....*Coelenterata.*
    b_2 Digestive tract distinct from body cavity.
      c_1 Tentacles present around the mouth and movable spines absent, or movable spines present and tentacles absent. Skeleton calcareous, composed of united plates or separately embedded spicules. Soft retractile structures (tube feet) often present and connected with

---

*T. J. Parker and W. A. Haswell—A Textbook of Zoology, Vol. 1; The Macmillan Company.

CLASSIFICATION OF ANIMALS 23

$\qquad$ internal hydraulic tubes (the water-vascular system)..............................Echinodermata.
a_2 Body bilaterally symmetrical.
$\quad b_1$ Without notochord, internal skeleton, or dorsal nerve cord.
$\quad\quad c_1$ Body unsegmented externally (except in tapeworms).
$\quad\quad\quad d_1$ With paired horizontal fins...........Chaetognatha.*
$\quad\quad\quad d_2$ Without paired locomotor appendages.
$\quad\quad\quad\quad e_1$ Body flattened..................Platyhelminthes.†
$\quad\quad\quad\quad e_2$ Body cylindrical or thread-like.
$\quad\quad\quad\quad\quad f_1$ Anterior portion either modified to form a protrusible introvert or bearing a retractile proboscis in front of mouth. Burrowing forms.
$\qquad$ Annulata (Gephyrea).
$\quad\quad\quad\quad\quad f_2$ Without introvert or anterior proboscis (except in case of certain internal parasites).
$\qquad$ Nemathelminthes.†
$\quad\quad\quad\quad e_3$ Body variously shaped, usually covered with calcareous or chitinous structures; generally attached, often colonial. Characterized by a ridge (lophophore) bearing ciliated tentacles....Molluscoidea.
$\quad\quad\quad d_3$ Organ of locomotion a single foot (sometimes modified into elongated arms, tentacles, or wing-like extensions). Body *usually* covered with a calcareous shell. Fundamental bilateral symmetry often obscured by spiral-shaped body............Mollusca.
$\quad\quad c_2$ Body partially or obscurely segmented.
$\quad\quad\quad d_1$ A retractile disk bearing cilia found on anterior end of body. Tail mobile or telescopically jointed. Microscopic....................Trochelminthes.†
$\quad\quad c_3$ Body externally segmented.
$\quad\quad\quad d_1$ Appendages if present, not segmented, body elongated...............................Annulata.
$\quad\quad\quad d_2$ Paired segmented appendages present....Arthropoda.
$\quad b_2$ With notochord or internal skeleton. Dorsal nerve cord present.................................Chordata.

Such keys can be constructed for every group of animals and even for the subspecies and varieties.

In conclusion, the purpose of keys and of the entire system of classification is to make our knowledge definite and

---

*By some authors ranked as a subphylum of *Vermes;* by others, a class of the *Nemathelminthes;* while many avoid the difficulty by calling it "a group of uncertain systematic position."

†Treated in the chapter "Lower Worms." Some consider these groups as phyla, others as subphyla of the phylum *Vermes*, which is often made to include *Molluscoidea* also.

available. If you wish to designate a certain animal, you must find its generic and specific names, just as you would use a code word or a catalog number in ordering goods from a supply catalog. The catalog number and the scientific name have the same purpose, namely, to avoid vague references and permit accuracy—to expedite business. The catalog number, however, is the arbitrary device of manufacturer or merchant, while scientific nomenclature is based on the natural relationships of living things so far as we understand them.

CHAPTER III

LIFE IN THE OCEAN

This book is concerned with the invertebrate animals found upon the shore, especially those forms that may be readily examined without a microscope or any other special equipment. Attention will be given to the kinds of animal life that are fitted to live within the narrow zone uncovered by the ebbing tide, and to a few species that live farther out but may be strewn upon the shore by the action of wind and waves. One naturally wonders if the conditions he observes upon the shore extend outward under the sea, and whether the same kinds of animals and plants exist on the floor of the ocean and its waters that may be found upon the margin.

Research along such lines is a difficult and expensive undertaking, calling for a large outlay for nets, dredges, sounding instruments, and other apparatus, besides the use of a seagoing vessel with officers, crew, and a trained scientific staff. Enough has been done, however, usually under the auspices of our own or European governments, to indicate the general conditions of marine life, and it has been found that the animals that live on the surface of the open sea and those in the great depths differ in many respects from each other and from the kinds we see along the shore.

At tide level the shore may be composed of rocks supporting a varied growth of marine plants and serving as a place of attachment for many lowly animals besides offering a place of refuge to others, or it may consist of sandy beaches into which such animals as clams and sand crabs can burrow for protection. The action of the breakers and the ebbing of the tide bring about a maximum exposure to the air. Wherever an anchorage can be obtained the giant brown kelp grows offshore, forming a fairly complete belt along the entire western coast and sheltering many animals besides the bright-colored fish that play between the streaming fronds. A fairly accurate idea of the kelp beds and the zones of seaweed in general may be obtained from the glass-bottomed boats at Avalon. But seaweed can not grow very far down because it must have light in order to produce carbohydrates for food and the amount of light rapidly diminishes with increasing depth.

In the comparatively shallow water immediately offshore, the bottom seems to consist mainly of sand and silt derived from the land, mixed with the hard parts of the animals that either lived there or in the water above. After a few miles of a gradual slope extending outward from the shore and termed the continental shelf, the bottom drops off steeply to great depths,—more than 12,000 feet on the average in the eastern Pacific. There is reason to think, that if we could see it, we should find the bottom of the sea much smoother and with more gentle contours than the surface of the continents, for the land surface is constantly being worn down by the agencies of nature, while under sea the fine material which is being deposited tends to collect in the hollows.

It has been known for many years that beyond the influence of the silt washed into the sea by the rivers, the bottom is chiefly composed of accumulations of shells, most largely of microscopic one-celled animals, the *Protozoa*, mixed with spicules or skeletons of sponges and the shells of pteropod mollusks to form a gray, oozy mud. This deposit is called globigerina-ooze from its most characteristic component, the calcareous shells of the protozoan, *Globigerina*, and its close relatives. It has been found that *Globigerina* does not live in the ooze but in the water near the surface and it is only after death that the minute bodies slowly settle down through thousands of feet of water to become part of the slimy

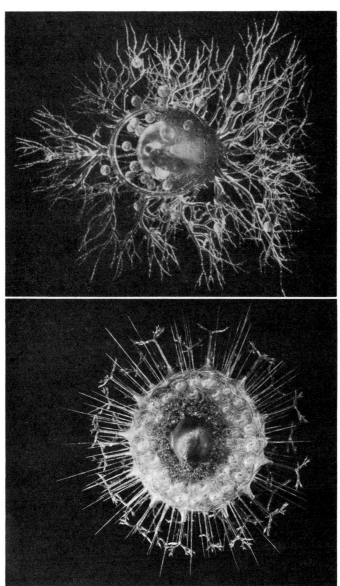

FIG. 9.—Glass models of *Radiolaria* in the American Museum of Natural History. Radiolaria are microscopic animals found in the sea. Photographs by courtesy of the American Museum of Natural History, New York.

deposit. In some places, where in the course of ages great elevations and extensions of the land surface have occurred, there are extensive beds of chalk. When examined with a microscope, samples of this rock are found to be composed of fragments of the same sort of tiny shells that make up the globigerina-ooze and geologists tell us that the chalk, limestone, and marble, which in the aggregate amount to a considerable proportion of our rocks, are all derived from this type of under-water deposit.

At depths greater than about 2,500 fathoms (15,000 feet) the globigerina-ooze is replaced by a reddish deposit, called red clay. This is without calcareous shells, owing to the fact that they are dissolved as they slowly sink through the miles of water, and consists largely of volcanic ash and meteoric dust, mingled with such hard organic remains as sharks' teeth and the ear bones of whales. The shells of one group of protozoans, the *Radiolaria*, being of a silicious, or glassy nature, are preserved and form a large part of the deposit (fig. 9). They are present in the globigerina-ooze also, but not to the same extent as the shells of the limy forms. In the Pacific, the red clay areas are much more extensive than those covered by globigerina-ooze. In deep water, the proportion of radiolarian shells increases, in some places, until we have a radiolarian-ooze, corresponding in its manner of origin with the globigerina-ooze, but silicious in character. There are regions where the bottom is covered by organic remains of other sorts, particularly in the coldest part of the Pacific where the silicious coverings of minute one-celled plants called diatoms predominate, but they are less extensive than the foregoing.

It has been found that the temperature decreases until at 2,000 fathoms it is in the neighborhood of 2° C. (35.6° F.). The great mass of the water is cold regardless of the temperature at the surface. Yet there are animals that inhabit these cold, unlighted depths and are adapted to the conditions there. Curiously shaped fish provided with organs for producing light are brought up by the dredge and many kinds of invertebrate animals belonging to the groups described in this book are taken from the ocean bottom. Most of them differ in some respects from the shore forms and belong to

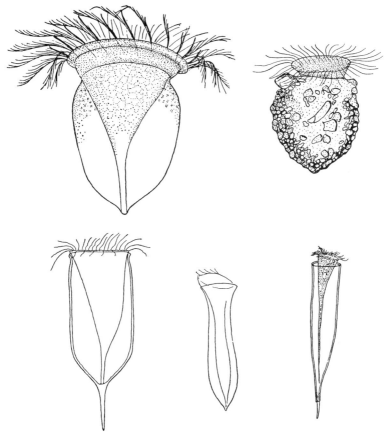

FIG. 10.—Marine ciliates (Tintinnidae) protected by chitinous, cup-shaped coverings. The cups, in some species, are encrusted with sand. The animals move rapidly by means of cilia, hair-like structures which serve as oars; greatly magnified.

different species but they are built on the same general plan. There are sponges, starfish, sea anemones, annelid worms, bryozoans, crustaceans, mollusks, and members of other groups there, particularly animals that creep about or live fastened to a substratum. The number of individuals taken by dredging in very deep water is, however, much less than

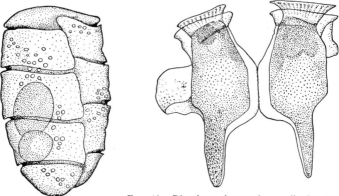

FIG. 11.—*Polykrikos* sp., an unarmored dinoflagellate; greatly magnified.

FIG. 12.—*Dinophysus homunculus*, a dinoflagellate. This one-celled animal has just divided to form two individuals, not yet entirely separated from one another; greatly magnified.

that obtained with the same number of hauls upon the continental shelf.

The intermediate depths and the surface could not be the haunts of fish alone, as is sometimes imagined, because like

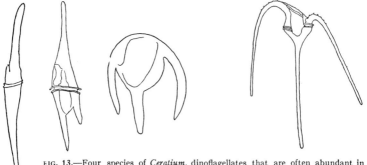

FIG. 13.—Four species of *Ceratium*, dinoflagellates that are often abundant in the plankton Each one has two whip-like flagella, one of which usually lies in the transverse groove which encircles the body; greatly magnified.

all other living things, fish require food. Animals, excepting a few one-celled species, can not manufacture food from inorganic substances with the aid of sunlight as most plants

LIFE IN THE OCEAN 31

do. Consequently animals are obliged to depend upon plants and the few simple animals that resemble plants in this respect, for their ultimate food supply. Of course large fish can feed upon little fish and they upon smaller ones but the chain is not complete and we find small fish feeding upon minute invertebrates and they in turn upon smaller ones or directly upon microscopic plants. As one looks at the water from the deck of a vessel these small organisms are invisible but their presence in enormous numbers is essential to the existence of higher animals in the open sea.

Collectively, the drifting population near the surface of the water is known as the plankton. In addition to the microscopic forms, the term includes such animals as the copepods, jellyfish, salpae, and floating mollusks, especially

FIG. 14.—*Prorocentrum micans*, one of the dinoflagellates that is often present when patches of "red water" appear and the sea is luminescent; greatly magnified.

FIG. 15.—*Peridinium* sp., one of the dinoflagellates; greatly magnified.

the pteropods. The more common of these are described in the following pages in connection with their shore-living relatives. It is the most minute kinds, the microplankton, that form the "floating meadows" of the sea. They consist of enormous numbers of one-celled plants, among them some with silicious coverings called diatoms, and equally enormous numbers of one-celled animals (figs. 9–17), distributed over great areas. Among the plankton live the shell-forming protozoans such as *Globigerina* and its relatives, that produce the limy shells which collect on the bottom to make up the gray ooze, and the radiolarians that have been mentioned in connection with the deposits of the deeper areas. Among the protozoans, the dinoflagellates are of great importance, owing to their abundance. At times they outnumber all other forms off our coast.

When large numbers of certain dinoflagellates appear we often see patches of "red water" by day and the luminous displays that accompany it at night. Many marine animals are capable of producing light but these protozoans are the cause of the diffused phosphorescence often seen in the

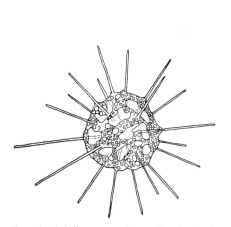

Fig. 16.—A heliozoan, a microscopic animal with ray-like pseudopodia and little power of movement; greatly magnified.

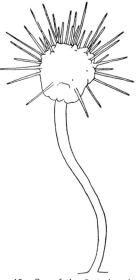

Fig. 17.—One of the *Suctoria*. An attached protozoan which may sometimes be found growing on hydroid stems; greatly magnified.

breakers during the summer and early fall. This luminescence is frequently spectacular and beautiful when seen on a dark night. The fish in the water shine with a blue-green light as they dart about, and the wake of a boat becomes a long trail of soft light. The surf is brightly illuminated and if one stamps upon the wet sand, sparkling points of light suddenly appear and disappear within a radius of several feet. By shaking some of the water in a bottle, rather

LIFE IN THE OCEAN 33

bright, sudden flashes can be produced, for the animals glow momentarily when disturbed instead of giving off a continuous light. While many species of dinoflagellates are luminous when stimulated, the most important ones on this coast are *Gonyaulax polyedra* Stein and *Prorocentrum micans* Ehrenberg. The dinoflagellates are so named because each one is usually provided with two flagella, or whip-like lashes, used in locomotion. Some kinds of dinoflagellates produce a yellowish-green discoloration of the water which resembles red water in being luminous when disturbed at night. An aftermath of extensive outbreaks of red water is the decay of inconceivable numbers of microscopic bodies stranded upon the beach, causing very offensive odors and poisoning the water sufficiently to kill animals such as sea cucumbers, crabs, and sometimes even fish, with the result that their bodies wash ashore and increase the stench.

There are many other marine protozoans and some differ greatly from the ones mentioned, but as their life-history and structure can not be studied without the help of the higher powers of the microscope and special technique, the reader is referred to text books. If more detailed information of our Pacific Ocean *Protozoa* is desired, the papers of Kofoid and Swezy should be consulted. Among them will be found some remarkable color plates of the dinoflagellates.

CHAPTER IV

Phylum—PORIFERA

(Sponges)

The sponge as we know it is only the skeleton of the sponge animal. When the sponges are alive, this skeleton is filled and overlaid with cells which line all of the numerous passageways as well as encase the outside of the framework.

Fig. 18.—Spicules of a sponge; greatly magnified.

The skeleton itself is usually made up of great numbers of curiously shaped, microscopic, glassy, or limy spicules (figs. 18–19). The commercial sponges, when alive, are said to show but little resemblance to the familiar bath sponge, the whole surface being covered with a thin slimy skin, usually dark brown or black and perforated to correspond with the openings of the canals. If we have one of our common shore sponges living in a dish of sea water, the only way we can tell that it is alive is by the fact that currents of water flow steadily in and out of the animal through the numerous openings and these openings sometimes open or close.

Aristotle, after studying the ordinary bath sponge found in the Mediterannean Sea decided that sponges are animals, though he admitted that they show many points in which they are like plants. In the years following Aristotle's time, however, opinions differed and many students classed sponges with plants so that their status was not definitely settled until the 19th century. The microscope then enabled observers to decide the question. They found the collar cells that bear flagella or tiny whip-like structures. These, by lashing to and fro, create the current that flows through the sponge. They also found that, instead of manufacturing

 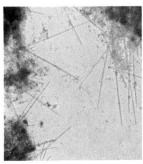

FIG. 19.—Two types of sponge spicules as shown by microphotographs.

their food as plants do, the sponges feed upon tiny plants and animals that are borne in on this current that flows through the canals.

The diagrams (fig. 20) show how the canals are arranged in different kinds of sponges. The body wall is made up of two layers of cells which are usually termed the dermal and gastral layers. The dermal layer makes up the greater part of the sponge and is mostly a gelatinous substance covered

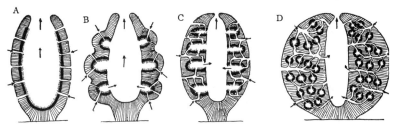

FIG. 20.—Various types of canal arrangement seen in sponges. Gastral layer solid black, dermal layer shaded. The arrows show the direction of the current of water. (Diagram modified from various sources.)

on all of its exposed surfaces with a flattened layer of epithelial cells. It contains the spicules of the skeleton, the cells that make them, and the wandering cells. The gastral layer is a single layer of the flagellate cells which line certain regions of the canals. These cells are called collar cells because of a peculiar flexible collar-like structure that surrounds the base of the flagellum. The current of water created by the flagella brings food and oxygen to the sponge. There are no special organs for digestion and respiration. The channels carrying the water pass through every part of the animal so that the exchange of gases in respiration can take place readily. The tiny animals and plants that are sucked in with the current of water are engulfed by the cells that line the canals. Digestion takes place within these cells and the others get their nourishment from them. Excretion takes place from the cells exposed to the current but wandering cells also engulf waste particles and carry them to the outside. These wandering cells probably also help in the transportation of food material.

Sponges become attached at an early stage in their history and remain so throughout their lifetime. The only movement of which they are capable is the closing of the pores by the contraction of the muscular tissue around the openings. If a sponge is injured, the pores near the injury may close but those in other parts of the animal may remain open. They have no active means of defense but the spicules naturally render them decidedly unpalatable and, in addition, they are often impregnated with substances that give them an unpleasant odor. However, they are not entirely without enemies, for certain nudibranchs have been found to feed upon them.

It is not always possible to say just what constitutes an individual sponge, especially in the case of some of the species which grow to a large size and whose canals branch and cross-connect in a bewildering fashion. Sponges of a single species may show great variation in form according to the situation in which they may be growing, and two sponges growing close together may fuse to form one large mass, or one variety may be grafted on to another.

Small portions may be detached from the mass and if kept

under favorable conditions will grow to full size. Individuals of a desirable kind are sometimes cut into small pieces and attached to stakes on a frame which is lowered into the water. After a few years the sponges are found to be of a good size and shape for marketing. If a sponge dies because of unfavorable conditions, a few cells may live over and grow into new sponges. Fresh water sponges, in the autumn, die and disintegrate, except for certain cells, the gemmules, which form hard coverings and live over until spring when they grow into new sponges. The gemmules may be dry for more than a year and then start to grow if put into water. They can also live through extremely cold weather.

There are no special sex organs but the ova and spermatozoa develop from the wandering cells of the dermal layer. The ova are usually fertilized in the canals and leave the body as ciliated larvae, which, after swimming about for a time, attach and develop into adults.

On our coast we find no sponges that are of any commercial value. Most of them are small or stiff with calcareous spicules and are irregular in shape. Many form whitish blotches or bright patches of red, orange, or sulphur-yellow color upon the rocks at the water's edge. Others live upon shells and there are forms that grow upon wharf piles and floats, on algae, and even upon the backs of crabs. Commercial sponges grow in the warm waters of Florida, West Indies, and the Mediterranean and Red Seas. The finest ones come from the Mediterranean region, but the catch from the Florida fishing grounds amounted in 1908 to 622,489 pounds, worth $548,876.

The sponges are taken with long-handled hooks which look like three-toothed rakes. One man rows the boat while his companion wields the hook, tearing the sponges from their place of attachment. This method can only be used in comparatively shallow water. Divers, with or without special diving apparatus, are also frequently employed to gather the crop. Dredges can be used but they are not favored because they destroy so many young sponges. The sponges that are gathered for the market are allowed to die on the deck of the vessel and then put into cages under water until the flesh becomes soft when they are beaten and washed under water and then strung in bunches and sold. Later, in the packing-houses, they are trimmed, sorted, and in some cases bleached, after which they are ready for the consumer.

As no commercial developments have centered around the sponges of the west coast of the United States, comparatively few species that inhabit this region have been described. Consequently we find it impossible to name and describe even our abundant shore species. Owing to the work of Lambe, on the West Canadian forms, a few of the northern species can be identified but even in this case it is not a simple task for any one but a specialist. Their external form is so dependent upon environment that individuals of the same species may differ markedly in appearance. The shape and kind of spicules are the reliable characters for use in classification. The spicules may be calcareous, composed of carbonate of lime; silicious, of a glassy nature; or they may be horny. Some sponges are described as non-spicular, in which case the skeletons are fibrous or cuticular.

The Phylum *Porifera* is commonly divided by the systematist into three classes; the *Calcarea*, or calcareous sponges; the *Hexactinellida*, or glass sponges; the *Demospongiae*, in which not only silicious spicules but horny fibres can serve as skeletal material. The following key from Pratt,* will serve to distinguish them.

a_1 Small marine sponges with calcareous spicules and large collar cells and mostly under 2 cm. in length...1. *Calcarea*
a_2 Usually larger sponges with silicious spicules or spongin fibres, or both, or without either.
b_1 Glass sponges; spicules usually six-rayed.2. *Hexactinellida*

---
*H. S. Pratt—Manual of the Common Invertebrate Animals; A. C. McClurg & Co.

PHYLUM—PORIFERA 39

b₂ Massive sponges without six-rayed spicules; skeleton of silicious spicules, spongin, or both, or wanting..........
3. *Demospongiae*

CLASS—CALCAREA
ORDER—HETEROCOELA
FAMILY—GRANTIIDAE

To this family belong the simple sponges of the genus *Grantia*, small cylindrical or urn-shaped forms which may be found on floats or wharves. In *Grantia* and its relatives the flagellated collar cells are found in chambers connected with the outside in the manner shown by fig. 20-A and 20-B and the central cavity opens to the outside with a single large excurrent pore or osculum situated at the upper end of the body.

CLASS—HEXACTINELLIDA

The glass sponges have received their common name from their skeleton of silicious spicules which in some cases are so joined together as to give the appearance of threads of spun glass. The best known glass sponge is the Venus' Flower Basket, the vase-shaped skeleton of a tropical sponge, *Euplectella*, that is obtained by dredging. Some members of the group do not have a continuous skeleton but are stiffened with solitary spicules. In all cases spicules are present and have a six-rayed structure. Lambe has described three species belonging to this class from the Straits of Georgia.

CLASS—DEMOSPONGIAE
ORDER—MONACTINELLIDA
FAMILY—ESPERELLIDAE

In the Puget Sound region nearly all of the large scallops have the upper valve of the shell covered by a yellowish-brown growth of sponge. *Esperella adhaerens* and *Myxilla parasitica*, both described by Lambe, are the incrusting species. The two differ largely in the structure of their skeletal spicules but superficially are much alike. In *Esperella adhaerens* the pores are very small and the excurrent openings or oscula are also small (0.33 mm. in width) and but very slightly raised above the surface. In *Myxilla parasitica* the incurrent pores approach the size of the oscula in the preceding species, reaching a diameter of about 0.27 mm., while the oscula are comparatively large, having an average diameter, according to Lambe,[*] of 5 mm.

FAMILY—SUBERITIDAE

Suberites latus Lambe, from Vancouver, is interesting as the living home of a hermit crab. The sponge, itself, is generally rather hemispherical in shape with a flattened base. The color is a uniform yellowish brown. A large proportion, Lambe[**] says four out of five, of these sponges are pierced from the base to the center with a spiral opening in which the hermit crab lives. Because of the hermit crab's activity, the sponge secures considerable free transportation. It may be conceived that the arrangement benefits both individuals, the crab being protected by the drab-appearing, passive sponge which few marine animals can eat, while the sponge profits from the disturbance created by the hermit crab in moving about and in feeding.

FAMILY—CHALINIDAE

Some sponges are widely distributed. The bread-crumb sponge, *Halichondria panicea* (Pallas), which occurs upon the beaches of British Columbia and Alaska, also inhabits the New England coast and European shores, and has been reported from places as far apart as Ceylon and Nova Zembla. It is a massive sponge of a grayish or yellowish color, stiffened by a mass of needle-like spicules and containing but few fibres of a horny nature.

Figure 21 shows an assortment of shallow water sponges from the California shore, illustrating the great diversity of form among the numerous species that inhabit the west coast.

---

[*]Lawrence M. Lambe—Sponges from the Pacific Coast of Canada; Transactions of the Royal Society of Canada; Vol. 11, p. 31.

[**]Lawrence M. Lambe—Some Sponges from the Pacific Coast of Canada and Behring Sea; Transactions of the Royal Society of Canada, Vol. 10, p. 72.

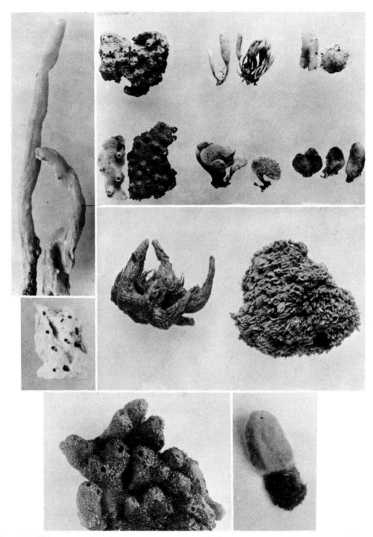

FIG. 21.—Thirteen species of shore sponges commonly found on the California coast. The four larger ones x ⅓; others x ⅔.

CHAPTER V

Phylum—COELENTERATA

(The hydroids, jellyfish, sea anemones, corals, and others)

Jellyfish, sea anemones, and corals, the coelenterates most often seen, seem to the casual observer to have very little in common. Closer observation, however, shows a fundamental plan which is the same for all of the group. Early zoologists called them plants or plant-animals (zoophytes) because of their flower-like appearance.

Literally older than the hills, the coelenterates have come down to us through unnumbered generations from the Cambrian seas, not without modification to be sure, but essentially the same as today. Fossil jellyfish of that age have been found that closely resemble species now living.

In complexity of form this group seems to stand between the sponges and the echinoderms. Typically, they have a central mouth surrounded by tentacles and a central cavity for circulation and digestion. They furnish an example of the radial plan of symmetry (fig. 8). The digestive tract is not distinct from the body cavity as it is in higher animals. The body is made up of two layers of cells with a jelly-like layer, called mesogloea, lying between them. Most of the species have stinging capsules upon the outer parts of the body which they can use as weapons of offense and defense.

Many of the coelenterates are colonial, a large number of individuals being connected to form a colony. Some of these colonies and many simple forms are fixed, while others are floating or free-swimming. In many of the species, alternation of generations is the rule. When this is the case the offspring differ in form from the parents but are, instead,

PHYLUM—COELENTERATA

like the grandparents. This is commonly seen in the hydroids. Regeneration, or the replacement of amputated parts, occurs to a remarkable degree in coelenterates. In this group are many species which are seen to be beautifully luminescent when disturbed at night.

Coelenterates are grouped in five classes; *Hydrozoa*, hydroids and medusae, the latter with a velum (fig. 22-B); the *Siphonophorae*, free-swimming colonies (plate 1); the *Scyphozoa*, hydroids and medusae, the latter often large and without the velum (plate 2); *Anthozoa* or *Actinozoa*, the corals, sea anemones (fig. 82), and gorgonians; the *Ctenophora*, free-swimming, solitary forms (fig. 88), sometimes called "comb jellies" or by the fishermen, "cat's eyes."

Class—HYDROZOA, HYDROMEDUSAE

(The hydroids and medusae)

The two forms, hydroids and medusae (fig. 22) are unlike in appearance but are two stages in the life-history of the same animal. The plant-like, colonial hydroids produce (by budding) free-swimming, solitary medusae and these umbrella-shaped medusae produce eggs which grow into the hydroid form.

The hydroids are more readily found than the medusae and may be seen at low tide as a fine seaweed-like growth on rocks, shells, kelp, or piling. Some kinds are so small that they appear to be a fuzzy growth on the supporting surface, other kinds are larger and more easily seen, one of the largest being the tubularians which often form matted clumps, the stems frequently reaching a length of two or three inches. One who has not given much attention to the group may confuse the hydroids with certain of the bryozoans, but examination with a hand lens will bring out considerable difference (compare fig. 42 and fig. 118). Hydroids may be recognized by the flower-like heads borne singly on a stalk or branching therefrom.

A colony of hydroids is usually made up of two kinds of individuals, the feeding individuals or hydranths which take in food for the colony and the reproductive individuals which

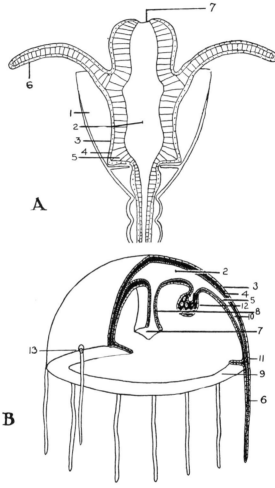

FIG. 22.—A. Diagram of a feeding individual, or hydranth, of the hydroid colony. B. Diagram of one of the *Leptomedusae*. (Modified from various sources.)

1. Hydrotheca.
2. Gastrovascular cavity.
3. Ectoderm.
4. Mesogloea.
5. Endoderm.
6. Tentacle.
7. Mouth.
8. Manubrium.
9. Velum.
10. Radial canal
11. Ring canal.
12. Gonads.
13. Lithocyst.

PHYLUM—COELENTERATA 45

produce medusae. These individuals which make up the fixed colony are also called polyps or zooids. The nutritive polyps are the flower-like ones with a crown of tentacles and a central mouth which leads into the stomach cavity. This cavity is in communication with the stalk and all the other individuals of the colony are likewise connected. The tentacles are delicate and sensitive and when a tiny creature swims against them they seize it by closing quickly, bringing the ends of all the tentacles to the mouth, into which the victim is taken. If the animal is an active one it probably calls forth a round of poison darts from the nettle cells, or nematocysts, of the tentacles. These poison darts numb the creature so that he falls an easy prey to the hydroid. The nettle cells are in the outer cell layer and differ somewhat in the different species of hydroids. A typical one consists of a cell containing a capsule which bears at one end a long thread continuous with the capsule wall. This thread is often spirally coiled within the capsule. When the animal is disturbed, the thread is shot forth, turning inside out and carrying the poison into the wound. The hydroids with these sting cells are able to penetrate the body covering of many of their enemies, and in many cases the animal is quickly paralyzed or even killed by the poison.

The reproductive polyps are frequently vase-shaped or club-shaped and contain a central axis from which are budded the medusae, which, when fully developed, escape from the end of the case and swim off into the water. In some kinds the medusae are borne on pendulous stems attached between the sets of tentacles.

The medusae are harder to obtain than the hydroids. They frequently appear in hauls taken with a fine meshed net, or if one has healthy specimens of hydroids in a dish of clean sea water, he usually will soon find medusae swimming about in the dish. The medusa is like a tiny umbrella, with a very short handle, the manubrium, at the end of which is the mouth. The mouth opens into a cavity in the center of the large part of the umbrella from which radial canals pass through the substance of the body to the outer rim where they join the ring canal, a circular canal in the margin of the umbrella. The medusae of this group have a velum, a narrow, thin shelf of tissue projecting inward from the margin of the umbrella (fig. 22-B). Tentacles provided with nettle cells hang from the margin of the umbrella and help the animal in capturing its food. Medusae swim in a jerky fashion, their movement being brought about by spasmodic contractions of the umbrella. The sexes are usually separate. The sperm cells are usually shed from the male medusae into the water and are borne by the currents to the female medusae where they fertilize the eggs. The embryos are set free from the medusae as tiny ciliated planulae which soon attach to some convenient support and grow into hydroid colonies.

Although the foregoing life history is typical, there are many variations from it. *Corymorpha palma* (fig. 29), for example, never sets free the medusae but keeps them attached to the parent where they pulsate in a futile fashion for days until, after their eggs are deposited, they finally wither away. The fresh-water hydra and some others have no medusae at all, the fixed form producing the eggs which grow into fixed individuals. Still other forms have lost the fixed form entirely and complete the life-history with the medusa stage alone. One can see that it is often difficult to follow the various stages in the life-history of a coelenterate. In many cases, medusae have been known for some time, but their fixed forms have never been recognized. In some cases, the fixed and free-swimming forms of a species were known a long time before the two were known to be connected. To establish the relationship, the hydroids can be kept until the medusae are liberated or medusae may be kept until the hydroids develop, but it is difficult to keep delicate pelagic forms like medusae in aquaria unless special equipment and an abundance of fresh sea water are available.

Order—GYMNOBLASTEA, ANTHOMEDUSAE

In this group the feeding individual forms no protective cup or hydrotheca around itself though the stems of the colony may have a firm protective covering. The medusae are either fixed or free-swimming, more or less bell-shaped, bear the reproductive organs on the manubrium, in contrast to the *Leptomedusae* in which the reproductive organs are attached to the radial canals (fig. 22), and possess ocelli. The ocelli are red or blue spots at the bases of the tentacles around the edge of the umbrella; their exact use is not fully known but they are thought to be light-perceiving organs.

Family—SYNCORYNIDAE, SARSIIDAE

Sarsia mirabilis L. Agassiz (fig. 24) is a small, transparent medusa about 7 mm. high and 4 mm. in diameter. There are

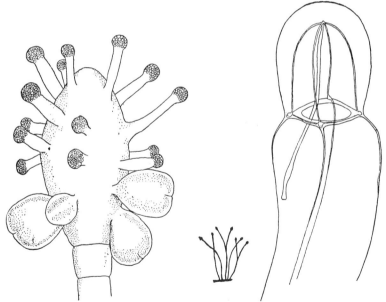

Fig. 23.—*Syncoryne mirabilis*; left x40, right x2.   Fig. 24.—*Sarsia mirabilis*; x5.

four contractile tentacles which may be four to six times the bell radius in length. Nematocyst cells are prominent particularly toward the distal ends of the tentacles, and there is a small ocellus on the outer side of each tentacle bulb. The manubrium is long and slender so that the mouth opening may hang far below the velum unless the manubrium is coiled within the bell as frequently happens. The hydroid (fig. 23) is *Syncoryne mirabilis* (Agassiz) which is attached to the support by a creeping stolon, grows to be 15 mm. high, and branches profusely. Polyps terminate the main stem and branches and develop medusa-buds on their sides just below the tentacles, each bearing 1–4 buds in different stages of development. The species is common on the New England coast from February to April and has been reported from the Pacific coast as far south as Chile. The specimen figured was taken at Friday Harbor, Puget Sound, in July.

FAMILY—BIMERIIDAE

Garveia annulata Nutting (fig. 25) has hydranths with a conical or dome-shaped proboscis, surrounded by a single whorl of about 16 thread-like tentacles. The reproducing polyps are fixed sporosacs on branch-like pedicels, and are not permanently surrounded by a perisarc, the outer covering of cuticle. The stalk is made up of a number of stems lightly held together. The whole colony may be 50 mm. high. The annulation of the stems may or may not be pronounced. The color is orange with the main stem and sporosacs a deeper color

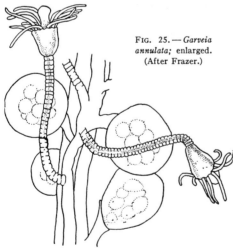

FIG. 25.—*Garveia annulata;* enlarged. (After Frazer.)

than the rest of the colony. The species has been reported from Alaska to Catalina Island.

FAMILY—BOUGAINVILLEIDAE

Stomotoca atra Agassiz (fig. 26) has a nearly transparent bell, two long contractile tentacles, and about 80 rudimentary tentacle bulbs. The reproductive organs are eight brown or black, linear cross-foldings on the sides of the stomach. The mouth has four lips. The hydroid is probably a tubularian, *Perigonimus*. The medusa is found in Puget Sound from June to September, often in considerable numbers. The bell is 20–25 mm. high and 20–22 mm. wide.

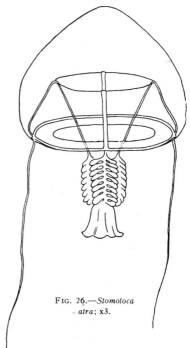

FIG. 26.—*Stomotoca atra*; x3.

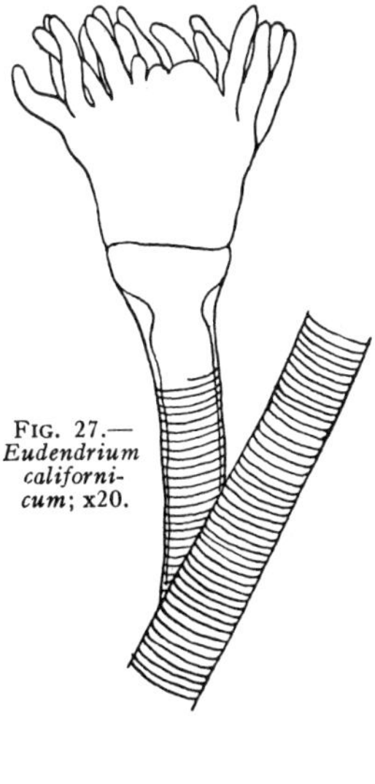

FIG. 27.—*Eudendrium californicum*; x20.

FAMILY—EUDENDRIIDAE

Eudendrium californicum Torrey (fig. 52-C) may easily be mistaken for one of the brown algae. The stem is made up of a single tube which has a stiff, brown outer covering. The flower-like hydranths are borne at the tips of the secondary branches. The stem has numerous narrow rings or annulations throughout its length (fig. 27). The reproductive zooids are borne on the hydranths just below the tentacles. The orange colored eggs give the female zooids an orange tint. The species is found from Pacific Grove to Puget Sound, and may reach 14 cm. or more in length.

Eudendrium rameum (Pallas) (fig. 52-D) resembles the last named species but is more irregularly branched and the main stem is thick, and composed of many dark brown tubes closely twisted together. The principal branches are compound below but run out to single tubes at the tips. The branches have 2-5 rings at their bases though these are not always clearly marked. The tentacles number 24-27 in one whorl. The colony may reach a length of 11 cm. This species is found from Puget Sound to San Diego.

FAMILY—CORYMORPHIDAE

Corymorpha palma Torrey (figs. 28 and 29), the fairy palm, is a delicate white or semi-transparent form which bears a single flower-like head with slender, graceful tentacles. It may reach a length of four inches and the tentacles may be nearly an inch long. It lives in shallow bays and may be found on the mud-flats when the tide is out. We have seen them in considerable numbers on the Balboa mud-flats and on the shore of Mission Bay. Dr. Torrey reported them as being numerous at San Pedro in 1902, but dredging operations or contamination of the water may have exterminated them for we have not seen them there.

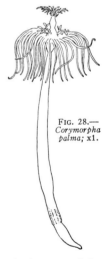

FIG. 28.—
Corymorpha palma; x1.

The stalk is thickest near the base where it is rooted in the mud by a tangle of fine processes. The head bears a whorl of 18-30 long tentacles and twice as many short ones, the latter close to the mouth and arranged in several whorls. A cross section of the stem shows that instead of a single central cavity there are a number of canals running lengthwise of the stalk near the outer surface, the portion between the canals being made up of gelatinous cells. There are no cilia on the tentacles, on the surface of the proboscis, nor on the body. Nettling cells are present on all the tentacles with a few on the surface of the body and the proboscis.

In the laboratory aquaria, when a current of water is flowing past the *Corymorpha*, they are usually erect with the plane of the tentacles horizontal or slightly tilted, but when the water is turned off, they soon begin a series of bowing operations. They bend double until the tentacles sweep the mud, and then rise to an erect position again. The long tentacles, in bringing food to the mouth, roll up at the ends first and the short ones comb the food off of them and pass it to the mouth. The animal bows about once in three minutes while it is feeding.

During both the summer and winter months, the medusoids, or imperfect medusae, are found in various stages of development. They remain permanently fixed to peduncles which arise at the base of the proboscis, within the large circle of tentacles. They have the umbrella and radial and ring canals, (fig. 22-B) but the manubrium is twice as long as the bell and without mouth or tentacles. In pulsating, they seem to be trying to free themselves, but so far as is known they are not successful. The eggs drop from the manubrium, adhere to whatever they first touch and develop without any free swimming stage.

Corymorpha shows great capacity for regeneration. If the hydranth, or flower-like head of the animal, is cut off and the animal kept under favorable conditions, a new hydranth will grow within a few days. The shorter the piece of stem left, the longer is the time it will take to replace the hydranth. Sections cut from the column will also regenerate complete individuals, the part of the section that was uppermost forming a hydranth and the lower part developing the holdfast. In the bibliography are listed some interesting papers that give in detail what has been learned about regeneration in *Corymorpha*.

Family—TUBULARIDAE

Tubularia crocea (Agassiz) is found growing in large clumps on floats or piles that are submerged most of the time, (fig. 30). The stems often become four or five inches long, are seldom branched, and are usually matted together in a bushy mass. The stems are yellowish, the tentacles transparent, and the mouth region coral pink. There are two sets of tentacles, the shorter ones 20-25 in number, in a circle close to the mouth, and the longer ones in a single whorl at the base of the head. Fixed medusoids are borne in clusters on peduncles attached between the two sets of tentacles. The tentacle processes of these medusoids are very small, flattened, 6-10 in number, small in the female, and often hardly visible in the male. The larva, when it escapes from the medusoid, has tentacles which help to keep it afloat for a time until it settles down to start a new colony. This species is found on the eastern coast as well as on our entire western coast.

In the laboratory it will be found that the animals drop their hydranths if the water becomes warm but if the water can be kept cool and well aerated the heads may regenerate in a few days. The stems with their firm covering seem well able to resist unfavorable conditions. *Morse reports *Tubularia* present throughout the summer at South Harpswell, Maine where the temperature seldom rises above 16° C. (61° F.) but at Woods Hole, Massachusetts he says they disappear when the temperature of the water reaches 20° C. (68° F.), appearing again in the fall.

Tubularia marina Torrey grows on rocks along the ocean shore where it is exposed to the breakers. The stems arise in clusters from a creeping stalk that adheres firmly to the

---

*Max Withrow Morse—Autotomy of the Hydranth of *Tubularia*; Biological Bulletin, Vol. 16, p. 181.

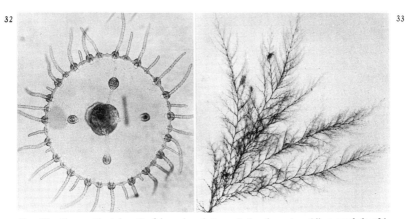

FIG. 29.—*Corymorpha palma*, the fairy palm, photographed under water while expanded; x⅔.
FIG. 30.—*Tubularia crocea*, photographed under water while expanded; x½.
FIG. 31.—*Clytia bakeri* on clam *Donax gouldii*; x1.
FIG. 32.—Microphotograph of the medusa, *Obelia* sp.; greatly enlarged.
FIG. 33.—Living branches of *Obelia longissima*. The dark objects along the branches are nudibranchs, *Galvina olivacea*; x⅔.

support. The stems, more or less annulated, and unbranched, are 30-50 mm. long. There are 20-26 tentacles in the outer and 26 in the inner circle. The female medusoids have four stout tentacles that are as long as the medusoids themselves (fig. 34). This latter fact and the different manner of growth will distinguish the species from *T. crocea* which it superficially resembles.

FIG. 34.—*Tubularia marina*, female medusoid; x20.

ORDER—CALYPTOBLASTEA, LEPTOMEDUSAE

In the hydroids of this group, the perisarc or transparent protective covering is not only found on the stems but also extends up around the polyps, forming, in the case of the feeding individuals, a protective cup or hydrotheca into which the polyp can withdraw (fig. 42-A) and in the case of the reproductive individuals, a cylindrical or vase-shaped capsule, the gonotheca, which envelopes the developing medusa (fig. 42-B). The medusoid generation is in some cases free swimming and in others fixed, remaining within the gonotheca. The medusae, called *Leptomedusae*, have lithocysts instead of ocelli for sense organs and bear the reproductive organs beneath the radial canals rather than on the manubrium (fig. 22-B). The lithocysts, also called statocysts, were formerly thought to be auditory organs and were called otocysts. The more recent view is that they are for perceiving the position of the body or the direction in which it is swimming. In the *Leptomedusae* it consists of a small vesicle in the margin of the umbrella (fig. 22-B 13), containing a tiny, hard, stony body called the statolith. Changes of position will cause, through the influence of gravity, slight changes in the position of the statolith against the sensitive lining of the vesicle.

Family—CAMPANULARIIDAE

The hydroids are either branched or simple stalks growing from a creeping rootstalk. The feeding polyps have bell-shaped protective cups and are usually borne on stalks. The mouth is on a trumpet-shaped proboscis. The reproductive polyps are large and produce either free-swimming medusae or embryos which develop into the next hydroid generation.

Campanularia gelatinosa (Pallas) has a fascicled stem and reaches a height of 200–250 mm. The small branches divide somewhat dichotomously (fig. 35) bringing a large number of hydranth

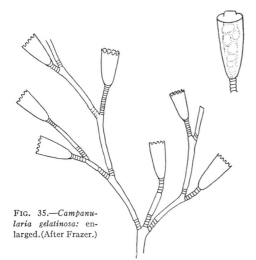

Fig. 35.—*Campanularia gelatinosa:* enlarged.(After Frazer.)

pedicels close together so that the colony has a more or less gelatinous appearance when it is in the water. The margin of the hydrotheca has about 10 teeth, each with two cusps. The sporosacs remain within the gonangia during the development of the planula, not being extruded into a sac at the summit of the gonangium as in *Gonothyraea*.

Clytia bakeri Torrey (figs. 31 and 36) grows on the living clam *Donax* as shown

Fig. 36.—*Clytia bakeri.* Nutritive polyps (above). Reproductive polyps (below); x20.

in the photograph. The clam lives on flat, sandy beaches and is often left partly uncovered at low tide. It buries itself in the sand with the broad end at the surface since the siphons through which it feeds are at that end of the body. The hydroids are also attached at the broad end of the shell so that they are exposed to the wash of the waves which brings food to the hydroid as well as to the clam. Frequently one would not suspect the presence of the clam in the sand if it were not for the tuft of seaweed-like hydroids growing on it.

The hydroid stalks are often 30–40 mm. long and usually unbranched, with polyps alternately arranged along the stem which is markedly annulated near the base, but has fewer annulations toward the tip. The stem is divided into internodes and the short pedicels or branches are borne on a sort of shoulder at the end of each internode. The hydrothecae are triangular in outline with smooth rims. The reproductive zooids are frequently in pairs, long, and with the large distal end covered by a membrane which breaks when the medusae are set free. The medusae are oval, thin-walled, with two long tentacles 3–4 times the length of the bell and two short, less developed ones.

Phialidium gregarium (Agassiz) is a medusa (fig. 37) which appears in great numbers at Friday Harbor during the summer months. The bell is from 12 to 20 mm. wide with 60 tentacles on the margin, each arising from a spherical bulb. A lens will show one or two lithocysts between each pair of tentacles. There are four radial canals with the reproductive organs located upon their distal portions. The stomach is small and the mouth is provided with four elongated, lobed lips.

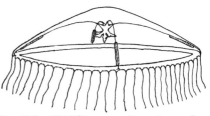

FIG. 37-A. *Phialidium gregarium*, the medusa form of the hydroid *Clytia inconspicua*; x3.

PHYLUM—COELENTERATA

Fig. 37-B.—*Clytia inconspicua*, the hydroid form of *Phialidium gregarium;* enlarged. Drawn from specimens raised at the Puget Sound Biological Station, Friday Harbor, Washington.

The hydroid (fig. 37-B) has been raised by Mrs. T. C. Frye and found to be *Clytia inconspicua* (Forbes). This species forms a small colony with the stem usually unbranched.* The polyp cup is small and bears seven blunt teeth. The short pedicel is annulated throughout or with a small smooth portion in the center. The reproductive individuals are on short annulated stalks. The terminal aperture is comparatively large. The height is about 1 mm.

Orthopyxis caliculata (Hincks) is a very small hydroid often found growing on seaweed (fig. 38). The stems bearing the hydrothecae are a quarter of an inch long and unbranched, springing from a creeping rootstalk. The pedicels may be regularly annulated throughout their length, or more rarely, may be almost smooth. The hydrothecae are variable in shape and in the thickness of their walls, but the walls are always thickened toward the bottom forming a shelf upon which the polyp rests. The reproductive polyp is not shown in the figure but Dr. Nutting describes it as oval, the distal end truncated, joined to the stem by a short peduncle, and with the walls coarsely and unevenly corrugated. The medusa is one sixteenth of an inch high with no tentacles but with four

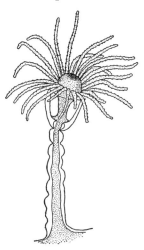

Fig. 38.—*Orthopyxis caliculata.* (Approximately x5.)

---

*L. H. Strong—Development of Certain Puget Sound Hydroids and Medusae; Publ. Puget Sd. Biol. Sta., Vol. 3, pp. 383-400.

minute pigmented bulbs, and eight lithocysts. The species is almost world wide in its distribution.

Orthopyxis compressa (Clark) resembles the last species but the pedicels are usually smooth except for an annulation just below the hydrotheca, and the reproductive polyps are laterally compressed.

Orthopyxis everta (Clark) also found all along our coast, resembles *O. caliculata*. It may be distinguished by the

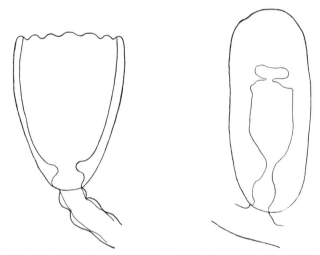

FIG. 39.—*Orthopyxis everta*, Hydrotheca (left), and gonangium (right); x50.

toothed margins of most of the hydrothecae. The reproductive polyps are regularly oval in outline, somewhat compressed (fig. 39).

Gonothyraea clarki (Marktanner-Turneretscher) has slender, branching stems with long internodes (fig. 40). The stems are sinuous, and dark horn-brown near the bases, but lighter distally. The margin of the hydrotheca has 10–12 sharply truncated teeth which give it a castellated appearance. The gonangium usually grows out of the axils of the branches but sometimes takes the place of the hydrothecae. There are four or five medusoids in each gonangium. This form has been found from Alaska to San Francisco Bay.

Obelia commissuralis McCrady (fig. 41) is reported as abundant in Oakland Harbor, San Francisco Bay. The stems are not fascicled, the margin of the hydrotheca is entire and polygonal in outline. The medusa has 16 tentacles at the time it is liberated and becomes a free medusa.

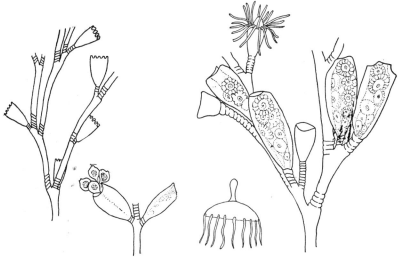

FIG. 40.—*Gonothyraea clarki;* enlarged. (After Frazer.)

FIG. 41.—*Obelia commissuralis;* enlarged. (After Nutting.)

Obelia dichotoma (Linnaeus) is one of the widely distributed hydroids found growing on piling, rocks, mussels, seaweed, and floating timber on nearly all coasts. The stems of the colonies may be an inch or more long. They are fine and thread-like, and form a dense, fuzzy growth. The colonies are often dichotomously branched with 5-8 annulations above the base of each branch (fig. 42). The hydrothecae often show slight folds or pleatings which make the margin of the cup somewhat irregular and sometimes give the effect of low, rounded, marginal teeth. The reproductive zooids are urn-shaped, and attached in the axils by a short, ringed stalk. The developing medusae may be seen through the wall of the gonangium thickly clustered around the central axis. If healthy colonies are kept for a few hours in a dish of clean sea water, medusae will usually be found in large numbers swimming around in the water. When set free, they are about 1 mm. in diameter and usually have 16 tentacles on the margin of the disk-shaped umbrella (fig. 42, compare fig. 32), a short, square manubrium, and four radial canals under which hang the

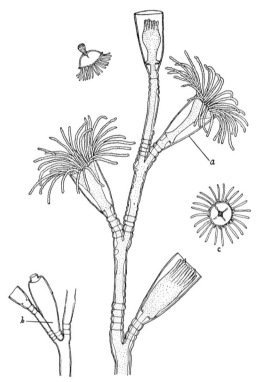
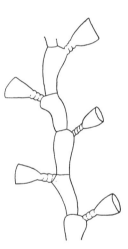

FIG. 42.—*Obelia dichotoma*, a hydroid that is abundant on wharf piles and rocks. A. Hydrothecae (feeding polyps). B. Gonangium (reproducing polyp). C. Medusa (free-swimming form). Drawn from life. (Approximately x20.)

FIG. 43.—*Obelia geniculata*, a common form of hydroid, enlarged. (After Nutting.)

ovaries. Under favorable conditions, the medusae increase rapidly as to size and number of tentacles.

Obelia geniculata (Linnaeus) is also a widely distributed species. Like the last named species in general appearance, the colonies grow from a creeping rootstalk but the stems are usually unbranched and zigzag in outline; the nodes are thickened below each bend and from this thickening the peduncle arises (fig. 43).

PHYLUM—COELENTERATA

Obelia longissima (Pallas) forms large colonies which may reach 50–60 cm. in height. The central, undulating stem is dark brown near the base and lighter toward the tip. The branches are alternate and are themselves branched (fig. 33). The pedicels vary in length and are annulated throughout (fig. 44). The margins of the cups are toothed but this is usually very hard to see. The reproductive individuals are in the axils and contain a dozen or more medusae. The medusae have 20–24 tentacles when set free.

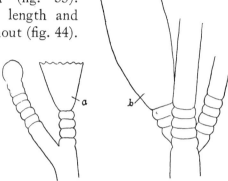

FIG. 44.—*Obelia longissima.* A. Hydrotheca. B. Gonangium; x40.

According to Frazer* this is the most abundant shallow water campanularian in the Puget Sound region. Its range extends northward from Puget Sound to Alaska.

FAMILY—SERTULARIIDAE

The feeding polyps are sessile, that is, they grow close to the stalk and are not on pedicels. They are arranged very definitely in more than one series along the stem. Usually there is a delicate transparent operculum made up of 1–4 parts which covers the mouth of the hydrotheca when the polyp is contracted. The reproductive individuals are larger than the hydrothecae and produce eggs or spermatozoa without an intervening free medusa form.

Abietinaria filicula (Ellis and Solander). The slender stems have alternating branches pinnately arranged (fig.

---

*C. McLean Frazer—Some Hydroids of the Vancouver Island Region; Transactions of the Royal Society of Canada, Third series, 1914, Vol. 8, p. 153.

52-A) and the branches may themselves be branched more than once. The flask-shaped hydrothecae are nearly opposite each other on the stem and have small, round openings (fig. 45). The gonangium is oval, smooth and has a small opening. This species is often found washed in with the seaweed and is an inhabitant of the Atlantic and Pacific coasts.

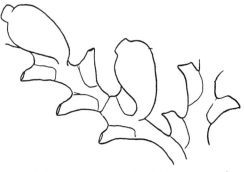

FIG. 45.—Branch of *Abietinaria filicula*, showing gonangia and hydrothecae; x20.

Sertularella turgida (Trask) forms small colonies with stout stems which are either unbranched (figs. 46 and 53) or have a few irregularly placed branches. The polyp cups are arranged alternately on opposite sides of the stem, and have three teeth on the margin. The reproductive individuals arise in the axils of the polyp cups, are elongated, with spines in varying numbers at the distal end. The species has been reported from many points in the Puget Sound region and ranges as far south as San Diego.

Sertularia pulchella (d'Orbigny) grows on eelgrass, the colonies with few or no branches grow from a creeping stolon sometimes becoming as much as an inch long. The hydrothecae are in

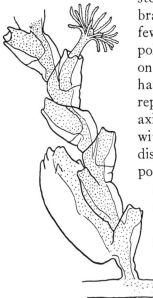

FIG. 46.—*Sertularia turgida*, showing one feeding polyp expanded, several hydrothecae, and the large gonangium; x20.

PHYLUM—COELENTERATA 63

pairs opposite each other, the two members of a pair being in contact on one side of the stem and some distance apart on the other side (fig. 47). There are two large mar-

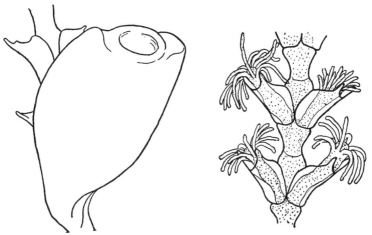

FIG. 47.—*Sertularia pulchella*, showing feeding polyps (right); x20; and a gonangium (left); x50.

ginal teeth and the hydrotheca has a large opening. The gonangia are at the lower part of the stem, large, oval, and compressed, with a short curved pedicel, a distinct collar and a large aperture. They may be found from Vancouver to San Diego.

FAMILY—PLUMULARIIDAE
(The plume hydroids)

In this group the colonies are plume-like (fig. 56), the polyps are only on one side of the branches, and without stalks. Nematophores are present, but there are no medusae formed.

The nematophores are different from anything found in the two preceding families. They are tiny thecae or cups (fig. 48), occupied by little polyps which are specialized for defense. In some groups the cup is free or jointed while in others it is sessile, but in any case the polyp is armed with either nematocysts (nettle cells) or adhesive processes or both. Dr. Nutting,* in quoting from von Lendenfield gives the following translation of a summary of his observations

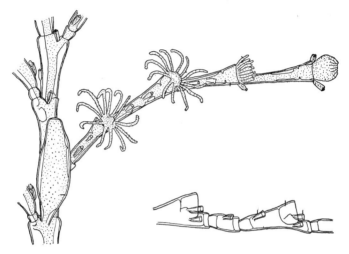

FIG. 48.—*Plumularia setacea*, showing expanded, flower-like, feeding polyps, tiny nematophores below and above the polyp cups, and one flask-shaped gonangium near the base of the larger branch; x45.

on how the living plumularian captures food. (The sarcostyle is the defensive polyp in the nematophore). He says, "The prey, coming in contact with a tentacle of the hydranth, is pierced by the tentacular nematocysts, which have a narcotizing effect. Next it comes in contact with one of the adhesive bodies at the end of the greatly produced sarcostyle. The adhesive cells adhere to the prey, and the body of the adhesive polyp quickly retracts, bringing the Zoea into contact with more of the globular adhesive masses, which hold it in spite of even the most violent struggles for liberty. It is thus brought again within the range of the tentacles and devoured. The adhesive cells are finally cast off, remaining attached to the victim, and the sarcostyles again retract.

"When a large animal, such as an Annelid, strikes the

*C. C. Nutting—American Hydroids I—The *Plumularidae*; U. S. National Museum Special Bulletin, 1900, p. 24.

tentacle, the adhesive threads immediately retract, as do also the tentacles, and the batteries of nematocysts on the other part of the sarcostyles are brought into play to repel the attack."

In the genus *Plumularia* the pinnately arranged branches of the colony are themselves unbranched, the hydrothecae have smooth margins and the nematophores are movable. The gonangia are sac-shaped or bottle-shaped.

Plumularia setacea (Ellis) forms delicate, unbranched, plume-like colonies sometimes three inches long (figs. 48 and 49). We have found it growing on the kelp. The pinnate branches are alternate, the hydrothecae shallow and cup-shaped, the nematophores are small, movable, and occur both on the main stem and on the branches, as shown in the figure. The gonangia are elongated with a narrow neck and situated in the axils of the pinnate branches. This form is found on both the west and east coast of North America.

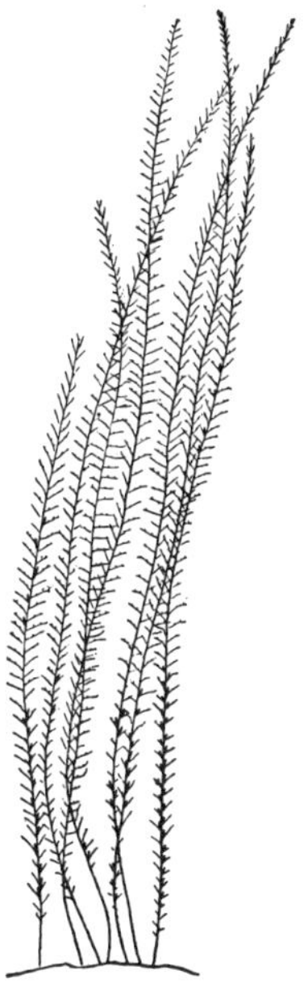

The genus *Aglaophenia* comprises hydroids with sessile nematophores, hydrothecae with toothed margins, and gonangia enclosed in a corbula formed by overlapping leaf-like structures.

Aglaophenia inconspicua Torrey (fig. 50), the small ostrich-plume hydroid, grows on algae or along with the larger hydroid next described. Though it resembles the next species it may easily be distinguished from it by the shape of the polyp cups, which

FIG. 50.—*Aglaophenia inconspicua.* A. Portion of branch with three hydrothecae; x30. B. Corbula, containing gonangia; x12. (After Torrey.)

FIG. 49.—*Plumularia setacea*; natural size. These colonies were attached to a large spherical float of kelp.

bear nine teeth on the margin, the shorter corbulae, and by the shorter stems. The latter are 30–40 mm. long. It is reported from Pacific Grove and San Diego.

Aglaophenia struthionides (Murray), the ostrich-plume hydroid, is one of the most common ones on the coast (fig. 56) being frequently cast up on the beach, often with bunches of seaweed which it greatly resembles. It occurs from Vancouver to San Diego. Each colony has regular, alternating branches from each side of the central stem, the whole often growing to a length of six inches. The branches are graded in length which makes the plume effect more pronounced. The hydrothecae are in one series on one side of the branch and each has eleven prominent teeth, and three nematophores (fig. 51). The middle nematophore is large and reaches the level of the margin of the hydrotheca, but the other two, one on each side of the hydrotheca, do not extend as far forward as its margin. There is a large triangular nematophore at the base of each branch. Some of the lateral branches of the colony are shorter than the others and bear swollen pod-like structures. These corbulae are made up of 8–13 pairs of overlapping leaves, each with a row of nematophores along its edges. At the bases of the corbula-leaves and enclosed by them are the gonangia. Free medusae are not formed but the egg cells develop into ciliated larvae within the gonangia.

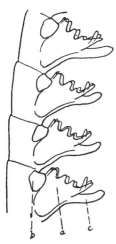

FIG. 51.—Hydrothecae of *Aglaophenia struthionides*, showing the toothed margin of the cups, or hydrothecae, the long bent nematophore below the cup, and the shorter nematophores on each side above the cup; x40.

FAMILY—THAUMANTIIDAE

Polyorchis penicillata (Eschscholtz) is a beautiful medusa which is found in the bays of the California and Washington

FIG. 52.—A. *Abietinaria filicula;* x½. B. *Aglaophenia inconspicua*, the small ostrich-plume hydroid; x½. C. *Eudendrium californicum;* x½. D. *Eudendrium rameum;* x½.

FIG. 53.—*Sertularella turgida*, growing on seaweed; about natural size.

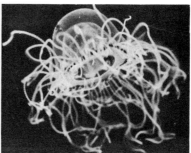

FIG. 54.—*Gonionemus vertens*. From a photograph of a living specimen in an aquarium; x1.

FIG. 55.—*Polyorchis pennicillata;* x⅔.

FIG. 56.—The ostrich-plume hydroid, *Aglaophenia struthionides;* about ⅛ natural size.

coasts (fig. 55). It is usually present from December to April and sometimes is found in great numbers. It is bellshaped, about 50-60 mm. high and 30-40 mm. broad. There are four short lips. The ring canal is unbranched but the radial canals each have 15-25 pairs of short branches which end blindly. There are 40-150 tentacles arranged in 2-4 rows. These can contract until they are only thick, short stubs or extend until they are twice the length of the bell. There is an eye spot at the base of each tentacle. The gonads, 4-8 in number, arise from each of the radial Canals between the sides of the stomach and the branched parts of the canals. The stomach, gonads, tentacle-bulbs, and radial canals are reddish brown to purple.

FAMILY—MITROCOMIDAE

Halistaura cellularia (A. Agassiz) (fig. 57) is an inhabitant of northern Pacific waters. It is exceedingly abundant in some parts of Puget Sound as at Friday Harbor where, together with species of *Aequorea* and *Phialidium*, it constitutes a large porportion of the pelagic life. The bell is flatter than a hemisphere, 45-90 mm. wide, and bears from 100 to 340 tentacles with swollen bases.

FIG. 57.—*Halistaura cellularia*, a luminous medusa of northern waters; x⅔.

This species is conspicuously marked by the reproductive organs which are linear in shape and extend nearly the whole length of the four radial canals, giving the appearance of a cross when the animal is seen from above. Like the medusae with which it is associated, this form becomes luminous if touched or jarred after dark. The hydroid is unknown.

There is considerable difference of opinion regarding the classification of this medusa and in many works it appears as *Thaumantias cellularia* (Haeckel).

Family—AEQUORIDAE

Aequorea aequorea (Forskal) (fig. 58 and 59) is a cosmopolitan medusa which occurs in our costal waters. At Friday Harbor it is the most conspicuous member of the summer swarms of medusae. If *A. coerulescens* Brandt of the California coast is really but a variety of this species as H. B. Bigelow* and Mayor† both consider probable, the range extends along the western coast of North America from Bering Sea to Lower California. It includes also the coast of Japan and the Atlantic coasts of Europe and North America as well as the Mediterranean. Throughout this extensive range there are many variations from the typical structure. Studies of the European variety have shown the hydroid to be a minute form known as *Campanulina*. *C. forskalea* Fraser, of our northern coast, may be the hydroid generation in the Pacific. *Aequorea* may have 100 or more radial canals upon which the reproductive glands are located. There are a large number of long tentacles (often more than 100) which can be coiled and retracted until they appear as a short fringe upon the bell or extended like threads for a distance excceding twice its diameter. Each tentacle arises from a swollen base and close observation will disclose numerous lithocysts upon the margin of the bell between them.

If disturbed after dark, *Aequorea* is brilliantly luminescent, giving off a soft greenish light (fig. 59). But when taken into a dark room during the daylight hours and stimulated immediately no light will be visible. That this is not due to lack of adjustment to the dark by the eyes of the observer may be shown by leaving the specimen in the dark room for an hour or two. Upon returning, it may be stimulated, either

---

*H. B. Bigelow—Medusae and Siphonophorae collected by the U. S. Fisheries Steamer *Albatross* in the Northwestern Pacific, 1906; Proceedings U. S. National Museum, Vol. 44, p. 36–41.

†A. G. Mayor—Medusae of the World, Vol. II, The Hydromedusae; Carnegie Institution Publication No. 109, pp. 326.

by jarring the dish of sea water containing it or by touching the medusa itself with some hard object, and it will become luminous. Similar results can be obtained with *Phialidium*

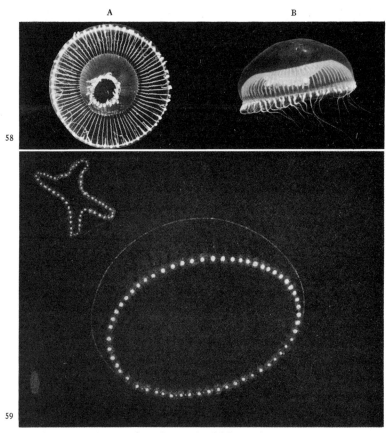

FIG. 58.—*Aequorea aequorea*, from photographs of living specimens. A. Viewed from below, to show the mouth and radial canals. B. Side view showing tentacles partly contracted; x½.
FIG. 59.—"Phosphorescence," or light production, in *Aequorea aequorea* (below) and *Phialidium gregarium* (above); x1. The illumination occurs at the bases of the tentacles.

gregarium, *Halistaura cellularia*, the ctenophores *Bolinopsis microptera* and *Pleurobrachia bachei* and, according to reports, with many other coelenterates. So far as we have been able to discover, light production in all these forms is practically

PHYLUM—COELENTERATA

the same as found by Dahlgren[*] in the Scyphozoan jellyfish, *Pelagia* (page 79), but in the medusae it shows at the bases of the tentacles and in the ctenophores it appears along the rows of paddle plates. Until recently this medusa has been known as *Aequorea forskalea* Peron and Lesueur but H. B. Bigelow has pointed out that by right of priority the name given by Forskal thirty-four years earlier should be used. As Forskal called it *Medusa aequorea*, and *Medusa* is no longer applicable while the genus *Aequorea* has long been recognized, the name now becomes *Aequorea aequorea* (Forskal).

Order—TRACHOMEDUSAE
Family—PETASIDAE

These are medusae with free or enclosed lithocysts, with the margin simple and not cut into lappets, and the gonads on the radial canals of which there are 4, 6, or 8.

Gonionemus vertens A. Agassiz (figs. 54 and 60), found all along the Pacific coast, is a little less than three-quarters of an inch across the disk and a little higher than it is wide. There are 60–70 tentacles, all similar, all projecting from the sides of the exumbrella above the margin, and all with pad-like adhesive disks near their outer ends, and bulb-like swellings on the umbrella margin below the insertion of the tentacles. There are 60–70 lithocysts alternating with the tentacles and four ribbon-like, sinuously folded gonads along the four radial canals. The manubrium is large. The mouth, which has fringed lips, is on a level with the velum. The bell is yellowish green, the gonads deep red, and the radial canals deep brown. The high bell, long slender tentacles, and deep red gonads distinguish it from other species. It is found in the kelp near shore in the summer time.

Dr. Mayor says that its habits appear to be similar to

---
[*]Ulric Dahlgren—The Production of Light by Animals; Journal of the Franklin Institute, 1916 and 1917.

those of *Gonionemus murbachii*, an Atlantic coast form which has been the subject of much study at the biological station at Wood's Hole. We take from Mayor,* the following quotation from Perkins concerning the habits of *G. murbachii*. "On cloudy days or toward nightfall the medusa is very active, swimming upward to the top of the water, and then floating back to the bottom. In swimming it propels itself upward with rhythmic pulsations of the bell-margin, the tentacles shortened and the bell very convex. Upon reaching the surface the creature keels over almost instantly, and floats downward with bell relaxed and inverted and the

FIG. 60.—*Gonionemus vertens:;* x2.

tentacles extended far out horizontally in a wide snare of stinging threads which carries certain destruction to creatures even larger than the jelly-fish itself.

"*Gonionema* continues this fishing, with little respite, all day long in cloudy weather. Occasionally it fastens itself to a blade of eelgrass or some other object near the bottom,

*A. G. Mayor—Medusae of the World, Vol. II, The Hydromedusae; Carnegie Institution Publication No. 109, p. 344.

PHYLUM—COELENTERATA

or stops midway in its course with tentacles extended. In this position it is well-nigh invisible, but a deadly foe to small fish or crustaceans which cross in its path."

The eggs are laid in the evening during the summer months. Twelve hours after the eggs are fertilized the ciliated larvae have developed. A cavity begins to form at the posterior end and the larva fixes itself to the bottom by the forward end. The mouth breaks through at the upper end and four tentacles appear, the first ones coming about three weeks from the time the egg was fertilized. Buds develop from the polyp just below the tentacles, become detached, and swim away, locating after a few days and growing into polyps like the one from which they were budded. The development of the medusa from this fixed form has not yet been observed but recent research on a related European species has shown that buds near the base of the polyp develop directly into tentacled medusea.*

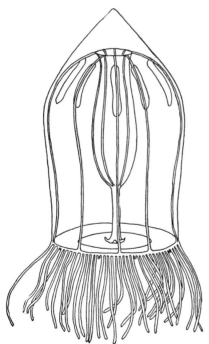

FIG. 61.—*Aglantha digitale*, a small, transparent medusa; x5.

FAMILY—TRACHYNEMIDAE

Aglantha digitale (Fabricius) Haeckel (fig. 61) is small and nearly transparent. At each contraction of the bell it darts swiftly through the water,

*H. F. Perkins—*Gonionemus*; Science, Vol. 63, 1926, p. 93.

in marked contrast to other medusae which swim in a rather leisurely fashion. Its shape reminds one of a rifle bullet with its pointed apex and nearly straight, somewhat rigid sides. There are 80–100 slender tentacles which are only moderately contractile and 4–8 lithocysts. A spindle-shaped manubrium hangs from under the umbrella and the 8 radial canals extend down it to the stomach. Eight sausage-shaped reproductive organs arise from the radial canals near the manubrium and hang down into the bell cavity. The surface of the bell is iridescent and the reproductive organs, stomach, and tentacles may show a faint pink color. It is found both in the North Atlantic and the North Pacific and reaches a height of 30 mm. and a diameter of 15 mm.

Order—SIPHONOPHORAE

The *Siphonophora* are beautiful, free-swimming, colonial animals which are found in the open ocean and are numerous in the warmer seas. Our most common representative of the group is *Velella* (Plate I) which is occasionally cast ashore on the California coast. Perhaps a more familiar example is the Portuguese Man-of-War which, though not often seen far north on the Pacific coast is known to many because of its being so often pictured both in scientific and popular books.

In any case the colonies are made up of many individuals of various sorts which are all connected. Some of the polyps are specialized for taking in food, some are armed with nettle cells which protect the colony and help it in capturing prey, some are swimming individuals which propel the colony through the water, and others reproduce the species.

The colonies swim about slowly in the sea and sometimes are many feet in length and contain thousands of individuals. Travelers in the warmer seas often speak of the beauty of

PHYLUM—COELENTERATA

these little argonauts which float hither and thither, blown by the wind and buoyed up by an air-filled float which often reflects the sunshine with beautiful iridescent colors.

Family—VELELLIDAE

Velella lata Chamisso and Eysenhardt is shown in the color sketch (plate 1). The raft-like portion, which sometimes reaches a length of four inches, contains air chambers which help to keep the colony afloat. The triangular sail which projects above the surface catches the wind and keeps the raft in motion. The individuals which make up the colony are attached to the under side of the float and hang down in the water. The individual polyps are easily visible with the unaided eye although, looking down upon the colony, one can only see the long contractile tentacles which make an outer double row around the polyps. In the center of the colony is a large feeding polyp called a gastrozooid, and between it and the tentacles are the reproductive individuals which also have mouths and are able to take in food, though they are much smaller than the central polyp.

The elliptical float is covered both above and below by the mantle which also covers both sides of the crest or keel, and extends even beyond, to form flaps along its margin. The keel runs diagonally across the horizontal float usually in the direction shown in the picture. The float and keel contain canals which surround a central point as concentric ellipses and send branches up into the keel. Air tubes extend throughout most of the float, keel, and liver mass. Parts of the liver system may be seen as brownish granular masses through the blue of the mantle. While the central polyp has the main feeding mouth the smaller ones surrounding it can take in food, though to a less extent; but all are connected at their bases with the canal system, and the nourishment they take in goes at once into the general circulation and the fluids are kept moving by the active cilia which line the canals, as well as by the activity of the polyps themselves. The polyps all bear nettle cells in great numbers and the tentacles are especially well provided with them. The reproductive polyps bear medusae near their bases in all stages of development. The medusae are small and become free-swimming. The float and keel are practically colorless but the mantle covering is deep blue, tinged in certain parts with green due to the presence of the liver mass within. The polyps and tentacles are a lighter blue. A number of species have been described from the Pacific but H. B. Bigelow* after studying a large series of specimens places them all under this one species. *Velella lata* is distributed over the warmer regions of the Pacific reaching as far north as Japan and Puget Sound; how far south it extends is not known. They are usually seen in schools, borne about by wind and tide and sometimes thrown up in considerable numbers upon the beach. Numbers of them are destroyed, for even moderate waves are likely to upset them and they die if they remain upside down for any length of time.

Specimens kept in dishes in the laboratory show an undulating movement of the mantle edge and strong contraction of the central zooids, but this seems to be the only independent motion of which they are capable. The shore collector will only find them when by chance they have been blown on shore and stranded there. If they have just been washed in they may be kept alive in clean sea water and watched for several days. It is well, however, not to put any other choice specimens, such as nudibranchs, into the same dish for *Velella's* nettle cells are very powerful and effective. We have never felt any effect from the nettle cells though we have handled small specimens

---

*H. B. Bigelow—The Siphonophorae. Memoirs of the Museum of Comparative Zoology at Harvard College, Vol. 38, No. 2, pp. 343-345.

freely, but we once put a rare nudibranch in with a *Velella* and came near having a tragedy. It took a day or two in favorable surroundings for the nudibranch to recover.

Class—SCYPHOZOA, SCYPHOMEDUSAE

(The jellyfishes)

The large jellyfishes usually seen floating on the surface of the sea or washed ashore as shapeless jelly-like masses belong to the *Scyphomedusae* and may be distinguished from the *Hydromedusae* by the absence of the velum (compare figs. 22-B and 64). The huge sea blubber, *Cyanea capillata* Eschscholtz, of the Atlantic coast, grows to be seven feet across the disk with tentacles one hundred and twenty feet long. We have no species as large as this, but one of the *Pelagia* group is sometimes two feet in diameter with tentacles two feet or more in length and the north Pacific variety of *C. capillata* reaches a diameter of one and a half feet. These animals are large and heavy but not more than 5% of the body is solid matter.

The Cambridge Natural History* says, "in China and Japan two species of *Rhizostomata* * * * are used as food. The jelly-fish is preserved with a mixture of alum and salt or between the steamed leaves of a kind of oak. To prepare the preserved food for the table it is soaked in water, cut into small pieces, and flavored."

Usually the *Scyphozoa* have an alternation of generations which reminds one of the life-history of the *Hydrozoa*. The medusa produces the egg which is set free as a ciliated larva, settles on and attaches itself to some support and develops tentacles at the free end. This fixed form, which somewhat resembles an unbranched hydroid, grows to be about half an inch long when constrictions appear which after a time divide it into a number of disks. These saucer-like disks

---

*S. J. Hickson—*Coelenterata* and *Ctenophora*, Cambridge Natural History, Vol. 1, p. 312; The Macmillan Company.

are set free, one at a time, as small medusae which grow rapidly under favorable conditions of food supply.

The nettle cells of the large jellyfish are usually powerful enough to penetrate a person's skin and be plainly felt. In the case of our large species of *Pelagia* and *Cyanea* the effect is much like that produced by ordinary nettles but is stronger and may last for half a day or more. The species that have large mouths can capture and take in good sized fish and crustaceans. One of the *Chrysaora* has been observed to select medusae, ctenophores, and pelagic worms. The *Rhizostomata* have such small mouths that their food must be minute. In spite of their carnivorous habits, fish, usually very small kinds, are frequently found associating with certain jellyfish, apparently for protection since they dart into the sub-umbrella cavity when they are alarmed.

The sense organs are small tentacle-like structures bearing statoliths at the tip and in some species there are one or two tiny eye spots. These organs are protected by a small hood or fold of the umbrella margin. The umbrella may be globular, disk-like, conical, cubical, or divided by a circular groove. In many species, triangular lips hang down from the mouth in long frills sometimes two or more times the length of the disk. The stomach cavity extends out toward the margin of the disk in a number of pouches. Sometimes the pouches are connected to a ring canal in the margin of the umbrella but in other cases the ring canal may be absent. The gonads, when nearly ripe, may be brilliantly colored, circular, horse-shoe shaped, or band-like organs on the lining of the stomach pouches. The eggs and sperm pass out through the mouth, the eggs are fertilized in the water and may remain in pouches in the manubrium for a part of their development. In *Pelagia* the development is direct, the egg growing into a free swimming medusa like the parent. In one group, the fixed generation only is present but most kinds pass through the alternation of generations already described.

PHYLUM—COELENTERATA

Passengers on our coastwise steamers often see the jellyfish as great balls of light in the sea after dark. We know of no studies on the light production in our own species but quote Dahlgren* who gives some interesting facts concerning the related *Pelagia noctiluca* of the Mediterranean Sea.

"When this animal is swimming freely at home in the sea at night, or in an aquarium in the laboratory with plenty of fresh sea-water, it gives no light at all. A direct stimulus is necessary to make it show its light. This stimulus may be mechanical, chemical, or electrical.

* * * * *

"A slight contact of the finger or a glass rod with the outer surface of the umbrella results in a spot of light at the point touched. This spot is local for only an instant, and then spreads out, but it may not cover the whole surface of the bell. The light appears in lines and streaks and occasionally in patches.

"When the contact is made more violently the light is brighter and spreads over the whole umbrellar surface. If violent enough, the luminosity does not spread, but appears on all the surface at the same time. In this case it is stimulated to appear by the mechanical jar that is communicated through the jelly.

"A very noticeable feature of this lighting power is the fact that when one touches the illuminated surface, however gently, the luminous material clings to the fingers and shines on them almost as long as on the umbrellar surface. Thus we have proof that the material is discharged from the cellular tissues (the covering epithelium). If one strokes that surface with moderate firmness, while the animal is resting and not lighting, and withdraws the finger instantly before the light appears, no light will be seen on the fingers, proving that it has not yet been thrown out of the cells.

---

*Ulric Dahlgren—The Production of Light by Animals; Journal Franklin Institute, February, 1916, pp. 2-4.

And again, if the outer cells are scraped off, they will illuminate, but the tissues beneath them will remain dark.

"These several experiments show that the light produced by the animal is the result of the bringing of a secretion, the luciferine, into contact with the free oxygen in the sea-water, when the luciferine is discharged from the cells that store it."

Order—DISCOPHORA

Suborder—SAEMAEOSTOMATA

Scyphomedusae are jellyfish without a circular furrow, with hollow tentacles, marginal sense organs, and a simple mouth from which hang long curtain-like or gelatinous lips.

Family—PELAGIIDAE

The purple-striped jellyfish, *Pelagia* sp. (plate 2), the largest of our jellyfishes, is common off the coast of California. It is often seen out at sea swimming near the surface or sometimes may be found washed up on shore. It has eight marginal tentacles alternating with the eight sense organs and between these are sixteen partially divided lappets. There are sixteen radiating stomach pouches but there is no ring canal.

The specimen from which our color sketch was made, was a foot in diameter. The tentacles ordinarily hung six or eight inches below the umbrella margin but could be extended so that they were much longer. The purple and red coloring makes these creatures quite conspicuous. If a person while surf bathing brushes up against the tentacles of one of these jellyfishes he gains a good notion of the action of the nettle cells. While not extremely painful, the parts of the skin that are stung, if not calloused, show a decided redness and the nettled effect may persist for an hour or more.

Chrysaora melanaster Brandt has eight marginal sense organs, twenty four tentacles (three tentacles between each pair of sense organs), and thirty two marginal lappets. The disk is a flattened hemisphere 10–12 inches in diameter and 4–6 inches high. The marginal lappets are of equal size and slightly narrower at the base than beyond it. The curtainlike lips taper from a wide base to pointed ends, the margins are much folded, and their length is as great as the diameter of the disk. The marginal tentacles are shorter than the disk radius. The general color is light bluish with 32 brown rays on the umbrella and 16 dark brownish or black radial streaks on the under side of the umbrella. The gonads are reddish brown and the tips of the tentacles red. This species is abundant in the north Pacific from Kamtschatka to California.

Chrysaora gilberti Kishinouye (plate 1) is distinguished from *C. melanaster* by having longer tentacles, often as long as the lips of the manubrium and by its semicircular lappets all similar in shape. The exumbrella is thickly sprinkled with nettle warts. The four oral curtains are broad near the mouth and have finely frilled margins. When contracted they are coiled. The umbrella is light brown, the tentacles and midrib of the mouth-arms darker brown. This variety is common in Monterey Bay in the summer time and is also found off southern California. The specimen figured was taken at Pacific Grove in February 1921.

FAMILY—CYANEIDAE

Cyanea capillata Eschscholtz, the "sea-blubber," is the giant jellyfish of the Atlantic with a disk measuring seven feet across and tentacles 120 feet long. One of the Pacific representatives, known as variety *ferruginea*, may reach nearly one and a half feet in diameter. The species is bright yellow or orange with brownish stomach and radial pouches. Some of the tentacles are reddish and some are nearly white. The colors show considerable variation. The disk is rather flat and the margin divided by 8 deep clefts into lobes, each with a median cleft and two short notches. At the bottom of each median cleft is a club-shaped sense organ. The bell is thick toward the center of the disk and thin at the edge. The tentacles number several hundred and can be extended until their length is about 25 times the bell diameter, reaching nearly 35 feet in the larger specimens. They are provided with powerful nettling cells as the unwary collector will soon know if he picks up a specimen with his hands. The four long mouth arms have greatly folded margins and are about the length of the bell diameter. The reproductive organs are in four pouches at the sides of the stomach between the radii. There is no ring canal. The variety *postelsii* is much like *ferruginea* but there are deep clefts on each side of the median cleft (with the sense organ) instead of short notches. Both varieties are found in the North Pacific.

FAMILY—AURELIIDAE

Aurelia aurita (Linnaeus), a jellyfish of almost world-wide distribution, frequently appears in large numbers (fig. 62).

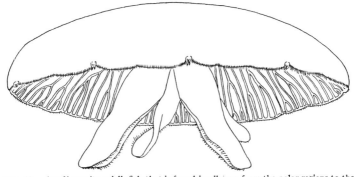

FIG. 62.—*Aurelia aurita*, a jellyfish that is found in all seas, from the polar regions to the tropics. It often appears in large numbers. Drawn from a preserved specimen that was somewhat contracted.

The body is disk-shaped and bears a fringe of small tentacles upon the edge. The disk sometimes reaches a diameter of 25 cm. Club-shaped organs which contain eye spots are located in eight marginal indentations. The mouth is surrounded by four long narrow lobes. The four reproduc-

PHYLUM—COELENTERATA

tive glands are horseshoe-shaped and arranged radially around the center of the animal. From the four-lobed stomach, narrow, branching canals extend outward becoming smaller until they join again in a circular canal just within the margin. This species is often nearly colorless, but sometimes is light violet or rose red. Mayor* says, "In the Tropics it lives very close to its heat death-temperature, and thus it is barely able to survive in the surface waters of the warmer seas in summer. Romanes found that specimens of this medusa from the British seas can withstand being frozen solidly into ice, and I find this to be true also of this medusa from Halifax, Nova Scotia. At Halifax, on the other hand, the medusa ceases to pulsate at 29.4° C., at which temperature it is most active at Tortugas, Florida. On the other hand, the Florida medusa is killed by being frozen into the ice. Thus the medusa becomes somewhat acclimated to the temperature of the waters in which it lives, and if accustomed to warm water it loses its resistance to cold and the opposite."

SUBORDER—RHIZOSTOMATA

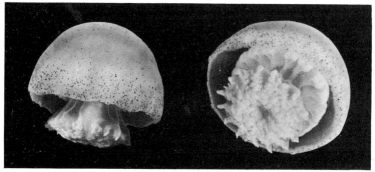

FIG. 63.—Two views of the jellyfish, *Stomolophus meleagris;* from photographs of a small, living specimen; x½.

*A. G. Mayor—Report upon the Scyphomedusae collected by the U. S. Bureau of Fisheries Steamer *Albatross* in the Philippine Islands and Malay Archipelago; U. S. National Museum Bulletin 100, Vol. 1, Part 3, pp. 204–205.

This order includes medusae without marginal tentacles and with numerous small mouths.

Stomolophus meleagris L. Agassiz (fig. 63) often appears in great numbers in San Diego Bay* in the summer time and is also a common species on the Atlantic Coast from the Carolinas southward and in the Gulf of Mexico. Mayor reports that "It often occurs in vast swarms, occupying an area which is sometimes over 100 miles in length." The bell is a hemisphere and the color Prussian blue, pale in small specimens and more pronounced in the large ones. The largest specimens are six to eight inches in diameter. The gelatinous substance is much firmer than it is in many of the jellyfish. There are no marginal tentacles, 8 marginal sense organs, and 130–150 marginal lappets. The mouth-tube projects downward below the umbrella, and the free ends of the mouth arms divide and flare out at the lower end. Besides the central mouth at the lower end of the manubrium there are 16 slit-like mouths on the knife-like projections on the upper part of the manubrium (fig. 64). Small, incessantly waving tentacles constantly drive food particles toward the mouths. The central stomach is wide and gives off 16 radial canals which in turn give off side branches and

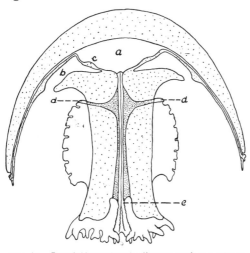

FIG. 64.—*Stomolophus meleagris*, diagram to show anatomy.
a. Stomach.   d. Lateral mouths.
b. Sub-genital pit.   e. Central mouth.
c. Gonad.   (After Mayor.)

*H. B. Bigelow—Note on the Medusan Genus *Stomolophus*, from San Diego; University of California Publications in Zoology, Vol. 13, pp. 239–241.

form an anastomosing net work connecting all of the radial-canals. There is no ring-canal. There are 16 semielliptical areas of circular muscles in the sub-umbrella and alternating with these there are triangular areas of weakly developed radial muscle fibres. The margin of the umbrella pulsates constantly and rapidly.

The Atlantic coast specimens are brownish with white or yellow spots in the margin of the umbrella but our specimens are light Prussian blue with dark blue spots around the margin of the umbrella. Dr. Mayor* reports the capture of immature medusae in which the disk was less than ⅛ inch in diameter. The bell was much flatter than it is in mature specimens and the umbrella surface was covered with warty clusters of nettle cells. According to Dr. Mayor the fact that the Atlantic and Pacific species are so much alike makes it probable that the medusae have remained unchanged since the closure of the Isthmus of Panama.

Class—ANTHOZOA or ACTINOZOA

(*Sea anemones, corals, sea pens*)

The flower-like sea anemones (fig. 82) are familiar to almost every visitor to the beach for they line the tide pools on all the rocky shores and thickly cover the piling of the wharves all along the coast. The corals reach their highest development in warmer waters than ours, but we have a few representatives here.

The individual animals of this group, whether solitary or members of a colony, have a flower-like circle of tentacles and a central mouth opening. Unlike the hydroids in which the mouth opens directly into the stomach, these animals have a gullet which leads from mouth to stomach cavity and the latter is partly partitioned off by a number of radial

---

*A. G. Mayor—Medusae of the World, Vol. 3, The Scyphomedusae; Carnegie Publication No. 109, pp. 710–711.

mesenteries into compartments which are all connected with the central part of the cavity (fig. 65). The hollow tentacles

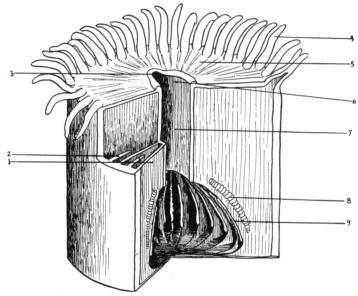

FIG. 65.—Diagram of a sea anemone.
1. Primary mesentery.
2. Secondary mesentery.
3. Siphonoglyph (one at each end of the slit-like mouth).
4. Tentacle.
5. Disk.
6. Mouth.
7. Gullet.
8. Gonads.
9. Mesenterial filament.

are usually provided with nettle cells. The edges of the mesenteries bear the digestive glands, and the gonads are on the sides or edges of the mesenteries. Muscle fibres running lengthwise on the mesenteries enable the animal to draw itself down and circular muscles can draw the disk together and cause it to close over the contracted tentacles.

Since anemones are to be found in abundance and are hardy animals that live almost indefinitely under favorable laboratory conditions, they have been the subject of numerous experiments. In this way their method of taking food has been learned. The normal expanded position of the sea anemone is with tentacles extended outward (fig. 65-4). If a

PHYLUM—COELENTERATA

little carmine (an inert substance which does not affect the sea anemone in any way) is dropped on the extended tentacles, the granules will be seen to gradually move from base to tip of the tentacle, finally being dropped off from the tip into the water. Microscopic examination shows that the tentacles are covered with tiny cilia which are constantly beating toward the tip of the organ. Thus the minute carmine granules are lashed along by the whip-like structures until they are finally pushed off. If the carmine is put on the disk between the mouth and the bases of the tentacles it will usually remain stationary. No cilia have been found on the disk. If the carmine is put on the mouth the granules that fall near the siphonoglyphs (fig. 65-3) are carried down the gullet while those that fall on the lips are carried off away from the mouth and on to the disk. The ordinary motion of the cilia then sends a current of water into the stomach cavity by way of the region of the siphonoglyphs and creates an outward current in the rest of the gullet, while the cilia on the tentacles are always beating toward the tip.

If we put a bit of food such as a piece of fresh oyster or mussel on the tentacles, the ones that are touched by the food contract over it and in so doing point toward the mouth. The cilia are still beating toward the tip of the tentacle but as it is pointing toward the mouth, the motion is now driving the food in that direction. The lips often seem to move over to meet the particle of food, they at least open slightly to allow it to pass down. It will be remembered that the cilia at the mouth, except for those on the siphonoglyphs are making an outgoing current. It has been found that the presence of food causes the cilia in the gullet to reverse the direction of their beating so that the food glides down the tube by the action of the reversed cilia. When the food has reached the stomach cavity the cilia once more resume their outward beat and the current of water again flows steadily in and out of the animal taking oxygen in with it and carrying off waste matter.

Most of the *Anthozoa* are fixed forms, the eggs developing into ciliated larvae, which swim about for a few days or hours and then attach to something and develop into the adult form, which may be either solitary or colonial. They are mostly shore and shallow water forms, not being found in great numbers below 50 fathoms.

The group is divided into two subclasses:—

Alcyonaria, in which the fully developed animals have only eight tentacles and eight mesenteries;

Zoantharia, in which the tentacles and mesenteries number 6, 12, 24, or more, but never 8.

Subclass—ALCYONARIA

The sea pens, sea fans, gorgonian corals, and sea pansies are in this group and though not abundant on the coast are either so odd or so beautiful that they never fail to attract attention. Nearly all of them are colonial, with complex groups of individuals united to a common stalk. The colony may be branching, or feather-like, or flat like a mat or a mushroom, but in any case the individual polyps are distributed over its surface in some special order, and the body cavity of each polyp communicates with canals which lead through the colony. The polyps of many of the *Alcyonaria* are dimorphic, that is, of two distinct types. The fully developed ones with eight pinnate tentacles and eight mesenteries are called autozooids in contrast to the siphonozooids which have no tentacles and only incompletely developed mesenteries. Cilia on the mesenteric filaments and in the gullet keep the water flowing in and out of both kinds of zooids and consequently of the canals of the stalk also, since all are connected. The skeletons of *Alcyonaria* are made up of spicules of either calcium carbonate or of a horny substance. The spicules may be warty clubs, spindles, or needles which lie loosely in the softer tissue or grow together to form a hard compact skeleton. The nettle cells are minute and although they probably can paralyze the

PHYLUM—COELENTERATA

tiny organisms which serve as food they cannot penetrate the human skin.

Order—ALCYONACEA

These are fixed colonial forms without an axis cylinder.

Family—ALCYONIDAE

Anthomastus ritteri Nutting (fig. 77). The red polyps of the colony grow out from a fleshy rounded head which is supported by a short, thick stem. Small siphonozooids closely cover the whole fleshy base, giving it a granulated appearance. One of the naturalists of the Albatross described the living specimen as resembling "an early rose potato stuck full of red cloves." The head is about 70 mm. long and 25 mm. deep. It has only been found in deep water off central and southern California.

Order—PENNATULACEA

The colonies in this order are free and consist of a stalk and an expanded portion called the rachis. The stalk is usually embedded in the sand or mud, while the rachis, lying at or above the surface of the mud, bears the individuals of the colony, called polyps or zooids, and may be expanded into a feather-like or plate-like structure. There is usually a horny axis which supports the colony, and the mesogloea may contain spicules. The polyps are connected by canals and are of two kinds; the autozooids, nutritive individuals, typical in structure, and the siphonozooids, without tentacles or gonads and serving to convey water into and out of the canals.

Family—PENNATULIDAE

Colonies are feather-like with the polyps borne each side of the axis on leaf-like structures.

Ptilosarcus quadrangularis Moroff (fig. 66) is a deep water form which is frequently brought in by the fishermen who get it on their trawl lines. The specimen shown was six inches long when contracted and when fully expanded was

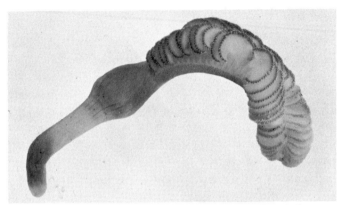

FIG. 66.—A sea pen from deep water, *Ptilosarcus quadrangularis*. (Approximately x⅓.)

about 15 inches. The "stem" is about one third of the whole length. The polyps are on the outer edges of semicircular leaf-like structures on each side of the stem. Each polyp has the eight pinnate tentacles characteristic of the group. Down the ventral side of the stem are two bands of small, red siphonozooids. These zooids have no tentacles but have mouth openings and body cavities which connect with the main canals which run lengthwise through the colony. The large polyps also connect with these canals by way of passages through the "leaves." The skeletal axis in the center, does not extend the full length of the colony, a specimen nearly 18 inches in length having an axis not quite 8 inches long. Spicules are in the edges of the leaves in a band about 2 mm. wide.

The great expansion of the colony seems to be brought about by taking in water through the mouths of the polyps and zooids or through an opening at the end of the stem. When fully expanded the rachis is 5–6 inches wide. The specimen, when photographed, was expanded almost to its fullest extent. The color of the stem is orange and white,

PHYLUM—COELENTERATA

red when contracted. The leaves are white, edged with red around the bases of the polyps which are pure white.

FAMILY—STYLATULIDAE

Colonies are long and slender with small "leaves" or pinnules supported by a plate of radiating spicules.

Stylatula elongata (Gabb). The sea pens (figs. 67 and 68)

FIG. 67.—The sea pen, *Stylatula elongata*, lives in the mud with the bulbous portion buried while the other end, which bears the zooids, extends up into the water; x⅓.

are slender, white colonies 10-12 inches long, found in the mud with their bulbous lower ends buried. The colony is supported by a slender stony axis and when it is contracted the tip of this axis is all that can be seen above the surface,

FIG. 68.—Part of a Sea Pen, *Stylatula elongata*, enlarged to show the polyps; x10.

but when the colony is fully expanded, the axis, covered with the filmy living tissue, extends several inches above the mud. The polyps appear to be arranged in whorls but are really on paired leaf-like structures which nearly surround the stem and overlap somewhat on the dorsal side. These "leaves" are supported below by 12–16 stiff, transparent, thorn-like stays which project outward from each leaf and inward nearly to the axis. There are about ten pairs of leaves to the inch with the vertical polyps standing on the upper edges of the leaves, closely crowded, 20–24 in a row. Siphonozooids are closely packed over the whole surface between and under the leaves but are not visible without dissection. This species is found from San Francisco to San Diego in the mud of shallow bays. We have found them in Mission Bay and Balboa Bay.

Stylatula gracile (Gabb) is much like the preceding species but the colonies are excessively slender, reaching two feet in length. The leaves are numerous, close together, and short, the largest being 6 mm. long. The siphonozooids on the sides between the leaves are readily visible and are in groups of 6–12. This species is found in Monterey Bay.

FAMILY—RENILLIDAE

The polyps grow from a flattened, heart-shaped rachis which has a short smooth peduncle without an axis (plate 4).

Renilla amethystina Verrill (fig. 70), the sea pansy, has a heart-shaped rachis which is violet or amethyst color with white transparent polyps. The tip of the short peduncle is also white. The autozooids may number two hundred or more and the siphonozooids are even more numerous. The colony can contract so that it is 40 mm. long and then become inflated until it is 80 mm. long. The polyps, fully expanded may be half an inch long. Dr. G. H. Parker[*] has some interesting accounts of experiments to determine the direction of the flow of the current. He says that the water flows into the colony mainly through the siphonozooids, down the peduncle to its tip and back up to and out of the axial siphonozooid which is on the upper surface of the colony near the center and at the end of a smooth tract of integument that starts from the root of the peduncle.

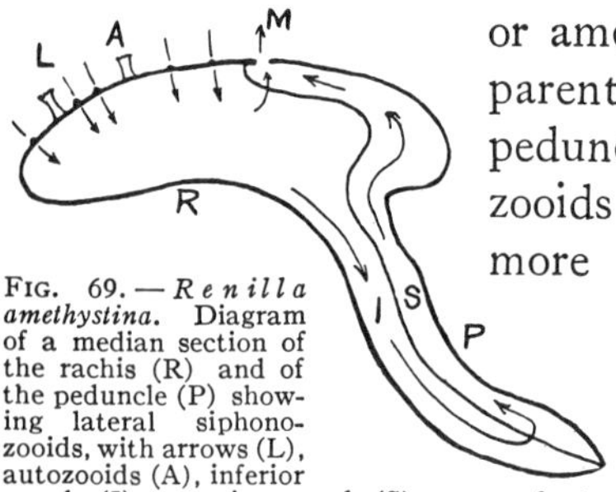

FIG. 69.—*Renilla amethystina*. Diagram of a median section of the rachis (R) and of the peduncle (P) showing lateral siphonozooids, with arrows (L), autozooids (A), inferior canal (I), superior canal (S), pore of the median siphonozooid (M). The direction of the current of water in a resting individual is shown by the arrows. (After G. H. Parker.)

---

[*]G. H. Parker—Activities of Colonial Animals. I. Circulation of Water in *Renilla*; II. Neuromuscular Movements and Phosphorescence in *Renilla*; Journal of Experimental Zoology, Vol. 31.

The diagram (fig. 69) of a lengthwise cut through the center of the colony shows this usual direction of flow of the current and also shows how the interior of the peduncle is divided into two canals, an upper and a lower one. The autozooids serve for the entrance of food and the discharge of ova and

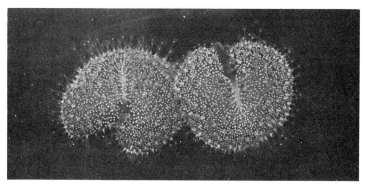

FIG. 70.—The sea pansy, *Renilla amethystina*, in sea water, fully expanded. Photographed by Frank W. Peirson, Pasadena. (Approximately x½.)

waste matter. Spawning has been observed by Dr. Parker in the month of August. Only under considerable pressure is water discharged from the autozooids, the lateral siphonozooids or the pore at the end of the peduncle.

We have found sea pansies at Mission Bay, near the Ocean Beach bridge, and at Newport along the west shore of the bay. One can get them at low tide close to the edge of the water or in the shallow water. They are usually contracted and partly covered with sand so that one must dig under the little horseshoe-shaped marks in the sandy mud.

This is an excellent species to keep alive in the laboratory for study. It bears captivity well and its peculiar undulating movements as it contracts or inflates various parts of the disk are most interesting. Its luminescence may be seen at night if one prods the animal with a blunt instrument. A wave of bluish light will run over the upper surface of the colony, starting from the point that was stimulated. If an animal is taken during the day time into the dark and stimulated, no light is seen, but if during the day time the animal is left in a dark room for several hours and then stimulated, the light will show clearly.

Order—GORGONACEAE

These are colonies having a distinct axis which, in the case of the branched colonies, extends through all the branches.

Family—PLEXAURIDAE

Euplexaura marki Kükenthal (figs. 71, 72 and 73) is a species that grows in deep water and is taken with the dredge. The photographs show the manner of branching, the colony as a whole being irregularly fan shaped. The color is coral-red and the living polyps are yellowish white.

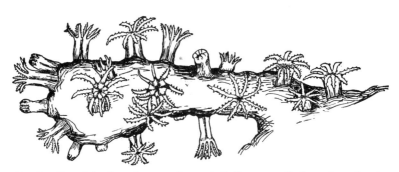

FIG. 71.—Expanded polyps of a gorgonian coral, *Euplexaura marki*: drawn from life; x7.

The cups from which the polyps project are low and rounded. The spicules are small, rough, warty spindles, many of them double, those on the tentacles and body of the polyps being few, more slender, and less warty than the others.

Now and then, after there has been a storm at sea, one may pick up on the beach living specimens of *Eunicea* sp. (fig. 74), a deep water gorgonian coral with a horny axis and a calcareous outer skeleton. The living polyps are

PHYLUM—COELENTERATA

yellowish white. The colony may reach 8-10 inches across.

Subclass—ZOANTHARIA

The sea anemones and the stony corals belong to this group. The species are diverse in form but they all differ from the *Alcyonaria* in that the tentacles are never pinnately branched but are usually simple, hollow, unbranched, and numerous. The gullet is flattened and the deep

Fig. 72.—Gorgonian coral, *Euplexaura marki*, photographed from life. The outlines of the small, expanded polyps can be seen along the sides of the branches against the light background. (About natural size.)

Fig. 73.—A gorgonian coral from deep water, *Euplexaura marki;* about ½ natural size.

grooves, one at each end of the slit-like mouth, are called the siphonoglyphs (fig. 65). The six pairs of primary mesenteries usually join the body wall and the gullet, the two pairs which join the body wall to the siphonoglyph being called the directives. There are usually a number of other mesenteries between the primary ones but most of them do not reach to the gullet. Nettle cells are usually present, sometimes in special short tentacle-like structures in the outermost circle of tentacles, and sometimes in long thread-like structures called acontia (fig. 83) which are discharged when the creature is disturbed, being shot out through the mouth or through the body wall at certain points.

Our species are included in the two orders; *Actiniaria*, sea

FIG. 74.—*Eunicea* sp., a gorgonian coral that grows in deep water; x½.

anemones, solitary animals without a skeleton; *Madreporaria*, corals, colonial animals with a calcareous skeleton.

Order—ACTINIARIA

The flower-like sea anemones are well known to anyone who has ever seen the rocky beaches of our coast. Many of the tide pools and rocks are fairly carpeted with the creatures. They extend from the shore to considerable depths and range in size from a fraction of an inch to a foot in diameter. Some kinds live in the sand, some in mud, and others attach to rocks, shells, seaweed, piles, or any support that is offered. Their colors are varied, even one species alone showing almost every possible color.

PHYLUM—COELENTERATA 97

Visitors at the beach are fond of poking the anemones in the tide pools, in order to see them contract and close. A collector who may be near a mass of anemones left uncovered and contracted at low tide, may be startled by jets of water squirted out from the mass. Some of the anemones have suddenly contracted still more, and have forced the water out from the body cavity through the mouth or through the body wall.

The order is sub-divided as follows:
  A. Colonial forms..............................*Zoantheae*.
  B. Solitary forms
    a. *Edwardsiae*—tentacles in one circle, body long, slender, with outer wall fluted along the lines of the eight mesenteries.
    b. *Hexactiniae*—tentacles numerous and varied in form and arrangement, six or more pairs of mesenteries (fig. 75).
    c. *Ceriantheae*—tentacles in two distinct circles.

Tribe—Edwardsiae

Edwardsiella californica McMurrich (plate 3) burrows in the sandy mud along the bay shores. We have found them

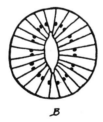
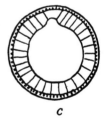

A   *B*   *C*

FIG. 75.—Diagrammatic cross sections of anemones to show arrangement of the mesenteries. Black spots represent muscles. A. *Edwardsiella;* B. *Harenactis;* C. *Cerianthus.*

at Balboa and at Mission Bay. The flower-like disk when expanded lies above the surface of the sand and is so nearly transparent that it is almost invisible. As much as half an inch of the column may extend above the mouth of the burrow. There are 16 tentacles, eight of which bend outward and eight are directed upward. Each tentacle is about twice as long as the diameter of the column.

Expanded specimens may be four inches or more in length and a quarter of an inch in diameter. The long, slender column is covered with a brown epidermis, except for a short basal portion and a part next to the tentacles. This clear part of the column is red-brown, darker on the ridges and lighter between. There are eight well marked ridges and furrows on the column corresponding with the mesenteries within. The mouth is somewhat raised above the disk in an expanded specimen and has an elongated oval opening with a red-brown rim. The tentacles are transparent with red-brown and white splotches. Specimens show considerable variability as to color. The typical arrangement of mesenteries is shown by the diagram (fig. 75-A).

Tribe—HEXACTINIAE

Our species belonging to this tribe are divided into the following families.

    A. Burrowing forms, not attached..............*Ilyanthidae*.
    B. Species usually attached by a pedal disk
        a. Nematocyst threads (acontia) absent........*Cribrinidae*.
        b. Nematocyst threads (acontia) present........*Sagartidae*.

Family—ILYANTHIDAE

Harenactis attenuata Torrey is a burrowing form which lives buried in fine sand or mud. When undisturbed the transparent

Fig. 76.—*Harenactis attenuata*, a burrowing species, as it appears with its tentacles extended on the surface of the mud at the bottom of a tide pool. Photographed from life; about ½ natural size.

Fig. 77.—*Anthomastus ritteri*; from a photograph of a preserved specimen, about ½ natural size.

tentacles fully expanded lie on the surface as shown in the photograph (fig. 76), but when disturbed the animal draws down into the hole, contracting to half its former length. The burrows are deep and as the animal may be extended to a length of 16 inches it is often difficult to dig deep enough to get a whole specimen. The base is sometimes spherical and sometimes a flattened disk.

There are 24 tentacles and 12 pairs of mesenteries (fig. 75-B). The body is dirty white and when contracted is about ¾ inch in diameter, has many transverse wrinkles and 24 shallow longitudinal furrows which mark the bases of the mesenteries. There are no acontia but toward the upper part of the column there are pores through which nematocysts may be protruded. We have taken this species at Mission Bay and at Balboa and it has been reported from San Pedro.

Family—CRIBRINIDAE

Cribrina xanthogrammica (Brandt) is the most common sea anemone of the coast (fig. 79 and plate 4). It is found

Fig. 78—*Epiactis prolifera*, a sea anemone that carries its young until they are partly grown. Note the circle of young anemones just below the tentacles of the parent. About natural size.

Fig. 79—*Cribrina xanthogrammica* is usually green if it has lived in a brightly lighted place, but specimens in shaded places may be white, pink, or lavender. About ½ natural size.

from Sitka, Alaska, to Panama. The disk may reach 6 inches in diameter with the column twice that length. The color is variable, the tubercles on the column serving better than color as a distinguishing mark. Dr. Torrey* states that this is "probably the only verrucose species occurring in the region (from San Francisco to San Diego) between tide marks." The column may be greatly extended or contracted and has longitudinal rows or tubercles to which pieces of shells often adhere by suction. The tubercles are largest near the tentacle bases where they have several lobes and contain numbers of nematocysts. The tentacles are in six cycles and there are about 24 pairs of perfect mesenteries.

Specimens living in the sunlight are usually green and much darker than those that grow in the shade. Dr. Torrey says, "The characteristic green color of the species is found only in individuals exposed to the sun. It is due to the presence of a unicellular alga in the endoderm of the column wall, mesenteries, and tentacles. Where sunlight does not penetrate, as under wharves (Calkins), or in caves, the algae, though present, do not develop so luxuriantly as in more exposed situations and the polyps are correspondingly pale." Although some anemones close when in a bright light and open when the light is less intense, this species opens in the day time and closes at night. If left in a room that is lighted day and night they remain open but if kept in a darkened room they stay closed. The paper by Dr. Wilson Gee,† listed in the bibliography, gives a detailed account of *Cribrina's* reaction to light.

The mouth disk has radiating stripes marking the position of the mesenteries. Specimens growing in the light are usually green, marked with purple, while those growing in shaded places are frequently white, or delicately tinted with pink or lavender.

This species is a hardy one and easy to keep alive in the

---

*H. B. Torrey—The California Shore Anemone *Bunodactis xanthogrammica;* University of California Publications in Zoology, Vol. 3, p. 44.

†Wilson Gee—Modifiability in the Behavior of the Shore-anemone *Cribrina xanthogrammica* Brandt; Journal of Animal Behavior, Vol. 3, pp. 305–328.

PHYLUM—COELENTERATA

laboratory. Unless one has a very large tank and quantities of sea water it is best to attempt to keep only small specimens an inch or two in diameter. Bits of mussel, clam, or oyster make good food for them but if these are not available they will usually take cooked white of egg. The Cambridge Natural History* states that a specimen of *Actinia* lived in an aquarium for 66 years and that specimens of *Sagartia*, still living, are known to be about 50 years old.

Epiactis prolifera Verrill (fig. 78) is a small sea anemone, the column in our largest specimen being about an inch in diameter. The point of special interest about it is its egg-pits on the outside of the body. It is thought that the eggs remain within the body until they have reached an advanced stage of development when they migrate to the pits where they complete their development. Specimens are frequently found with a complete circle of little ones around the middle of the column.

The color is usually red or red-brown, sometimes greenish. The column is marked by vertical lines of lighter color and the pedal disk often spreads out at the base as shown in the photograph giving the margin a fluted appearance. The tentacles number 90 in a large specimen. The mesenteries are in sixes and there are 4–5 cycles. The species has been reported from Puget Sound to San Pedro. At Friday Harbor it is abundant on the eelgrass at low tide. We have found it on the under side of the rocks in tide pools at Point Firmin and at La Jolla.

Family—SAGARTIDAE

The *Sagartia* shown in plate 3, figure 5 is undescribed, but greatly resembles *Sagartia luciae* Verrill of the Atlantic coast. The color is usually more of an olive-green than the figure shows, with a few narrow, orange or yellow, vertical

---
*S. J. Hickson—*Coelenterata* and *Ctenophora*; Cambridge Natural History, Vol. 1, p. 375; The Macmillan Company.

stripes on the column. The tentacles, 25-50 in number, are long, slender, and arranged in poorly defined whorls. Our specimens are small, about one-fourth inch in diameter and a little higher than wide. The oral disk is light with darker radial lines. Frequently there is a white diameter line across the disk or it may be only a radius, if the animal has divided recently. Acontia are freely given off when the animal is disturbed. *Sagartia luciae* is found all along the Atlantic coast and has been reported from San Francisco. We have found this similar form to be very common at certain points in San Diego Bay and Mission Bay.

This is an excellent species in which to study reproduction by division. The animals will be found on sticks, eelgrass,

FIG. 80.—A group of sea anemones, *Sagartia* sp. growing on old oyster shells. Photographed from life, about ½ natural size.

FIG. 81.—*Metridium dianthus;* about ½ natural size.

shells, etc. After being put into sea water they will move to the bottom or sides of the dish and begin dividing without delay. The basal disk elongates and finally divides into two parts, the tear extending upward until both the column and disk have been torn in two. The torn edges of each half come together and heal so that we have two anemones each about half the size of the original one.

The *Sagartia* shown in figure 80 is probably undescribed but resembles *Sagartia davisi* Torrey. It is usually found in the mud, attached to clam or oyster shells or to rocks. The anemone extends its tentacles above the layer of soft mud which may have drifted over the object to which it is

PHYLUM—COELENTERATA 103

clinging, being able to extend its thin-walled column a distance three or four times greater than the length of the contracted column. The diameter is about 10-15 mm. and the spread of tentacles reaches 25 mm. in large specimens. The foot disk is extensible, the body wall thin, brown or flesh-colored, and semitransparent so that the mesenteries are visible through it. There is a well defined collar a little below the level of the tentacles and the column above this point is slightly narrower than it is below. The disk is light-colored, often with radiating or circular markings, the mouth a slit, often bordered with orange or salmon color. When the animals are disturbed, acontia are freely shot out

FIG. 82.—Puget Sound anemones in an aquarium. We do not know the name of the largest specimen. The others are *Metridium dianthus;* about ⅓ natural size.

from near the base and higher up on the column, which makes the species an excellent one in which to study the nature of acontia and sting cells (fig. 83).

Specimens in the aquarium move about quite freely and will often be found creeping along the surface of the water, hanging upside down from the surface film. They will readily reproduce by dividing and also by what is called "pedal laceration." The basal disk expands, spreading out in a thin film covering twice the usual area. When the animal resumes its normal shape some of the outer part of the film is left. These fragments break up into a number of spherical portions each of which soon will have developed tentacles and mouth. Many of these new individuals are hardly the size of a pin head and there are usually great numbers of them produced, especially if the water in the aquarium is allowed to get slightly stale.

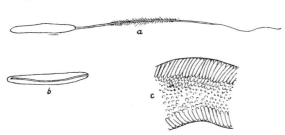

FIG. 83.—Nematocysts and acontia of *Sagartia* sp. (Fig. 80). a. Nettle cell discharged; b. nettle cell not discharged; c. portion of nettle thread magnified to show great number of nettle cells within, greatly enlarged.

This species may be found at San Pedro, Balboa, and San Diego Bay. In a slough off from San Diego Bay at National City we have seen the floor of the channel (under the 8th Street bridge) fairly carpeted with these anemones. When one gets down to the spot, the animals may have all contracted so that none can be seen but if some of the rocks and clam and oyster shells are gathered the anemones will usually be found on them in great numbers.

Metridium dianthus Ellis (figs. 81 and 82) may be recognized by the short, numerous tentacles which extend over nearly the whole of the greatly expanded and frilled oral disk. The color may be brown, salmon, orange, or white. The species is one of the largest and most common anemones on the Atlantic coast from New Jersey to Labrador and is found on our coast from Alaska to Monterey. *Metridium* reproduces sexually and nonsexually. The latter may be by fission or dividing into two nearly equal parts, by budding, or by basal fragmentation as described for the preceding species. The size is given by Pratt thus, "length up to 10 cm., width 7 cm." The photograph (fig. 81) shows an individual with a greatly expanded base. Specimens in glass dishes in the laboratory often spread out in this fashion. The anemone shown in plate 3, figure 2 is undescribed. It is a sand dweller and is often found along with *Harenactis attenuata*. It lives buried in the sand with the tentacles spread out at the surface, the disk remaining in an expanded state even after the tide has left it uncovered. The specimen figured was 12 mm. in diameter, the spread of tentacles being 20 mm. or more. The color of the disk and tentacles may be seen in the color sketch. The brown is often replaced by orange and the mouth may have more white on the lips. The tentacles may show more black and the white radiating lines are more noticeable in some specimens. The tentacles are in 5 series and there are about 100 in all. Those in the outer row are usually short and held back against the column. The column is white. Circular muscle bands are prominent but longitudinal ones are not visible from the outside. In retraction the tentacles do not wholly disappear. No acontia have been seen though the animals have been prodded diligently. In one case we found a specimen attached to a shell two inches below the surface of the sand. Specimens have been collected at Mission Bay and under the piers on the bay shore at Balboa.

Corynactis sp. (plate 3) is a beautiful little anemone found at Monterey. It is usually in crevices between rocks or under overhanging portions of rock along with the coral *Balanophyllia*. The anemone is almost blood-red with white tentacles which are markedly capitate, that is, enlarged and rounded at the tips. It is a small species commonly not over 15 mm. in diameter.

Tribe—CERIANTHEAE

Family—CERIANTHIDAE

Cerianthus aestuari Torrey and Kleeburger (fig. 84) is a burrowing form found on mud flats at Mission Bay along with *Harenactis attenuata* and *Corymorpha palma*. The body projects a little from the burrow and the tentacles are spread out on the sand sometimes making a circle of 4–5 inches in diameter. There are two sets of tentacles, those of the outer circle are long and the inner ones, close to the mouth, are short. There are about 30 in each set with the same number of mesenteries. The column is streaked and

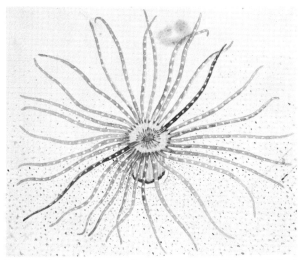

FIG. 84.—*Cerianthus aestuari*, a burrowing anemone which has a circle of short tentacles around its mouth, as well as its conspicuous circle of long tentacles; x⅔.

mottled with brown, and the transparent tentacles are delicately banded, a single pair being often more deeply colored than the others. There are no acontia. There is a terminal pore through the body wall at the basal end. The animal surrounds itself or lines the burrow with a felt-like sheath of nematocysts.

Order—MADREPORARIA

The skeletons, at least, of the coral animals are well known to everyone although one seldom sees the living animals. The creature is like a tiny sea anemone except that it has the

PHYLUM—COELENTERATA 107

habit of depositing a plate of carbonate of lime at its base and between its mesenteries. The skeleton is formed by the outer layer of cells, the ectoderm, and is external to the body. The basal part is built up at the outer margin in many species until it forms a distinct cup into which the animal can draw itself when it is disturbed.

There are both solitary and colonial corals. In the colonial ones, the body of an individual extends beyond the calyx, or cup, as if it overflows it and produces successive layers of skeleton in the spaces between. New polyps are also budded off from this part of the animal. The exact arrangement and method of producing these new polyps and the skeleton between them and the general form of the whole colony is different in the different species.

The reef-forming corals have flourished best in the warm, shallow waters of the world but other corals may be found outside of this habitat. One species has been found off the coast of Alaska and another off the Cape of Good Hope. Though our shore region now boasts only a few scattering species of solitary corals, there was a time in the distant past when corals flourished here, since fossil coral reefs have been found in a number of places in California. One of these ancient reefs is on the Colorado Desert and runs across Coyote Mountain.

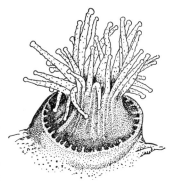

FIG. 85.—*Astrangia* sp., a small coral found on the southern California coast. Drawn from life; x7.

Investigations of coral reefs have shown that the greatest depth at which the reef coral animals are alive is 30–50 fathoms. Many of the tropical islands of the Atlantic and Pacific have been built up largely by the growth of coral polyps. At one time a number of borings were made into a typical coral island and the material thus obtained was examined and found to be of the same sort from top to bottom though the borings reached over a thousand feet in depth. A lime-secreting seaweed, such as frequently lives among coral reefs, predominated, with the skeletons of one celled animals and corals next in importance. These results would indicate that the islands have been built up through long ages by the growth of the corals and seaweeds and the deposition of the skeletons of the one celled animals. The fact that the greatest depth at which the reef coral animals are able to live is 30–50 fathoms, coupled with the fact that corals were found in the borings at a depth of 1000 feet would indicate that the island and probably the whole region has been sinking for a considerable period of time. We refer the reader to the writings of Darwin, Lyell, and Dana and the more recent work of Vaughan for further discussion of coral reef formation.

MADREPORARIA IMPERFORATA

Family—ASTRANGIDAE

Astrangia sp. (fig. 85) is a coral that may be found growing on the rocks in the tide pools near La Jolla. When alive it

Fig. 86.—The cup shaped skeletons of a small California coral, *Astrangia* sp.; x3/2.

Fig. 87.—The empty cup of the coral, *Balanophyllia elegans*, which is abundant along shore at Pacific Grove; x3/2.

is orange or coral-red with lighter tentacles. There are 36 tentacles with white, blunt tips. The mouth is a slit-like

PHYLUM—COELENTERATA

opening. The height of the column is 1/4 inch and the spread of tentacles is 1/2 inch. The septa radiate in a fan-shaped arrangement (fig. 86) and the cup is nearly circular in outline.

Family—EUPSAMMIDAE

Balanophyllia elegans Verrill is a beautiful orange-red or flame-red coral that is fairly abundant at low water at Monterey. It is usually found adhering to the under side of large stones. The color sketch (plate 3) shows the appearance of the living animal and the photograph (fig. 87) shows the cup with its characteristic star pattern. The cup is 5–10 mm. high and has a diameter of 7–10 mm. This species has been reported as far north as Puget Sound.

Class—CTENOPHORA

(The comb jellies)

The ctenophores are spherical, lobed, thimble-shaped, or band-like, usually transparent and gelatinous, and are found floating at the surface of the ocean. Their transparency makes them almost invisible and they are usually only taken with the tow net. Some of the spherical or thimble-shaped Atlantic coast species are 4–6 inches in length and the ribbon-like forms reach 3 feet in length and 3 inches wide.

The name *Ctenophora* comes from a word meaning comb and the name comb jellies is a common one given the group. The "combs" are eight rows of paddle-like organs which are arranged like lines of longitude on a globe. Each little paddle is made up of a row of cilia which are fused together at the base. These paddles are quickly raised and slowly flattened down again. As the plates from one end of the meridian to the other go through these motions in a rhythmic fashion it looks like a series of waves traveling down the ribs and drives the animal through the water oral end first. The sense organ consists of a hard statolith supported by four tufts of fused cilia and usually covered by a dome which is believed to be formed by a fusion of cilia. One ctenophore has both male and female sex organs which are located on the outer wall of the meridian canals. The ova and sperm reach the outside by way of the mouth. The development is direct, there being no alternation of generations. They often occur in schools and many of them are highly luminescent at night.

The various kinds of ctenophores are grouped as follows:
 A. *Tentaculata*, with either tentacles or oral lobes;
 B. *Nuda*, with neither tentacles nor oral lobes.
  The *Tentaculata* are divided into
   a. *Cydippida*, body globose or cylindrical, with long tentacles.
   b. *Lobata*, body globose or cylindrical, no tentacles.
   c. *Cestida*, body ribbon-like.

SUBCLASS—TENTACULATA

ORDER—CYDIPPIDA

FAMILY—PLEUROBRACHII-DAE

Pleurobrachia bachei A. Agassiz (fig. 88). These ctenophores are called "cat's eyes" by the fishermen. They are remarkably transparent and are colorless except for the tentacles and the esophagus which are sometimes tinted with reddish. The sphere is 13 mm. or more in diameter and the tentacles can extend more than five times the diameter of the body.

FIG. 88.—*Pleurobrachia bachei*, a ctenophore or "cat's eye;" x3/2. If the water of the aquarium is undisturbed, the tentacles may extend until they are much longer than they are shown in the figure, but if they are disturbed they contract quickly.

The combs, as they move in rhythmic succession, often show iridescent colors. The tentacles which are fringed with pinnae are very sensitive and contract quickly when they touch anything. They are provided with adhesive cells which help in capturing the prey. The esophagus is short and the funnel tubes open from it at a point about ¼ the distance from pole to pole. They are found from Puget Sound to San Diego.

Order—LOBATA

Family—BOLINOPSIDAE

Among the schools of jellyfish which occur near the surface at Friday Harbor, Washington, during the summer months, one frequently finds a soft, bilobed ctenophore with the body markedly compressed. The creature is so delicate that it is only by the exercise of considerable care that it may be taken up without injury. It is a species of *Bolinopsis*, probably *B. microptera* (A. Agassiz), originally described from the Gulf of Georgia. There are eight rows of swimming plates, four longer than the others. Figure 89 shows only the plates upon one half of the animal, the features upon the opposite side being omitted for the sake of clearness, though the animal is so transparent when alive that all of them can be distinctly seen from any aspect.

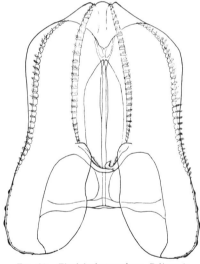

Fig. 89.—The lobed ctenophore, *Bolinopsis microptera*: x2.

Subclass—NUDA

Family—BEROIDAE

Accompanying swarms of medusae and ctenophores, one frequently finds the delicate *Beroe forskalii* Milne Edwards (fig. 90). It may be distinguished from the other ctenophores by its vase-like shape, and by its translucency, often like that of ground glass, in contrast to the marked transparency of other forms. Torrey described the very young individuals as colorless and the half grown ones as rosy, with brilliantly iridescent rows of swimming plates. A conspicuous feature is the network of ramifying canals which unite with the gastric canals. There are eight rows of ciliated plates extending from the apex of the body about two-thirds of the distance to the mouth. There are no tentacles. It is carnivorous and said to be exceedingly voracious, often feeding on other ctenophores nearly as large as itself. This species of *Beroe* has been recorded from the California coast, the Antarctic, and a number of other places.

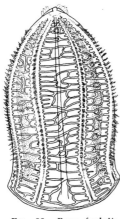

FIG. 90.—*Beroe forskali;* somewhat reduced. (After Torrey.)

CHAPTER VI.

THE LOWER WORMS

The lower worms which make up this phylum are flat worms, round worms, "wheel animals," arrow worms and others. Worms are readily recognized by their slender elongated bodies and the lower worms may be distinguished from most of the annelid worms discussed in another chapter by the fact that their bodies are not divided into segments or rings, or if segmented, have no appendages for locomotion.

These lower worms are not often seen by the casual collector but may be found in abundance if one knows where to look for them. If a mass of mussels, anemones, and goose barnacles are torn from their attachments on the piling of some wharf, and examined, flatworms will often be found. If at low tide, one turns over the stones in the tide pools, usually a number of flatworms will be found on the under side. A person must be quick to see them, however, for they lose no time in disappearing into the nearest hiding place. They so resemble their surroundings that many of them are overlooked by collectors. Some species of the flatworms are parasitic. A few round worms are to be found swimming at the surface of the sea but the great majority of the group are parasites. The "wheel animals" or rotifers are microscopic and move by means of cilia so arranged that when they are in motion the animal seems to have revolving wheels on the anterior end. The arrow worms or chaetognaths are minute, nearly transparent, pelagic forms which are taken only in fine-meshed tow nets.

The worms of this group are alike in being bilaterally symmetrical, in fact they are the lowest group to exhibit such symmetry.

114 SEASHORE ANIMALS OF THE PACIFIC COAST

The following subdivisions of the group are represented in the marine life of the coast. These subdivisions are variously ranked as phyla or subphyla by different writers.

Platyhelminthes—body flattened, leaf-like, many species are parasitic.
Nemathelminthes—body round, sometimes thread-like, many species parasitic.
Trochelminthes—minute, aquatic animals, with crown of cilia at the anterior end.
Chaetognatha—minute aquatic animals without cilia.

Phylum—PLATYHELMINTHES

(*The flat worms*)

The bodies of these worms are flattened and well adapted for hiding under stones and among algae or, in the case of the parasites, for living within the tissues of their hosts. Some of them are broad and leaf-like while others are long and ribbon-like. They are variously colored, usually resembling their surroundings, but some of the free living ones are bright orange or pink or gaily striped with several colors. The body is in most cases unsegmented, with no definite head, and no paired appendages.

There are four classes of *Platyhelminthes:*—

Turbellaria—free-living, flat, leaf-like, unsegmented, the surface of the body covered with cilia.
Trematoda —(the flukes) parasitic, small, unsegmented.
Cestoda —(tapeworms) parasitic, body usually made up of many proglottides, no mouth or intestine.
Nemertea —free-living, long, and ribbon-like.

Class—TURBELLARIA

(*The ciliated flatworms*)

These are free-living, soft-bodied flatworms, usually less than two inches long, which live in secluded places under rocks or among mussel attachments or algae. They usually resemble their surroundings in color, this color sometimes being due to the material that the worms have eaten. On the other hand, some species are brightly colored and even gaily striped. Most flatworms are smooth but, in some, the dorsal surface is covered with papillae or spines. Some have nettle cells but these they have acquired from coelenterates which they have eaten. The largest specimen recorded is a free-swimming form found off Ceylon, which measured six inches in length and four in breadth, but most of our species are much smaller.

They crawl with a wave-like gliding motion, due to cilia which cover the surface of the animal. A secretion produced by gland cells envelops the surface of the body and the cilia work in it and are thus enabled to carry the creature along, no matter what may be the nature of the surface over which it is moving. This movement of the cilia sometimes creates a slight turbulence in the water around the animal, hence the name of the group. Some of the turbellarians have suckers with which they can adhere to surfaces.

Turbellarians are probably all carnivorous and take in their food with the pharynx which is thrust out through the mouth, located at or near the middle of the body on the under side. The pharynx may be bell or trumpet-shaped or may have a wide, frilled margin. The intestine is either tubular or branched, usually ending blindly, so that the waste must be ejected through the pharynx. There is an excretory system which consists of one or more longitudinal canals. These receive the excretion from tiny capillaries which have their beginnings in "flame cells," the excreting cells.

Eyes of a simple or primitive sort are usually present in large numbers and are sometimes grouped on the anterior margin but are more often farther back. The brain is some distance from the anterior margin, usually near the eyes, and from it a pair of ventral nerve cords leads to the posterior part of the body as well as to the eyes and the sensitive anterior region of the animal.

Turbellaria are usually hermaphroditic. The ova after fertilization receive a shell coating and are deposited on the rocks as plate-like masses or spirals, the eggs in the mass being united by a gelatinous secretion. The ova are usually cross-fertilized since the male organs develop before the ovaries do, but the periods of maturation of the two elements overlap so that self-fertilization is possible.

Unfortunately, comparatively few of the marine free-living flatworms of our coast have been named. The principal account is that of Heath and McGregor (1912) in which sixteen species from the southern shore of Monterey Bay are described. However, the small number of publications does not indicate any scarcity of individuals, for collectors usually find them plentiful on rocky beaches.

TRIBE—ACOTYLEA

(*Turbellarians without suckers*)

FAMILY—PLANOCERIDAE

Planocera burchami Heath and McGregor (plate 4) is found under stones uncovered by an ordinary tide. The known range extends from Bird Rock, near San Diego, to Monterey. A little way behind the anterior margin are a pair of conical elevations (nuchal tentacles) that can be retracted. Minute eyespots are located about their bases and other eyespots are scattered over the area between them. The mouth is in the middle of the body on the lower surface. The markings vary in color but are distinct. The internal organs appear white when seen through the lower surface. We have found this worm to be quite active, swimming about freely in the laboratory aquarium. A large specimen is about 26 mm. long.

Planocera californica Heath and McGregor is found rather abundantly in the vicinity of Monterey. It agrees with *P. burchami* in the possession of nuchal tentacles, in the position of the mouth, and in most external features. In the grouping

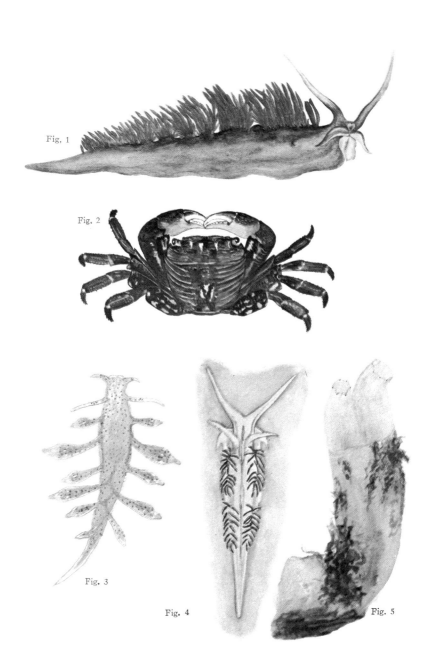

FRONTISPIECE.

FIG. 1 (p. 503).—A marine slug or mollusk, *Flabellina iodinea;* x 2. FIG. 2 (p. 395).—The striped shore crab, *Pachygrapsus crassipes;* x ⅔. FIG. 3 (p. 502).—A mollusk, *Galvina olivacea;* x 7. FIG. 4 (p. 502).—A mollusk, *Hermissenda crassicornis;* x 6 (small specimen). FIG. 5 (p. 594).—*Ciona intestinalis,* a cosmopolitan "sea squirt" or ascidian found on floats and piling in harbors; about ⅔ natural size.

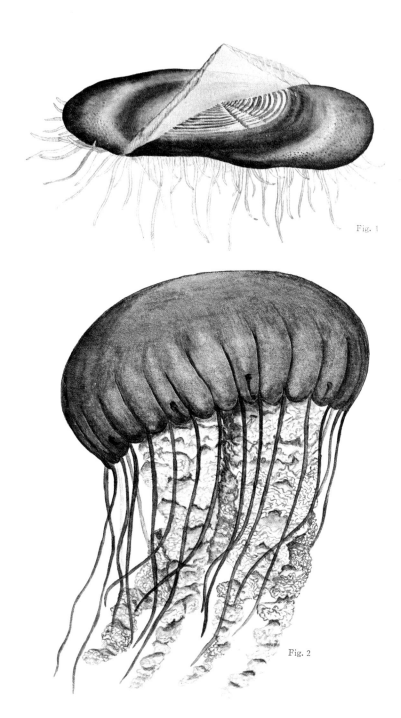

PLATE I.
FIG. 1 (p. 75).—*Velella lata*, from a color sketch of a small, living specimen; x 2. FIG. 2 (p. 81).—*Chrysaora gilberti*, a jellyfish of the California coast; x ½.

PLATE II.

The jellyfish, *Pelagia* sp.; x ⅓. Although jellyfish like this are large and heavy, the body tissues contain less than five per cent of solid substance. (p. 80).

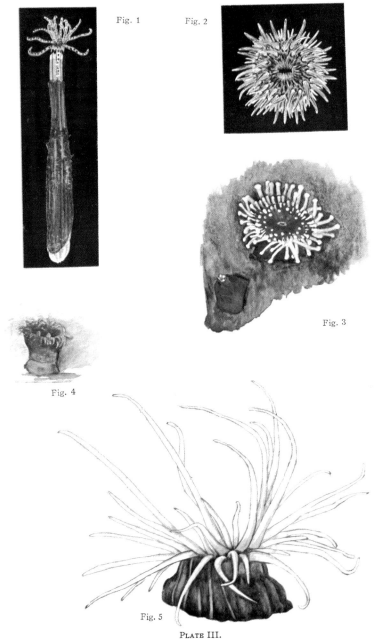

PLATE III.

FIG. 1 (p. 97).—A burrowing anemone, *Edwardsiella californica*. Color drawing from life; x 4/3. FIG. 2 (p. 105).—An anemone that lives with its column buried in the sand and its tentacles and disk at the surface of the sand. Color sketch from living specimen, fully expanded, slightly enlarged. FIG. 3 (p. 105).—*Corynactis* sp., a sea anemone with capitate tentacles. One specimen expanded, the other contracted. About natural size. FIG. 4 (p. 109).—*Balanophyllia elegans*, a coral that is abundant alongshore at Pacific Grove. Drawn from a living specimen in an aquarium. About natural size. FIG. 5 (p. 101).—The striped sea anemone, *Sagartia* sp.; drawn from a living specimen fully expanded; x 10.

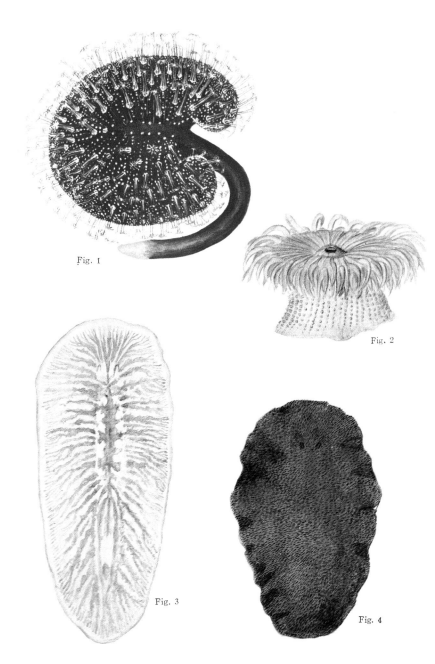

PLATE IV.

FIG. 1 (p. 92).—The sea pansy, *Renilla amethystina*, a large colony with rachis and zoöids expanded, viewed from above. The peduncle, when buried in the sand, serves as an anchor. The zoöids extend above the surface of the sand when the tide is in and the colony is expanded. (About natural size.) FIG. 2 (p. 99).—*Cribrina xanthogrammica*. A small individual, about natural size. (From a color sketch by C. E. von Geldern.) FIG. 3 (p. 116).—A flatworm, *Planocera burchami*; x 3. FIG. 4 (p. 117).—A flatworm from Puget Sound; x 2 (a small specimen).

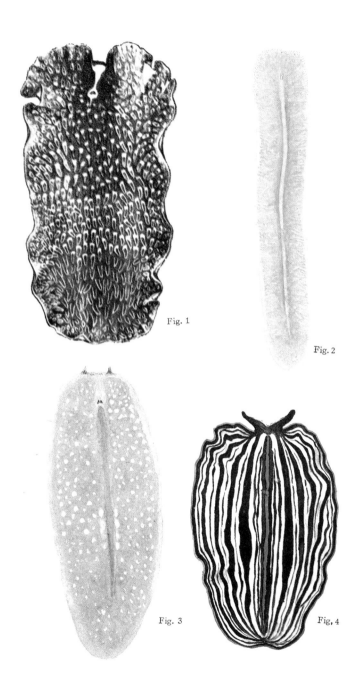

PLATE V. FLATWORMS.

FIG. 1 (p. 118).—*Thysanozoon* sp. Drawn from a specimen taken at La Jolla, California; x 2. FIG. 2 (p. 119).—*Prosthiostomum* (?) a southern California species; x 3. FIG. 3 (p. 118).—*Eurylepta aurantiaca;* x 5. FIG. 4 (p. 118)—The striped flatworm. Drawn from a specimen found at La Jolla; x 2.

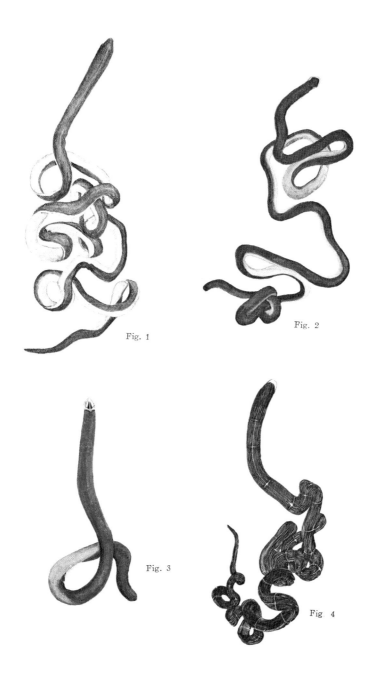

PLATE VI. NEMERTEAN WORMS

FIG. 1 (p. 124).—*Emplectonema gracile;* x 3. FIG. 2 (p. 124).—*Paranemertes peregrina;* x 1.
FIG. 3 (p. 125).—*Amphiporus bimaculatus;* x 1⅓. FIG. 4 (p. 126.).—*Lineus pictifrons;* x 2.

Fig. 1

Fig. 2

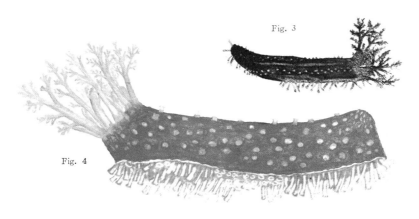

Fig. 3

Fig. 4

Fig. 5

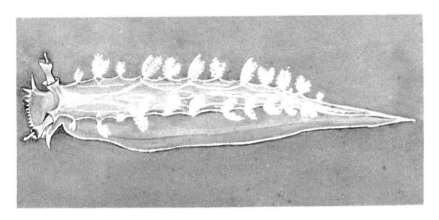

PLATE VII.

FIG. 1 (p. 169).—*Eudistylia polymorpha*. The gills of the worm project from the canals of a sponge in which the worm has made its tube; x ⅔. FIG. 2 (p. 170).—*Serpula columbiana*. The brightly colored gills project from the limy tube in which the worm lives; x 1⅓. FIG. 3 (p. 244).—*Cucumaria curata*. This sea cucumber broods the eggs under the ventral surface; x 1⅔. FIG. 4 (p. 243).—*Thyonepsolus nutrians*, a sea cucumber that carries the young upon its back. The structures shown on the back of this specimen are tube feet. (Small specimen, x 3⅛.). FIG. 5 (p. 491).—A sea slug, *Tritonia festiva*; x 2⅔.

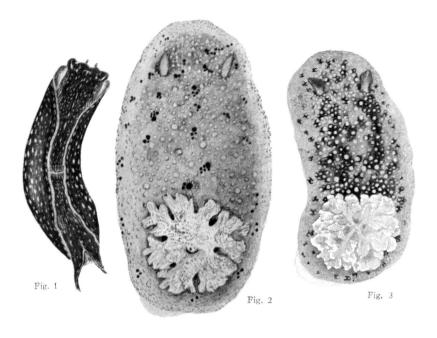
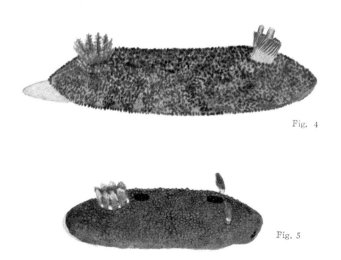

PLATE VIII. NUDIBRANCH MOLLUSKS OR SEA SLUG.
FIG. 1 (p. 485).—*Navanax inermis*, a sea slug often found in muddy bays as well as in rocky tide pools; x ⅔. FIG. 2 (p. 491).—*Archidoris montereyensis;* x 1⅓. FIG. 3 (p. 491).—*Anisodoris nobilis;* x ⅔ (small specimen). FIG. 4 (p. 492).—*Rostanga pulchra*. Most specimens show much more red than this specimen did; x 6. FIG. 5 (p. 492).—*Aldisa sanguinea;* x 2⅗.

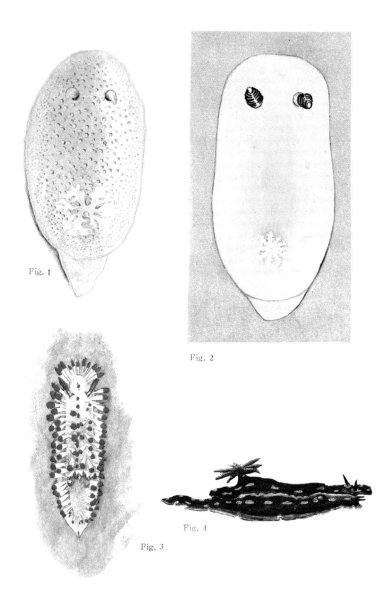

PLATE IX. NUDIBRANCH MOLLUSKS OR SEA SLUGS.
FIG. 1 (p. 493).—*Cadlina marginata*; x 1⅓. FIG. 2 (p. 493).—*Cadlina flavomaculata*; x 6. FIG. 3 (p. 494).—*Laila cockerelli*; x 2. FIG. 4 (p. 494).—*Chromodoris californiensis*; x ⅔.

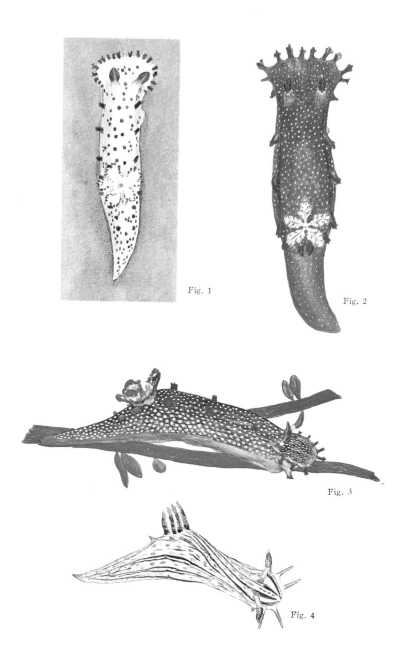

PLATE X. NUDIBRANCH MOLLUSKS OR SEA SLUGS.

FIG. 1 (p. 495).—*Triopha carpenteri*; x 4/3. FIG. 2 (p. 495).—*Triopha maculata*, a small specimen; x 4. FIG. 3 (p. 495).—*Triopha maculata*, a large specimen showing bluish white spots not seen in young specimens; x 2. FIG. 4 (p. 496).—*Polycera atra*; x 3.

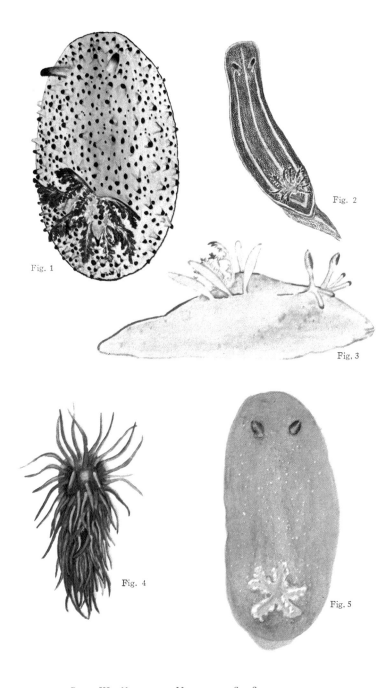

PLATE XI. NUDIBRANCH MOLLUSKS OR SEA SLUGS.
FIG. 1 (p. 496).—*Acanthodoris rhodoceras;* x 6. FIG. 2 (p. 494).—*Chromodoris macfarlandi;* x 1.
FIG. 3 (p. 497).—*Ancula pacifica;* x 5. FIG. 4 (p. 497).—*Hopkinsia rosacea;* small specimen, x 2.
FIG. 5 (p. 498).—*Doriopsis fulva;* x 4/3.

THE LOWER WORMS 117

of the eyespots, however, it differs from the former species. The eyes which are in clusters near the tentacles are arranged is somewhat elongated areas, diverging anteriorly (fig. 21), while the other eyes tend to occupy two narrow parallel areas, beginning between the tentacles and extending forward. The body is too opaque to show the internal organs. Heath and McGregor* in discussing the habits of the species, say: "*Planocera californica* occupies sites farthest removed from low tide mark. Under stones or in crevices of the rocks it finds a hiding place and a food supply consisting of small animals together with scant quantities of diatoms. Throughout the greater part of the year its egg masses, forming more or less circular patches from two to six millimeters in diameter, appear like encrusting plant growths concealed in crevices of the rocks or attached to the under surfaces of boulders scattered on the beach." It is said to reach a length of 24 mm. and a width of 14 mm.

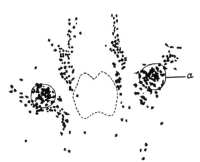

FIG. 91.—Eyes of *Planocera californica*: greatly enlarged: a. A tentacle cluster. (After Heath and McGregor.)

Puget Sound is the home of a large flatworm (plate 4) which seems to be related to the foregoing. It possesses a pair of nuchal tentacles which can be completely retracted into the body. The mouth is a little in front of the center on the lower side. The upper surface is tawny, rather uniformly marked with numerous elongated dark specks. The lower surface is unpigmented. The body is of a very firm consistency and almost leathery in appearance. This species is fairly abundant under rocks in some localities near Friday Harbor and its large size, sometimes three inches or

---

*H. Heath and E. A. McGregor—New Polyclads from Monterey Bay, California; Proceedings of the Academy of Sciences of Philadelphia, Vol. 64, p. 455.

more in length, makes it a conspicuous member of the shore fauna.

Tribe—COTYLEA

(Turbellarians with a ventral sucker)

Family—PSEUDOCERIDAE

Plate 5, figure 1 represents a flatworm of remarkable appearance which we have found only twice, at low tide near La Jolla. It is probably more abundant than our collections would indicate, for it is a rapid swimmer and is able to escape readily when the rocks under which it hides, are moved. The tentacles are folds of the anterior margin of the body. The entire dorsal surface is covered with long papillae which have light colored bands at their bases. This worm belongs to the genus *Thysanozoon*, which is widely distributed in tropical and sub-tropical seas. Our specimens were about an inch in length.

Family—EURYLEPTIDAE

Eurylepta aurantiaca Health and McGregor (plate 5) may be found under rocks at low tide. It is a sluggish form and becomes nearly circular when at rest. A ventral sucker helps it to cling tenaciously to the rocks. The mouth is less than one-sixth of the length of the body from the anterior margin. The tentacles, placed upon the margin, are speckled with numerous eyespots, while two clusters of eyespots occur above the brain region. The species is fairly common both at Monterey and in the San Diego region.

From the standpoint of color, one of the most conspicuous little animals we have seen is the one shown in plate 5, figure 4. On the ventral side it was porcelain white; above, as shown in the drawing, there were stripes of black on a white background, with a stripe of orange along the axis of the body and a border of orange and black upon the margin. The tentacles, mainly black, were marked with white, tinged

THE LOWER WORMS

with red. Altogether the color pattern was one of severe contrasts. We have taken but one specimen and that was somewhat injured in collecting. It was found under a stone north of the caves at La Jolla. The length is about one inch. Superficially, this closely resembles the species of the genus *Prostheceraeus* Schmarda, some of which have been described from the South American and east Asian coasts.

Family—PROSTHIOSTOMIDAE

Quite distinct from the preceding species is the long, narrow form plate 5, figure 2. It is common at La Jolla. The remarkable feature of this worm is the occurrence of numerous eyespots on the anterior margin of the body in addition to the two groups in the brain region. It is without tentacles. When extended, the dimensions of a large individual are 25 mm. by 3.5 mm.

Class—TREMATODA

(*The flukes*)

The trematodes are parasitic flatworms and their manner of life has influenced their form to some extent. The adults are not ciliated, the body being everywhere covered with cuticle. They are provided with suckers and sometimes with hooks, as well, by which they can adhere to their hosts. Most of them are leaf-shaped though some are long and slender. The mouth is at or near the anterior end of the body and opens into a muscular pharynx used to pump in the food which includes blood and other juices of the prey.

The ectoparasitic trematodes, the ones that live on the outside of the body of the host instead of within it, usually require only one host in order to complete their life history. The adults lay eggs which hatch into small worms like the parent. They often attach to the gills of fish since the blood there is so close to the surface that it may be obtained easily and the worms are somewhat protected by the gill covers.

The endoparasitic trematodes live within the host and usually require two or more hosts in order to complete their life cycle. Because these parasites are very small and go through such complicated transformations, the life histories of many species are not known.

A typical life history follows: the egg passes out of the body of the host, usually a vertebrate, and develops into a ciliated larva which finds the intermediate host, often a mollusk, within whose tissues it develops, passing through one or more stages and usually forming great numbers of young individuals, the cercariae, which swim away and find other hosts. If the intermediate host is now devoured by the vertebrate, the latter will become the host, the worms taking up their abode in the intestine, liver, bladder, or some other organ of the body and completing the cycle by producing the eggs.

Comparatively few marine trematodes have been studied extensively. One, however, that interests us lives as an adult in a marine fish while the intermediate host is the eastern oyster. Doubtless many of our mollusks are infested. We examined the common clam *Donax* a few years ago and found in the gonads a great many trematodes in the sporocyst stage (fig. 92-A) containing developing embryos. The clam is, in this case, evidently the intermediate host. The other host might be a marine fish, a shore bird, or some other vertebrate that eats the clam. At the time *Donax* was found thus infested, people were gathering great numbers of them to eat and ran little chance of becoming infested since, even if the worms were not killed in the cooking, the chances are, they would not live in human beings if the usual adult host is a bird or a fish. However, some kinds of parasites can adapt themselves to different environments so it is safer to cook mollusks before one eats them.

Class—CESTODA

(The tapeworms)

The tapeworms are soft bodied, flat, parasitic worms. They always live within the body of the host and usually require two hosts in order to complete their life history. They have neither mouth nor intestine and absorb their food through the body wall. The adult form lives in the intestine of some vertebrate and the larvae usually live in the muscles or other tissues of an intermediate host. This intermediate host serves as food for the final host. The larva is not affected by the digestive juices of the new host, but attaches itself, usually within the digestive tract, takes in the partly digested food of the host by absorption and grows to adult form.

The body of the adult worm is made up of the scolex, which is sometimes called the head, and the strobila. The scolex bears suckers and sometimes hooks as well, which enable the worm to fasten itself to the host. The strobila lengthens as the animal grows, sometimes becoming several feet long, and divides into segments each of which contains reproductive organs which form great numbers of eggs. The segments are cast off and leave the intestine. Many of the eggs never reach the intermediate host but those that do, develop into the larval stage there. Fish are often infested with tapeworms which may take up their abode in man if he eats the fish without cooking them.

Somewhat different from the typical cestode form just described is *Gyrocotyle fimbriata* Watson, which is unsegmented and has an elliptical, leaf-like body with fluted margins. It has a sucker at one end, a rosette at the other, and lives in the spiral valve or intestinal tract of *Chimaera colliei*, the ratfish. The intermediate host is not known. It is probably some invertebrate but as the fish eats almost all kinds of invertebrate animals indiscriminately, it will be difficult to find which one is the intermediate host of the parasite.

Mollusks and worms are known to serve as intermediate hosts for many cestodes. Figure 92-B is a cestode from the

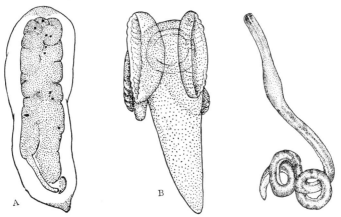

FIG. 92.—Two parasites from the little bean clam, *Donax*. Both are *Plathyelminthes*, or flatworms. A. A trematode; x120. B. A cestode or tapeworm; x60.

FIG. 93.—A yellow nemertian, *Lineus flavescens;* x2.

clam *Donax*, showing the four suckers by which the worm will attach itself when it reaches its final host. Here again the host is not known but may be a sea bird, fish, or some other vertebrate that eats *Donax*.

Class—NEMERTINEA
(The ribbon worms)

The nemerteans are long, slender, flatworms which are often brilliantly colored and always very contractile. Some of them are able to shorten the body to one-tenth of its ordinary length. Some are minute, while others may be several yards long. One species, reported from California but not described in this book, is as slender as a thread and reaches a length of 25 yards.

Nemerteans have a long thread-like proboscis which they sometimes shoot out if they are roughly handled. This proboscis may be as long as the body and is in many cases armed at the tip with a tiny stylet and serves as a weapon of offense and defense. The body of the prey can be pierced by the stylet and at the same time can be held by a tenacious viscid secretion of the lining of the proboscis chamber. A nemertean that is parasitic on the gills of the crab uses the stylet to puncture the gills of the host. The blood and other body fluids that exude are then taken in by the worm. In worms that have no stylet, the proboscis coils around the enemy and secretes the tenacious mucus.

The proboscis is enclosed in a sheath which lies above the digestive tract, and may open either in connection with the mouth or separately. The mouth is on the under surface of the body near the anterior end and the anus is at the posterior end of the body. The body is not segmented, is covered with cilia, and has many gland cells which secrete mucus. The intestine often has paired pockets throughout its length which may sometimes give the body a segmented appearance. There are three principal longitudinal blood vessels with connecting branches which also communicate with large blood spaces. The blood is circulated only by the contractions of the body and follows no regular course through the vessels. There is a well-developed, four-lobed brain, and from it run longitudinal nerves which extend to the hind end of the body and are connected at intervals. Most nemerteans have ocelli in the region anterior to the brain and some forms have them scattered along the sides of the body, often in great numbers. The excretory canals are similar to those of turbellarians.

Some nemerteans are hermaphroditic, but more are unisexual. The reproductive organs lie along the intestine and open directly to the surface of the body. Some are viviparous, some develop directly into the adult form, while others go through several metamorphoses before they take on the form of the parents.

Most nemerteans are marine and live among algae, under stones, or in burrows in the sea bottom or near shore. A few kinds live as commensals in the mantle chambers of clams or ascidians, and others live as parasites among the gills or the egg masses of crabs.

Order—HOPLONEMERTEA

The proboscis usually bears stylets, the mouth is in front of the brain, usually opening along with the proboscis.

Emplectonema gracile (Johnston) is 20-50 cm. long and about 2 mm. wide, blue-green, or yellowish green above and nearly white beneath (plate 6). The ocelli are in two groups on each side of the head, a row of 8-10 anteriorly and a cluster of 10-20 farther back. The central stylet has a long slender basis with a short stylet which is curved like a scythe and only $\frac{1}{3}$–$\frac{1}{2}$ as long as the basis (fig. 94). There are

Fig. 94.—*Emplectonema gracile*, central stylet and one of the accessory pouches; x50. (After Coe.)

Fig. 95.—*Paranemertes peregrina*, a central stylet; x300. (After Coe.)

two accessory pouches near the central stylet, each containing 5-7 curved stylets similar to the central one in shape and size. This species is found from San Diego to Alaska, being very abundant in its northern range. One finds them among the mussels and seaweed sometimes coiled together in great masses, also under stones in muddy localities.

Paranemertes peregrina Coe (plate 6) is long and slender. The head is variable in shape but somewhat flattened and bordered with the lighter color of the ventral surface. The color is variable, ranging from dark brown to orange-brown or purple-brown while the under side is white or yellowish. Sometimes the light color occupies only the middle third of the under surface. Specimens have been reported up to 400 mm. long but 150 mm. for extended specimens is a more common size. The stylet of the proboscis is about the same length as the basis and has a remarkable braided appearance (fig. 95). The reserve stylets, grouped in 2-4 pouches, are similarly braided.

Amphiporus bimaculatus Coe (plate 6) is short, flattened, and broad for a nemertean. The whole dorsal surface back of the head is deep brownish red and there are two spots of deep red or black on the head in the center of a pale area. These spots, in the specimen figured, are decidedly angular but Dr. Coe* says they are often oval. The eyes number 25-30 on each side in an irregular row outside the dark spots. The specimen figured also fails to show the usual dark and inconspicuous line on a paler stripe down the dorsal side of the body. Beginning at the head, this line extends about ⅙ the length of the body. The under surface is flesh color or yellowish and pale. Two pinkish spots mark the position of the brain. The proboscis is large with a central stylet and usually four pouches of accessory stylets. The central stylet is long and slender with a short basis. This species has been reported from Monterey Bay to Alaska and may be found on wharf piles or among the algae on rocks at low tide. Individuals range from 40-150 mm. in length.

ORDER—HETERONEMERTEA

The proboscis is without stylets, the mouth is posterior to the brain, the intestinal caecum is absent and the muscular walls of the body are in three main layers of which the inner is longitudinal.

Lineus flavescens Coe (fig. 93) is usually pale yellow, tinged with orange anteriorly and with ocher posteriorly. Some are ocher anteriorly with a distinct white line running longitudinally in the mid-dorsal region. Some specimens are dark buff and others golden brown. The brain region is usually more rosy in color and the side and anterior margins

*W. R. Coe—Nemerteans; (Harriman Alaska Series, Vol. 11), Smithsonian Institution Publication, 1998, p. 44.

pale or white. It is 8–40 mm. long, but deep-water specimens are often three times that length and 2–3 mm. wide. This species is a hardy one and will live for some time in the laboratory. It is reported from San Pedro in annelid tubes and crevices of rocks. Our specimen was taken at La Jolla in a similar location.

Lineus pictifrons Coe (plate 6) is dark brown usually with fine markings of pale yellow. The body is soft and the shape changeable though usually somewhat flattened, often snarled and twisted. Head and posterior end are narrower than the middle portion of the body and the slits at the side of the head are unusually long. Ordinary specimens are 5–15 cm. long and 3–4 mm. wide but sometimes they are longer. They often have a velvety sheen. The longitudinal and transverse markings and the margin of the head usually show more yellow than the figure represents though in some specimens the markings are almost lacking. The dorsal surface is corrugated with longitudinal flutings, accentuated by the fine longitudinal lines. The narrow border of white on the head encloses two oval, orange colored spots embedded in an area of lemon yellow. The orange spots may be indistinct and in some specimens the yellow color is lacking, but the white border is always present. The flesh colored proboscis is long and slender. The worm is found in rock crevices, under stones between tides, and among the tunicates on the wharf piles along the southern California coast.

Phylum—NEMATHELMINTHES

(The round worms)

The round worms are unsegmented, have round, slender bodies and are often called thread or hair worms. The body is without cilia or paired appendages and in most species has no suckers or bristles. Most of the species are unisexual. Many of them are parasites within plants or animals but some live a free life in water or moist soil. Familiar

THE LOWER WORMS 127

examples of species of round worms parasitic in man, are the trichina and hookworm.

The collector of marine animals may occasionally find free-living round worms in hauls made with a fine tow net or among the algae on rocks along shore. We have frequently found them in fish. They are usually among the viscera but sometimes may be among the muscles. The body usually tapers at both ends and is covered with cuticle.

Phylum—TROCHELMINTHES

(The wheel animalcules and others)

The animals of this group are minute and transparent and resemble the larvae of the annelid worms described in a later group. Although they are many-celled animals with digestive, nervous, and excretory systems well developed, they are no larger than many of the one-celled animals, so it is no wonder that they were placed among the *Infusoria* by those who first became acquainted with them. The body is not segmented, but often has external rings which make it appear to be segmented.

Most of the *Trochelminthes* belong in the class *Rotifera* and have been given the name wheel animalcules because of the ciliated discs at the front end of the body which look like revolving wheels when the cilia are in motion. In many of the rotifers, the central portion of the body is encased in a shell called the lorica. The hinder part of the body, called the foot, is usually retractile and ends in one, two, or more toes. An adhesive substance is secreted by which the animal can attach its foot to an object temporarily.

The mouth is in the middle of the ciliated region and opens into a pharynx provided with a pair of large jaws which work vigorously when the cilia are in action. There is an esophagus, glandular stomach, and sometimes a dorsal anus, a brain with nerves extending from it, and two kidney tubules which open into a contractile vacuole which connects with the intestine. Sexes are distinct and the males are usually smaller than the females though males of some of the species have not been found.

CHAETOGNATHA

(The bristle-jawed worms or arrow worms)

The chaetognaths are slender, transparent, torpedo-shaped animals which swim near the surface of the sea. The largest reach 70 mm. in length, but 20 mm. is more usual. The mouth is armed with bristles which help in seizing the prey. The animal darts swiftly about by muscular contractions of the body and is guided somewhat by the lateral and tail fins which seem to act as balancers.

The mouth is surrounded by a sort of hood within which are 2–4 rows of short spines and outside of these are two rows of sickle-shaped hooks. The mouth is usually terminal and the digestive canal runs straight through the body, opening to the exterior by the anus at the junction of the trunk and tail. There are no special respiratory, circulatory, or excretory organs. The brain is well developed and there are a pair of small eyes and an olfactory organ. Chaetognaths are hermaphroditic, the ovaries being in the trunk and the male organs in the tail. Eggs are laid in the water and usually float at the surface. When the young hatch from the egg they resemble the adults in everything except size.

There are incredible numbers of the chaetognaths during certain seasons of the year. They must form a large part of the food supply of many of the animals that live on the plankton. The fact that they lay eggs all the year round and develop rapidly accounts in some measure for their great numbers. They eat larvae of crustaceans, one-celled animals and plants, and probably the small larval fishes.

Interesting studies were made upon the daily migrations of chaetognaths by Mr. E. L. Michael* at the Scripps Institu-

---

*E. L. Michael—Classification and Vertical Distribution of the *Chaetognatha* of the San Diego Region; University of California Publications in Zoology, Vol. 8, p. 144.

tion. He found that *Sagitta bipunctata*, the most common species of the region, is rarely found below 75 fathoms and is most abundant between 20 fathoms and the surface. The species reaches its maximum abundance at the surface within an hour after sunrise and its evening maximum within an hour after sunset, being found in abundance at considerable depths at other than the hours mentioned. Why they make these vertical migrations is not known.

Sagitta bipunctata Quoy et Gaimard (fig. 96) has two pairs of lateral fins and two paired rows of teeth. A collarette

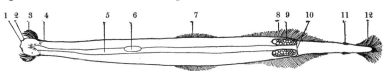

FIG. 96.—The chaetognath, *Sagitta bipunctata*; x7.
1. Anterior teeth.
2. Posterior teeth.
3. Seizing jaws.
4. Collarette.
5. Intestine.
6. Ventral ganglion.
7. Anterior fin.
8. Posterior fin.
9. Ovary.
10. Tail septum.
11. Seminal vesicle.
12. Tail fin.

is present which never extends more than half way to the ganglion, and at least one half of the posterior fin is in front of the tail septum, the anterior fin never extending to the ventral ganglion. The tail is not over one-fourth of the total length, the constriction at the tail septum being evident. The jaws are 7–8 in number. The length is 9–17 mm.

CHAPTER VII.

Phylum—MOLLUSCOIDEA

(The moss animals and lamp shells)

In the phylum *Molluscoidea* are included the *Bryozoa*, or moss animals (fig. 109), the *Phoronida*, worm-like forms which live in tubes in the sand, and the *Brachiopoda* or lamp-shells (fig. 127). Although few of these names are familiar ones to most people, the *Bryozoa* are exceedingly numerous on almost every beach. *Phoronida* are forms so uncommon that few students of zoology have had the privilege of seeing them. Brachiopods are not very plentiful at the present time, and are found mostly in deep water, but ages ago, judging by their fossil remains, they were more numerous than the mollusks, so that those we have now are the scattering survivors of a once prosperous family.

The three groups are alike in possessing a true body-cavity within which is suspended an alimentary canal. The mouth and anus are close together, and overhanging the mouth are two structures, offshoots from the body cavity. These are the epistome, a short process and the lophophore, a tentacle-bearing ridge which is usually horseshoe-shaped.

Class BRYOZOA or POLYZOA

(The moss animals)

Although common enough, the *Bryozoa*, on account of their small size and peculiar habit of growth have been entirely overlooked by the majority of visitors to the seaside, and when seen they are mistaken for delicate seaweeds or,

PHYLUM—MOLLUSCOIDEA

in some cases, for corals, but observed with a microscope, or even with a hand lens, they show a surprising variety of form and color and many of them are no less beautiful than the hydroids and corals which they superficially resemble. They are colonial forms made up of a large number of individuals, each one possessing a circle of tentacles. The latter can be withdrawn into a little chamber of chitinous or calcareous substance that surrounds the animal and serves as a protecting skeleton. This group has two names. *Polyzoa*, meaning many animals, is a term referring to their colonial nature, while the name *Bryozoa*, refers to the moss-like appearance of many of the species.

They are found in all the oceans of the world from the tropics to the cold waters of the Arctic and Antarctic Oceans and a few kinds inhabit fresh water. While many of them are to be seen in shallow waters, they often occur at great depths, certain forms having been taken in three thousand fathoms of water. Geologists tell us that they are a very old group as their fossil remains have been found in most rock formations since the Ordovician. Several thousand kinds are known to be living today and they differ markedly among themselves in shape, size, and means of protection.

Some kinds cover the surfaces of shells or seaweeds with patches of delicate lace work made up of hundreds of limy capsules containing the individuals of a flat colony. These chambers, though small, are frequently profusely ornamented and often bear protecting spines. Other species hang from the fronds of kelp and rockweed or grow in crevices between the rocks, taking the form of miniature shrubs. Such colonies are common in our bays and harbors where they grow luxuriantly on the piles and floats in common with hydroids, tunicates, and algae. Possibly the most remarkable of our Pacific coast bryozoans are the tall, calcareous ones that form coral-like masses, two or three inches in height. Because of their habit of growth this last type shares with certain of our algae the name of corallines. By way of contrast, there are also a few species of a fleshy or gelatinous consistency, often forming large branching colonies. The microscope shows us a very complicated structure. As we have just said, the individual or polypide, as it is called, is contained within a calcareous or chitinous capsule or box, the zooecium. It is attached to the inner wall of this chamber, but is able to protrude or retract the circle of tentacles through an opening near one end of it. The tentacles are provided with cilia which by their constant action produce currents in the water, bringing diatoms and other minute organisms to a mouth located within the circle. The mouth of each polypide communicates with a stomach and intestine, a condition very far in advance of that characterizing the hydroid colonies, where we find the cavities of the different hydranths connected with each other and a true alimentary canal lacking. The nervous system is rather simple, consisting usually of a central ganglion and a few strands of nervous tissue. There is no heart or vascular system.

The polypides are exceedingly sensitive and retract their tentacles quickly if touched. If not further disturbed, the tentacles are more slowly pushed outward in a compact group, later expanding into the typical funnel (fig. 116). It is great fun to watch a colony of the common incrusting species, *Membranipora membranacea* (fig. 118), perform under the microscope. There seem to be countless little tentacles arranged in small funnels standing out above the capsules. The movements are very quick and jerky; first one tentacle and then another gives a little twitch as the tip turns inward, then, perhaps in an instant, the whole circle closes up so that the tentacles lie parallel to each other and vanish as they are withdrawn into the chamber below, leaving only the general outlines to show dimly through the sheath. They come back in reverse order and when the circle is fully expanded again we can distinctly see the mouth, a circular opening at the bottom of the funnel. Little particles floating in the water are suddenly drawn into this funnel, either disappearing entirely or being shot out again with considerable force. These motions are due to a current brought about by the beating of minute cilia which project from the surface of the tentacles.

In looking at colonies of *Bryozoa*, with the aid of a lens, one usually discovers numerous dark brown spots within the zooecia. They are characteristic of the group and have received the name of brown bodies. From time to time the polypides break down, the tentacles can no longer be protruded, and all that remains is the dark mass we have just mentioned. A bud is formed which now develops into a new polypide, the brown body either remaining within the capsule or passing to the exterior by means of the alimentary canal.

Surprising as is this feature of the life of these small organisms, there are other peculiarities quite as remarkable. If a living colony of *Bugula* (fig. 117) be examined with the microscope there will be found, in addition to the moving circles of tentacles, some motile appendages of a very different sort. These are shaped like the head and beak of a bird of prey and are often attached to the zooecium by a short stalk which resembles a neck, the whole structure looking exceedingly bird-like, whence its name, avicularium. The avicularia have a fixed upper and a movable lower jaw worked by means of muscles and capable of closing up with a snap. They are constantly in motion and serve possibly to capture but probably to frighten away small animals which wander over the surface of the colony. As they cannot convey their catch to the mouth their use is somewhat problematical, probably they are only defensive or they may help to keep the surface of the colony clean. Many *Bryozoa* are entirely without avicularia and in certain other species they rarely occur. On the other hand, some colonies may bear several varieties, as is the case with the members of the genus *Scrupocellaria* found along our coast.

Besides the avicularia, some *Bryozoa* bear long whip-like

appendages called vibracula (fig. 108), which continually beat through the water. Their use is not entirely clear, either, but it has been suggested that they may assist in keeping the colony clean by preventing the attachment of foreign particles.

The size of the colony is increased by a process of budding but new colonies are brought about through the formation of egg cells which are subsequently fertilized and develop into free-swimming, ciliated, larvae. In some species the eggs are passed directly into the water but in many cases development takes place in a special structure, to which the name of ovicell has been given (figs. 100 and 120). It has been discovered that in some forms, as in certain members of the genus *Crisia* living on the Pacific coast, the embryo divides into secondary embryos and these, in turn, divide again, the resulting embryos developing into larvae (fig. 97). The larvae soon settle down and begin the formation of a new colony, the number of individuals being increased by budding.

FIG. 97.—Bryozoan larva found in the plankton at La Jolla. The species is unknown; greatly enlarged.

The group is divided into two subclasses, the *Entoprocta* and the *Ectoprocta*, the great majority of our forms belonging to the latter division. The first is characterized by having the anus, or intestinal opening, within the circle of tentacles, whereas it is outside of the circle in all of the *Ectoprocta*. The latter are generally considered the more highly organized group.

Subclass—ENTOPROCTA

Pedicellina echinata Sars. White knot-like zooecia are borne on stems that rise from creeping stolons. The upright stems are provided with short spines. The species is found along the California coast on shells and among hydroid colonies. The tentacles roll up instead of being retracted in the usual fashion.

PHYLUM—MOLLUSCOIDEA

Subclass—ECTOPROCTA
Superorder—GYMNOLAEMATA

The *Ectoprocta* are divided into two superorders, the *Phylactolaemata* and the *Gymnolaemata*. The former are exclusively fresh-water forms, and, as such, fall without the limits of this volume, while most of the latter group are marine. Three orders of the *Gymnolaemata* occur in the California area.

Order—CYCLOSTOMATA

The zooecia are more or less tubular and are calcareous; as a rule the opening is circular and they differ from the next suborder conspicuously, in the absence of an operculum to close the orifice. The embryos develop in ovicells which appear as large swellings.

Family—CRISIIDAE

Crisia geniculata Milne Edwards (figs. 98 and 99). This species ranges from San Francisco Bay to San Diego and is abundant in many localities. Like the other species of *Crisia* it has an erect habit of growth, forming delicate little tufts on other *Bryozoa* and seaweed, and may be left exposed at low tide. Although the tube-like zooecia are very stiff and calcareous, freedom of motion is gained by numerous flexible joints. The zooecia are straight or very slightly curved and divided into two parts by the joint, which is light brown in the younger parts of the colony and dark brown or black in the older parts. The ovicells are rather narrow but some-

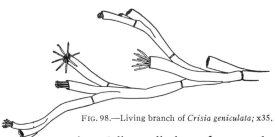

FIG. 98.—Living branch of *Crisia geniculata*; x35.

what inflated above and provided with an internal horizontal partition near the base.

Crisia maxima Robertson (figs. 100 and 112), is found upon the southern California coast and ranges from the zone uncovered by the lower tides to a depth of thirty or thirty-five fathoms. It forms large tufts with a superficial resemblance to *Bugula*. The colony is stiff and coarse and may be an inch or more in height. The ovicell is large enough to be seen with the

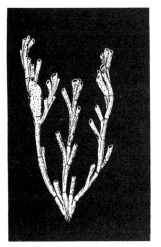

FIG. 99.—*Crisia geniculata;* x10.

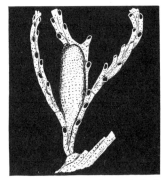

FIG. 100.—Zooecia and an ovicell of *Crisia maxima;* x10.

naked eye, if searched for carefully, and may be of an orange color, in contrast to the white branches about it. The curved, tubular zooecia are arranged in a double series, and form a stem with occasional chitinous joints. There is no horizontal partition across the interior of the ovicell.

Crisia pugeti Robertson is found in Puget Sound and resembles the foregoing species in most features. It differs principally in respect to the tube which rises from the ovicell and opens to the exterior. In *C. pugeti* this is long and bent sharply forward and toward one side, while in *C. maxima* it is short and inconspicuous.

Crisia edwardsiana (d'Orbigny) which is found between the tide marks from San Diego to Alaska, may be distin-

PHYLUM—MOLLUSCOIDEA

guished by the possession of long curving spines rising from the zooecia and small ovicells which open to the exterior by short tubes on the dorsal surface some distance below the summit.

Crisia occidentalis Trask. This is considered by some as identical with *C. eburnea* of the Atlantic coast which it very closely resembles. The ovicell is smaller than in *C. maxima* and opens by a small pore at the end of a short, inconspicuous tube, directed upward, and placed near the summit but on the dorsal side. It is abundant in San Francisco Bay and has been found at San Pedro and on the coast of Washington.

FAMILY—TUBULIPORIDAE

Tubulipora pacifica Robertson (figs. 101 and 123-3) is abundant upon the fronds of kelp (*Nereocystis*) that drift ashore along the coast of southern California. It is an incrusting form and makes flat fan-shaped or nearly circular patches upon the algae. The long zooecial tubes radiate outward and

FIG. 101.—A young colony of *Tubulipora pacifica* from a frond of kelp; x12.

FIG. 102.—*Tubulipora pulchra*. Camera lucida drawing of a specimen found upon a frond of kelp; x12, the smaller figure a little less than natural size.

turn upward at the tip, sometimes interlacing, one of the

series turned to the right, the other to the left, and so on. The ovicell is very large and extends over the space between the series of zooecia. The opening is funnel-shaped or slit-like and larger than the opening of a zooecial tube.

Tubulipora pulchra MacGillivray (fig. 102) forms delicate and graceful colonies. The characteristic shape is shown in the small habit sketch. There are usually three or four elongated, fan-shaped lobes growing out of a circular disk. Enlarged thirty or forty diameters, tooth-like projections upon the disk and sides of the colony can be seen. The projections and the shape of the opening of the ooecium will serve to distinguish this species from any other along our coast. An additional point is the shape of the zooecial tubes which are more slender than in our other species. *T. pulchra* forms a part of the "sugar coating" of the kelp on the coast of southern California. We have found it abundant at La Jolla on the fronds of kelp which drift ashore.

Tubulipora flabellaris Fabricius is found on kelp in the region of Puget Sound and its range extends southward to San Pedro, California. This species is very much like *T. pacifica*. The opening of the ovicell occurs at the end of a small tube which rises freely from the ovicell wall and is smaller than the aperture of a zooecium.

With respect to beauty the species of this family are worthy of note. If they were larger they would be as much desired for curios as the corals and cowries but the largest colonies seldom exceed a quarter of an inch in diameter. A hand lens, however, will show their structure with its graceful curves, and the translucent effect of the stony parts.

Idmonea californica d'Orbigny (figs. 103 and 122). Seen with the unaided eye, a colony of this species rather closely resembles a gorgonian coral. It is erect, branched, with the zooecia opening only on the ventral side. The branches are without joints and are flattened, often anastomosing or growing together into a network of stems. The zooecia are arranged in transverse rows and are bent outward sharply

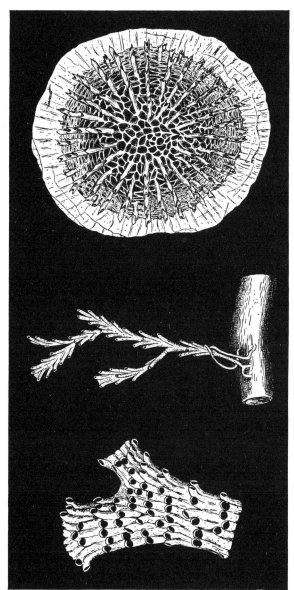

Fig. 103.—A portion of the stem of *Idmonea californica*; x12.

Fig. 104.—*Crisulipora occidentalis*, habit drawing to show form of colony; x6.

Fig. 105.—*Lichenopora radiata* forms white patches on the kelp that resemble tiny coral skeletons; x12.

near the end. It is this feature that brings about the resemblance to the fan corals as the tubes remind one of polyps along the branch, and they are just large enough to be noticed without a microscope. They are calcareous, as is true of all of the cyclostomes, and in this respect this species resembles the true corals. It is sometimes known as a coralline but the name, unfortunately, is confusing as it is also borne by a group of seaweeds that are extremely common along our shores.

While this species is abundant along the coast of southern California, it is less common on shore than in water several fathoms deep. The width of the branches is usually greater in specimens from deeper water, while the shallow water specimens show a much more complex network.

Crisulipora occidentalis Robertson (fig. 104). This very closely resembles the species of *Crisia*. The colony figured was a very young one, without an ovicell. This structure is characteristic and resembles that found in the genus *Tubulipora*, appearing as an inflation of the surface which surrounds and partly buries a number of zooecia. It is coarsely punctate and has a tubular aperture. The colonies are anchored by a few rootlets or arise from creeping stems, and are composed of segments united by flexible chitinous joints. *C. occidentalis* is common along the southern coast of California.

FAMILY—LICHENOPORIDAE

Lichenopora radiata (Audouin) (figs. 105 and 123-4). This incrusting species forms little white, or grayish white patches on the fronds of kelp which remind one somewhat of the thecae, or cups, of coral polyps. This resemblance is due to the fact that the zooecia, which radiate outward from the center in rows, are each raised slightly at the end so that they point upward and outward, giving the row the appearance of an elevated septum. The center of the colony is occupied by the ovicell which extends between the zooecia and has an aperture in the central area near the base of the first ring of tubes. The surface is marked off by a network of ridges. *L. radiata* is rather common on kelp which is washed ashore. When viewed with a binocular microscope on a black background the colony has the appearance of a bit of exquisite, sculptured lace work. The species is found principally south of Point Conception.

Order—CHILOSTOMATA

In this suborder, the wall of the zooecia may be either calcareous or chitinous. When the polypide is retracted, the opening, or orifice, is closed by an operculum, or movable lip, and is frequently surrounded by protective spines. Again, the frontal wall of the zooecium is usually composed of calcite assuming most delicate patterns. The ovicells, when present, overhang the orifice. Many of the chilostomes are incrusting forms but some of them make erect, and often plant-like, colonies.

Family—AETEIDAE

Aetea anguina (Linnaeus) (fig. 106). The colony is composed of creeping branches which adhere to the substratum at least for a part of their length and throw up short, erect, and somewhat club-shaped tubes. The ventral side of the expanded portion is occupied by a membranous wall, at the upper end of which is the operculum. This species has a world-wide range and is not uncommon from San Pedro to San Diego, where it is found on shells, eelgrass, fronds of algae, hydroid stems, and on other bryozoans.

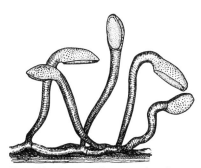

Fig. 106.—*Aetea anguina;* approximately x25.

Family—CELLULARIIDAE

Menipea occidentalis Trask (fig. 107) is an abundant species along the western coast from the Queen Charlotte Islands to San Diego. It forms bushy tufts on kelp or rocks, with the branches bending outward in such a manner as to bring all the polypides on the lower surface. It is white or yellowish and armed with numerous spines. The zooecia are elongated, tapering somewhat below. The aperture, which is closed by the membranous sheath, occupies nearly half the front and is surrounded by about six spines. A scutum, or shield, may be found extending over the aperture from the margin but it is often very small, sometimes a mere spine. Avicularia occur on this species attached to the sides of some of the zooecia. They are of the sessile type. The ovicells are globose. Rootlets coming from chambers above the lateral avicularia pass downward and form an anchor for the colony, twisting together into a cable-like stalk. The figure shows a young colony with polypides retracted. To show detail it has been made to stand upright; ordinarily from this viewpoint it would nod toward the observer.

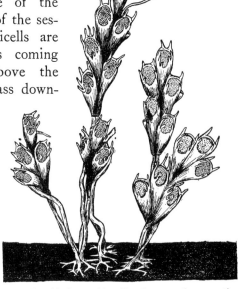

FIG. 107.—*Menipea occidentalis*. A young colony, greatly enlarged, to show the appearance of the stems and zooecia; about x25.

South of Point Conception the sub-species *catalinensis* occurs (fig. 123–1). It

PHYLUM—MOLLUSCOIDEA 143

is similar to the typical form but the scutum is fan-shaped and divided into many lobes. Some of the spines, also, are inclined to fork at the tip.

Menipea ternata (Ellis and Solander) is found from Puget Sound and points further north to the vicinity of San Francisco. The habit of growth is different from that of *M. occidentalis*, the colony being made up of straggling branches, spreading widely. The scutum varies considerably but is frequently broad and the zooecium is proportionately longer. The two species can be distinguished thus: the scutum in *M. occidentalis* arises within the lower third of the margin of the aperture and in *M. ternata* it is half way down the margin; the root fibres in *M. occidentalis* are directed downward to afford an anchorage and are not developed on the upper portions of the colony, while *M. ternata* gives rise to anchoring rootlets in the lower part of the colony and to tendril-like, climbing ones in the upper portion.

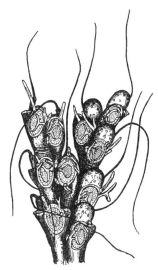

FIG. 108. — *Scrupocellaria diegensis*. Tip of branch, x33. (After Robertson.) This species has both avicularia and waving vibracula.

Scrupocellaria diegensis Robertson (figs. 108 and 115) is provided with vibracula and with three different kinds of avicularia, not to mention spines and joints and other features of considerable complexity. Some of the avicularia are lateral and sessile, resembling those found on species of the genus *Menipea*. Between the zooecia are smaller sessile avicularia which have the beak opening upward. The zooecium at the forks of a branch usually bears a very large avicularium just below the aperture, which stands out prominently from the front of the structure. The vibracula come from a chamber on the dorsal side of each zooecium and are about as long as three zooecia.

This species is calcareous and is well provided with spines and vibracula so that the clumps which it forms are rough to the touch. It is abundant in San Diego Bay on rocks, piling, floats, etc., and is obtained at San Pedro, but is found only in small quantities in San Francisco Bay and disappears farther north.

The three species of *Scrupocellaria* which occur along our coast may be distinguished by characters based on the number and length of the vibracula. *S. varians* Hinks agrees with *S. diegensis* Robertson in having a vibraculum on each zooecium, but in *S. varians* it is only slightly longer than the zooecium. *S. varians* is characteristic of Puget Sound and the region north of it, although it reaches San Diego to the south. *S. californica* Trask has fewer vibracula than zooecia, the upper part of the colony usually having none, and those present being shorter than a single zooecium. This species is found along the shore of northern California and grows sparingly along the southern coast.

Family—BICELLARIIDAE

Bugula neritina (Linnaeus) (figs. 109 and 116) consists of brown, or purple tufts which may reach a length of nearly

FIG. 109.—*Bugula neritina*, the purple *Bugula*; natural size. This form is often mistaken for seaweed.

FIG. 110.—*Zoobotryon pellucida*; x⅔.

four inches and is commonly mistaken for seaweed. The zooecia are elongated, and somewhat rectangular. The aperture occupies nearly all of the front. The ovicells are spherical and are attached by a short, broad stalk. Figure 116 shows a portion of a branch with ovicells and an expanded polypide. The operculum is a part of the membrane which forms the front wall and so is not conspicuous. In the figure a part of the zooecia are shown without the polypide or the membrane of the aperture, leaving only the rigid structure. Most members of the genus have avicularia but this species lacks them. *B. neritina* occurs both in the Atlantic and the Pacific Oceans but along our coast has not been found further north than Monterey. It is a common form in the harbor at San Pedro and along the southern coast of California.

FIG. 111.—An encrusting bryozoan *Membranipora tehuelcha*, on a piece of seaweed; natural size.

Bugula pacifica Robertson (fig. 117) ranges from Bering Sea to San Francisco Bay, growing more abundantly and to a larger size in Alaskan waters. The colony has a rather spiral form and varies in color from a purple tint, to white, yellowish, or even greenish. It reaches a height in northern waters of 75 mm. The zooecia have three spines at the end. The avicularia are large and attached by a stalk. The ovicells are very small and the embryo extends downward into the zooecium, the polypide having degenerated into a brown body to make room for it.

Bugula murrayana (Johnston) (fig. 113). This distinctive species with its broad, frond-like branches is circumpolar in its range. Its southern limit in the Pacific is Puget Sound. The zooecia are arranged in many rows to form thin fronds and bear a varying number of spines, curving inward. Avicularia are plentiful and stalked.

FAMILY—MEMBRANIPORIDAE

Membranipora membranacea (Linnaeus) (fig. 118) forms thin, flat, incrusting colonies upon the floats and fronds of

FIG. 112.—*Crisia maxima*; x1. FIG. 113.—*Bugula murrayana*; x1. FIG. 114.—*Thalamoporella rozieri*; x1. FIG. 115.—*Scrupocellaria diegensis*; x1.

kelp. The zooecia radiate outward from the older part of the colony, forming a more or less irregular patch which seems to the naked eye to be filled with small rectangular spaces marked off by narrow white ridges. The ridges are formed by the elevated calcareous walls of the zooecia. The aperture is membranous and transparent, showing the polypide within. A short, blunt, hollow, chitinous spine occurs at each anterior angle. There are no ovicells. The

PHYLUM—MOLLUSCOIDEA

range includes our entire Pacific coast. It is found in Atlantic waters also.

Membranipora villosa Hincks is a very spiny form. In addition to the spines found in the angles as in *M. membranacea* there are numerous spiny processes of small size pointing inward above the aperture and there are a few spines flaring outward. This species occurs from Puget Sound to San Diego in the same situations as *M. membranacea*.

Membranipora tehuelcha (d'Orbigny) (figs. 111 and 119) is another incrusting variety covering stems of rockweed (*Fucus*) with a deep white network. This species is marked by thick, blunt spines which are so arranged as to guard the aperture above and below. The zooecia are rectangular as in *M. membranacea* but shorter in proportion to width, and also have heavier rims than the latter species. *M. tehuelcha* was first described from Patagonia and has an extensive range. It is found from San Francisco southward on our coast and is common south of Point Conception.

Fig. 116. — A somewhat diagrammatic drawing of *Bugula neritina*. One polypide has the tentacles extended, the others have withdrawn within their zooecia. Three zooecia are represented as vacant, but with the hard parts intact. The membrane of the aperture (stippled) may be seen on those zooecia that are represented as alive. It covers the polypide when retracted. The spherical bodies are ovicells or special structures for the development of the young.

Fig. 117. — *Bugula pacifica*. Some of the avicularia are open, some closed, and one is clamped upon a part of a small worm which it has caught. (After Robertson.)

Family—THALAMOPORELLIDAE

Thalamoporella rozieri (Audoin). For beauty and complexity this species yields to none (fig. 120). Here the zooecia are divided into two chambers by a horizontal calcareous partition or septum. The upper one is covered by a transparent membrane through which the septum can be seen, and which carries the operculum. It is filled with fluid. The lower chamber contains the polypide and is flask-shaped, owing to the fact that the edges of the horizontal partition bend downward at the sides of the orifice. This leaves two openings lateral to the constriction. The orifice is covered in part by the operculum and in part by the transparent membrane of the upper chamber. Avicularia are present in some colonies, absent on others. When present they are large, have a triangular "beak," and are scattered over the surface of the colony. The ovicells are large

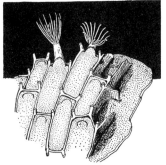

FIG. 118.—A somewhat diagrammatic representation of the cosmopolitan *Membranipora membranacea*, showing two polypides extended. This species is very abundant; x20.

FIG. 119.— *Membranipora tehuelcha*. A common species found from California to Patagonia. A surface view with polyipdes retracted; about x20.

and divided into two lobes. *T. rozieri* sometimes covers stems of seaweed, etc., as an incrusting form, or may adopt an erect branching manner of growth with flexible joints at intervals (fig. 114). At times a colony may be partly incrusting and partly erect. Dr. Robertson mentions* one large tuft found on the beach at San Pedro "consisting of a roll of intertwining branches twenty centimeters long and seven and a half centimeters in its greatest diameter." It is found from San Pedro to San Diego between the tide marks and occurs plentifully on gulfweed (*Sargassum*) that drifts ashore in that region.

---

*Alice Robertson—The Incrusting Chilostomatous *Bryozoa* of the West Coast of North America; University of California Publications in Zoology, Vol. 4, p. 278.

Family—HIPPOTHOIDAE

Hippotha hyalinoa (Linnaeus) (figs. 124 and 123-2). The colonies form circular incrustations on kelp and other seaweeds. The ovicells are just large enough to appear to the eye as opaque white spots. This species varies so much that it is difficult to describe. Frequently, the zooecia are placed some distance apart and connected by a net work of calcareous ridges. At other times they are crowded together and occasionally are piled on top of each other in two or more irregular layers. The shape of the zooecia is not constant, that of the orifice, in particular, varying widely. In some specimens two well defined calcareous processes point forward from the sides of the orifice, in others they are lacking. This form is found abundantly from Alaska to San Diego.

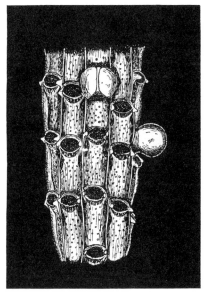

Fig. 120.—A portion of a colony of *Thalamoporella rozieri*. The globose structures are ovicells; about x20.

Family—RETEPORIDAE

Retepora pacifica Robertson (fig. 121). This species forms an incrusting base and from it grow thin perforated sheets. It is widely distributed, being found along the coast between the tide marks from Puget Sound to San Diego and southward. It lives, also, in deeper water, having been taken by dredge in thirty-five fathoms. The habit of growth is sufficient to distinguish it from others on the Pacific coast but the size and shape of the perforations vary considerably in different colonies. Large colonies, an inch

FIG. 121.—*Retepora pacifica*, a "coralline" bryozoan; natural size.

FIG. 122.—A "coralline" bryozoan. *Idmonea californica;* natural size.

to an inch and a half in height, are not uncommon and are frequently mistaken for coral. The term "lace-coral," often given to *Retepora*, while misleading with respect to the true relationships of the species, has the virtue of expressing at

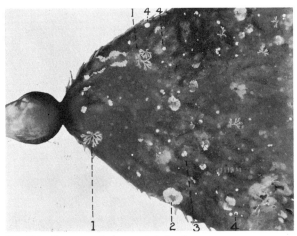

FIG. 123.—Bryozoan colonies on a kelp frond.
1. *Menipea occidentalis catalinensis*.
2. *Hippothoa hyalina*.
3. *Tubulipora pacifica*.
4. *Lichenopora radiata*.
(About three-fourths natural size.)

once the delicacy and the calcareous nature of the hard parts of the colony.

FIG. 124.—Zooecia and an ovicell of *Hippothoa hyalina*. This species varies greatly and the calcareous lace-work is not always present between the zooecia; about x33.

ORDER—CTENOSTOMATA

In this group the body wall is soft and the cavity into which the tentacles are withdrawn is closed by an operculum of setae which resembles a comb. This gives the order its name (ctenos, comb).

FAMILY—VESICULARIIDAE

Zoobotryon pellucida Ehrenberg (figs. 110 and 125). Figure 125 shows the forks of a branch of this form and a single zooecium with the polypide partly expanded. The collar may be clearly seen. The zooecia are arranged in rows along the sides of the stems which have transverse partitions at the nodes. The soft, flexible colonies form great masses in San Diego Bay.

CLASS—PHORONIDA

This class is represented on our coast by *Phoronis pacifica* Torrey, a form which resembles some of the marine worms and lives in a sand encrusted, chitinous tube which is straight and cylindrical. The length is 90 mm. and the diameter 2 mm. They may be found on sand and mud flats left uncovered by the tide and have been reported from Puget Sound, Humboldt Bay, and Laguna Beach.

The anterior end of the body bears the horseshoe-shaped lophophore which consists of a double ridge produced into a spiral coil at each side, and bears two rows of tentacles. The anus and mouth are close together in the middle of the lophophore but are separated by an extension of the body wall. The U-shaped digestive tract is made up of esophagus, stomach, and intestine. The kidneys lead from the body cavity to the outside, the openings being near the anus. The sex products pass off through the kidney openings.

CLASS—BRACHIOPODA

(*The lamp shells*)

The brachiopods (figs. 126–131) so resemble the clams that they were at first classed with the mollusks, but a study of the shell and the internal organs of the animals shows them to be more closely related to the worms. The bivalve shell is in some cases calcareous and in others horny. The animal is attached by the peduncle, which is a stout muscular stalk, a prolongation of the hinder end of the body. When the valves are opened, the most conspicuous parts of the animal's anatomy are the pair of coiled ridges bearing tentacles. Along the surface of each of these ridges is a ciliated groove, on one side of which is the row of ciliated tentacles. This whole structure is

FIG. 125.—*Zoobotryon pellucida*, showing a portion of the colony stem to illustrate the method of branching and the disposition of the polypides, also one of the polypides partially extended; about x5.

PHYLUM—MOLLUSCOIDEA

called the lophophore. The cilia sweep minute organisms into the mouth which lies at the base of the ridges. The digestive tube is made up of esophagus, stomach, and intestine, with digestive glands opening into the stomach. In some groups there is an anus, but in others the intestine ends blindly. The nervous system consists of a ring of nerve tissue around the esophagus connecting dorsal and ventral nerve centers. There is a structure present which has been called a heart, but for the animals of the first group, at least, it has been shown that cilia are responsible for the circulation. Excretory organs open from the body cavity into the mantle cavity and the eggs also find their way out through these openings.

Order—ATREMATA
Family—LINGULIDAE

Glottidia albida (Hinds) (fig. 127) is a small, clam-like brachiopod with a thin, white, horny shell and a long contractile peduncle which anchors it in the sand. It has been found from San Diego to Monterey from low water mark to sixty fathoms depth. We have taken living specimens in the sandy mud of the bay shore at Balboa. Living specimens were taken on the shore of Mission Bay before dredging had changed the shore line. The stalk is buried with the shell partly extending and when it is disturbed it withdraws suddenly, leaving a slit-like hole in the mud. Dr. E. S.

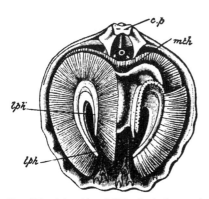

FIG. 126.—A brachiopod (*Magellania flavescens*) with the ventral valve removed to show the lophophore; c. p. cardinal process; 1ph. arm of lophophore; 1ph.' its coiled process having the tentacles removed on the right side; mth. mouth. (From Parker and Haswell after Davidson.)

154 SEASHORE ANIMALS OF THE PACIFIC COAST

Morse, in collecting them, used a large hand-dredge in the form of a dust-pan with a closely perforated bottom. With this, one can scoop up the sand and sift it at the same time. A few items are taken from Dr. Morse's account of his observations on living brachiopods.* He kept specimens alive in a small bowl for six months, and says, "the vitality manifested by them is almost beyond belief. * * One cannot help associating this remarkable vitality in these genera with their persistence through geological horizons from the Cambrian to the present day almost unchanged in character.

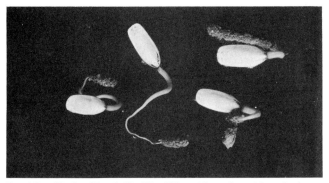

FIG. 127.—The brachiopod, *Glottidia albida*, with sand particles adhering to the peduncle; x1.

Living as they do in shallow seas, the gradual elevation or subsidence of the coast-line would in no way affect their condition. Temperature alone has probably caused their disappearance from the more northern regions." The shells are horn-like in texture and the valves are of equal size, pointed behind and truncated in front. On the edges of the mantle are long, stiff setae which project beyond the shell. The dorsal valve has two sharp internal ridges diverging from the beak and extending about a third of the length of the shell. The ventral valve has a median ridge extending forward from the beak. The valves are connected by muscles only, having

---

*E. S. Morse—Observations on Living Brachiopods; Memoirs of the Boston Society of Natural History, Vol. 5, p. 316.

PHYLUM—MOLLUSCOIDEA

no posterior hinge. The shell is 14 mm. long by 6 mm. wide and the peduncle is about three times the length of the body.

Dr. Morse describes various methods of burrowing observed in a closely related species of *Glottidia*. One "individual traveled a distance of two inches on one side of its body, the setae plowing into the sand and forcing the body forward; other specimens penetrated the sand by means of the peduncle and, in a vermiform way, dragged the body out of sight." He saw one specimen, whose peduncle was broken off close to the body, bury itself by going down head first. He says that when the animals are at rest, the body is half out of its burrow, the shells partly open, and the bundles of setae brought together in such a way as to form "rude channels" for the water which flows in and out in definite currents.

The dorsal shell is often oscillated from side to side in a peculiar sliding fashion, while the setae move rhythmically. Dr. Morse was of the opinion that circulation of the blood

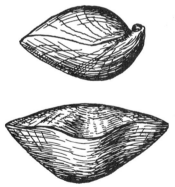

FIG. 128.—*Terebratalia transversa*, the red lamp-shell; x2. Side view (above). Front view (below).

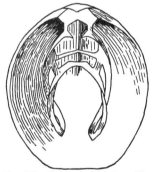

FIG. 129.—*Laqueus californicus*. The interior of the dorsal valve, showing the loop; slightly reduced. (After Davidson.)

is due to ciliary action alone, as he saw circulation in the peduncle even after the body had decayed, and saw the blood flowing both ways in a blood channel, the two currents, going in opposite directions, being separated only by a ciliary ridge.

156   SEASHORE ANIMALS OF THE PACIFIC COAST

Order—TELOTREMATA

Our representatives of this order are typical "lamp shells" and the appropriateness of the name is seen if the shell is viewed from the side (fig. 128). The shells are calcareous and the lower one has a beak at the hinder end which is perforated for the peduncle. Calcareous arms or loops project from the inner side of the dorsal shell and form a support for the tentacular ridges or lophophore (figs. 129 and 126).

Family—TEREBRATULIDAE

Terebratulina unguicula (Carpenter), the Snake's head lamp shell (fig. 130), is calcareous, ovate, longer than wide, and possesses a hinge. The surface has numerous delicate radiating ribs which increase in number toward the margin

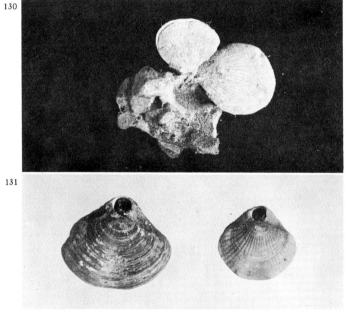

Fig. 130.—Two specimens of the snake's head lamp-shell, *Terebratulina unguicula*, attached to a rock; x5/4.   Fig. 131.—*Terebratalia transversa*, dorsal view, showing the foramen through which the peduncle passes; x4/3.

of the shell. Concentric striations cross the ribs. The calcareous loop on the dorsal shell is short and simple. The height is 12 mm., length 10 mm., and diameter 5 mm.

Family—TEREBRATELLIDAE

Terebratalia transversa (Sowerby), the red lamp shell (fig. 131), a large form found from Alaska to southern California, from low-water mark to 100 fathoms, is the most common brachiopod on the coast. It is usually wider than long, and broadest near the hinge line which is nearly straight. The calcareous loop within the dorsal valve is long and doubly attached. The valves have numerous radiating ribs and the edge has several scallops, the central one of which dips downward in contrast to that of *T. occidentalis*, a less common form whose central scallop bends upward. A variety, *T. transversa caurina* (Gould) is often found. Dall* says, "The typical *transversa* which is smooth or nearly so, grows to a much greater size than the wide strongly ribbed *caurina*, which is on the whole more southern in distribution. The former is generally of a grayish color, the latter tends to reddish." A good sized specimen of *transversa* measures 40 mm. long, 30 mm. high, and 20 mm. in diameter.

Laqueus californicus (Koch) has a large, thin, longitudinally oval, inflated shell (fig. 129). The margins are slightly sinuous and the dorsal valve uniformly convex. The beak curves inward and has a small circular foramen for the peduncle. The surface is smooth with inconspicuous concentric lines of growth. The color is yellow brown or light reddish brown. It lives in deep water off the coast from Alaska southward to southern California. An average specimen measures 50 mm. in height.

---
*W. H. Dall—Annotated list of the recent *Brachiopoda* in the Collection of the U. S. National Museum; Proceedings U. S. National Museum, Vol. 57, p. 340.

CHAPTER VIII.

Phylum—ANNULATA

(The segmented worms)

The annulates are elongated worms with bodies divided into numerous segments. The first or head segment contains the mouth and, in many cases, also bears eyes and tentacles. The remaining segments are similar and frequently bear bristles and other appendages which are not segmented.

Annelids live in the water or in moist places on the land and are numerous and varied in form and habit. Most of them swim through the water at times, but many of them build tubes within which they live and which they rarely leave. Some of the tube dwellers live in the sand and form tubes of mucus and sand, others form tubes of mucous in matted clumps of seaweed, while in other cases the tubes are of a tough parchment-like material. Some forms build tubes of limy material, on rocks, shells, or kelp. Some kinds have plate-like scales on the body. Many of the worms are highly colored, having brilliant red or orange-colored gills on the anterior part of the body. When these are expanded as in Plate 7 the worm looks like a gorgeous flower. One may get a glimpse of the worm in the bottom of a tide pool only to have it suddenly disappear into its tube at the least disturbance of the water.

Class—CHAETOPODA

Order—POLYCHAETA

Most of the marine annelids belong to this order. As we have already said, the worms are usually beautiful when alive and most interesting in their habits. The preserved

PHYLUM—ANNULATA 159

worms are uninteresting to most people, but even the worm expert can rarely identify a living worm for you off hand. He must usually get its various bristles and head appendages under the microscope before he will even venture to give you the name of the genus. The student who insists upon having the scientific names for all the specimens of his collection will have to be prepared to do some patient microscopic examination of bristles, jaws, tentacles, etc., and probably to make a weary search through the scattered literature on the worms of the Pacific coast. When he has done all this, it is quite likely that he will find his worm has not yet been described.

The polychaets have parapodia, leg-like appendages, that are made up of dorsal and ventral branches, both of which may bear setae and cirri (fig. 133). In some forms the parapodia are modified to form "fins" for swimming. The cirri, which are sensory, may in some cases have an extra blood supply, be much enlarged, and serve as gills. The bristles are useful in locomotion and probably for defense also.

Cuticle forms the outer layer of the body wall and the large body cavity is usually divided by thin partitions into as many spaces as there are body segments. The pharynx, into which the mouth leads, can usually be thrust out through the mouth to form a proboscis. Frequently there are jaws, which vary in number and arrangement in the different families. The intestine is usually straight with the anus at the posterior end of the body. The blood tubes are well developed, there being a dorsal and a ventral longitudinal one with transverse ones connecting them. Many worms have red blood and the body cavity contains a circulatory fluid which may or may not be connected with the blood tubes. The color is due to haemoglobin, as in the case of the blood of higher animals, but here it occurs not in corpuscles but in the liquid part or plasma. Usually there is a pair of kidney tubules in each segment. Respiration is carried on through the entire body wall and also in special organs on the parapodia or the head. The brain is a small, paired ganglion connected with the ventral nerve cord which has a ganglion for each segment. Eyes may be present on the prostomium and sometimes on other segments or on the branchiae. Many worms are luminescent.

Polychaetes are usually unisexual though a few genera are hermaphroditic. In *Spirorbis* the anterior abdominal segments are female and the posterior ones male. The eggs and sperm cells are usually discharged into the sea through the kidney openings or through the body wall. The fertilized egg develops into a free-swimming larva called the trochosphere. Some worms attach their eggs to the body. In *Spirorbis* the operculum may serve as a brood pouch, and a few species are viviparous.

In the *Syllidae* there is sometimes asexual reproduction. Eyes and feelers begin to develop at intervals along the body of the worm and later the worm breaks up into a corresponding number of small individuals. In certain worms (*Eunice*) the eggs are in the posterior end of the body and this whole portion is cast off. The cast off portion swims about at the surface for some time, contracting in such a way as to squeeze all of the eggs out, while the anterior part sinks to the bottom and regenerates a new posterior portion.

Some of the polychaets are carnivorous and feed on small animals of all sorts, some eat seaweeds, but others are like the earthworm in that they burrow in sand and mud, which they swallow, digesting the organic parts. Other forms feed on minute organisms which are taken in by means of the cilia on the tentacles. Certain worms which swarm abundantly at the spawning season, are eaten and regarded as great delicacies by the natives of Samoa and Fiji Islands who eat them either alive or baked.

Family—SYLLIDAE

This group of worms includes some forms that become luminescent periodically. A species in the Bermudas swarms at intervals of about a month during the late summer.* The numerous females as they swim on the surface begin to glow dimly, suddenly becoming more distinctly luminous. They swim rapidly through the water, making small light circles about two inches or more in diameter. Around the circles is a luminous halo which is possibly caused by the escaping eggs or body fluids. The eggs continue to be slightly luminous for a time. The male appears as a glint of light a few feet away coming up obliquely from deeper water. They dart for the center of the light circle and male and female rotate together in wider circles scattering eggs and sperm cells into the water. This lasts only for a few seconds, when the worms lose their luminescence and so disappear from sight. The swarming usually begins at sunset and lasts from half an hour to an hour. Luminous species on our coast have been found to differ only in minor details from the one described.

Odontosyllis phosphorea Moore, a luminous form which has often been observed on our coast, is found from Puget Sound to San Diego. It is about 23 mm. long and is stout, being widest in the middle. The body is pale yellow, crossed by narrow, black, transverse, intersegmental lines. There is a conspicuous spot on the prostomium. The swarming was observed at Departure Bay,† British Columbia on the following dates:

1912	1913	1914
Aug. 18	Aug. 21, 22	July 30, 31
Sept. 5	Sept. 6	Aug. 14, 19, 29
		Sept. 5, 21
		Oct. 6, 8

*T. W. Galloway and P. S. Welch—Studies on a Phosphorescent Bermudan Annelid, *Odontosyllis enopla* Verrill; Transactions of the American Microscopical Society, Vol. 30, pp. 13–39.

†C. McLean Frazer—The Swarming of *Odontosyllis;* Transactions of the Royal Society of Canada, Series III, Vol. 9, p. 43.

162 SEASHORE ANIMALS OF THE PACIFIC COAST

Comparing these dates with the swarming time of the Bermuda species, one can see that closely related species may resemble each other in behavior though they may be far apart geographically.

Family—APHRODITIDAE

Aphrodita, the sea mouse (fig. 132), is an unmistakable form. The convex, dorsal surface of the elliptical body is covered with long felt-like fibers through which project

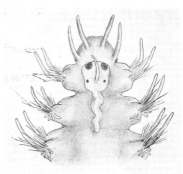

FIG. 132.—*Aphrodita* sp. Photograph of preserved specimen dredged off San Pedro; about one-half natural size. FIG. 133.—The head region of *Eurythoe californica*: x8.

stronger bristles. The worms reach 3 inches in length *Aphrodita* lives only in deep water but sometimes after severe storms may be found washed on shore. Some of the species have beautiful iridescent lateral fibers.* Cuvier† wrote of one, "From its sides spring bundles of flexible bristles shining brilliantly with all the splendor of gold and changing into all the hues of the rainbow. They do not yield in beauty either to the plumage of the humming bird or to the most brilliant of the precious stones." On our coast *Aphrodita refulgida* Moore is the species to which such a description

*C. Essenberg—Some New Species of *Aphroditidae* from the Coast of California; University of California Publications in Zoology, Vol. 16, p. 402.

†G. Cuvier—Animal Kingdom (1834, London, Henderson).

PHYLUM—ANNULATA 163

might apply. In most of the other kinds the lateral fibers are less brilliantly colored.

FAMILY—AMPHINOMIDAE

Eurythoe californica Johnson (fig. 133) reaches 106 mm. in length, 5 mm. in width, and tapers at both ends. A cross section of the body is squarish, the segments very distinct. On the head are two pairs of cirri and one median one. The somites number 60–93. The living worms are flesh color to dark brown and, when mature, the eggs shine through the body wall and give the female a purple tinge. Mature males are red and there is often a purple and green iridescence on the under side.

FAMILY—POLYNOIDAE

The *Polynoidae* (fig. 135) have overlapping dorsal scales and the body is flattened and comparatively short or oblong.

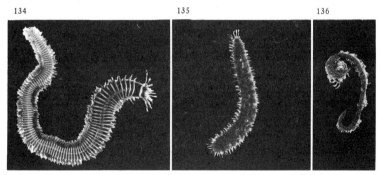

FIG. 134.—A specimen of *Nereis* sp. from Friday Harbor, Washington. The living worm was photographed in the water; x⅔. FIG. 135.—One of the *Polynoidae:* about two-thirds natural size. FIG. 136.—One of the *Cirratulidae*. Photographed from life. The thread-like gills shown in irregular or spiral coils may be extended until they are almost as long as the worm; about one-half natural size.

Over fifty species have been described from this coast. Individuals differ greatly in shape, color, and size according to the environments in which they are found. For example, a specimen of *Halosydna pulchra* (Johnson) which is found on or in the common sea cucumber is likely to be colored red-brown like the cucumber, while individuals of the same species taken from under the mantle of the key hole limpet, *Megathura crenulata*, are either without pigment or dark with conspicuous black rings on the elytra, or scales.

Halosydna lordi Baird is commensal with the limpet *Diadora aspera* in which it nestles under the mantle. It has also been found in the gill groove of the large chiton *Cryptochiton stelleri*.

Halosydna insignis Baird is abundant from Alaska to San Diego. It is often free-living, found under stones, among tunicates, mussels, and seaweeds, but is also found as a commensal in the tubes of certain other worms. In the latter situation it may grow to be very long and slender. The cirri or tentacle-like structures on the head are marked by a dark band just below the bulbous enlargement near the tip. The length is 57–75 mm., width 9–11 mm. There are 37 somites and 18 pairs of scales. Each scale has a definite white spot near its center, the remainder of the scale being mottled.

Halosydna californica (Johnson) is much like the above but ranges from Humboldt to San Diego. It also is either free living or commensal in worm tubes. The scales increase in size up to the 12th pair which is usually the largest. The upper surfaces are pitted with tiny cavities. A large specimen measures 48 mm. long and 7.25 mm. wide.

Family—NEREIDAE

The nereids are elongated worms with two small tentacles, two palps, four eyes, and four pairs of cirri or long tentacle-like structures on the head (fig. 134). The proboscis has one pair of large jaws. The "feet" are well developed and bear numerous tufts of bristles. There are many species of *Nereis*, the characters that distinguish them being slight differences in the relative lengths of prostomium, cirri, and tentacles, and in the types of the bristles and lobes of the parapodia, or "feet."

Nereis vexillosa Grube (fig. 137) is one of the most common species on the coast, found from Alaska to San Diego. Fishermen know it and other allied species as pile worms, for it is found among the barnacles and mussels of the wharf piles. It is also found on gravelly beaches and under rocks and is largely used for bait. The color is dark brown or blue green, often beautifully iridescent. The segments number about 118, the prostomium or anterior part of the head, is longer than broad, the tentacles are shorter than the prostomium and well separated at the bases, the palps are large and reach beyond the tentacles. Only the longest of the cirri of the head reach beyond the palps. Microscopic details of bristles and feet complete the distinguishing characters.

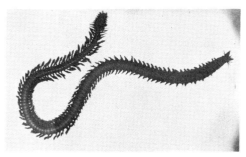

Fig. 137.—*Nereis vexillosa*. Photographed from life; x⅔.

Family—LEODICIDAE

In the *Leodicidae* the jaws in the proboscis are complicated and the cirri of the anterior feet form branched gills. Usually the worms produce a permanent parchment-like tube in which they live.

The specimen figured (fig. 138) was from deep water and was washed in after a storm. The branched gills were bright red and, beginning a short distance back from the head, were located on the segments. The worms were in tubes to which sand and broken shells adhered.

Similar tubes are numerous along shore on the mud flats at least from central California southward. It is difficult to secure perfect specimens because the worm contracts as soon as one begins to dig around its tube. We have often failed to get the entire worm even though we may have secured a foot or more of the tube intact. The upper end of the tube usually projects an inch or two above the surface

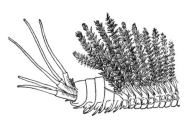

FIG. 138.—Anterior end of *Diopatra* sp.; x2.

FIG. 139.—A small burrowing worm found in great numbers on some sandy beaches. The sharp point at the posterior end is used for digging into the sand. The burrows are about the diameter of a pin-head; x2.

and lies on the mud. When it is undisturbed the worm projects the anterior segments beyond the elliptical mouth of its tube so that the waving tentacles may be seen, but it is not often that it emerges far enough to show the plume-like red gills.

Diopatra californica Moore is one of the species of *Leodicidae* most commonly collected. There are five principal tentacles, the basal fourth of each having 12–14 annulations. The palps are prominent, and their bases come together in the middle line. There is a pair of cirri on the peristomium, the posterior portion of the head. The branchiae begin on the fourth parapodium and increase in size to the sixth or seventh, then gradually decrease in size until they finally disappear by the 60th parapodium.

FAMILY—CIRRATULIDAE

The cirratulids are cylindrical worms which taper at the ends, the head is without a proboscis, the segments are similar throughout, and many of the rudimentary feet bear long thread-like gills which are often more numerous near the head (fig. 136). When expanded the threads may be

PHYLUM—ANNULATA 167

as long as the worm itself. The worms usually live in burrows and may be found under the rocks or in the sand.

FAMILY—TEREBELLIDAE

The terebellids are cylindrical and usually larger in diameter near the head than at the tail end. The anterior part of the head or prostomium forms a sort of upper lip which has a series of tentacles. There are in many cases 1–3 pairs of gills on the anterior segments. The foot lobes are small and bear bristles. The lower surface of the first few segments is thickened by glands or shields which secrete the mucus used in building the tube. The mucus tubes are usually covered with adhering foreign material. As the figure shows (fig. 140), the tentacles may be extended from the tube to a distance much greater than the length of the worm and are kept constantly in motion. The tentacles of the species photographed are flesh-colored, and the slender thread-like gills, which do not show in the picture, are bright red and continually in motion, coiling and twisting about.

FIG. 140.—One of the *Terebellidae* photographed in an aquarium. The numerous tentacles with which the worm snares its prey are shown partly extended, and the tube from which the worm was removed is shown at the left; x⅔.

The worm shown in figure 139 is, so far as we know, undescribed. It may belong to the family *Chlorhaemidae*. It makes tiny "pinholes" in the sand of the sandy beaches near La Jolla, and probably at other places. The worm is red, 20–30 mm. long and 1–2 mm. in diameter. The bulb-like posterior end is very muscular and the pointed part is used for digging when the worm burrows in the sand, tail end first. Although the worm does not live in a permanent tube, mucus secreted by it causes the grains of sand to adhere and form a partial covering. The burrows are usually 2 mm. in diameter and close together in a zone 1-2 feet wide about half way between high- and low-water marks.

FAMILY—MALDANIDAE

A sandy bay shore at low tide is often covered with hundreds of projecting worm tubes such as are shown in figure 141. These are about one-fourth inch in diameter and

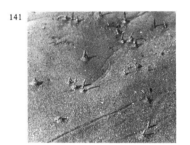
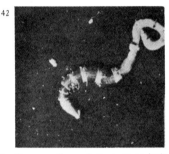

FIG. 141.—The tubes, in which the *Maldanidae* live, usually project about an inch above the surface of the sand. FIG. 142.—*Clymenella rubrocincta*.

extend for an inch above and 8–10 inches below the surface of the sand. Many of them are well above extreme low-tide mark but all are covered at high tide. Such worm tubes are likely to be the dwellings of members of the family *Maldanidae*.

Clymenella rubrocincta Johnson (fig. 142) is a common form. It has 22 segments, and is widest in the region of somites 10–12. Segments 15–17 are longest and the most slender, while the 21st is the shortest. The head plate is oval, concave, and nearly bisected by a median ridge. Fine bristles begin on the second somite. The length of a large individual is 162 mm., diameter 3.5 mm. The 5th to 8th segments are marked with a broad Indian-red band with a narrower-lighter band in front. The hinder end is funnel-shaped and has a border of 18–30 cirri, surrounding the anus.

PHYLUM—ANNULATA

FAMILY—ARENICOLIDAE

The burrowing worm, probably *Arenicola claperedii* Levinson (fig. 143) is known to fishermen as the "lug-worm."

FIG. 143.—A burrowing worm, probably *Arenicola claperedii* Levinson. (From Puget Sound.) The large specimen was 6 inches long.

It shows three distinct body regions. The anterior end is thick and blunt, the middle portion is more slender and has branching gills above the rudimentary feet. The specimen photographed was from False Bay, San Juan Island, Puget Sound. At low tide the worms were crawling about over the exposed surface of the mud. The large specimen was 6 inches long. The peculiar rough, black surface is well shown in the photograph.

FAMILY—SABELLIDAE

Eudistylia polymorpha (Johnson) (plate 7) lives in a translucent tube which adheres to rocks or piling and is often encrusted with sand and bits of shell. The specimen figured was living in a large sponge, the tube extending far down into its canals. The species is reported from Alaska to San Pedro and is especially conspicuous at Pacific Grove. The color of the feathery gills is variable, being purple or wine color, whitish or tawny. One color only may be present or the gills may be banded with two colors. The thoracic segments are distinct from the tapering abdomen. The gills number about 30 on each side and are in spirals of 2–3 turns each. Each gill has 2–10 conspicuous black eyespots. An average specimen measures 95 mm. long and 6 mm. wide.

In a quiet tide pool the extended gills look like gaily colored flowers but quickly disappear when one reaches for them. In fact, it often requires but the shadow of one's hand passing over the tiny black eyespots, to cause the worm to retreat into its hole. The tubes usually extend far down into crevices between the rocks so that one needs the help of a crowbar if he is to succeed in securing a whole specimen.

Family—SERPULIDAE

Serpula columbiana Johnson (plate 7) is also to be found in tide pools or under stones near low-water mark from Puget Sound to San Diego. It forms a white calcareous tube more or less coiled. A large specimen may be 55 mm. long and 7 mm. wide at the widest point. The specimen figured is a small one with fewer branchiae than are found on the large individuals. A funnel-shaped operculum closes the opening when the worm withdraws into its tube. The collar, gills, and operculum are variously banded with scarlet. Sometimes they are almost wholly red, but in other cases white predominates. The operculum is on the right side and has a notched border made up of one hundred or more ribs, its mate on the left side is short and rudimentary. The branchiae number 54 and the abdominal segments 250 or more.

At least two species of *Eupomatus* closely resembling figures 144 and 146 are found from San Pedro to San Diego, one species extending northward, at least as far as Pacific Grove. The photograph shows the appearance of the small calcareous tubes which may be found in great numbers on rocks and old shells. The gills are feather-like and often banded with dull colors. The peduncle has a circle of horn-colored spines. Growth of specimens in the laboratory

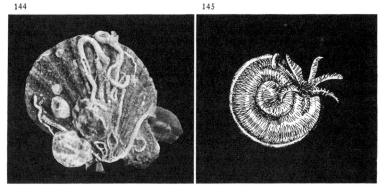

FIG. 144.—*Eupomatus* sp. forms calcareous tubes on old shells and rocks; about natural size.
FIG. 145.—*Spirorbis* sp. greatly enlarged to show the gills.

was noted as follows; one grew 7 mm. in 2 months, another grew 7 mm. in 4–5 months and made two bends in its tube, a third took 4–5 months to make 5 mm. of straight tube.

There are at least two species of *Spirabranchus* common on the coast. Figure 147 shows one of them with the gills

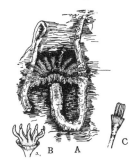

FIG. 146.—*Eupomatus* sp., (A) with gills extended (slightly enlarged); (B) with operculum opened (x6); (C) with operculum contracted.

FIG. 147.—*Spirabranchus* sp., (A) with gills extended (x1½); (B) with gills partly contracted (x4/5).

extended. The tube is calcareous, large, and firm, with a distinct median keel which projects in a point at the anterior margin. The base of the tube spreads out on the supporting rocks and is often overgrown with other tube worms and

172 SEASHORE ANIMALS OF THE PACIFIC COAST

corallines. They are usually on the underside of rocks in the tide pools. The gills, which remind one of feather dusters, are brown, yellow, white, or lavender. Some are all white near the tip while others are banded with the different colors.

There are many species of *Spirorbis*, most of them small and hard to distinguish. Figure 148 shows a group of them

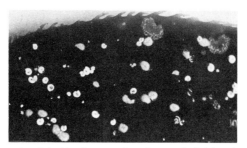

FIG. 148.—*Spirorbis* tubes on a piece of seaweed; slightly reduced.

on a piece of seaweed and figure 145 shows a single one much enlarged. The coiled tubes are either calcareous or glassy, the coil being 1–4 mm. across and 1–4 mm. high. The branchiae are feather-like and there is an operculum which closes the opening of the tube when the animal is contracted. The species are distinguished by the character of the tubes and by the form of the bristles and the number of the segments of the worm. A form that is common at Pacific Grove has conspicuous red gills which are in striking contrast to the white tubes. In some species the embryos are developed in the tube and the operculum is used only to protect the animal by closing the tube. In other species the operculum becomes a thin-walled pouch or cavity in which the embryos develop.

Family—HERMELLIDAE

Sabellaria californica Fewkes (fig. 149) forms enormous masses of tubes on the rocks along shore left exposed at low tide. The characteristic appearance of these masses is shown in the photograph. The tubes are made of a mucus-like substance, impregnated with mud, sand, and pieces of shells. The species is found from San Francisco to San

FIG. 149.—The large rock is encrusted with masses of tubes of the worm *Sabellaria californica*. Such rocks as this are uncovered only at low tide; much reduced.

Diego. A similar species, *S. cementarium* Moore, is reported from Alaska to San Diego, the tubes of this species usually being found farther out and singly or in small groups attached to shells and stones. The worms are about 3 inches long.

ORDER—OLIGOCHAETA

To this order belong worms with neither parapodia nor head appendages. Most of them, like the familiar earthworm, are terrestrial, or live in fresh water. A very few oligochaetes live in salt water, and so far as we have been able to discover, no marine species on our coast have been described.

CLASS—GEPHYREA

The worms in this group are mostly burrowing forms which live in the sand, in the matted roots of eelgrass, or among the masses of animals that encrust the rocks and wharf piling.

When removed from their burrows and placed in a dish of sea water, they actively extend and retract the fore part of the body which is called the introvert. When the animal is fully extended, a circle of more or less prominent tentacles is visible at the extreme anterior end. When it contracts, these tentacles first disappear within the introvert which next gradually rolls in, or invaginates, until it also has almost entirely disappeared inside of the posterior portion of the body after which the introvert gradually rolls out until the tentacles again appear. This action is repeated constantly.

The worms swallow great quantities of sand and mud and use as food the organic material contained therein. The mouth is within the tentacular fold or circle and the digestive canal extends the length of the body and back to the dorsal anus which is usually in the anterior third of the body (fig. 150). The intestine is much coiled. A pair of nephridia are usually present on each side of the anus a short distance from it. A cerebral ganglion lies on the dorsal side of the esophagus and is connected by a pair of nerves with the ventral nerve cord which extends to the hinder end of

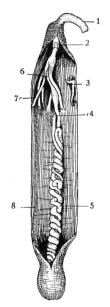

FIG. 150.—*Sipunculus nudus*, dissected.
1. Introvert.
2. Cerebral ganglion.
3. Nephridium.
4. Anus.
5. Intestine.
6. Esophagus.
7. Retractor muscles.
8. Ventral nerve cord.

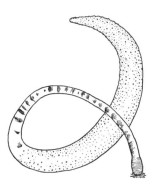

FIG. 151.—Sipunculid worm with introvert extended. From the pier at Venice; x1½.

the body. The longitudinal and circular muscles are well

developed, sometimes being arranged in continuous sheets and sometimes divided into bundles. Two or four retractor muscles extend from the anterior part of the introvert back to the wall of the hinder part of the body and serve to bring about the infolding of the introvert. The body cavity is filled with a blood fluid which contains corpuscles. A ring canal which passes around the esophagus, tentacular canals extending into the tentacles, and one or two contractile caeca make up the blood vascular system. This circulation probably brings about the expansion of the tentacles and provides for respiration. Gonads develop at the bases of the retractor muscles. Eggs and sperm find their way out through the nephridia. The young worm passes through a trochophore or free-swimming stage. In the adult form they show no trace of segmentation.

ORDER—INERMIA

Sipunculus nudus Linnaeus (fig. 152-c) is the largest sipunculid we have found on the coast. Our specimens, when contracted, measure 130 mm. in length and 12 mm. in diameter. We have found them deep in the sandy mud along the shore of Balboa Bay when we were digging out specimens of *Harenactis*. Mr. Barnhart of the Scripps Institution has found them at False Bay near the north end of the Ocean Beach bridge.

The skin is white or flesh color, somewhat iridescent, and the surface is marked off into small rectangular areas by the bundles of circular and longitudinal muscles. The introvert, except for a space next to the tip, is covered with tiny papillae which are somewhat triangular in shape with the tips pointing away from the mouth; they are so close together that they overlap somewhat like the shingles on a roof. The tentacles are not well marked as in our other species, there being only a horseshoe-shaped tentacular fold.

There are 30–32 longitudinal muscle bands and four

176 SEASHORE ANIMALS OF THE PACIFIC COAST

retractor muscles of the introvert. The two nephridia open some distance anterior to the anus and are usually free. The external opening of the anus is prominent. The coiled intestine and ventral nerve cord are attached throughout their length.

Figures 151 and 152-A show a small sipunculid frequently found among masses of mussels and other animals

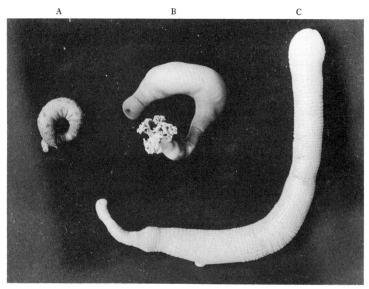

FIG. 152.—A group of sipunculids; slightly reduced. The large one is *Sipunculus nudus*, the others apparently have not yet been named and described.

taken from piles of the piers at Venice and Newport. The length when the animal is extended is 70–90 mm. and the diameter 4–5 mm. Contracted, it is 40 mm. long and 5–6 mm. in diameter at the largest point. The animal is straw color, with tiny brown spots evenly distributed over the surface of the trunk, with a few larger light brown blotches on the skin. The introvert is lighter, with black circular bands, which are not regularly spaced, and do not completely encircle the introvert tube. Near the tip of the introvert, extending about 5 mm. from the bases of the tentacles, are 17 rings of tiny dark colored, curved setae which point toward the distal end of the introvert. The single line of 16 tentacles is arranged in horseshoe-shape, the tentacles being simple, finger-like, and surrounded by a collar-like fold. There are 22 bands of longitudinal muscles; the nephridia are free in the body cavity and open at the level of the anus. The two retractor muscles are joined for $\frac{2}{3}$ to $\frac{1}{2}$ of their length, and attach to the posterior third of the body wall. The coiled intestine is attached at the posterior end of the body, and the nerve cord is also attached toward the posterior end, though apparently not attached all the way. The skin and muscular layers are comparatively thin.

Figure 152-B shows a sipunculid common in the eelgrass at La Jolla, great numbers of them often being obtainable at Bird Rock. If one pulls up a clump of the eelgrass, the thick, light colored bodies of the worms will be found dangling from the sod. Worms that are extended, measure 105 mm. in length and contracted ones measure 65 mm. The diameter, after contraction, is 10 mm. They are grayish or dirty white, nearly smooth, but with fine circular lines or a finely pebbled surface. After preservation in formalin, they are red-brown. The tentacles branch in tree-like fashion, the branches being in four large and two smaller clumps. The introvert is whitish and only slightly narrower than the anterior part of the trunk, which is thickest toward the posterior end and pointed at the tip. There are no bristles nor hooks on the introvert. The longitudinal and circular muscles are not divided into bundles, and there are two retractor muscles of the introvert, which originate in the posterior third of the body. The intestine is spirally coiled and not attached behind. The pair of nephridia are free in the body cavity and open on a level with the anus. A contracted specimen is 30 mm. long from anus to posterior tip of body and 17 mm. from anus to base of tentacles.

Order—ARMATA

Echiurus is a thick-bodied, cylindrical worm. The specimen figured (fig. 153) was dug from the mud above Moss Landing, north of Pacific Grove. It lives in surroundings similar to those of the ghost shrimp, *Callianassa*, and at about the same depth. A specimen kept in an aquarium is most interesting to watch. The flesh-colored body contracts rhythmically, a given part alternately swelling and contracting. The drawings give an idea of the forms assumed by an active specimen. The one figured would extend until it was 140 mm. long, its greatest width being 25 mm. There are 10 anal bristles and 2 hooks on the proboscis.

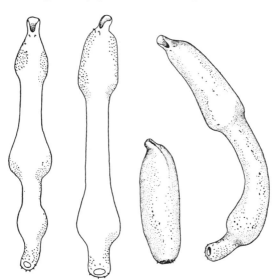

FIG. 153.—*Echiurus* sp. The four views show the same animal in different states of contraction; x½.

The method of burrowing in the mud is described by C. B. Wilson* for *Echiurus pallassii*, an east coast species. He says the animal rests upon its ventral surface, thus bringing its two large anterior setae in contact with the mud. The proboscis is turned upward and backward and remains thus with its edges somewhat rolled in and takes no part in the burrowing. The anterior end of the body is made wedge shaped, the under surface flattened and extending forward

---

*C. B. Wilson—Our North-American Echiurids; Biological Bulletin, Vol. 1, p. 166.

with the two setae projecting from its anterior edge. These are thrust into the mud and then the hinder part of the body is drawn forward. This is anchored by the anal bristles and the anterior end is pushed still further down into the mud. The process is described as being very slow, taking 40 minutes to get the body out of sight below the surface. The burrow goes down diagonally 10–18 inches, then horizontally 6 inches to 2 or 3 feet and then vertically to the surface. Once at the surface, the anterior end comes out far enough to free the proboscis and get it in its normal position, and then the body is withdrawn. Mud washes in to the unoccupied parts of the burrow, but an opening the size of a lead pencil remains, through which the proboscis is thrust in search of food when the tide is in. The proboscis may extend to a length of 5 or 6 inches. When a piece of food is located, the proboscis rolls to form a tube, and the inner surface being ciliated, a current is produced which carries the food into the mouth.

CHAPTER IX.

Phylum—ECHINODERMATA

(*The starfishes, sea urchins, and their relatives*)

The four groups of echinoderms are widely different in their appearance and habits. The name means "spiny-skinned" and most of the species have a granulated or spiny outer covering, while the few that seem smooth to the eye will be found to have little calcareous plates hidden in the skin. They differ from the coelenterates in having a distinct body cavity and digestive tract and may be distinguished from all other groups by their radial symmetry. Most animals are bilaterally symmetrical, the organs and appendages of one side of the body duplicating, or at least very closely resembling those of the other, but in the coelenterates and echinoderms the similar structures radiate outward from the central axis of the body. This is most easily seen in the starfish, which has its rays attached to the central disk as the spokes of a wheel are joined to the hub.

The water-vascular system also distinguishes the echinoderms from other animals. It is a system of tubes extending throughout the body and connecting with the tube feet, the organs of locomotion peculiar to echinoderms.

Echinoderms are found only in the sea. As a whole they are bottom dwelling animals and comparatively few kinds reach the shore line. Nearly 8,000 species are known.

The phylum is divided into five classes. The following brief key adapted from Pratt[*] will serve to distinguish them.

---

[*]H. S. Pratt—Manual of the Common Invertebrate Animals, p. 618; A. C. McClurg & Co.

PHYLUM—ECHINODERMATA

a_1 Arms present.
   b_1 Arms with small branches called pinnules; mouth directed upward.
        1. Crinoidea
   b_2 Arms without pinnules; mouth directed downward.
      c_1 Oral surface of arm with a deep longitudinal (ambulacral) groove.
        2. Asteroidea
      c_2 Ambulacral grooves absent.
        3. Ophiuroidea
a_2 Arms absent.
   b_1 Body covered with movable spines; no tentacles around the mouth.
        4. Echinoidea
   b_2 Body without spines; tentacles surround the mouth.
        5. Holothurioidea

Class—ASTEROIDEA

(The starfishes)

Starfish are found in all the oceans of the world in deep or shallow water. They are bottom dwellers and creep about on the sand, rocks, or mud, with a slow, gliding motion. The name is derived from their radial structure which takes the shape of a star as conventionally drawn. There is a disk or central portion (figs. 154 and 158) from which extend a number of rays or arms. Usually there are five of these but some kinds have more than twenty. On the under surface there is a mouth which occupies the center of the disk and from which a furrow or ambulacral groove, as it is called, radiates outward on each arm. The organs of locomotion, the tube feet, project from this furrow (figs. 167 and 155).

The animal is given a certain degree of rigidity by a skeleton composed of a loose meshwork of calcareous plates or rods. This preserves the shape of the body and protects the vital organs from injury but is loosely constructed so that the rays are more or less flexible. If they were not, it would be difficult for a starfish to right itself when turned

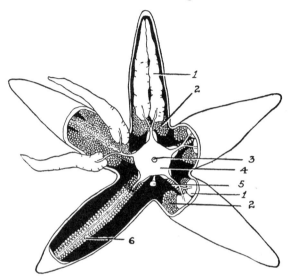

FIG. 154.—Dorsal view of a starfish dissected to show the internal organs. (Modified from various sources.)
1. Pyloric caeca.
2. Reproductive organs.
3. Anus.
4. Stomach.
5. Madreporic plate or sieve plate.
6. Ampullae.

over. Projecting from the skeletal meshwork there is generally a multitude of spines, often movable, which are evidently of protective value. In the majority of forms, peculiar organs known as pedicellariae occur. They are minute pincer-like structures which have the power of opening and closing (fig. 156).

Some of them are scattered over the body while others are close together in wreathes or rosettes about the spines. The presence or absence of pedicellariae and their shape and distribution help in classification. Their function is to keep the animal clean and to protect the papulae or breathing organs. There is evidence, also, that they aid the creature in obtaining food. Professor Jennings* made a study of the

---

*H. S. Jennings—Behavior of the Starfish *Asterias forreri* de Loriol; University of California Publications in Zoology, Vol. 4, pp. 53–185.

common "soft starfish" of southern California shores, *Astrometis sertulifera* (fig. 166) and the following account is taken largely from his report. When the starfish is undisturbed, pedicellariae are found embedded in fleshy rings at the base of the spines. If they are stimulated by a small animal crawling over them or by a bristle in the hands of the experimenter, which serves as well, the ring moves upward and hides the spine. The same thing occurs on neighboring spines. Now the pedicellariae which are attached

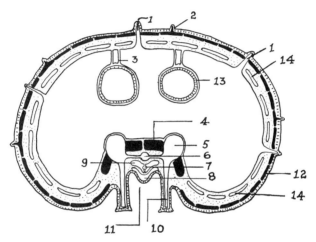

FIG. 155.—Cross section of a ray of a starfish (diagrammatic). Modified from various sources.

1. Branchiae or papulae.
2. Spine.
3. Mesentery supporting the caecum.
4. Ambulacral ossicle.
5. Ampulla of tube foot.
6. Radial ambulacral vessel.
7. Radial septum.
8. Radial nerve.
9. Perihaemal canal.
10. Tube foot.
11. Ambulacral groove.
12. Ossicles or calcareous plates of body wall.
13. Pyloric caecum.
14. Perihaemal spaces.

by short stalks extend outward as far as possible and open widely and the spines themselves are bent over toward the spot where the disturbance takes place. Consequently, an animal which ventures to crawl over the surface of a starfish provided with these structures soon finds itself confronted with a small army of Lilliputian antagonists. The gaping jaws snap shut upon contact, seizing the creature in their merciless grip and holding it for days, if need be, until death and decay take place.

It would seem impossible for such minute pincers to catch or hold any but the smallest creatures but Professor Jennings observed the capture of sand crabs (*Emerita*) and of true crabs an inch or more across. The small size of the jaws prevents them from grasping the limbs or claws but crabs are so plentifully provided with little bristles that it is possible for a great number of pedicellariae to obtain a grip. The united effort makes them very strong. It was found possible to lift the starfish out of water after allowing the pedicellariae to fasten on the hairs of the back of the hand. Indirectly they help to feed the starfish since crustaceans and other animals caught by them can be gradually transferred to the mouth by bending the rays into a convenient position and making use of the tube feet.

Each pedicellaria of the type discussed above is a complicated structure (fig. 156). It consists of two movable calcareous blades or jaws, toothed on the inner margin and crossing each other below, like a pair of pliers. These are attached to a third, or basal piece, and the whole is supported by a flexible stalk. Such crossed, or forcipiform, pedicellariae are peculiar to a single order of starfish, on this account called the *Forcipulata*. Another kind of pedicellaria found on *Astrometis* is much larger and differs in that the two jaws do not cross but are opposed throughout their entire length. They present a tiny counterpart to the open mouth of the alligator awaiting its prey. Professor Jennings is reminded of the hippopotamus in the circus posters, a resemblance which is increased by the comparatively large teeth upon the jaws. This kind of

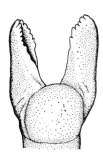

FIG. 156.—Pedicellariae from *Coscinasterias sertulifera*. Large forciciform type and small forcipiform type from the spine wreath; greatly enlarged.

pedicellaria is also stalked. It is a solitary form distributed over the surface between the spines. Technically it is known as the forficiform or "straight" pedicellaria. Crossed pedicellariae are limited to the one order and members of the other groups have sessile kinds with a varying number of jaws or valves. In some species they are set in pits, in others they are placed immediately over a pore.

Every living creature requires oxygen and if the supply ceases it dies. It must also eliminate carbon dioxide. The particular way in which the starfish does this is worthy of note. A membrane covers the spaces between the skeletal framework and on this there are many small finger-like processes. These are the structures which act as gills. They are hollow and open directly into the body cavity. There is a fluid filling this cavity which also fills the hollows of the gills or papulae as they are more properly called (fig. 155). Hair-like, microscopic structures, called cilia, line the interior and cause the fluid within the gills to circulate. Cilia are located externally also, and produce a current which moves toward the tip of the gills and continually renews the water in the immediate vicinity. The papular membrane is thin and permits an interchange of gases between the moving fluid within and the sea water without. Although the structures are so different, respiration in the starfish is nevertheless fundamentally the same as in man where the interchange of gases takes place through a thin, somewhat moist membrane between the blood and the atmosphere. When one thinks of the defenseless condition of these breathing organs and their exposed position, the value of pedicellariae and protecting spines becomes apparent.

The ambulacral grooves have already been mentioned. In cross section they have somewhat the shape of an inverted letter V and the sides are strengthened by a series of calcareous structures called ambulacral ossicles (fig. 155). A series of movable spines are usually attached to them and help to protect the tube feet. The latter are the organs of locomotion. They are hollow cylinders of thin skin ending in a point or a sucking disk. They are capable of considerable extension or retraction. Individually, they cannot be very strong, but they occur in large numbers.* Verrill has counted more than 22,000 of them on a half grown specimen of the great twenty-rayed starfish of our northern coast and he estimates that 40,000 might be present on a large individual. In our five-rayed species the number would, of course, be much less but still considerable. An attempt to pull a large animal off of a rock on which it has fastened itself will convince one of their collective strength.

Watching starfish climb up the glass of an aquarium, one gets the impression that they move by taking hold with the sucking disk and then retracting the tube feet, thus pulling themselves along. Where climbing is unnecessary this does not seem to be the case. A starfish can travel over the sand with facility and sand yields too readily to allow firm attachment and subsequent pulling by the tube feet. Moreover, some of our most active forms have no suckers on the tube feet. The actual method of locomotion seems to be as follows. The tube feet are extended in the direction toward which the animal is moving regardless of the position of the various rays. The tips are placed firmly on the substratum, and if conditions favor, are sometimes attached to prevent slipping. The tube feet are now used as levers to shove the body forward in much the same way as the legs of the higher vertebrates, except that they curve throughout their length instead of bending only at joints.

The connection between the tube feet and the water-vascular system has been mentioned, but not in detail. On the upper side of the disk, excentrically placed, is the sieve plate, known technically as the madreporite (fig. 154). It is a calcareous plate that is perforated with minute pores,

---

*A. E. Verrill—Monograph of the Shallow-water Starfishes of the North Pacific Coast from the Arctic Ocean to California; Harriman Alaska Series, Vol. 14, Smithsonian Institution Publication 2140, pp. 4, 6.

part of which communicate with the water-vascular system and part with the body cavity. Connecting with the sieve plate is the stone canal which runs to the ring canal around the mouth. The name is suggested by a crust of lime which covers the tube. Both the stone canal and the tiny pores in the madreporic plate are lined with cilia. The ring canal encircles the mouth and opens at intervals into the radial canals that extend outward along the ambulacral grooves of all the rays (fig. 155). It will be seen that the radial canal occupies a position outside of the skeleton and that it is connected with a pair of tube feet by means of side branches. The latter are provided with bulbs, the ampullae, located within the body. Just where the side branches enter the tube feet are valves which allow the fluid to pass into each tube foot but prevent its return. The bulbs are contracted by a set of circular muscles running around them and this forces the water into the tube feet and extends them. On the other hand, an enlargement of the bulbs coupled with the action of the longitudinal muscles of the tube feet will serve to shorten them. It is thought that the tube-feet are not quite water-tight and that they draw upon the radial canal to make up the loss of fluid through their thin walls. The radial canal in turn obtains its supply from the ring canal, and the ring canal is filled with sea water by means of the stone canal and the madreporic plate. The radial canal ends at the tip of the arm in a closed tube called the tentacle which is provided at the base with two eye spots.

Starfish are voracious feeders. They are carnivorous and feed largely on shell fish and barnacles, which are fixed or slow moving, though crustaceans and other more active animals are occasionally captured with the aid of the pedicellariae. In close quarters, as in aquaria, they often capture fish. They will devour dead fish or invertebrates, thus acting as scavengers, and they exhibit canabalistic traits by feeding on small members of their own group.

The starfish have no hard jaws to break open shells yet

they live largely on mussels and clams. Small mollusks may be taken into the stomach entire and the empty shells discarded later, but larger ones are eaten in quite a different manner. Mention has been made of the fluid which fills the body cavity. When the muscles of the body wall contract, pressure is brought to bear upon this fluid which in turn forces the stomach out through the mouth. The prey is held by the tube feet, and may readily be surrounded by the protruding stomach and digested without entering the body. Some species open the shells of large bivalves by fastening on them with the tube feet and pulling steadily, meanwhile humping up the body until only the tips of the rays are left attached to the substratum. The raising of the body permits the tube feet to pull at right angles to the valves of the shell. After the meal, the stomach is pulled back into place by a group of retractor muscles and the starfish glides on in search of other prey. This manner of feeding is not characteristic of the entire group since many kinds, including the sand star, have no suckers on the tube feet and cannot fasten on to a shell. They are limited to prey which they can swallow whole.

Starfish are very destructive to oysters and mussels and are injurious to all shellfish industries. As an example of the damage done to the oysters, it has been estimated that in 1888 starfishes destroyed $631,500 worth of oysters on the beds of Connecticut alone. They sometimes appear suddenly in great numbers and moving slowly along destroy everything edible they find in their path. On our coast this is particularly true of *Pisaster ochraceus*, the ochre star, which is always abundant.

The stomach is a baggy, pouched sack. Above it is another sack, of smaller proportions, the pyloric sack, from which tubes radiate outward into the rays where they lead into plume-like masses of small pouches, called the pyloric caeca. It is the function of the pyloric caeca to secrete digestive fluids. In most cases a short intestine or rectum leads to a small opening, the anus, in the dorsal wall of the disk. The anus is absent in some species, among them our common *Luidia*.

There are no organs in the starfish which quite take the place of the kidney and it is believed that nitrogenous waste matter is eliminated by way of the body cavity and the papulae. There are cells in the body-fluid, called amoebocytes, which somewhat resemble the protozoan, *Amoeba*, in their movements and the white corpuscles of higher animals in their function. If indigo-carmine, a dye stuff which is eliminated by the kidney when injected into the bodies of higher animals, is put into the starfish it will be taken up by the amoebocytes and carried to the papulae. The thin walls of the latter are often ruptured by masses of amoebocytes which creep through to the outside and carry away their freight of indigo-carmine. Amoebocytes can readily be observed with a compound microscope if some of the body-fluid is placed on a slightly warm slide.

The starfish carries on a series of complicated activities which could not be performed without the cooperation of many structures. Examples of this are the group reactions of the pedicellariae in which all of those organs in the affected area (as well as the spines) take a part, and the movements of the tube feet in locomotion and in food-getting. Moreover, if a starfish is turned over on its back a succession of movements takes place which calls for a considerable degree of coordination between the different arms and the tube feet (fig. 167). This would evidently be impossible without a nervous system of some kind. However, the starfish has no brain or spinal cord. The main part of the nervous system consists of a central nerve ring which lies just below the ring canal, and radial nerve cords which lie between the tube feet and below the radial water canal (fig. 155). Nerve cells and their fibrils are to be found widely scattered over the body. The radial nerve cords end at the base of the finger-like tentacles which terminate the radial canals of the water-vascular system. There they form structures known as the "eyes" because they are capable of distinguishing light from darkness. The double canals shown in the diagram to lie between the nerve cord and the water canal are known as the perihaemal canals. They are connected with a circular canal in the mouth region and communicate with the body cavity. Within them there is circulation of fluid.

The sexes are separate in starfish. The gonads are located at the base of the arms and open to the exterior through minute pores in the inter-radii. The eggs and sperm cells in most species are extruded into the sea water and fertilization takes place there. The larva is a free-swimming form which changes later into a tiny individual with the adult shape. At this stage they are most helpless and immense numbers are eaten by fish. There are starfish, however, including a number of those found within our limits, that carry the eggs in a hollow, formed about the mouth by arching the disk, and retain the developing larvae until they grow into the adult shape.

Asexual reproduction also occurs in a few species. Rays that have been broken off from the starfish may regenerate a disk and new rays and the lost part of the original animal may be restored by new growth. Self-mutilation and subsequent replacement of the missing part occurs in several species in which the lost part does not have the power of reproducing the whole animal. Regeneration of rays lost by accident is a common phenomenon throughout the group (fig. 157).

Sea stars are commonly divided into three main groups or orders, all of which are represented along the Pacific coast by shallow water species. The members of the order *Forcipulata* have stalked pincer-like pedicellariae, commonly of two kinds known as the forcipiform (crossed) and forficiform (straight) (fig. 156). On the upper part of the body the skeleton forms a network of rectangular or irregular meshes and the marginal plates are inconspicuous. The tube feet are provided with sucking disks.

Sea stars of the order *Spinulosa* agree with those of the *Forcipulata* in the small size of the marginal plates and the

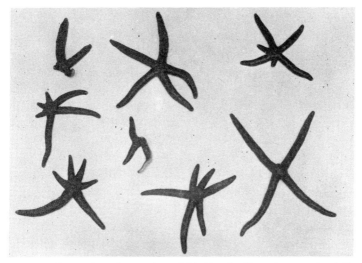

FIG. 157.—One nearly perfect specimen of *Linckia columbiae* and seven other individuals that were regenerating parts of the body which had been lost; greatly reduced. Photographed by Frank W. Peirson.

presence of sucking disks upon the tube feet but when pedicellariae occur, which is but rarely in this group, they are not stalked and do not have the forcipiform shape.

The *Phanerozonia* is not as clearly set apart from the *Spinulosa* as the *Forcipulata*. There are large and conspicuous marginal plates, usually arranged in two rows and in contact (although there is but one distinct row in the genus *Luidia*), so that they make a marginal frame for the dorsal area. The upper surface may be covered with paxillae (fig. 179) or made up of flat plates which may be smooth, granulated, or spined, in some cases covered with membrane, in others, naked. Pedicellariae are generally present and are either sessile or set in pits. The tube feet may have sucking disks, as in the other two groups, but in many species they are merely pointed or furnished with a small pointed knob at the tip.

In the descriptions which follow, an attempt has been made to eliminate technical terms wherever possible. Difference in the size and shape of the pedicellariae and many other characters of importance to specialists have been purposely omitted because their inclusion would confuse rather than aid the beginner. For the same reason no mention has been made of the varieties which are recognized in some of the species.

The greater number of starfish are deep water forms that are collected only by dredging. With one exception those we have selected for description are inhabitants of shallow water and most of them may be found at times upon the beach or among the rocks at low tide. It should be kept in mind that our list includes only a small proportion of the total number of different kinds present in our region, and that specimens brought up from the depths are not likely to be mentioned. However, the collector on shore will be able to identify most of his catch from this list.

The northern Pacific coast is peculiarly rich, both in number of species and of individuals. Many of the forms reach a large size, and one of them, *Pycnopodia helianthoides*, the twenty-rayed star, is one of the largest starfish known. According to Professor Verrill,* there are at least nine species which reach a diameter of over two feet. The southern part of the coast is poorer in individuals and species, but a number of interesting forms extend their range into this region from Mexican waters. Among them is *Linckia columbiae*, remarkable for its variability and its power to regenerate lost rays. Some kinds of starfish which are uncovered by the tide in the north can only be obtained by dredging in southern latitudes.

---

*A. E. Verrill—Starfishes; (Harriman Alaska Series, Vol. 14), Smithsonian Institution Publication, No. 2140, p. 11.

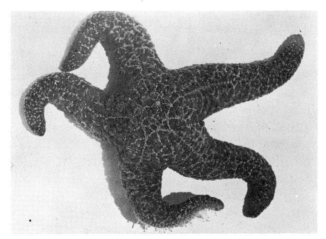

FIG. 158.—The ochre sea star, *Pisaster ochraceus*, from a photograph of a living specimen in a shallow pan of sea water; x¼.

FIG. 159.—The ochre starfish, *Pisaster ochraceus*. From a dried specimen; x¼.

Order—FORCIPULATA
Family—ASTERIIDAE

The most abundant starfish of our rocky shores is the large *Pisaster ochraceus* (Brandt) (figs. 158 and 159). The color varies from yellow through orange and brown to purple, the spines being somewhat lighter than the body color. The disk is thick and broad and there are five, occasionally six, stout, tapering rays. The short dorsal spines are arranged in close-set rows forming a distinct network. On the disk they make a well-marked pentagon. Some of the larger specimens exceed twenty inches in diameter. *P. ochraceus* may be found in shallow water and upon the rocks at low tide from the San Diego region to southern Alaska. The scientific name may be translated as the ochre starfish though the animal has too many color phases to be aptly characterized by the name of any pigment. To many people, it is simply *the* starfish because it is so much more in evidence than other forms.

Pisaster giganteus (Stimpson) may be found from Vancouver Island to south of Monterey. The spines do not form a network and there is no pentagon on the disk. They are smaller, more numerous, and set closer together than in the following species. The rays, also, are more tapering and less blunt at the tip than in *P. giganteus capitatus*. In size, this species ranks with *P. ochraceus*, reaching a diameter in some cases of 20 to 22 inches.

Accompanying the above in the southern part of its range is *Pisaster giganteus capitatus* (Stimpson) (fig. 160). The dorsal spines of this species are less numerous than in *P. ochraceus* but most of them are larger. They stand further apart, too, and do not form a well defined network; even the pentagonal figure on the disk is difficult to trace or entirely lacking and in its place there is a central tubercle or group of tubercles. This species occurs at low tide mark, and in shallow water as far north as Point Conception.

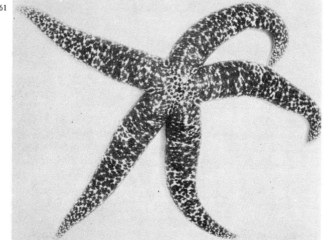

Fig. 160.—*Pisaster giganteus capitatus*; x¼. Fig. 161.—*Evasterias troschelii*, from a photograph of a living specimen under water; greatly reduced.

FIG. 162.—*Pisaster brevispinus* forma *paucispinus*; x¼.

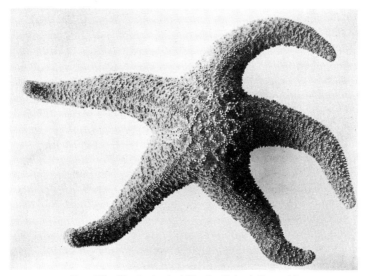

FIG. 163.—The short-spined *Pisaster*, *P. brevispinus*; x¼.

PHYLUM—ECHINODERMATA

Another species of *Pisaster* that lives in shallow water on sandy bottoms is *P. brevispinus* (Stimpson) (fig. 163). As the term *brevispinus* implies, the spines upon the upper surface are very short in comparison with those of the three preceding species. They are rather numerous, blunt, and are usually so placed as to describe a pentagon on the disk but in very large specimens this is occasionally obliterated by the growth of other spines. An irregular median band is usually present on the dorsal surface of the rays. According to Verrill, the known range is from Monterey Bay to Departure Bay, British Columbia.

As its name indicates, *Pisaster brevispinus* forma *paucispinus* (Stimpson) is notable for the scarcity of its dorsal spines (fig. 162). They are arranged in a few distinct rows upon the rays, and upon the disk they form a pentagon which usually includes a single spine or cluster of spines. This is not a very common form. The specimen photographed was taken at Tomales Bay, California, at low tide. It also occurs in Puget Sound and was originally described from that region.

Evasterias troschelii (Stimpson) is abundant along shore at Puget Sound. Several varieties have been described but there are many intergradations and not a few supposed hybrids, making their classification difficult. It is a large starfish often reaching two feet in diameter. It has a comparatively small disk and five long tapering rays. The dorsal spines are numerous, arranged in a network or clustered together, and there are five or six rows of short upward-pointing spines occupying broad zones beginning at the margin of the groove and extending up on the sides of each ray. The skeletal framework is firm. Figure 161 shows a living specimen under water. When they have been out of water for a short time the papulae are withdrawn, leaving deep pit-like areas between the skeletal network and altering the appearance of the animal.

Leptasterias hexactis (Stimpson) (fig. 164) is a small six-

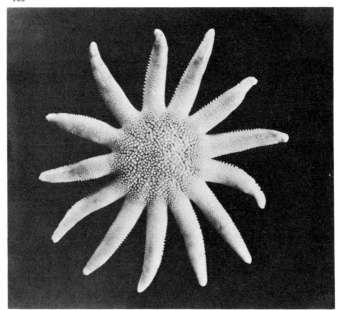

Fig. 164.—*Leptasterias aequalis* (left) and *Leptasterias hexactis* (right); x3/2. Fig. 165.—*Solaster dawsoni*; from a preserved specimen.

rayed starfish that may be found, often abundantly, on the shore and in shallow water from Monterey to Sitka. Like other members of the genus to which it belongs, this species probably has no free-swimming larval stage, but carries its young in clusters attached to the body in the region of the mouth until they have reached the true starfish form, and are able in some degree to shift for themselves. Many individuals carrying eggs were found at Monterey in February. As these starfish were often only 35 mm. across, the species must mature when quite small. Large specimens reach a diameter of about two and one-half inches. The color is variable but most of the specimens are light, mottled with pink or orange. Gray specimens are sometimes found. The surface, rather than being granular, like that of the next species, has the appearance of basket work.

Leptasterias aequalis (Stimpson) resembles the preceding species but the rays are comparatively short and blunt (fig. 164). The dorsal spines are shorter, thicker, more numerous, and evenly distributed, giving a more even, granular surface than is found in *L. hexactis*. This form commonly reaches 60 mm. in diameter but may be even larger. It is often uncovered by medium low tides and is common on the coast from Monterey to Puget Sound, its extreme range extending from San Diego to Vancouver Island.

Orthasterias columbiana Verrill has the dorsal spines in five or more indistinct radial rows, in contrast to the next species in which they form three indistinct radial rows. The arms are long and flexible, tapering gradually to a small tip. The color is usually gray with bright red markings but varies somewhat. The species is often found along shore at Puget Sound and the range extends northward along the southern coast of Alaska. Specimens often are more than a foot in diameter.

Astrometis sertulifera (Xantus) (figs. 166 and 167), the soft starfish, is a red-brown form with a small disk and five

Fig. 166.—*Astrometis sertulifera*. The animal was photographed in a shallow tide pool among its natural surroundings; x⅓.

Fig. 167.—A living specimen of *Astrometis sertulifera* was placed wrong side up in a dish of water and photographed when it was endeavoring to right itself; x½.

PHYLUM—ECHINODERMATA

long rays armed with tapering spines set well apart and wreathed at the base with circles of pedicellariae. The body is covered with a soft integument. It is one of the common kinds on the California coast.

The enormous, twenty-rayed sunflower-star, *Pycnopodia helianthoides* (Brandt) (fig. 169) is common from central California to Alaska. It is often exposed by low tides and is usually conspicuous by reason of its large size and large number of rays. Very young specimens may have as few as six rays. Ritter and Crocker* investigated the manner in which new rays are added and found that they appear in pairs between the older ones. The first ones bud out in the interspaces at the base of a certain ray, and the next two, in turn, come out between these and the two original or primary rays nearest them. Succeeding pairs arise bilaterally in the interspace between the youngest pair and the same two primary rays. (See fig. 168). Ray production proceeds until there are from twenty to twenty-four; generally there is an even number but in a large series specimens may occasionally be found with an odd number of rays.

FIG. 168.—*Pycnopodia helianthoides* (young) to show the sequence of budding new arms. a. First pair added; b. second pair added. (After Ritter and Crocker.)

This starfish has a broad, soft disk and the gracefully tapered arms radiate outward in a fashion suggesting the conventional image of the sunflower (hence the name *helianthoides*, helianthus-like). Like the disk, they are

---

*W. E. Ritter and G. R. Crocker—Multiplication of Rays and Bilateral Symmetry in the 20-rayed Starfish, *Pycnopodia helianthoides* (Stimpson); Proceedings of the Washington Academy of Sciences, Vol. 2, pp. 247–274.

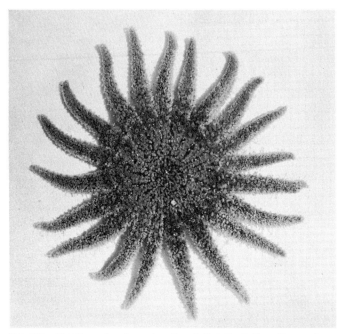

FIG. 169.—The 20-rayed sunflower star, *Pycnopodia helianthoides*. From a photograph of a small living specimen in a shallow pan of sea water; greatly reduced.

FIG. 170.—The red starfish, *Henricia leviuscula*; x⅔.

covered above with a soft skin. The whole dorsal surface is sparsely beset with spines which have wreaths of pedicellariae about them, and thick clusters of pedicellariae are scattered over the disk. Like many other starfish *Pycnopodia* varies greatly in color; yellow and orange, bright red and purplish, or gray specimens may be collected. It frequently exceeds two feet in diameter and there is a possibility, according to Verrill, that some specimens have reached a diameter of four feet, which would make this the largest species of starfish known. It is limited to the Pacific coast.

ORDER—SPINULOSA

FAMILY—ECHINASTERIDAE

From central California to Alaska, ebb tides may expose a neat five-rayed starfish with the dorsal surface covered with numerous groups of short spinelets (or pseudo-paxillae) arranged in a fine-meshed network with little spaces intervening. It has a small disk and the slender, strongly arched rays taper gradually. The color of the dorsal parts may be tan, orange, orange-red, or even purple, often mottled or banded with darker shades. The lower surface is lighter, generally yellowish or orange. Pedicellariae are absent. This form is *Henricia leviuscula* (Stimpson) (fig. 170). The species is of moderate size, commonly less than six inches in diameter. It varies greatly in form, color, and shape of spines, giving rise to a number of sub-species and varieties, and for that reason has proved quite a puzzle to descriptive naturalists. At Pacific Grove, Dr. Fisher[*] has found small specimens bearing eggs in January. They are orange-yellow and are carried in a depression around the mouth made by arching the disk. He notes that brooding mothers are generally found hidden under or between rocks and in darkness.

[*] W. K. Fisher—*Asteroidea* of the North Pacific and Adjacent Waters, Part 1, *Phanerozonia* and *Spinulosa;* U. S. National Museum Bulletin 76, p. 283.

FIG. 171.—*Solaster stimpsoni*, from a photograph of a living specimen under water.

FIG. 172.—The rose star, *Crossaster papposus;* from a dried specimen; x½.

Family—SOLASTERIDAE

The sun-stars, as the members of the genus *Solaster* may be called, have a number of rays but may be readily distinguished from *Pycnopodia* by the cluster of paxilliform spinelets covering the dorsal surface like sheaves of little spines, in contrast to the soft integument of the latter species. Two kinds may be encountered at low tide along the northern portion of the Pacific coast. Stimpson's sun-star *Solaster stimpsoni* Verrill, has a broad disk with ten slender rays (fig. 171). In one specimen, the radius of the disk is 20 mm. and the ray 82 mm. It is found from Oregon to Bering Sea. In the specimens we have found at Friday Harbor, Washington, the center of the disk and a broad band extending almost to the tip of each ray is a dull bluish gray. This is bordered by yellow-ochre or orange. The lower surface is light with a narrow blue-gray stripe on each side of the groove.

Solaster dawsoni Verrill (fig. 165), Dawson's sun star, has a wider range but is less frequently seen at tide level. In one specimen, the radius of the disk is 48 mm. and a ray is 163 mm. It is recorded from Monterey Bay to the Aleutian Islands and a subspecies occurs in the Arctic Ocean. The number of rays varies from eight to fifteen and is most commonly eleven to thirteen. These two species of *Solaster* may be distinguished by the fact that *S. dawsoni* has larger, flat-topped groups of spinelets (pseudopaxillae) and longer spinelets along the grooves. The latter are nearly as long as the actinal spines and form a "bristling fringe" along the edges of the grooves. In *S. stimpsoni* the spinelets of the ambulacral grooves are much shorter than the fringing spines, which are themselves shorter and less bristling. The marginal plates are more conspicuous in *S. dawsoni*.

While very rarely, if ever, found upon the beach, the rose-star, *Crossaster papposus* (Linnaeus), is common in relatively shallow water in the north. Fig. 172 shows the general form. On the upper surface the skeleton forms an open network supporting tall paxillae. The latter are rather evenly spaced, not crowded, and consist of a pedicel, or stalk, bearing a large number of slender spinelets. The color pattern is somewhat variable, as in most starfish, but the Puget Sound specimens we have seen alive were marked above, either with two broad concentric rings of a reddish color, one upon the disk and the other produced by a band across the middle portion of all the rays, or with a broad circular spot upon the disk and the band across the rays. According to Fisher, the number of rays may vary from eight to fourteen.* The rose star is circumpolar in its distribution and is found in the North Atlantic, Arctic, and North Pacific oceans, reaching Puget Sound and the west coast of Washington.

Family—ASTERINIDAE

One of the most characteristic starfish of the California shore is *Asterina miniata* (Brandt) (fig. 173). It can be recognized by the thick, inflated disk and the short triangular rays, usually five in number. The dorsal plates are crescent-shaped and roughly granulated as are the depressed areas between them. The concave side of the crescent faces the center of the disk. It is a handsome form, commonly bright red or scarlet above and yellowish below, though dark red, purple, or even greenish individuals are found. *Asterina miniata*† occurs from Lower California to Alaska both on rocks and sandy bottoms.

---

*W. K. Fisher—*Asteroidea* of the North Pacific and Adjacent Waters, Part 1, *Phanerozonia* and *Spinulosa;* U. S. National Museum Bulletin 76, p. 325.

†A. E. Verrill—Starfishes; (Harriman Alaska Series, Vol. 14), Smithsonian Institution Publication, No. 2140, p. 263. Professor Verrill believes that the genus *Asterina* should be divided and that the name should be confined to species closely resembling the European *A. gibbosa* which is the type of the genus. Consequently he calls this species *Patiria miniata* (Brandt).

FIG. 173.—*Asterina miniata* may be bright red, purple, or tinged with green; x3/4

FIG. 174.—The leather star, *Dermasterias imbricata*; x2/5.

Order—PHANEROZONIA
Family—GONIASTERIDAE

Figure 175 shows *Ceramaster leptoceramus* (Fisher), an almost pentagonal form. As it lives in deep water, this species is included mainly on account of its peculiar shape. It is frequently exhibited in curio shops, and attracts attention because it looks quite different from the ordinary starfish. It is very thin but is slightly inflated in the radial areas leading to the short rays. The marginal plates accentuate the contour of the animal. The dorsal area is covered with low granulated plates regularly spaced but somewhat close together near the margin. Living specimens are a brilliant vermilion above and yellowish below. This species is found from San Diego to Point Conception.

An allied species of similar shape, *C. arcticus* (Verrill) may be found upon the beach in Alaska. Both kinds reach a diameter of about five inches.

Mediaster aequalis Stimpson (fig. 176) resembles the preceding forms superficially but is more star-shaped. The disk is broad and flat and the rays are moderately long and tapered. The marginal plates are granulated and well developed and the dorsal area is covered with well spaced paxilliform plates. This is a very brilliant little starfish. In life the upper side is vermilion or deep red, the under parts scarlet, salmon, or orange, and the tube feet frequently red. Large specimens are about six inches across. While the range extends from Lower California to Alaska, it is found at tide level only in the north and then not frequently. It appears to be common in water twenty or more fathoms deep.

Family—ASTEROPIDAE

Dermasterias imbricata (Grube), the leather star (fig. 174), is in sharp contrast with the other members of the *Spinulosa* listed here, as the entire animal is covered with a

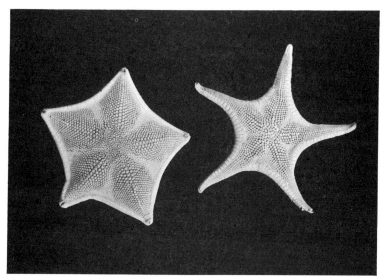

FIG. 175.—*Ceramaster leptoceramus*; x2/5.   FIG. 176.—*Mediaster aequalis* is vermilion or deep red; x2/5.

FIG. 177.—*Astropecten armatus*, the spiny sand star (below); x2/5. *Astropecten ornatissimus*, an immature specimen of the ornate sand star (above, left); x2/5. *Astropecten californicus*, the California sand star (right); x2/5.

thick, soft membrane concealing the spines except along the grooves. This membrane contains hidden spicules of lime in the shape of rods and peforated plates, and may be covered with mucus. Sessile pedicellariae are usually present dorsally, but are lacking in many individuals. There are usually five rays. It is red, orange, or lead-blue mottled with dull red. This smooth-skinned starfish is fairly abundant from the vicinity of Monterey to Sitka, Alaska, and may be found at low tide. In one specimen, the radius of the disk is 45 mm. and one ray is 120 mm. long.

FAMILY—OPHIDIASTERIDAE

One of the most variable of all starfish is *Linckia columbiae* Gray (fig. 178). It repairs broken rays with remarkable facility and may even regenerate the whole animal from part of one ray. Miss Monks* made a careful study of this species and found that *Linckia* is capable of breaking off the rays of its own accord, and that the broken ray may form a new disk, mouth, and rays in about six months. For this reason a series of specimens shows great differences in shape. Figure 157 shows a few individuals collected at La Jolla and gives an idea of the extreme variability of the species. The color varies, also, but is generally grayish mottled with dull red or red-brown. The body is covered above and below with coarse granules which become larger in the vicinity of the groove.

FIG. 178.—*Linckia columbiae*. One ray is in process of regeneration. About ½ natural size.

Linckia reaches its extreme northern limit on the coast of

---

*Sarah P. Monks—Variability and Autotomy of *Phataria;* Proceedings of the Academy of Natural Sciences of Philadelphia, Vol. 56, pp. 596–600.

southern California, being common among the rocks and in the tide pools at least as far north as San Pedro. In the south it is reported from Panama, Columbia, and the Galapagos Islands. A closely allied species of this genus with similar habits lives in the Atlantic.

Family—ASTROPECTINIDAE

The members of the genus *Astropecten* are not shore species but are found in shallow water. They inhabit sandy areas and might be called sand stars because of their habit of concealing themselves in the sand. The flat upper surface is covered with paxillae, and two prominent rows of marginal plates are always found along the edge. There are two rows of pointed tube feet without suckers. It is a curious fact that these starfish have no true intestine, although vestiges of this organ exist. Most species are also without pedicellariae.

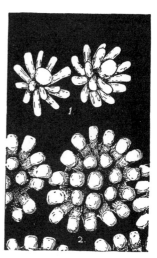

Fig. 179.—(1) Paxillae of *Astropecten armatus*; (2) Paxilliform plates of *Mediaster aequalis*; x28.

Three species are found in California waters and further south (fig. 177). *Astropecten armatus* Gray may be distinguished from the others by the larger size of the marginal plates and the greater length of the lateral spines that form a bristling fringe around the animal. Another marked characteristic is the presence of short vertical spines on the upper row of marginal plates. The usual shape is well shown by the photograph.

The California sand star, *A. californicus* Fisher (fig. 177), has longer rays and smaller marginal plates and spines than the preceding form. The upper marginal plates lack spines or tubercles. It also occupies about the same range as the preceding form.

A. ornatissimus Fisher, the ornate sand star (fig. 177), is much like *californicus*. It has larger paxillae, bearing longer and more slender spinelets. The paxillae are more distinctly spaced, while in the California sand star they are smaller and crowded closely together in the central area. The specimen figured is immature. The species occurs from San Pedro to Mexican waters.

The *Astropectens* are all of moderate size. *A. armatus*, the largest (fig. 177), seldom reaches a diameter of ten inches, though a few individuals have been found nearly a foot across. The California sand star may attain a diameter of eight inches, but five and six inch specimens are more frequent. The ornate sand star is four or five inches across.

FAMILY—LUIDIIDAE

Luidia foliolata Grube (fig. 180) resembles the species of *Astropecten* in many ways. Internally, it is like them in the absence of an intestine, externally, in the presence of paxillae and pointed tube feet. However, only the lower row of marginal plates is in evidence; the upper row is composed of paxilliform plates. The rays are long with slender tips and the animal has the power of breaking them off when roughly hand-

FIG. 180.—*Luidia foliolata*.

led. The tube feet are very large. This starfish is found in shallow water from British Columbia to San Diego, and is common in some localities. It reaches a diameter of fifteen to sixteen inches. The dorsal parts are mottled with gray and the lower surface is yellowish. This is a very active starfish.

CLASS—OPHIUROIDEA

(Brittle stars or serpent stars)

The ophiurans are commonly known as brittle stars or serpent stars. The first name comes from their habit of breaking off their arms when seized or roughly treated, the second is applied because of writhing movements of the arms and the superficial resemblance of the plates to the scales of a serpent. The scientific name of the class is derived from three Greek words meaning "snake-tail-form."

They resemble starfish in general form but the rays are very long and slender, the disk is sharply marked off from the arms, and there are no longitudinal furrows or ambulacral grooves on the lower surface. There is a water-vascular system with the madreporic plate on the ventral side. Tube feet, located on the arms, are modified into suckers which are useless for walking, but are supposed to be of value as sense organs. Beside the base of each ray or arm there are two slits which open into pouches called egg sacks, more technically, genital bur-

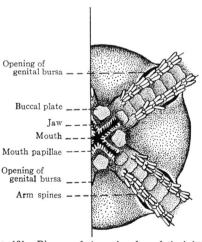

FIG. 181.—Diagram of the oral surface of the brittle star, *Ophioderma panamensis*. One of the buccal plates is also the madreporite. The extensions of the mouth between adjacent jaws are known as mouth-angles.

sae (fig. 181). Since the water continually flows in and out of these cavities, it is thought that they act as breathing organs as well as receptacles for the egg masses.

The possession of buccal plates (fig. 182) is peculiar to this group of echinoderms. The sieve plate is situated on one of

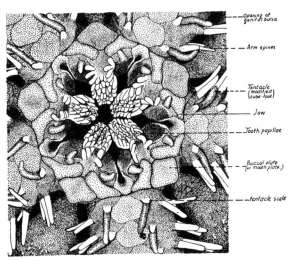

FIG. 182.—Mouth parts of *Ophiopteris papillosa*; greatly enlarged.

them. The plates surrounding the mouth form a rigid structure, called the mouth frame. The five projections are known as "jaws." They give the star-like shape to the mouth. Between them are the "mouth angles." At the tip of each jaw is a plate bearing a vertical row of spines or "teeth." A group of spines, which are borne on the edge of the apex of each jaw, are called the "tooth papillae" (fig. 182). The words "tooth" and "jaw" are not used in their usual sense as they are not organs of mastication. They can be moved somewhat, however, and when turned downward the mouth opening is enlarged. When turned upward they are said to serve as strainers to prevent the entrance of large particles.

Internally, the stomach almost fills the disk. It can not be extruded. The alimentary canal ends blindly, without caecum, intestine, or anus. The tube feet have no ampullae. The radial canals end in tentacles, as in the starfish, but there are no eyespots. The nervous system resembles that found in the *Asteroidea* but is contained within the skeleton. Neither pedicellariae nor dermal gills are found on serpent stars.

Some species may be found under stones and in other protected situations at low-tide mark. They are more active than starfish and move by wriggling the arms. A number of forms are exceedingly fragile and mutilate themselves, apparently with the greatest ease, not only breaking off pieces of the arms but frequently throwing off the entire arm, when caught. The injury so produced is not permanent as they have the power of regenerating the lost portions, and brittle stars are often found which are in process of reproducing the missing parts. In some forms the disk undergoes division and two individuals result. The usual method of reproduction is by the extrusion of the eggs from the genital bursae. When fertilized, they give rise to a free-swimming larva, known as a pluteus. This transforms into the adult form.

A series of fossil species which have been found in the rocks of Silurian and Devonian ages are intermediate in structure between serpent stars and starfish, and probably represent the ancestral forms from which both have descended.

Though the *Ophiuroidea* is said to be the largest class of echinoderms, with more than a thousand recognized species, the great majority of them live on the sea bottom in deep water. Very little has yet been published about the ones that live near shore on our Pacific coast, and their habits are not well known.

The class *Ophiuroidea* is divided into two orders by systematists. The order *Ophiurae* is characterized by unbranched arms, generally with distinct plates. It contains the serpent stars, proper. The other order, the

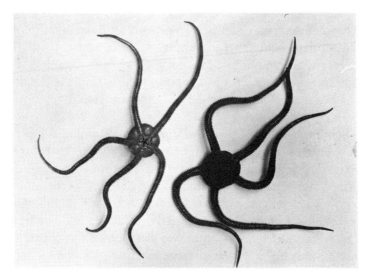

FIG. 183.—*Ophioderma panamensis*, the Panama brittle star; x2/5.

FIG. 184.—*Ophioplocus esmarki;* about natural size.

Euryalae, contains the basket stars. The arms are usually branched and roll up toward the mouth. The superficial plates of the arms are wanting, or poorly defined, and the body is covered with a thick skin.

Order—OPHIURAE

Family—OPHIODERMATIDAE

The generic name *Ophioderma* means "snake-skin" and in our species, *O. panamensis* Lütken (fig. 183), the resemblance to the scaly skin of a serpent is strongly marked. The shape is well shown by the photograph. The disk is finely granulated and there is a notch in it over each arm. The radial shields are almost hidden by the granulations. In large specimens the disk may be 25 mm. in diameter. The mouth papillae and teeth are small and the tooth papillae lacking. The openings into the egg sacks are in two parts so that there are four slits between each pair of arms, the two innermost beginning at the outside of the mouth shield. The plates of the arm are distinct and those at the sides bear flattened spines which lie parallel to the surface. The color is dark brown, lighter below, with alternating light and dark bands around the arms.

As its name suggests, this is a tropical brittle star and the southern coast of California marks its northern limit. At San Pedro and La Jolla it is the largest and one of the most abundant species of the class. It may be found under rocks or moving about in the pools when the tide is low.

Family—OPHIOLEPIDIDAE

Ophiura sarsii Lütken occurs in the North Pacific and is also found in European waters and on the American side of the North Atlantic. The disk is covered with several large, irregular scales surrounded by smaller ones. There are only two slits between each pair of arms. As in the case of the preceding species there are no tooth papillae, and the disk is notched over the base of each arm. In this indentation the disk bears a fringe of well spaced little spines, called the arm comb. The arm plates are regular, and the lateral plates bear close lying spines seldom exceeding the length of a joint.

Dr. H. L. Clark examined more than 20,000 specimens from the Pacific coasts of Asia and North America.* He makes the following comment about their size and color variations: "The smallest specimens have the disk about 3 mm. in diameter, while in the largest specimens it exceeds 32. Most of the specimens are uniformly gray in color, but there is great variety of shade, some being very dark, others very light, others decidedly yellowish, and others more or less brown. Some specimens, generally young ones, have the arms banded with yellowish or whitish, and in a few cases there are whitish spots or markings on the disk. Several specimens are distinctly spotted with black." The latter, he tells us, came from the coast of Japan.

Ophiura lütkeni (Lyman) is a closely allied species found from Alaska to the San Diego region. It may be distinguished from *O. sarsii* by the nature of the arm comb. In the latter, the papillae are so spaced as to resemble a comb, while in the former, they are short and broad and so closely crowded together that the resemblance vanishes. The disk shows a tendency to a pentagonal, instead of the usual circular shape. The ground color is gray, sometimes marked with blackish and whitish spots on the disk. The lower surface is very light in color—nearly white. This is not common along shore but is found in water of moderate depth.

Ophioplocus esmarki Lyman (fig. 184) is a shore form that lives from Monterey Bay southward. The disk is covered with irregular swollen scales which almost hide the radial shields, and give it a pebbled appearance. It is flattened, as

---

*H. L. Clark—North Pacific Ophiurans in the Collection of the U. S. National Museum; U. S. National Museum Bulletin No. 75, p. 44.

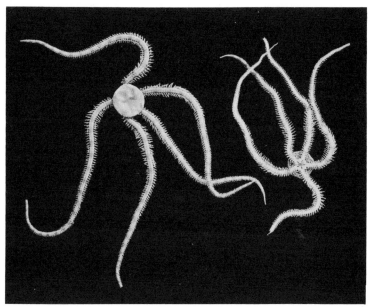

FIG. 185.—The banded serpent star, *Ophionereis annulata;* about natural size.

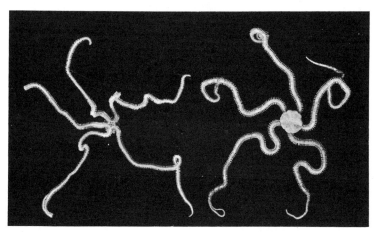

FIG. 186.—*Amphiodia occidentalis;* about natural size.

are the arms, and may attain a diameter of 30 mm. The arms are rather short for a serpent star, their length being a trifle less than three times the diameter of the disk. The arm spines are in groups of three, blunt and short, about two thirds the length of the arm joints. The genital openings are short slits reaching only half way to the margin of the disk and beginning some distance from the mouth.

Family—AMPHIURIDAE

Another shore species, that is rather abundant in the southern California region, is *Ophionereis annulata* LeConte (fig. 185). It reaches its northern limit near San Pedro, and ranges at least as far south as Central America. This form differs from all of the preceding species, and resembles all to be hereafter described, in having the spines perpendicular to the axis of the arm. The disk is small, 10 to 13 mm. in diameter, and bulges out between the arms. It is covered with small overlapping scales that almost hide the radial shields. Mouth papillae and teeth are present but there are no tooth papillae. The banded serpent star, as this may be called by translating its technical name, is gray with dark bands around the arms. It may be found at low tide inhabiting pools between the rocks or hiding beneath stones. It is an active species much given to twists and contortions of the slender arms.

The long-armed brittle star shown in figure 186 is *Amphiodia (Amphiura) occidentalis* (Lyman). It may be taken at low tide along the central California coast and is reported to occur as far north as Kodiak, Alaska. The small disk, 5 mm. in diameter in our larger specimens, is yellowish or greenish in color and covered with overlapping scales of almost microscopic size. The narrow radial shields are separated proximally by a few scales, but toward the edge of the disk they are joined. There is a notch in the margin of the disk between the radial shields and the base of the arm. At times, the portions of the disk between the arms become very much swollen with reproductive cells, altering the appearance greatly. The arms are very long, eight to ten times the diameter of the disk, and distinctly flat. The blunt arm-spines are in groups of three, and each spine is about the length of an arm-joint. These blunt, compressed spines are characteristic of the species and distinguish it from similar, but less common, members of the genus.

Like many other serpent stars, this one is easily broken and regenerates readily. We found one specimen in which four arms, broken a little way from the disk, were regenerating at the same time. The new portions each consisted of many tiny joints, but were exceedingly narrow compared with the old stumps, so that the transition was abrupt.

Ophiopholis aculeata (Linnaeus) Gray (fig. 187) may be found on rocky shores in the Puget Sound region. It is a small, reddish species, often curiously streaked and mottled with lighter colors. In light colored specimens the rays may be banded with red stripes. Sometimes a dull greenish color forms a part of the pattern. The disk is small with five conspicuous lobes located between the rays, and, in large specimens, appearing somewhat inflated on the lower surface. It is covered with small conical spines, which appear longer and sharper upon the outermost parts of the lobes, and a series of three or more rounded plates in a row from the central part of the disk to the base of each ray. There are five spines on the lateral plates of the rays, the middle one the longest.

Family—OPHIOCOMIDAE

Ophiopteris papillosa (Lyman) (fig. 188) is common in the San Diego region and on the coast of Mexico. We do not know the northern limit of this species but we have taken large specimens at Pacific Grove. The disk reaches a diameter of 11 mm., and has a soft skin covered with granules. The radial plates are entirely concealed. The arms are

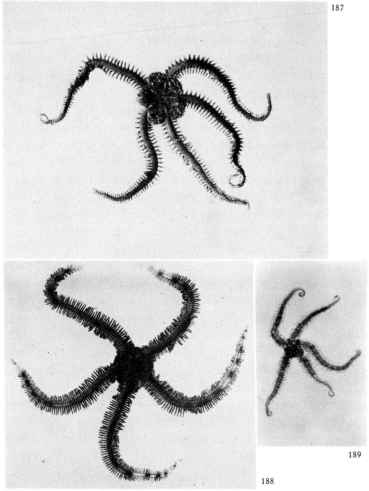

FIG. 187.—The daisy serpent star, *Ophiopholis aculeata*; slightly enlarged.
FIG. 188.—*Ophiopteris papillosa*; about one-half natural size.
FIG. 189.—*Ophiothrix spiculata*; x5/6.

comparatively short and support a fringe of long flat spines standing at right angles to them. Those of the upper row have small accessory spines overlapping them at the base. Altogether, there are five longitudinal rows. Tooth papillae are present and are arranged in four or five vertical rows. The color is brown with narrow, darker bands around the arms.

Ophiothrix spiculata LeConte (fig. 189) is found as far north as Monterey Bay. In this species the disk is thickly covered with small spines except on the radial shields, where they are less numerous. As a result, the radial shields appear to be depressed. They are long and triangular. The egg sacks are conspicuous, bulging out between the arms. The slender arm spines are placed in seven rows. When viewed with a microscope they will be seen to be roughened with minute projections. Mouth papillae are absent; tooth papillae present. This neat brittle star is found at low tide in the pools and is often seen on kelp holdfasts. *O. spiculata* is common in many places and is easily obtained. It shows great variation in color. McClendon* gives the color characters as follows: "Color greenish-brown, sometimes yellowish. Arms interrupted with orange bands. A cluster of orange spots near base of each arm on upper side of disk and internal to these the disk is speckled with orange (in one case the central area was white while the rest of the animal was colored normally). Mouth region whitish." Five specimens were taken at La Jolla from a single kelp holdfast washed ashore. Two were large with disks of about 15 mm.; three were smaller with disks not over 7 mm. across. They are given here in the order of their size.

1. The disk had a greenish brown border around the edge within which it was orange. The arms had a bluish tint with brown along the outer parts, and an orange tint down through the center.

---

*J. F. McClendon—The Ophiurans of the San Diego Region; University of California Publications in Zoology, Vol. 6, p. 50.

2. Resembled No. 1. The central part of the disk was a bright orange, the margin of brown narrower.
3. Disk with a brown margin and a greenish star-figure in the center. Rays with a greenish edge, and yellow center.
4. Disk brown, with a slightly lighter star-figure in the center. Arms light brown, crossed with orange bands.
5. Brown mottled disk, lightest toward center. Arms and spines reddish.

The fallacy of depending too much on color characters is clearly shown in this case, and many other species are similarly variable. Structural characters vary also, but, as a rule, less widely than color.

Order—EURYALAE

Family—ASTROPHYTIDAE

The basket fish, or basket star, *Gorgonocephalus caryi* Lyman (fig. 190), is found in relatively shallow water and

FIG. 190.—A living basket star, *Gorgonocephalus caryi*, in a small aquarium. The star was held so that the mouth and disk would show in the photograph. The tendril-like ends of the rays were continually in motion, alternately curling and straightening; x⅙.

arouses considerable interest because of its shape. The five arms each branch about twelve times beginning at the disk, and can be rolled in toward the mouth. The axial plates of the arm are double. The disk is covered with a soft skin in place of the characteristic ophiuran scales, but the surface is granulated. It is thick, and the dorsal portion is marked with radiating ridges. The creature is flesh-colored with orange markings.

When observed in an aquarium, the tiny end branches are found to be very active at times, curling up and stretching out in a graceful, waving movement. One specimen we watched, clung to the kelp with the smaller end branches of the rays. Part of the time it was motionless, but when a fish came too close, all of the branches on that side became very active and waved back and forth for several minutes. When the animal is at the bottom of the tank, it appears to be standing on its toes as it supports itself on the tip of the branches with the disk held high. It is said to capture fish by raising its basket of branches in the fashion of a purse-net.

According to Clark,* this basket star is found from California northward to Bering Sea, and thence southward along the coast of Asia to southern Japan. It ranges in depth from 8 to more than 500 fathoms.

The western species is closely allied to an Atlantic form, *G. eucnemis* Müller and Troschel, and possibly should be considered as a local race of the latter, from which it differs only in the amount and distribution of the granulations (in particular, of close rows of granules about the genital slits), and the length of the tentacle scales. Many northern echinoderms are found to be circumpolar, occuring in the colder waters of the Arctic, Pacific, and Atlantic Oceans.

Class—CRINOIDEA

Crinoids are characteristically deep sea forms, none on this coast occurring in shallow water, but their peculiarities

---

*H. L. Clark—North Pacific Ophiurans in the collection of the U. S. National Museum; U. S. National Museum Bulletin No. 75, p. 287.

of structure and the fact that they make up one of the principal divisions of the *Echinodermata* entitle them to consideration. The word, *Crinoidea*, is translated as "lilyform," and the name is well chosen. They are commonly known as sea lilies and feather stars.

Imperfectly as we know the animals of the past through their fossil remains, there are about 2,000 fossil species of crinoids listed, and several thousand species of the present day forms have already been described. The fossil forms are commonly known as stone lilies and encrinites. They are found abundantly in the rocks of many localities in the United States and are familiar objects to the geologist the world over.

The sea lily superficially resembles a basket star turned upside down and attached to the bottom by means of a jointed stalk. The mouth is directed upward and is in a depression, for the disk (called the calyx in this group) is cup-shaped. The anus is in an elevation near the mouth. The arms are branched and bear short side branches, called pinnules, which give them their feathery appearance. Ambulacral grooves are found on all the arms and their branches, except the pinnules. Some of the species retain their stalks as adults while a more numerous group are without them when adult, though they are attached in their younger stages. The members of the latter group retain a stump from which processes, or cirri, are developed that serve as fixing organs or temporary holdfasts.

The food is said to consist of small organisms brought to the mouth by means of cilia, in the ambulacral grooves and on the tube feet, which produce a current in the water. The arms have been compared to "a net spread out in the water to catch swimming prey." The mouth leads to a stomach which is curved horizontally within the calyx and opens into a short intestine. There is a water-vascular system and a nervous system much like those of the starfish and sea urchin, though differing in detail. The tube feet are

PHYLUM—ECHINODERMATA

probably used as organs of respiration. They are also used as tactile organs, and excretion is carried on through their thin walls. Several hundred species live at various depths in the Pacific.

Class—ECHINOIDEA

The sea urchins, sea eggs, or sea porcupines, as they are variously called, are spiny, globular, or disk-shaped creatures without arms. Only a few species live within tide limits but the small number of forms is offset by the great numbers of individuals.

The skeleton of the sea urchin (fig. 192) is composed of closely joined calcareous plates forming a case about the vital organs. People often refer to this as a "shell" but it should not be so considered. A typical shell arises by secretion from the outer parts of an animal, while the plates of the sea-urchin are formed by connective tissues beneath the surface. The structure is more accurately called a test.

Empty tests with the spines lost can usually be found along the shore in localities where sea urchins are abundant, or they can be prepared by putting the dead animals in water to which some lye has been added and boiling until the soft parts can be removed with a pair of forceps. Equally good results can be obtained by boiling in water to which some soda is added, in fact, the last method preserves the color of the skeleton much better. An examination of the test will give evidence of radial symmetry, not of arms, as in the case of the other groups of echinoderms we have discussed, but of zones, which extend from the mouth region to the opposite extremity. These zones may be compared to those bounded by the meridians of longitude which extend from one terrestrial pole to the other. Five double bands of perforated plates, each with a narrow zone of rounded knobs down the center, are separated by five broader areas covered with similar knobs. The perforations are for the passage of the minute canals which connect the tube feet with their ampullae and the water-vascular system, consequently, the double bands are known as ambulacral areas, and the regions between them are termed interambulacral areas. The flexible part surrounding the mouth is the peristome, and at the opposite pole are a number of small plates comprising the periproct. The sieve plate is one of them. The spines are joined to the knobs or tubercles. They are concave at the base and are able to slip over the rounded projections. Motion is produced by the contraction of muscle fibres attached to them.

There are several highly developed types of pedicellariae among the spines and on the peristome. As a rule, they have three valves and a calcareous basal piece within the stalk (fig. 191). One kind, the gemiform type, found in

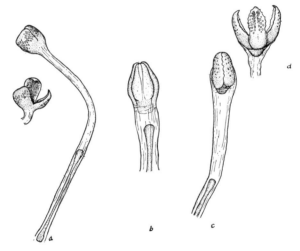

FIG. 191.—Pedicellariae of *Strongylocentrotus purpuratus*, greatly enlarged. a. Trifoliate; b. Gemiform (poisonous); c. Ophiocephalous (snake-headed) type, closed; d. Ophiocephalous type, open.

some of our species, has poison glands. It is stated in the *Cambridge Natural History** that a European investigator found the "bite" of a single gemiform pedicellaria would

*E. W. MacBride—*Echinodermata;* Cambridge Natural History, Vol. 1, p. 509; The Macmillan Company.

cause a frog's heart to stop beating. This is paralleled by a West Indian form which has poisonous spines that penetrate the skin and then break off. Our western species are not difficult to handle, though the spines are rather sharp. All the *Echinoidea*, except the heart urchins, are provided with a remarkable system of hard jaws and teeth for use in mastication. The great Greek philosopher and naturalist, Aristotle, described the structure and pointed out its resemblance to the ancient kind of lantern. To this day it is known as Aristotle's lantern. For a detailed account the reader is referred to any of the general textbooks in our list of citations. The essential features are five long teeth, like so many chisels, meeting practically in the mouth opening, where they may be seen in living specimens, each one held in place by a pair of calcareous rods joined together to make a structure shaped like a letter V. This is the jaw. The jaws are joined together by other ossicles so that a rigid framework results. It is the function of the frame work to support the teeth and afford attachment to a complicated system of muscles that move them up and down, and to another set that causes the entire lantern to protrude or withdraw. Beach-worn tests seldom contain the lantern, but it can be preserved in good condition by boiling the specimen as described above.

The alimentary canal is a baggy tube that winds once around the body cavity, then folds back and runs in the opposite direction to the anus. The lower portion is called the stomach. One peculiarity which deserves notice is the presence of an accessory tube or "siphon" which branches off near the mouth, parallels the stomach, and enters the canal farther back. It is lined with cilia and is thought to keep a current of water in motion through the alimentary canal.

The water-vascular system resembles that of the starfish in having a sieve plate, ring canal, stone canal, and radial water vessels but differs in detail. The ring canal is situated above the lantern and not immediately above the nerve ring, as in the *Asteroidea*. The stone canal is not calcified but is a simple, soft-walled tube so the word "stone" is a misnomer in this case. Each tube foot is connected with its ampulla by a pair of minute canals that are lined with cilia. The beating of the cilia creates a current through one canal into the tube foot, the return current flowing through the other tiny canal back to the ampulla. Such an adaptation makes the tube feet efficient breathing organs. Oxygen can pass through the thin walls into the moving water within them, and thence through the walls of the ampullae into the fluid of the body cavity. In most species, our common purple urchins, for example, the tube feet are used for locomotion (fig. 192-c) and are provided with suckers. They are organs of touch and respiration in all cases. There are other ways of breathing. The intestine helps by means of its accessory tube which keeps up a current of water, and the majority of species have gills upon the peristome. Excretion takes place by means of amoebocytes, as in the starfish.

The nervous system consists of a ring about the mouth and radial branches in the ambulacral areas. It lies within the skeleton. Nerve fibres are widely distributed. The use of the tube feet as sense organs has been mentioned. The sphaeridia are another kind of sense organ. They are little spherical masses of hard material found in the skin of the ambulacral areas. Their function is not definitely known, but, since they contain nerve cells, it is thought that they are organs of some special sense. Possibly they are concerned with the sense of balance. The pedicellariae are sensitive and if the spines are touched, the muscles which move them are set to work—another nervous reaction.

Sea urchins pass their eggs into the sea water where they are fertilized. The egg develops into a ciliated larva called a pluteus which looks much like that of the serpent stars. The larvae are very small but must exist in enormous numbers as it is claimed that one female of the European

PHYLUM—ECHINODERMATA

edible sea urchin will produce 20,000,000 eggs in a single season.* Probably, only an extremely small proportion of the fertilized eggs succeed in reaching maturity, and many are not even fertilized.

Professor Jacques Loeb and others have performed some interesting experiments with the eggs of our California sea urchins. When an echinoderm egg is penetrated by the sperm cell, it instantly forms a membrane about itself called the fertilization membrane which prevents the entrance of other sperm cells. Next, the nuclei of the two cells unite and the egg is said to be fertilized. Following fertilization, the egg nucleus divides and subsequently the entire cell is divided into two cells. The two cells divide and become four, in the same way, and this process continues until the larva is produced. Dr. Loeb found that by treating the eggs with certain chemicals he could produce the fertilization membrane and cell division.† Thus he was able to obtain larvae from unfertilized eggs.

The sexes are separate and the mature individuals have five large gonads in the upper part of the body cavity that open through pores in the periproct. The color of the male organs is whitish while the ovaries are a deep yellow or orange. They are utilized for food in many parts of the world and have considerable commercial importance in some places. Pratt states,‡ "over 100,000 dozen sea urchins are yearly brought into the fish markets of Marseilles." On the Pacific coast they have not become an article of commerce, as they are used mainly by immigrants from Europe, who collect their own supply. At Pacific Grove the eggs ripen from December to July.

The sea urchins can be observed in tide pools and their

---

*E. W. MacBride—*Echinodermata;* Cambridge Natural History, Vol. 1, p. 529; The Macmillan Company.

†Jacques Loeb—On Artificial Parthenogenesis in Sea Urchins; Science, N. S., Vol. 11, p. 612.

‡H. S. Pratt—A Manual of the Common Invertebrate Animals; p. 641, A. C. McClurg Co.

manner of living studied. When active, the sea urchin extends great numbers of tube feet from among its forest of moving spines. They stretch out and attach to the substratum by means of little suckers. The spines are also used as aids to locomotion. The food is largely vegetable and consists of algae and of small organisms found in the sand. Some species swallow great quantities of sand, digesting and utilizing the organic matter in it. This helps to grind the particles and make them finer and rounder, and in some places, as in the Bay of Fundy on the North Atlantic coast, where the green urchin lives in enormous numbers, they have greatly influenced the nature of the sands.

On the Pacific coast they are more plentiful on and among the rocks than in sandy areas. The purple urchin, in particular, lives in enormous colonies in some places, and literally honeycombs the rocks with small cavities in which it is secure from the pounding of the surf or destruction by the larger fishes. It is difficult to understand the way in which they excavate the pits in rock which seems harder than their own teeth and spines. Some observers maintain* that they chisel away the rocks with their teeth by constantly turning round and round, and by continued effort gradually increase the size of their cave. They may become prisoners within the cave if the entrance is not made larger as they increase in size. In more quiet water they do not "dig in" in this fashion, though the nature of the rock might be favorable, but live in pools and crevices between the ledges.

Besides the sea urchins, proper, the heart urchins and sand dollars belong to the *Echinoidea*. The heart urchins do not have the lantern of Aristotle and the sand dollars are compressed into the form of a disk or shield. There are other differences in structure and habit, but they may be considered as modified sea urchins. The class includes about five hundred known species distributed in all seas from the shore line to great depths.

*A. F. Arnold—Sea-Beach at Ebb-Tide, p. 221; The Century Company.

FIG. 192.—The common purple sea urchin, *Strongylocentrotus purpuratus*. A. Ventral view of test with spines removed. B. Dorsal view of test. C. Dorsal view of a living animal in motion with spines intact and tube feet extended; x½.

FIG. 193.—The green sea urchin, *Strongylocentrotus drobachiensis*, photographed in a tide pool; x⅓. FIG. 194.—The purple sea urchin, *Strongylocentrotus purpuratus*, from a photograph of a living specimen in motion in a shallow tide pool; about x⅔.

FIG. 195.—Sea urchins in tide pool at Pacific Grove. The one large individual is *Strongylocentrotus franciscanus* and the numerous smaller ones are *S. purpuratus*; approximately x1/5.

Order—CENTRECHINOIDA

Family—STRONGYLOCENTROTIDAE

The green urchin, *Strongylocentrotus drobachiensis* (Müller) (fig. 193) occurs on the Pacific coast from Washington to the Arctic, and on the Atlantic coast from the Arctic to New Jersey. The color is greenish, sometimes with a violet tinge; often the spines are deep green and the test violet. Occasionally, the entire animal is a dull brown. The body is about two inches in diameter, and the spines. are of moderate length and pointed. This species deserves some notice, if only for the fact that it bears one of the longest technical names applied to any animal on this coast. The generic name means "round-spined" and that of the species is given in honor of the Norwegian fjord of Dröbak.

The purple sea urchin, *Strongylocentrotus purpuratus* (Stimpson) (fig. 192) is the most abundant one on our coast. It is one that burrows into the rocks of the California shore, and it also occurs in large numbers in the tide pools. It ranges from Mexican to Alaskan waters, overlapping the area occupied by the preceding species. It is an inch and a half or two inches across, with thick, rather short, fluted blunt spines. The general color is a decided purple but very young specimens are usually greenish and are apt to be taken for a new species.

Strongylocentrotus franciscanus (A. Agassiz) (fig. 195) is the large sea urchin of this region, the test alone often exceeding five inches in diameter. It occupies about the same range as the common purple sea urchin, but usually inhabits only the deeper tide pools. Unless one observes carefully, he will think it only a very large form of *S. purpuratus* but the test is a little higher and the spines relatively longer. There are two color phases, red-brown and dark purple.

Order—CLYPEASTROIDA
Family—SCUTELLIDAE

The sand dollar *Dendraster excentricus* (Eschscholtz) (fig. 196) has a very flat body, somewhat circular in outline, which is so thickly covered with minute spines that it has a velvety appearance. The ambulacral areas form a figure on the upper surface that looks like a flower with five petals. Two of the zones are shorter than the other three which gives the effect of being off center (excentric), and this effect is increased by the straightened posterior edge of the disk. The anal opening of the intestine is on the lower surface near the margin.

They move by means of the tube feet and spines. With the microscope the latter are seen to be blunt and inflated with longitudinal and cross ridges. Some are longer than the others, especially at the edge of the disk and underneath. All the spines move together in long wavy lines, reminding one somewhat of a field of grain with the wind blowing over it, but bending more slowly and stiffly. The tube feet of a living specimen can be seen extending beyond the spines. There are, in fact, two sorts of tube feet. Those which occur on the ambulacral areas are used in respiration. The locomotor tube feet are smaller and are found on both surfaces.

The sand dollars live on top of the sand or bury themselves in it. They abound in many lagoons or sandy stretches along the coast, and the white tests are common objects on the beaches. The living animals are seen less frequently, but can be found in the lagoons when the tide is low. One July morning we saw a large group of them in Mission Bay, near San Diego. They were at the bottom of a tidal channel, and were completely covered with water, which was clear and did not interfere with our observation. There were literally hundreds of them there, and each one was half buried in the sand in an oblique position with the upper surface slanting in the direction the current was moving. They were grouped in a haphazard fashion, but so uniform was the degree of inclination, that they appeared to have a military formation. To what extent their position in this case was due to the force of the current, it would be hard to say, but probably it accounts, at least, for the uniformity. We have noticed the same thing at Balboa Bay several times. In other places, and at other times, they were found lying flat on top of the sand or mud or at various angles beneath it. The mouth side, however, is always kept downward, at least so far as we have observed. In this connection it may be noted that when placed on a hard surface with the

FIG. 196.—The sanddollar, *Dendraster excentricus*. A. Dorsal; B. ventral views of the test after the spines have been removed. A little more than ⅙ natural size. FIG. 197.—The heart urchin, *Lovenia cordiformis*. A. Test with spines in place. B. Test with spines removed; about ⅓ natural size.

ventral side uppermost it is impossible for a sand dollar to right itself. We had difficulty in keeping living specimens any length of time until we accidently ran across the following plan. A sand dollar, a mussel, and some old shells with tube worms on them were left together in a dish partly filled with sea water. There was a little eelgrass and some mud and debris on the bottom of the dish. After a week or ten days we were surprised to find the animals still alive, and so we kept them to see how long they would last. Enough salt water was run into the dish from time to time to remove the dust that gathered on the surface of the water. Sometimes the animals would be undisturbed for a month. The sand dollar was still alive when, after more than six months of life in the laboratory, it was returned to the ocean.

Evidently the sand dollar has dwelt in this region for a long time, for fossil remains of Pleistocene age are abundant in the deposits at the edge of the water at Mission Bay. Except for their fossilized condition, the tests cannot be distinguished from those picked up on the beach today, and they belong to the same species.

The sand dollar is purple when alive, but this changes later to a greenish hue. Verrill* says that the fishermen prepare an indelible ink by grinding the spines and skin and mixing the mass with water. *D. excentricus* is found from Alaska to Lower California, overlapping the range of our other sand dollar.

Echinarachnius parma (Lamark) is another sand dollar common to the North Pacific and the North Atlantic Oceans. On the Atlantic side of the continent it abounds as far south as New Jersey, and in our waters, it is reported to live as far south as Puget Sound. It differs from the other species in several ways. The petals of the ambulacral system are approximately equal in length, and center at the apex of the test instead of back of it. The posterior intestinal opening is located on the margin, instead of the lower surface. Finally, the test is a trifle flatter than in *Dendraster*.

---

*A. E. Verrill—Report of U. S. Fish. Commissioner, 1871–1872, p. 362–363.

Order—SPATANGOIDA
Family—SPATANGIDAE

Lovenia cordiformis A. Agassiz (fig. 197) belongs to the group commonly called the heart urchins. It is a Mexican species which reaches the southern coast of California, but does not inhabit very shallow water, and on that account, is little known. It deserves the name of sea porcupine, often applied to our sea urchins, for its general shape and long bristling spines, pointing backward, strongly suggest this rodent. The animal has its mouth near the forward end of the body, and the anus is in the posterior portion, making it somewhat bilaterally symmetrical. Aristotle's lantern is absent. When the spines are removed the test is seen to be roughly heart-shaped and marked above with five petaloid areas. As it is a burrowing creature and not found on shore, *Lovenia* is taken with dredging apparatus.

Class—HOLOTHURIOIDEA
(*Sea cucumbers*)

The sea cucumbers are elongated and cylindrical with a tough, leathery, often warty skin. One form is worm-like with a smooth translucent body covering. To judge from external appearance, the group has lost all trace of a skeleton, but if one uses a microscope he will find, buried in the skin, tiny calcareous granules in the shape of plates, buttons, tables, wheels, or anchors (figs. 199, 206), all that is left of the very complete skeleton possessed by the other members of the family.

The mouth is at the anterior end of the long body, and is not directed downward as it is in many of the other members of the group. A ring of tentacles surrounds it, as is the case with coelenterates, but the holothurians, unlike the coelenterates, keep the tentacles continually in motion when the animal is not contracted. Each tentacle

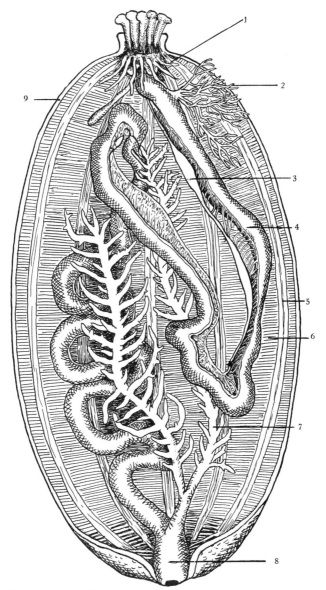

FIG. 198.—Sea cucumber dissected to show internal organs.
1. Ampullae of tentacles.
2. Reproductive organs.
3. Blood vessel.
4. Intestine.
5. Longitudinal muscles.
6. Circular muscles.
7. Respiratory trees.
8. Cloaca.
9. Polian vesicle.

expands to its full extent so that it is at right angles to the body and then bends in until its tip touches the mouth. This motion goes on constantly, one tentacle being turned outward when the two next to it on each side are turned inward (fig. 205). The tentacles are connected with the water-vascular system, and are furnished with ampullae so that the flow of water through them as they move is thought to assist in respiration.

The water-vascular system is provided with an internal, rather than an external, madreporite. Since the body cavity is always tensely filled with fluid, a plentiful supply of water for the tube feet is obtainable from within. The tube feet are much less numerous than in the starfish and in some species are absent altogether. These latter forms move wholly by expanding and contracting the body by means of the strong longitudinal and circular muscles. The tentacles also help in locomotion at times, but all the members of the group are rather sluggish in their movements. The food canal is long, nearly three times the length of the body, and ends in the cloaca (fig. 198). This cloaca, or rectum, is a good sized chamber and in addition to conveying the waste from the food canal is used to pump water into the respiratory trees. These "gills," which are not present in all species, are long, thin walled, branched organs lying in the body cavity. Water taken into the rectum is forced into them, and at the thin walled ends of the branches, some of the water diffuses into the liquid of the body cavity, thus giving it oxygen.

The reproductive organs open to the outside just back of the tentacles. In most of the sea cucumbers the eggs are scattered in the water and fertilized there but, in some species here described, the young are carried by the mother, and in one kind develop in the body cavity of the mother.

Holothurians are found in all seas, but occur in greatest numbers in the eastern oceans. They live from shallow water down to great depths, and range from a few millimeters to a meter in length. The shore collector usually finds sea cucumbers in tide pools along rocky beaches, or he may occasionally find some on kelp holdfasts that have been recently washed ashore. The worm-like ones are found under rocks in sand or buried in the mud. Holothurians all live on organic material that may be found in the sand and mud or that may be growing on the rocks. Sometimes they may get some of the small animals that swim about in the water. The food is conveyed to the mouth by the tentacles.

The collector of sea cucumbers is often surprised to find his cucumbers in a sad state when he gets them to his laboratory. If they are handled roughly or if the water becomes stale, they are apt to eject a large share of their internal organs. Under favorable conditions these organs would be completely regenerated by the animal. *Leptosynapta* has a peculiar habit of constricting off parts of its body so that a specimen kept in captivity may become shorter and shorter until little more than the tentacles are left.

In the Malay Archipelago certain species of sea cucumbers are taken in great numbers and prepared for use as food. They are cooked in sea water, dried in the sun, boiled in fresh water repeatedly until the salt is extracted, then dried and marketed. This dried product is called "trepang" and regarded as a great delicacy by the Chinese who use it in making soup.

The classification of the holothurians is in a more or less chaotic state because those who first described and named species made much of size, shape, color, number of tentacles, and other characters which show such great individual diversity as to make them worthless in distinguishing species.

Family—SYNAPTIDAE

Leptosynapta inhaerens (O. F. Müller) is a small white or brownish worm-like animal (fig. 204) which lives buried in sand or mud at tide line, or in deep water. Those commonly found are about 90 mm. in length though specimens 250 mm. long have been reported. The tentacles are ten to twelve in number with 5-7 branches on each side. The body wall between the prominent muscle bands is translucent showing some of the organs within.

When specimens are handled they will be found to adhere to one's fingers, and fine whitish particles may be seen in the body wall even without the aid of the microscope. When examined under the microscope these particles will be found to be tiny perforated plates and anchors with toothed arms (fig. 199). The anchors help to keep the animal from slipping as it burrows through the sand. When the circular muscles are contracted the anchors are elevated and hold against the sand. The contraction begins next to the posterior end and moves forward pushing the head end onward. The anchors in the anterior end lie flat in the skin and are short, whlie those at the back are longer and more numerous. The tentacles help somewhat in locomotion by loosening and separating the grains of sand as the creature contracts and extends the body. These actions may be seen in the laboratory if the animals are in a small dish containing sand with the sea water. If fresh sea water is not supplied they soon begin to constrict off parts of the body until only small portions are left which soon die. Under favorable conditions, when cut in two, the anterior end will live and grow if it still has the mouth and a part of the digestive tract. The posterior part does not grow again, probably because it has no way of taking food.

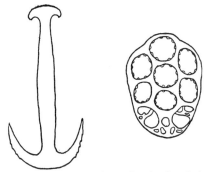

FIG. 199.—Calcareous plate and anchor from body wall of *Leptosynapta inhaerens:* x200.

This species has a wide distribution, being found from Maine to South Carolina; Puget Sound to San Diego; in the Mediterranean, Eastern Atlantic, North Sea, and Arctic Ocean.

FAMILY—CUCUMARIIDAE

Psolus chitinoides Clark (fig. 200) somewhat resembles a chiton in form when it is contracted. The dorsal surface is covered with many firm, granulated plates. The lower side is flattened to form a creeping sole and the tube feet run in rows around the edge of the sole and lengthwise down the center of it. The sole contains many knobbed, perforated plates. The color is generally orange but the crown of tentacles and the cylindrical neck are a brilliant crimson. It is occasionally found at tide line at Puget Sound. The length is 30-65 mm. and the width is about one-third the length.

Thyonepsolus nutriens Clark (plate 7) is found on rocks and kelp holdfasts along shore at Pacific Grove. It is red, and only 15-20 mm. long and 8 mm. wide. The ventral surface is flat and the dorsal surface flattened slightly. The tube feet are numerous, scattered irregularly all over the dorsal surface but arranged in three longitudinal series on the ventral side. There are 8-10 tentacles which may be tinged with yellow or purple. The deposits in the skin are reticulated cups and irregular perforated plates. The most interesting thing about this holothurian is the way the young are carried on the back imbedded in the soft skin. The young are small and ellipsoidal with a pair of tube feet near the rear end of the body, which seem to rest upon the stiff skin of the mother's back.

Cucumaria lubrica Clark (fig. 201) is found at Puget Sound, is pale brownish gray, 50-100 mm. long and cylindrical. The calcareous deposits in the body wall are irregular, thick, knobbed plates, or buttons. In the tube feet, deposits are few and in the tentacles there are a few large supporting rods. The tube-feet are in two rows in each radius. The calcareous ring about the esophagus is made up of plates having short anterior prolongations, the radial ones being notched. The skin is white with brown dots, the dots sometimes almost covering the surface.

Cucumaria curata Cowles (plate 7) is a very small black holothurian found at Pacific Grove. We found numbers of them in rocky tide pools on the point north of Carmel Bay. They are 15–20 mm. in length, and very active when kept in the laboratory. The deposits are rather thick, smooth buttons with 1–9 holes, and in the ventral wall a few much larger plates with larger holes. The eggs are cared for by the mother, being brooded under the ventral surface. Collectors are apt to overlook these sea cucumbers because they resemble the bits of tar that are often seen on the rocks.

Cucumaria miniata (Brandt) Ludwig (fig. 202) is abundant along shore and in deep water at Puget Sound. It is large, reaching 200 mm. in length. The tube feet are large and arranged along the radii, though not in regular rows. The dorsal side may have the papillae or pedicels between the radii. There are ten much branched tentacles which are long and of equal size. The calcareous deposits are not numerous, being rod-like in the tentacles while in the body wall there are irregularly perforated, more or less knobbed plates with serrated margins. The calcareous ring is delicate and the pieces are notched posteriorly. The color is salmon-red to dark brownish purple.

Cucumaria chronhjelmi Théel (fig. 203) is found from Pacific Grove northward, being common at tide line at Puget Sound. The tube feet are numerous and long, forming a double row along each of the radii. There are ten short, much branched, tuft-like, yellow, or orange tipped tentacles, the two ventral ones being smaller. The calcareous plates are many and varied in form, being like cups or baskets in the body wall, plate-like in the tube feet, and rod-like in

FIG. 200.—A living specimen of *Psolus chitinoides*. The animal has attached itself to a rock in the aquarium and is fully extended. The tentacles were continually in motion so some of them are not clearly outlined in the photograph; x½. FIG. 201.—*Cucumaria lubrica;* from life, x⅔.

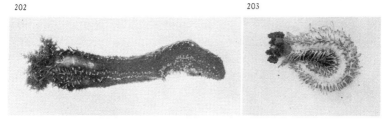

FIG. 202.—*Cucumaria miniata* with tentacles extended; from life, x½. FIG. 203.—*Cucumaria cronhjelmi;* from life, x⅔.

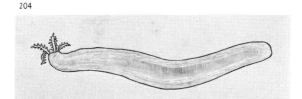

FIG. 204.—The burrowing holothurian, *Leptosynapta inhaerens;* x1.

the tentacles. The calcareous ring is large and the radial pieces have pointed posterior prolongations. The color is white or yellowish, and the length is 70–150 mm.

Thyone rubra Clark, reported from Pacific Grove, is a small species, 20 mm. in length, reddish pink above and white below. The body is almost cylindrical and uniformly covered with numerous tube feet which are seldom arranged in rows at any point. The tentacles number ten with the two ventral ones much smaller than the others. The deposits in the body wall are symmetrical, four-holed plates in the deeper layers, and tables with low spires in the outer layers. On the dorsal side they are larger than on the ventral and are knobbed. The radial pieces of the calcareous ring have prominent posterior prolongations.

This species is viviparous, the embryos developing in the body cavity. According to Clark,* one specimen contained ten young from 2–7 mm. in length. The young are pure white. If they possess developed tube feet these are in rows on the radii. The color and the numerous irregularly arranged tube feet seem to be acquired as the animals grow older.

Family—HOLOTHURIIDAE

The sea cucumbers most commonly encountered along the California coast are the large red-brown ones belonging to the genus *Stichopus*. There are three species closely allied and overlapping in their range, in fact, Clark says that future study may show that all are forms of a single variable species. The drawing (fig. 205) may represent any of the three, as they do not differ in external form. They reach eighteen inches in length and are yellowish or chestnut-brown, paler below with tentacles tipped with yellow and tube feet tipped with black. In some cases, probably *Stichopus johnsoni*, the tentacles and tube feet are tipped

---

*H. L. Clark—The Holothurians of the Pacific Coast of North America; Zoologischer Anzeiger, Band 24, p. 167.

PHYLUM—ECHINODERMATA

with red. The tube feet are numerous and in rather irregular rows on the ventral side. The dorsal side bears large warts

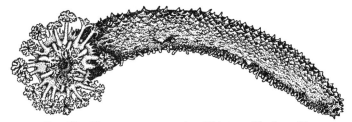

FIG. 205.—The common sea cucumber, *Stichopus californicus*; x½.

and papillae which are often tipped with black spine-like processes. The calcareous deposits are tables and numerous long, thin buttons (figs. 206 and 207). A few knobbed or

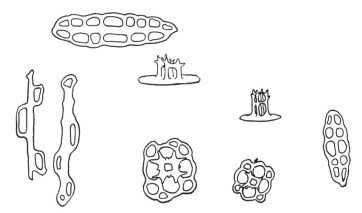

FIG. 206.—Calcareous plates and tables from the body wall of *Stichopus californicus*; x200.

FIG. 207.—Calcareous plate and tables from body wall of *Stichopus parvimensis*; x200.

branched rods are found at the bases of the tentacles and in the skin of the oral disk. The calcareous ring is well developed. The following series of measurements and data from Clark† will serve to distinguish the three species (the unit of measurement = 0.001 mm.).

†H. L. Clark—Echinoderms from Lower California, with Descriptions of New Species; Bulletin American Museum of Natural History, Vol. 32, p. 234.

	Buttons		Tables		Spire		Type Locality
	Length	Holes	Length of Disk	Perforations	No. of Teeth	Diameter	
Stichopus parvimensis Clark.	90	3–4 pr.	45	4+	8–10	20—	Lower Calif.
Stichopus johnsoni Théel.	165–190	10–16	120–170	25–40	20–25	50—	Santa Barbara
Stichopus californicus (Stimpson.)	140–165	10–12	50–90	8–18	12+	25	Pacific Grove & Puget Sound

CHAPTER X.

Phylum—ARTHROPODA

The arthropod group is the most successful one of the animal world in point of numbers and species. If we could name over all the species of barnacles, water fleas, crabs, shrimps, lobsters, sand hoppers, insects, spiders, and myriapods, we would have a list of about 400,000 names, all arthropods. Probably one reason for their success lies in the fact that they have adapted themselves to every sort of environment. They are found in the sea, in fresh water, in burrows underground, flying in the air, living on vegetation of all sorts, and sometimes unfortunately for us, parasitic on man and other animals. According to Pratt,* four-fifths of all known species of animals are arthropods.

Arthropods are easily recognized, since all of them possess a hard, chitinous or horn-like, outer body-covering that is divided externally into somites or segments, and these segments have jointed appendages attached to them. The hard covering affords a place of attachment for the muscles, and maintains the shape of the body, but does not permit growth. Arthropods shed this hard exoskeleton periodically, and before the new skeleton has formed, the creature has grown rapidly, so that the new covering is markedly larger than the old one, and remains the same size until the next moult (fig. 306). Some decapods emerge from the old exoskeleton through a crack along the dorsal part of the cephalothorax, but crabs and lobsters emerge through a dorsal slit between the carapace and abdomen. Since

---
*H. S. Pratt—Manual of the Common Invertebrate Animals; A. C. McClurg & Co., p. 325.

the skeleton is formed as a secretion from the underlying membrane, the animal is able to loosen it while a substitute is being prepared beneath. Of all the arthropods we have kept in aquaria, shrimps seem to be the most ready to demonstrate this moulting habit. If a few tide-pool shrimps are kept in small aquaria for a day or two, some of them are almost sure to moult.

Within the group of arthropods, there are, on one hand, the air-breathing forms, insects, spiders, and myriapods, and on the other hand, the crustaceans, arthropods that breathe by gills.

Class—CRUSTACEA

Most marine arthropods are crustaceans, so called on account of the heavy armor with which the body is covered. Nearly every one is acquainted with crabs, shrimps, lobsters, and barnacles, but few people realize how many kinds of crustaceans there are, or how useful they are to mankind. Some are used for food, others nourish the fish we eat, while some, again, are parasitic upon other sea creatures. Barnacles grow upon the bottoms of ships and make expensive cleaning operations necessary. On the other hand, some of the natives of Hawaiian and other Pacific Islands have found the barnacle an acceptable article of diet. Many crustaceans are scavengers, eating dead animal or plant substances, and thus helping to keep the places in which they live free from decaying matter.

Crustaceans are fitted to live in a variety of situations, and diversity of habitat and food requires a similar diversity of structure. The flat bodies of the porcelain crabs, *Petrolisthes* (fig. 299), enable them to find shelter within the crevices of rocks. The narrow body and elongated legs of *Pinnixa longipes* (fig. 348), enable it to slip into narrow worm tubes with ease. The sand crab, *Emerita* (fig. 290), uses its antennae to sift its food from the sand and running water, while its antennules form a breathing tube to convey pure sea water to its gills. Every animal is adapted to live within a certain environment and possesses organs fitted to cope with its surroundings. If the environment changes too rapidly, or the species is too imperfectly adapted, it is soon exterminated in the struggle for existence, and its place taken by others.

In *Crustacea*, breathing is carried on by means of gills which are attached to the legs, formed from abdominal appendages, or, more rarely, attached to the body wall. The appendages are typically divided into two branches, though in some cases one branch may be lost or poorly developed. There are two pairs of antennae. These are whip-like "feelers" bearing sense organs and serve as organs of touch, and perhaps as organs of smell. The first pair are known as antennules and the term antennae is applied to the second pair, which are usually larger. The eyes are often compound and in the higher forms are borne on stalks.

The mouth parts vary considerably but one generally finds a pair of short, hard mandibles, the biting parts, which work from side to side as in the insects. Next outside of these are the leg-like maxillae and sometimes maxillipeds as well, which help to hold the food. In some of the orders it is possible to group the segments into body regions; and we speak of the head, thorax, and abdomen. In others the head and thorax are fused, forming a division termed the cephalothorax.

The chitinous covering of the body may contain calcium carbonate in addition to the chitin and becomes a very tough armor in the larger crabs; but in many forms, especially the smaller ones, it is thin and comparatively delicate, often rather transparent. A good example of the latter is found in the burrowing shrimp, *Callianassa* (fig. 276). In this crustacean the form and movements of the internal organs can sometimes be seen.

The brain is a double knot of nervous tissue in the head above the esophagus. It is connected with two nerve cords that pass around the esophagus and then extend backward along the mid-line on the lower side (figs. 271 and 309). Branch nerves communicate with this ventral nerve cord and other nerves pass from the brain to the antennae and eyes. As a rule, there is a heart in the dorsal part of the body which keeps the blood in circulation, but in some of the smaller crustaceans it is not present. As in most invertebrates, the blood is nearly colorless.

The digestive tract consists essentially of a tube running through the body. In the higher forms, stomach, intestine, and liver can be distinguished, the latter being a gland giving off a secretion used in digesting the food.

The crustaceans are usually divided into two subclasses, the *Entomostraca* and the *Malacostraca*. In the latter are the crabs, shrimps, lobsters, mantis shrimps, and other crustaceans usually possessing abdominal appendages. They are highly organized and often large forms. The first subclass contains the barnacles and many small or ninute species without any abdominal appendages. Shipley and MacBride say: "This group may be regarded as a lumber room for all Crustacea which are not included in the well-defined division Malacostraca, and the only character which can be attributed to all the members is that of not possessing the marks of Malacostraca." The name, *Entomostraca*, comes from two Greek words meaning "cut in pieces" and "a shell." The first larvae are typically nauplii, minute, free-swimming forms having three pairs of appendages, the last two biramous. Most of the *Malacostraca* pass through a nauplius stage while still within the egg and the course of development may include several other stages such as the zoea and megalops of the crabs (figs. 304 and 305).

Subclass—ENTOMOSTRACA
Order—COPEPODA

The *Copepoda* (fig. 208) exist in the ocean in such countless numbers that although they are very small, they form the principal food supply of many much larger marine animals. A fine tow net will usually capture large numbers of them, or they may often be found in dishes of sea water that have contained collections of algae, or bryozoans. They are delicate and nearly transparent, and swim about in a rapid, jerky fashion. The females of some species carry a pair of egg sacs or brood pouches attached to the first segment of the abdomen (fig. 209). The males are often smaller than the females.

The body is large anteriorly, and tapers rather sharply posteriorly. Most species have only a single median eye. On the head are two pairs of antennae, the first unbranched, a pair of mandibles, two pairs of maxillae, and a pair of maxillipeds. The thorax bears five pairs of branched (biramous) swimming legs which are without gills. Each leg is joined to its mate on the opposite side by a transverse movable ridge so that the two move together. The abdominal segments bear no appendages, but the last segment bears two processes forming a forked tail with long caudal bristles. These bristles are remarkably developed on some of the pelagic copepods. The female in certain species carries the egg sacs about during the breeding season. Each sac contains a number of eggs which are glued together with a cement-like substance. Some of the pelagic copepods are brilliantly colored.

Dr. Esterly* of the Scripps Institution at La Jolla has made a study of the occurrence of copepods of various species in the different hauls that have been made in the waters along the southern California coast and he finds that the copepods of the species studied make daily vertical migrations. Generally they are found in deep water, mostly at about 250 fathoms, during the day, and at higher levels, even up near the surface, at night. The greatest

---

*C. O. Esterly—The Occurrence and Vertical Distribution of the *Copepoda* of the San Diego Region; University of California Publications in Zoology, Vol. 9, p. 335.

number of one species, for example, was found in the daytime at about 200 fathoms, in the evening at 100 fathoms, at midnight at 5–10 fathoms, and from 4–6 A.M. at 100 fathoms. These figures indicate that they begin to travel upward when darkness begins to fall, but start down again before daylight. Further researches are being made to determine, if possible, some of the conditions which may account for this migration.

Many species of copepods are parasitic upon fish, or other animals, and are called fish lice. Some of these

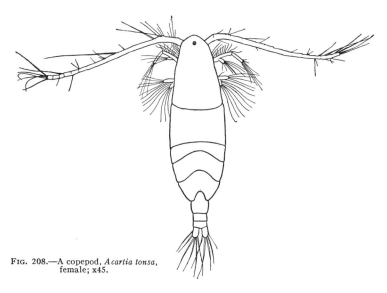

FIG. 208.—A copepod, *Acartia tonsa*, female; x45.

parasitic forms are strangely modified in structure on account of this habit. In many, the appendages have been reduced, the eyes have disappeared, the antennae have become modified to form hooks with which to hold on to the host, and the mouth parts adapted for sucking.

More than 2200 species of copepods have been described, of which nine-tenths are marine. Over one hundred species have been described from the San Diego region. Descriptions are based on the contour of the body, on the propor-

PHYLUM—ARTHROPODA

tionate lengths and the number of the joints of appendages, number of bristles or spines, and other minute details of structure.

Acartia tonsa Dana is a typical free swimming copepod (fig. 208) often taken with the net a short distance off shore. The female reaches 1.5 mm. in length and the male 1 mm. They are without color and transparent. The female does not carry the eggs in sacks as is the habit in many other families of copepods.

Oithona sp. (fig. 209) is a smaller form taken in hauls along with the last named species. The females of this family carry the egg sacks until the young are set free.

Pseudomolgus navanaci Wilson is parasitic upon *Navanax inermis*, the striped tectibranch mollusk (plate 8). The copepods can wim freely about in the water, but when they are disturbed they speedily attach themselves to the lining of the mantle cavity of the host. The females carry salmon-colored egg sacs. Length about 2 mm.

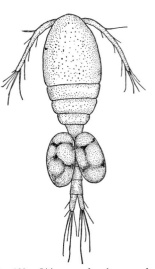

FIG. 209.—*Oithona* sp., female copepod carrying egg sacs; x40.

Order—OSTRACODA

The body of an ostracod is made up of relatively few segments and is completely enclosed in a carapace which resembles a bivalve shell. In practically all species the segments are so closely united that no traces of segmentation may be discovered. Seven is the usual number of appendages though a few species have less. Owing to their small size (averaging about one millimeter) the ostracods are little known to laymen and even to most scientists, but they are a group that is rich in species, no less than 1700-1800 species having been described to date.

They are found in both fresh and salt water, but the greater number of them are marine. Most of the marine species live on the bottom, ranging from the shore down to very great depths (some thousand meters). Comparatively few are pelagic and these range from the surface to great depths.

Family—CYPRIDINIDAE

Cylindroleberis sp. belongs to the same group as *C. oblonga* (Grube), characterized by an oblong, cylindrical shell. It lives in the mud and sand of the bottom, but is occasionally taken in the tow net. It sometimes appears in the hauls taken from the pier at the Scripps Institution. It is about 1.5 mm. long with an oblong carapace (fig. 210). The living animals are orange-red and swim about actively. The beating of the heart can be plainly seen through the carapace and the pair of lateral, black eyes are very well developed. The first antennae are mainly sensory. The second antennae are the only swimming organs and can not only be moved up and down but sideways as well, like oars. A furrow or excavation in the anterior end of the carapace permits the extension of the antennae for these oar-like movements.

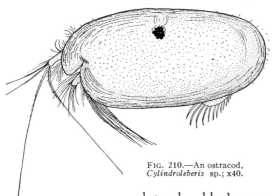

Fig. 210.—An ostracod, *Cylindroleberis* sp.; x40.

Order—CIRRIPEDIA

The *Cirripedia* are commonly known as barnacles, and are found in great numbers in all oceans. Any solid object that has been in the salt water for a time is usually encrusted with them, and the wharf piles below high water mark are often entirely covered. The bottoms of ships, especially of those that sail the warm seas, have to be scraped, now and then, to rid them of the accumulation of barnacles. The fixed *Cirripedia* on this coast are of two chief kinds, the goose and the acorn barnacles. The former (fig. 212) are stalked, the peduncle sometimes being several inches long. These frequently form great masses made up of individuals of all ages and sizes, some of them attached to the rock or floating piece of wood and the rest attached to each other. The acorn barnacle is sessile, fastened directly to some object such as a rock or shell, a piece of wood or kelp, the carapace of a lobster, or even the skin of a whale. Some species of cirripeds are parasitic on other crustaceans. A cirriped that has this habit loses all trace of resemblance to its relatives, becoming a mere sac with a system of branching roots which reach throughout the body of the host.

Early students classified barnacles with the mollusks on account of their hard shells; in fact it was not until 1830 that the early stages of the life-history were known and their relationships thus made clear. The young barnacle leaves the mantle cavity of the parent as a free swimming nauplius which looks much like the nauplius stage of other crustaceans. As the nauplii swim freely at the surface for a short time, they may be scattered far from the place of their birth. After moulting, the young barnacle resembles an ostracod, having a two-piece shell, and hence is called a cypris larva. This swims about for a time until it finds a suitable place to attach. At the base of each first antenna is a large cement gland which secretes an adhesive substance by which the cypris attaches itself to its support. The

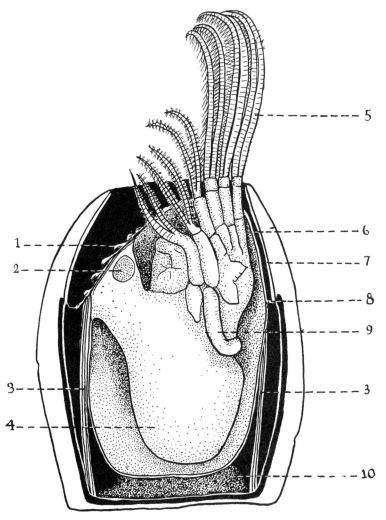

Fig. 211.—Diagram of a lengthwise section of a sessile barnacle.

1. Scutum.
2. Muscle (cross section).
3. Muscle (lengthwise).
4. Stomach region.
5. Cirri.
6. Tergum.
7. Sheath.
8. Opercular membrane.
9. Penis.
10. Ovary.

swimming legs are cast off and the two-branched cirri (fig. 211) of the adult appear. The larval bivalve carapace is shed and the shell-like plates begin to form from the outer surface of the mantle. In the goose barnacle, the head region elongates greatly and becomes the peduncle. The body thus lies free in the mantle cavity and when the valves are closed is protected by the hard shell and the mantle which lines the cavity. When the barnacle feeds, the cirri, or thoracic appendages, are thrust out through the mantle opening, and between the valves, and sweep a current of water, carrying food, in toward the mouth, where the particles are masticated by the mandibles. This sweeping or grasping movement of the cirri takes place with clock-like regularity, and suggests a hand clutching for something in the water and then hastily withdrawing. In large species of *Balanus*, this extension of the cirri occurs from 18–23 times a minute. Huxley* has very aptly said that a barnacle is a crustacean which is fixed by its head and kicks food into its mouth with its legs.

Most barnacles are hermaphroditic, the ovaries being located below the mantle cavity. A pair of oviducts open into the mantle cavity on each side of the head. The pedunculate barnacles have no branchial organs but the acorn barnacles have two branchiae formed from folds of the inner side of the mantle. The hard covering of the cirri and of the body, and the lining of the shell are shed at each moult, but the shell itself is permanent and is enlarged as the animal grows. Dr. Heath† found that both the sessile and pedunculate barnacles reach sexual maturity at the age of eighty days.

Suborder—LEPADOMORPHA

This group includes the goose barnacles, *Cirripedia* having flexible peduncles provided with muscles. The valves

---

*T. H. Huxley—Anatomy of the Invertebrated Animals; D. Appleton and Company.
†California Fish and Game, Vol. 3, p. 35.

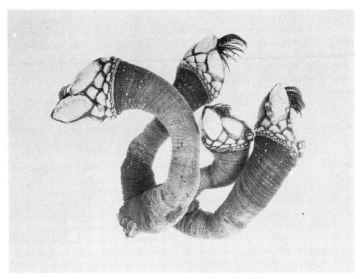

FIG. 212.—*Mitella polymerus*, a goose barnacle, slightly reduced.

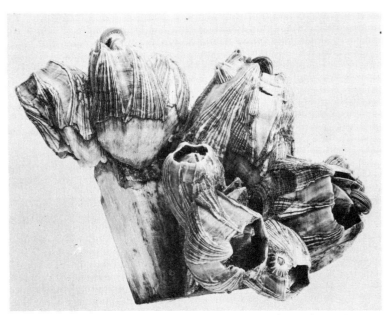

FIG. 213—*Balanus tintinnabulum californicus*, an acorn barnacle that is abundant on the coast; x1.

PHYLUM—ARTHROPODA

or plates resemble scales and are not grown together solidly. Species are occasionally found which have no plates at all. One of these, *Alepas pacifica*, is pelagic and is carried about attached to a large jellyfish.

An interesting picture in an old zoology book shows a group of the goose-barnacles borne on a stem which is extending from the water.* The individuals of the group show the supposed transformation from barnacles to geese. Heads of little geese may be seen extending from two of the barnacles and other geese are swimming about in the water. One student of natural history as late as 1676 described what he thought were little birds which he found enclosed in the barnacle shells.

California Fish and Game (1916) states† that the goose barnacles have recently come to be appreciated as having food value, and are being shipped to the San Francisco markets. When properly prepared, they are said to taste much like lobster. To prepare them, we are told to wash them thoroughly, clean with a small brush, boil in strong salt water until the barnacles shrink free from the shell, then remove the heavy skin from the "necks." The barnacle may then be prepared as salad, or in other ways. We have not tried this recipe yet, but may be reduced to it some day.

Family—SCALPELLIDAE

In this family, there is a basal whorl of plates below the five principal ones and the peduncle is scaly. There are a number of species of this group that are dredged from deep water off this coast but the species given is the only one the ordinary collector will be likely to find. For the meaning of the technical terms used, see figure 214.

Mitella polymerus (Sowerby) (figs. 221 and 212) has eighteen or more plates on the capitulum, with numerous

---
*One of these old figures is reproduced in *Organic Evolution* by M. M. Metcalf; The Macmillan Company, p. 5.
†California Fish and Game, Vol. 2, p. 150.

irregularly arranged scales at the base, and the peduncle is covered with fine scales. The general color is red or reddish brown to yellowish brown, except for the plates, which are white. The peduncle may be three inches or more in length and the capitulum nearly an inch across. This is the common species of goose barnacle on the coast, and is found from British Columbia to Lower California on piles, driftwood, and ships. It has also been found growing on the humpback whale. The barnacles in no way injure their support, but great numbers of them attached to a whale or a ship can considerably impede progress.

Family—LEPADIDAE

Members of this group have five plates which fit closely together. They are common in all seas on floating objects. As any of the five American species may be picked up along the west coast, the following key, adapted from Pilsbry† is given. Descriptive terms are shown in figure 214.

    a. Carina ends below in a flat external disk; valves or plates thin and papery. *L. fascicularis*

    a.a. Carina ends below in a fork; valves hardened with limy material.

        b. Valves radially furrowed or strongly lined on the surface.

            c. The outer margin of the scutum is arched, protuberant. *L. anserifera*

            c.c. The outer margin is close to the ridge from umbo to apex. *L. pectinata*

        b.b. Valves are smooth or have minute radial striations.

            c. Valves smooth or delicately lined, an internal tooth at the umbo on the right hand scutum. *L. anatifera*

            c.c. Valves not radially lined; no internal tooth at the umbo of the scuta. *L. hilli*

Lepas anatifera Linnaeus is the most abundant of the goose barnacles. It greatly resembles *L. hilli* (fig. 214) but has radial striations on the valves, and there is an internal tooth on the umbo of the right hand scutum, but none on the left. The carina is often toothed. The capitulum is 21 mm. long.

†Henry A. Pilsbry—The Barnacles (*Cirripedia*) contained in the Collections of the U. S. National Museum; U. S. National Museum Bulletin 60, p. 79.

Lepas hilli Leach is also abundant on floating objects. The specimens photographed (fig. 223) were on a kelp stem. A fifteen foot piece of kelp was brought ashore, and throughout the whole length it was covered with the barnacles as thickly as in the piece shown in the picture. Among the

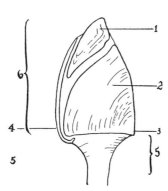

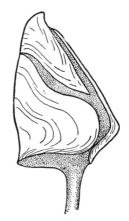

FIG. 214.—*Lepas hilli*, a goose barnacle; x3.
1. Tergum. 4. Carina.
2. Scutum. 5. Peduncle.
3. Umbo.   6. Capitulum.

FIG. 215.—*Lepas fascicularis*, a goose barnacle; x3.

barnacles were hydroids, bryozoa, and caprellas in great numbers. How numerous these forms must be in the ocean when a favorable foothold is appropriated by so many kinds of creatures! The valves in this species are nearly smooth (fig. 214) and there is no internal tooth at the umbo. The capitulum may reach a length of 35 mm.

Lepas anserifera Linnaeus. The valves have distinct radial grooves. The back of the scutum is strongly arched.

Lepas pectinata Spengler chiefly flourishes in warm seas. The valves are distinctly grooved and often spiny. The tergum is notched to receive the tip of the scutum.

Lepas fascicularis Ellis and Solander (fig. 215) has thin paper-like plates and the carina is bent at a decided angle expanding into a flat basal disk. The umbo flares out somewhat at the base. They are 14 mm. in length exclusive of the short peduncle.

Suborder—BALANOMORPHA

This group is made up of the sessile barnacles called acorn barnacles which encrust the rocks, piles, and floating wood, and are often found on living animals that have a tough or hard integument. They vary from 5 to 60 mm. or more in diameter, and by growing upon each other may form large masses.

Family—BALANIDAE

This family includes species in which the rostrum is provided with radii or wings which overlap the next lateral plate or compartment. (figs. 216 and 219)*

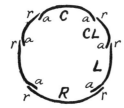 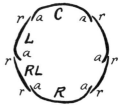

FIG. 216.—Diagram to show arrangement of compartments in *Balanus* (left) and *Chthamalus* (right).

a, a. Alae.
r, r. radii.
C. Carina.
CL. Carino-lateral compartment.
L. lateral compartment.
RL. Rostro-lateral comparment.
R. Rostrum.

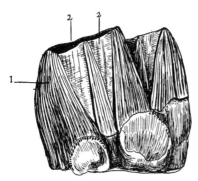

FIG. 217.—*Balanus tintinnabulum californicus:* x4/3. 1. Parietes; 2. radii.

The species of *Balanus* have six compartments or plates in the shell; there are usually radii except on the carina. The scutum and the tergum which make up the beak-like covering of the contracted cirri are interlocked.

Balanus (*Megabalanus*) *tintinnabulum californicus* Pilsbry (figs. 213 and 217)

*Henry A. Pilsbry—The Sessile Barnacles; U. S. National Museum Bulletin 93.

PHYLUM—ARTHROPODA

is one of the largest and most conspicuous barnacles. The parietes of the shell are rather finely striate, the lines being white on a red ground. The radii are indistinctly striated transversely. The parietes contain thin walled vertical pores, and the walls of the radii have pores which run parallel with the base. The form and markings of the scuta and terga are shown in the figure (fig. 218). When the cirri are extended, the mantle encircles their bases like a

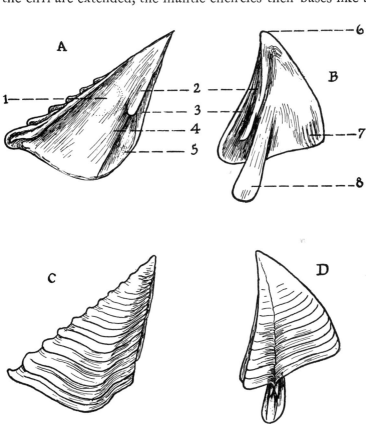

FIG. 218.—*Balanus tintinnabulum californicus*, showing scutum and tergum; x4.
    A. Scutum, inside.   C. Scutum, outside.
    B. Tergum, inside.   D. Tergum, outside.
    1. Attachment of adductor muscle.   5. Attachment of depressor muscle.
    2. Articular ridge.   6. Apex.
    3. Articular furrow.   7. Crests for depressor muscle.
    4. Adductor ridge.   8. Spur.

gorgeous collar. On the dorsal side, the collar is red with blue and light spots and edged with white, while on the ventral side it is blue, edged with red. This collar is covered over by the scuta and terga when the cirri are contracted. The greatest diameter is 57 mm., height 34 mm. It has been found throughout the southern California region.

Balanus crenatus Bruguière is extremely variable in form. The photograph (fig. 224) shows the smooth sort, but there is a rough form, described as rudely plicate, which externally looks quite unlike those in the figure. The ribs on the inner walls of the compartments are continuous with the septa or partitions of the walls. The scutum and the tergum have strong articular

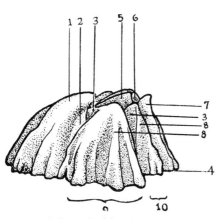

FIG. 219.—*Balanus glandula*: x4.
1. Rostrum.
2. Radii.
3. Alae.
4. Basis.
5. Scutum.
6. Tergum.
7. Carina.
8. Parietes.
9. Lateral compartment.
10. Carino-lateral compartment.

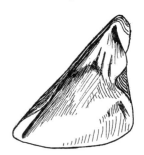

FIG. 220.—*Balanus crenatus*, inside view of scutum; x4.

ridges (fig. 220). The species has a wide range, which includes the Arctic Ocean, North Atlantic, Bering Sea; and North Pacific south to Santa Barbara, and northern Japan. The diameter at the base may be 16–18 mm.

Balanus glandula Darwin (figs. 219 and 222) is one of the smaller barnacles which is found chiefly between low and high-tide mark on rocks, mussels, etc., all the way from the Aleutian Islands to southern California. It is found in great numbers and usually along with *Chthamalus fissus* in

FIG. 221.—Mass of *Mitella polymerus* attached to a rock on the shore. FIG. 222.—Rock covered with barnacles *Balanus glandula* (larger species) and *Chthamalus fissus* (smaller species); about ½ natural size. FIG. 223.—The goose barnacle, *Lepas hilli*, on a piece of kelp, photographed from life; ×½. FIG. 224.—*Balanus crenatus*, an acorn barnacle found from the Arctic, south to Santa Barbara; ×½. FIG. 225.—*Balanus cariosus*, an acorn barnacle that is abundant on the Washington and Alaska coast; ×½.

the south. *Balanus glandula* is the larger of the two species and is dirty white, while the other is olive-brown. Large specimens measure 14 mm. in diameter and 7–8 mm. in height, are usually conical, deeply ribbed, and sometimes, in crowded areas, are much elongated. The pores are small in young specimens and nearly filled up with lime in larger ones. The species can be identified by the small pit or depression near the center of the inside of the scutum.

Balanus cariosus (Pallas) is a common shore form (fig. 225) on the Washington and Alaska coast. At first glance it resembles *Tetraclita* (fig. 227) with its membraneous basis, porous wall, and thatched appearance. It has many deeply cut, irregular ribs, which have occasional projecting points. These ribs which overlap each other make the shell appear to be thatched with straw.

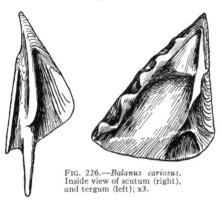

Fig. 226.—*Balanus cariosus.* Inside view of scutum (right), and tergum (left); x3.

The radii are narrow and usually cannot be well distinguished except in the cylindrical forms. The wall is thick, formed by several rows of unequal sized pores. The tergum is narrow with the apex beaked and the spur sharply pointed (fig. 226). The cirri are nearly black. The young barnacles are regular and starlike in form but the number of ribs and pores increases with age so that the old individuals are irregular and variable in form. Speci-

Fig. 227.—*Tetraclita squamosa rubescens*, the thatched barnacle; x1.

mens may vary from 56 mm. in diameter at the base and 32 mm. high, to 35 mm. in diameter and 100 mm. high.

Tetraclita squamosa rubescens Darwin, the thatched barnacle, (fig. 227) is found on the rocks that are exposed at low tide from the Farallone Islands to Lower California. It is roughly conical with only four compartments which often are not distinctly marked. The walls are permeated by pores which form several irregular rows in the large specimens. The color is a dull red, green, or grayish, and the surface is usually much eroded and roughened like thatch. The margin of the mantle is dull red, bordered in places with yellowish white. The largest specimens are about 40 mm. in diameter.

Family—CHTHAMALIDAE

The *Chthamalidae* includes species in which the rostrum has alae (figs. 216 and 219) and the walls are not porous.

Chthamalus dalli Pilsbry is a small conical barnacle with an oval opening. The compartments show a few unequal ribs. When the surface is much worn it is pale gray with a narrow basal band of buff where the cuticle remains. The interior is whitish with flesh tints and the opercular valves are of a like color. The scutum and tergum distinguish it from the next species. In *C. dalli* the tergal margin of the scutum is two-thirds the length of the base while in *C. fissus* it is one-half. *C. dalli* also has a strong adductor ridge and has crests in the scutum for the attachment of the lateral depressor muscle. The species is taken in Alaska and on the Washington coast. The basal diameter is 6.2 mm., height 2.3 mm.

Chthamalus fissus Darwin (fig. 222) is a very common, small acorn barnacle abundant on the rocks, mussels, and limpets on the southern California coast. The small olive-brown or gray, encrusting shells are almost the color of the rocks which they thickly cover. They extend well up above low-tide mark so that they are often left for several hours untouched by the waves and look dry and dead. If they are put into sea water they will quickly open, thrust out the cirri, and begin driving the food bearing current of water in toward the mouth.

Isolated individuals are low with broad bases, but when great numbers are crowded together they become elongated so that the height is 2–3 times that of uncrowded individuals. Ordinarily they are 5–7 mm. in diameter, and 3 mm. high with an opening 2–3 mm. long. The walls are somewhat ribbed

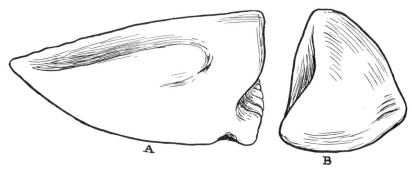

FIG. 228.—*Chthamalus fissus;* x27. A. Scutum; B. Tergum.

or folded near the base, smooth in the younger stages and covered with an olive-brown cuticle when they are not worn by the waves. The tergum and scutum (fig. 228) distinguish it from *C. dalli,* the northern form which it strongly resembles.

Subclass—MALACOSTRACA

With the exception of the barnacles, most of the common shore forms of crustaceans belong to this division. In most cases some of the thoracic appendages are modified to help in handling the food and are known as maxillipeds. The ones not used for defense and taking of food are used principally for locomotion. Technically, these appendages are known as periopoda, or walking legs, in distinction to

PHYLUM—ARTHROPODA 271

the pleopoda or abdominal appendages, also called swimmerets. The periopoda are often terminated by prehensile claws which may have a chelate structure or belong to the subchelate type (fig. 229). The swimmerets may be used in locomotion, as their name implies, or they may serve for the attachment of eggs. In some instances they are modified for use as breathing organs.

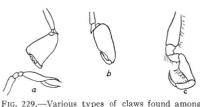

FIG. 229.—Various types of claws found among arthropods. A. Chelate. B. Subchelate. C. Simple.

ORDER—AMPHIPODA

(The beach fleas and their relatives)

The amphipod most commonly known is the sand hopper (fig. 233), so often seen feeding on the kelp along sandy beaches, or jumping about far up on the sand. The amphipods nearly all have laterally compressed bodies, that is, their bodies are flattened from side to side, in contrast to the isopods and some other forms, which are compressed from above downward. They have no carapace. The segments are free and covered with shiny, flexible cuticle and the gills are borne on the under side of the basal joints of the legs.

Three suborders of amphipods are represented on the Pacific coast. The members of the first suborder, the *Hyperiidea* are sometimes found living within the bodies of animals such as medusae, ctenophores, and salpae, but several species are to be found in myriads leading a free pelagic life, and these form an important part of the food of whales. As they are exclusively pelagic they are not often seen except in museum collections. The second suborder, the *Gammaridea* are the common sand hoppers or beach fleas. If a handful of algae or bryozoans is taken from the rocks or piling and placed in a dish of sea water, the little gammarids will be seen swimming about in all directions. Specimens of the third suborder of amphipods, the *Caprellidea* may be present also, but will be harder to see since both their color and their form make them almost indistinguishable. They look like animated sticks as they cling to the seaweed with their hind legs, and execute, with the anterior part of the body, a series of bowing and swaying movements through the water.

Suborder—GAMMARIDEA

The sand hoppers or sand fleas (fig. 233) are usually found in great numbers on every sandy beach especially if there is a good deal of seaweed present. They burrow in the sand, sometimes to a considerable depth, and some species are said to go far inland if the soil is moist. Although called sand fleas, they do not bite people as do their namesakes, the name having been given on account of their jumping habit. The small gammarids are numerous near shore in the tide pools where the rocks are covered with algae. Some kinds that live in the algae, spin tubes within which they lie in wait for their prey. One species sews up the margins of the kelp frond to make a tubular nest for itself. The thread for this weaving and spinning is secreted by glands that are situated on the first two periopods, and the fine strands of web may be seen passing out from the tip of the claw when the amphipod is weaving its nest.

The male gammarids are usually larger than the females and have the curious habit of carrying the females about under their bodies at the time the eggs are laid. Several different species have been observed at this time and probably the habits of most of the members of the group are much the same. The male swims about, grasping the female and holding her under his body. He lets her go when she moults but takes her again soon after and deposits the spermatozoa on the under surface of the thorax. The eggs are deposited about half an hour after the placing of the sperm and the male continues to carry his burden about after this for a short time. Gammarids have theoretically the normal number of segments for *Malacostraca*. The eyes are not stalked. The antennules may have an accessory branch while the antennae have a single branch. The first two pairs of appendages of the thorax are usually modified into chelate or subchelate grasping organs known as gnathopods, the remaining five are legs and of these the anterior two are directed backward and the posterior three forward. In the female, there are plate-like structures on certain of the legs which form a brood pouch for carrying the eggs. The first three pairs of appendages (pleopods) on the abdomen are two-branched and are useful for swimming, while the posterior three pairs (uropods) are stiff and used for jumping.

The classification of the *Gammaridea* is beset with many difficulties as any one who has worked with them will testify. Rev. T. R. R. Stebbing,* who has published a monograph in which he describes several hundred species, says, "the creatures themselves have conspired in various ways to make the path of knowledge thorny and fatiguing. Genera, the species of which have different habits, and which are separated by the unlikeness of the males, are in the females scarcely distinguishable. * * A great increase in the number of known species brings to light the missing links. * * Characters which at one time distinguished large groups, or were valid for the whole family, are gradually nibbled away by exceptions here and exceptions there till all the neatness and completeness of the arrangement they provided are muddled away and spoiled."

There are, undoubtedly, great numbers of species of *Gammaridea* on the Pacific coast that have never been

---

*T. R. R. Stebbing—*Amphipoda* from the Copenhagen Museum and other Sources; Transactions of the Linnean Society of London, Second Series, Zoology, Vol. VII, p. 395.

described since no one has made a complete survey of the group here as yet. A few only of the most common of the described species are given. If the differences that separate the species seem slight and trivial to the reader he must remember the difficulties that attend the classifying of the group.

Family—LYSIANASSIDAE

Aruga oculata Holmes (fig. 230) is one of the small amphipods. It reaches about 14 mm. in length. The eyes are large and oblong and the angles at the side of the head are produced into an acute triangular lobe. The antennae are short and of about equal length. In the first antennae the peduncle is stout, the first joint is nearly as long as wide, and the third joint is short. The first joint of the flagellum is long, and the secondary one has about five joints. The first gnathopods are simple and the second almost chelate. In the first and second walking legs the merus is produced at the lower anterior angle and the third has the merus produced at the lower posterior angle. These are found in shallow and deeper water along the southern California coast.

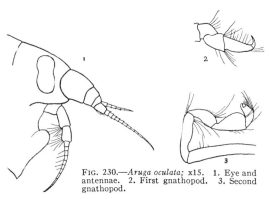

Fig. 230.—*Aruga oculata;* x15. 1. Eye and antennae. 2. First gnathopod. 3. Second gnathopod.

Family—GAMMARIDAE

Members of this family usually live in fresh or brackish water, but some of them are found in the ocean. The antennae are usually slender and the first usually has a small accessory flagellum. The posterior pair of abdominal appendages are longer than the two preceding pairs.

Melita fresnelii (Audouin) (fig. 231) is sometimes picked up on kelp holdfasts. It can be recognized by the one large

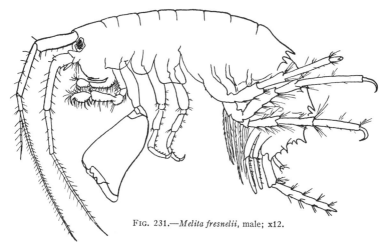

FIG. 231.—*Melita fresnelii*, male; x12.

pincer claw of the male. In the female the first two thoracic appendages bear the usual moderately developed subchelate hand in which the claw folds up like the blade of a jackknife, but in the male the palm of the second appendage on the left side is very large and is a real pinching or chelate claw. The antennae are long and slender and the dorsal segments of the abdomen are toothed on the posterior margin. This species has a wide distribution, being reported from points on the Atlantic, Pacific, and Indian oceans.

FAMILY—TALITRIDAE

In this family the mouth parts project strongly below. The side plates are rather large, the fifth one being bilobed. The first antenna is usually much shorter than the second and without an extra flagellum. The last pair of abdominal appendages usually have but a single branch each. The second thoracic appendage is either a poorly developed pincer or is subchelate.

Orchestoidea californiana (Brandt) (figs. 232 and 233-A) is common on the California beaches. The first antennae are

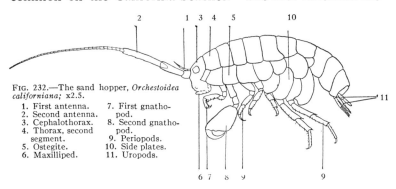

FIG. 232.—The sand hopper, *Orchestoidea californiana*; x2.5.
1. First antenna.
2. Second antenna.
3. Cephalothorax.
4. Thorax, second segment.
5. Ostegite.
6. Maxilliped.
7. First gnathopod.
8. Second gnathopod.
9. Periopods.
10. Side plates.
11. Uropods.

short, reaching only to the middle of the next to the last joint of the peduncles of the second antennae. The second antennae are longer than the body, with the flagellum longer than the peduncle. Specimens reach a length of 30 mm. They are usually white with antennae and part of the dorsal surface bright orange.

This species is usually seen feeding on the kelp that is left stranded on the beach by the receding tide. When the animals are abundant one has but to lift up one of these clumps of seaweed to see hundreds of the creatures go hopping away, and often when the tide is coming in, great numbers of them may be seen, apparently fleeing from the incoming waves. One evening we watched them digging into the moist sand. They went in head first and then remained below the surface with just enough covering to hide them from sight. On the same stretch of beach one may sometimes see the hoppers in countless numbers while at other times there may be none visible. Often when there may seem to be none, one will find them by digging down six inches or a foot into the sand at high-water line.

Orchestoidea corniculata Stout (fig. 233-c) is a large form

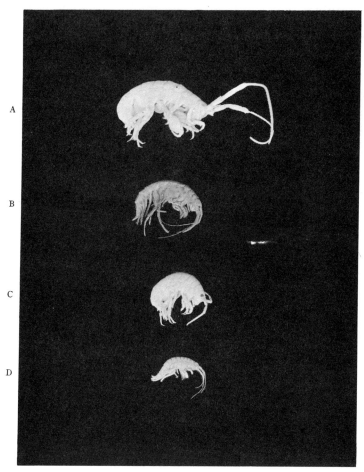

FIG. 233.—A. *Orchestoidea californiana.* B. *Amphithoe humeralis.* C. *Orchestoidea corniculata.* D. *Paragrubia uncinata;* x5/4.

resembling *O. californiana*, but is pink or mottled with red and brown according to Miss Stout.* The length of the body is 10-15 mm., and the second antennae of the male are about half the length of the body, while those of the female are about one-fourth the length of the body. The peduncle of the second antenna is longer than the 14-18 jointed flagellum in contrast to *O. californiana*, in which the flagellum has about thirty joints and is longer than the peduncle.

Orchestia traskiana Stimpson is found among the decaying algae at high-water mark or in the mud sloughs. It differs from the last two species in having the first thoracic appendage subchelate where the others have simple ones; also the second antennae are short, being only one-third the length of the body. These amphipods are grayish brown or green with bluish appendages. The males are about 15 mm. long and the females somewhat smaller.

Family—AMPHITHOIDAE

In this family the side plates are regular and the first antenna has a short third joint. There may or may not be an accessory flagellum. The second thoracic appendage usually has the larger hand. The uropods (the three posterior abdominal appendages) have two branches each, those of the last one being very short with distinct little hooked spines on them.

Amphithoe humeralis Stimpson (figs. 233-B and 234) is found from Puget Sound southward. The antennae are more than one-half the length of the body. The flagellum of the first antenna is two to two and one-half times the length of the peduncle. The second antennae are stout. In the gnathopods the fifth joint is longer than the sixth joint, or hand, and is lobed behind. In the first and second

---

*V. R. Stout—Studies in Laguna *Amphipoda* II; Zoologische Jahrbücher, Abteilung für Systematik, Vol. 34, pp. 647-650.

periopods, the fourth joint is widened and lobed. The
third periopod is very short and the fifth one longer than

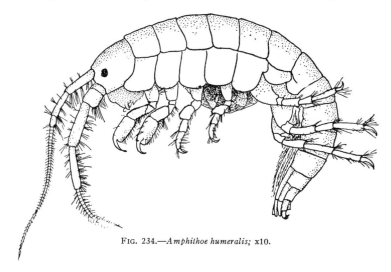

FIG. 234.—*Amphithoe humeralis;* x10.

the fourth. The third abdominal appendage does not reach
beyond the second, and its outer branch has two hooks.

Paragrubia uncinata (Stout) is a tube dweller (fig. 233-D)
and will be found now and then on kelp that is washed on
shore. The length is 8-23 mm., the body orange-green,
the antennae are ringed with white, and the eyes bright
rose color. The first antenna is longer than the second with
a 40-60 jointed flagellum and a tiny accessory flagellum as
long as the first four joints of the other flagellum. The
second antenna is shorter, the peduncle being longer than
the flagellum which has thirty-one joints. The appendages
are thickly covered with setae and the first and second
gnathopods are both subchelate.

SUBORDER—CAPRELLIDEA

Caprellids (fig. 235) are grotesque forms with elongated,
stick-like bodies. The pictures of the animals are curious

enough, but to see the living ones bowing this way and that, as they cling to the seaweed with their hind legs, or looping along in measuring-worm fashion is most entertaining. Some of our associates have dubbed them "Aunt Fannys."

The first thoracic segment is united with the head. The abdomen is rudimentary and gills are found on the third and fourth thoracic segments. The appendages of the first and second thoracic segments have prominent subchelate claws called gnathopods. The legs just back of these two gnathopods are missing in many of the species, and the abdominal appendages are rudimentary or wanting. The posterior thoracic appendages are prehensile.

Caprellids are found among masses of algae and among bryozoan and hydroid colonies either close in shore or in deep water. They grasp the supporting material with their prehensile hind legs and the gnathopods on the free, fore part of the body reach out for food as the animal bows and bends up and down and from side to side. Those found on the Pacific coast vary in length from 5–30 mm. Most of them are reddish or yellow-brown. The males are usually larger than the females. The latter have a brood pouch situated below the third and fourth segments.

Caprellids are not often seen by collectors because their color and outline make them look so much like their surroundings, and many of the species are too small to be noticed unless they are moving about. Some of the genera include specimens in which legs are present on the third and fourth segments, though they may be rudimentary structures. Other genera include individuals with four instead of six prehensile legs. In the genus *Caprella*, however, we have the gills borne on the third and fourth segments with no legs on these segments. There are six prehensile legs.

Caprella aequilibra Say (fig. 235) is commonly from 12–25 mm. long. The body is smooth and there is no spine on the head. The figures show the proportions and outlines of various parts in both male and female specimens. In young males the first two segments are shorter than they are in the old ones. A sharp spine-like projection between the bases of the second gnathopods (not shown in the figure) and an outward pointing spine on each side by the basal joint are characteristic of this species.

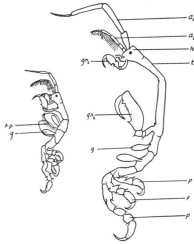

FIG. 235.—*Caprella aequilibra*. Male (right). Female (left); x3. (*a*) antenna; (*h*) head; (*t*) thorax; (*gn*) gnathopod; (*bp*) brood pouch; (*g*) gill; (*p*) periopod.

Caprella scaura Templeton (fig. 236) was taken from kelp holdfasts washed up on the beach. It is reported from points along the coast from Humboldt Bay to San Diego and from Japan and South America. The males are 18-20 mm. long and the females 10-12 mm. Both are very slender, the hand and first two body segments being especially elongated. There is a long sharp forward-pointing spine on the head, but otherwise the body is smooth except for a few blunt prominences. Some varieties of this species have been described which have numerous spines and tubercles on the body. This is especially true of female specimens. The palm of the large hand is also variable in outline and length, but it usually is as long as the first joint of the appendage. The

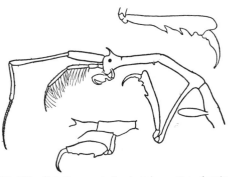

FIG. 236.—*Caprella scaura;* x3. Anterior portion of male. Second gnathopod of female (below). Second gnathopod of male (above).

long slender antennae are characteristic. In the female the body segments, the palm and first joint of the second gnathopod, and the first antennae are all shorter than those of the male.

Caprella kennerlyi Stimpson (fig. 237) is a large species which has been found in Alaska, Puget Sound, and along the California coast, at least as far south as Santa Barbara. Male specimens reach a length of 30 mm. and females may be half that long. There is a pair of small forward-pointing spines on the head. The male has a number of dorsal and ventro-lateral tubercles or spines, especially on the posterior segments, and the female has a very rough appearance, the spines being found on all the segments. The proportions and outline of parts are shown in the figure. The prominent prongs on the basal joints of the legs are characteristic. The flagellum of the first antenna has 15–18 joints.

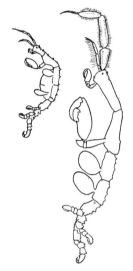

FIG. 237.—*Caprella kennerlyi;* x1.5. Male (right). Female (left).

Order—ISOPODA

The sow bug or pill bug so often found under boards in damp places is a familiar isopod. The group consists, for the most part, of small and inconspicuous forms. The species show a great variation in habitat and are widely distributed. They are not limited to salt water, but many of them may be found in inland lakes and streams or even living upon the land, as in the case of the sow bug just mentioned. Indeed, some isopods are hardy enough to inhabit the hot waters of thermal springs in some parts of the country. Not a few are parasites and infest the bodies of larger crustacea—the shrimps, schizopods, and crabs, while species of fish serve as hosts to other kinds.

PHYLUM—ARTHROPODA 283

The shore collector will have little to do with these, however, and we have, therefore, confined our descriptions to the comparatively few forms that may commonly be found within his province. A search of the crevices of rocks uncovered by the tide and of the algae will usually reveal some of these isopods, and they are frequently found upon the kelp which drifts in to shore. Several species frequent sandy beaches, and others dwell within the piling of old wharves.

In common with the amphipods many of the isopods feed upon the debris in the water, and others are predaceous. They are eaten by the larger animals and serve man indirectly by forming a part of the food of the fish which he eats.

The isopods have the body compressed in a dorso-ventral direction. Typically, there is a head, seven thoracic, and six abdominal segments. One or two of the segments of the thorax may be united with the head. Frequently the segments of the abdomen are firmly united so that it is not always possible to distinguish the six parts. The thoracic limbs, usually six or seven pair, are used in locomotion and are generally very much alike. This fact gives the group its name. The members of one group possess a cheliform anterior pair of appendages. The abdominal appendages are of two sorts, the pleopoda, which are occasionally used in swimming but usually function as breathing organs, and the uropoda. The latter may be in the form of broad blades laterally placed or they may be attached terminally to the abdomen. In a large proportion of the common species frequenting our shore they are turned under the abdomen forming two flaps, and serve to encase and protect the pleopoda (fig. 248). There are two pairs of antennae, a mouth with a pair of mandibles, a pair of maxillipeds, and two pairs of maxillae. The eyes are compound and are not on stalks.

The development is direct, there being no complicated

larval transformations. The eggs and young are carried in an incubatory pouch, composed of overlapping scales on the under side of the female (fig. 248).

Superfamily—TANAIOIDEA (CHELIFERA)

Representatives of the *Tanaioidea* live among the algae in tide pools, on sponges, barnacles, and in similar situations. They are small forms which can be distinguished from the members of all other groups of isopods by the claw-like (cheliform) structure on the first pair of legs. The pleopoda are used in swimming and the uropoda are attached to the end of the abdomen.

Tanais normani Richardson (fig. 238) may be found by washing out masses of brown algae from the tide pools. It

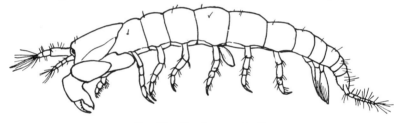

Fig. 238.—*Tanais normani;* x20.

reaches a length of 4 mm. and is large enough so that the cheliform legs can just be distinguished with the unaided eye. There are three pairs of pleopoda and the abdomen is composed of six distinct segments. The head and the first segment of the thorax are united to form a short carapace. The incubatory pouch is formed by two plates attached to the base of the fifth pair of legs. The uropoda are composed of six segments. The species has been taken at Monterey Bay and La Jolla.

Superfamily—CYMOTHOIDEA (FLABELLIFERA)

The uropoda are attached to the sides of the abdomen and extend laterally, giving a fan-shaped appearance. The pleopoda help in swimming.

Cirolana harfordi (Lockington) is a common form (fig. 239) on the Pacific coast, occurring from British Columbia to Lower California. It may be found between the tides, under rocks, or among beds of mussels. The colors are not constant but, in general, the ground color is light and the pattern is formed by minute black dots. We have taken specimens that were a very dark brown, almost black, and have found other specimens that were nearly colorless, in

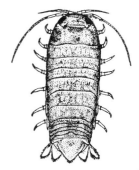

FIG. 239.—*Cirolana harfordi;* x2.6.

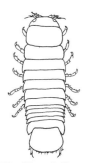

FIG. 240.—*Limnoria lignorum*, the gribble; x12.

the same locality. They often are tawny or orange-yellow at the sides and marked with gray or brown above. The pattern is variable. The uropoda have two branches and the abdomen which is composed of six segments bears a number of spines upon its posterior border. Epimera are present upon the last six thoracic segments. Length 8.5 to 15 mm.

The gribble, *Limnoria lignorum* (Rathke) (fig. 240), has a wide range and is a great pest wherever found. It destroys wooden piling by boring holes in it with its mandibles. It may cut into the wood at a rate of one-half to one inch per year, doing an enormous amount of damage. It is found on the Atlantic coast of North America and Europe, and in the cooler portion of the Pacific. Along the California coast this isopod is associated with three species of "ship worms" or boring mollusks that burrow into wooden structures with great rapidity. The ship worm and *Limnoria* together are stated to have destroyed piles a foot in diameter in less than two years. Attempts have been made with some measure of success to protect the wooden pile by metallic sheathing, by copper paints, by treating with creosote under pressure, or by surrounding it with a concrete jacket, but the slightest breach in the defenses is readily seized upon by the borers with disastrous effects.

Concerning the life-history of this form Professor Kofoid[*] says, "On the under side of the body between the legs the female has a brood pouch in which the eggs are carried during their development. These eggs are large, nearly one-fourth the width of the body in diameter, and when the young are hatched out they are ready at once to begin the burrowing process near the parent. There is no free-swimming larval stage, no period of forced migrations, and little risk of destruction. Few young are produced and these are ready at once to dig in for themselves. For this reason, the spread of *Limnoria* is like the growth of a lichen or the spread of the mange or itch. The colony grows at the periphery and burrows constantly deeper into the wood.

"We have no precise data as to their breeding seasons and length of life. Young and eggs are found throughout the year, though it is probable that reproduction is more rapid during the warmer season. The two sexes are much alike in appearance and differ only in slight anatomical details. We have found about the same number of males and females. A single square inch of infected Douglas fir from Tiburon heavily attacked by *Limnoria* was found to contain 79 females, 33 of them with eggs attached, 82 males, and 221 young in various stages, a total of 382 individuals."

Limnoria is a small creature, having a body length of 3 mm. and a width of 1 to 1.5 mm. It is convex dorsally and tends to roll up into a ball. The first segment of the thorax is nearly twice as long as any of the others. The last abdom-

---

[*]C. A. Kofoid—The Marine Borers of the San Francisco Bay Region; Report on the San Francisco Bay Marine Piling Survey, 1921, pp. 51-52.

PHYLUM—ARTHROPODA 287

inal segment is large, broad, and flattened above. The outer branches of the uropoda are hardly distinguishable. The two pairs of antennae are short and almost equal in length. *Sphaeroma pentodon* Richardson (fig. 241) also shows a

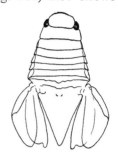

FIG. 241.—*Sphaeroma pentodon*, the rock-boring isopod. Dorsal view (left); x4. Side view showing typical attitude (right).

FIG. 242.—*Exosphaeroma amplicauda*; x7.5.

strong tendency to curl up like an armadillo, if disturbed, thus protecting the ventral surface of the body and its appendages. The legs and antennae are apt to be hidden when the animal is viewed from above. The body is short, oval, and convex dorsally. It can be recognized by the prominent ridge in front of the eyes, the double row of small tubercles on the abdomen, and the ridge near the end of the abdomen. The color is a dark, dull brown, lighter in some individuals and darker in others. The surface of the body is minutely granular, but this feature can not be seen without a lens. It reaches a length of 9 mm. The young are more slender than the adult, with slender uropoda.

This species is closely related to *Sphaeroma destructor*, of the coast of Florida, which destroys piling in a manner similar to that of *Limnoria*. The western species burrows in wood, also, but is not so destructive as *Limnoria*. It occurs in San Francisco Bay and we have found it in large numbers in a piece of decaying timber near a mud bank in the inner harbor at San Pedro. At Chula Vista wharf, San Diego Bay, it may be found in holes in the hard adobe-like mud at the high tide line. The holes are from 6 to

13 mm. deep and the animals are found curled up at the bottom of them. In certain locations on the shore of San Pablo Bay, Dr. A. L. Barrows* found it occupying burrows in reefs of soft sandstone and tuff. Some of the holes were 35 mm. deep and of sufficient diameter to fit very closely around the bodies of the isopods that lived in them. He took some specimens into the laboratory and watched them burrow into pieces of chalk. The boring was done with the mandibles. The fragments bitten off were passed back by the palpi and the feet until beneath the abdomen and were then washed out of the hole by a current of water produced by the swimmerets. The animals seldom left their burrows but frequently rested at the mouth of the bores with the posterior end of the body outward, retreating down the holes if alarmed. *Sphaeroma pentodon* is also reported in brackish water at Antioch. It may be found as far north as Alaska.

Exosphaeroma oregonensis (Dana) has a flatter body than *Sphaeroma pentodon* and the head is set further back into the first thoracic segment. The last segment of the abdomen is shorter and flatter, and is without tubercles. It is unlike *Sphaeroma pentodon*, also, in having no teeth on the outer branch of the uropods, but it is necessary to use a lens to determine this point. It is found on the beach at low tide, under stones, or in mud, from Alaska to Monterey Bay. Both species occur in San Francisco Bay.

Exosphaeroma amplicauda (Stimpson). Superficially, this form shows little resemblance to the last, though related to it (fig. 242). This is due to the large size of the uropoda, both branches extending to the tip of the abdomen. They are very wide and as the body gradually increases in width toward the abdomen the specimens exhibit a peculiar triangular appearance when seen from above. The inner

---

*A. L. Barrows—The Occurrence of a Rock-boring Isopod along the Shore of San Francisco Bay, California; University of California Publications in Zoology, Vol. 19, p. 307.

branch of the uropoda is fastened rigidly to the abdomen, but the outer one is movable. There are small tubercles on the dorsal surface of some of the thoracic segments. These isopods are grayish white in color with some brown spots on the thoracic segments. Length 5 to 8 mm. *Exosphaeroma amplicauda* is found from Alaska to the San Diego region.

SUPERFAMILY—IDOTHEIDAE or VALVIFERA

The species belonging to this division have the uropoda valve-like, turned under, and covering the pleopoda on the ventral surface of the abdomen. The pleopoda are used in respiration. Most of the *Idotheidae* are flat, narrow-bodied crustacea.

Idothea rectilinea (Lockington) is a common form along the coast of southern California and is found as far north as Humboldt County (fig. 243). It may live among rocks below high-tide level. The color varies from a dark to a light brown, the male being lighter. *Idothea*

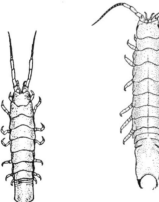
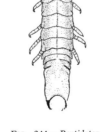
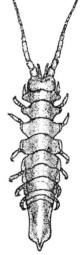

FIG. 243.—*Idothea rectilinea*, female; x1.5.   FIG. 244.—*Pentidotea resecata*; x1.   FIG. 245.—*Pentidotea aculeata*; x2.

rectilinea may be recognized by the shape of the final segment of the abdomen. It reaches a length of 20 mm.

Idothea urotoma (Stimpson) is found in the Puget Sound

region. It is about the same size as *Idothea rectilinea*, but may be distinguished by a shorter and broader abdomen with a shield-like contour when viewed from above.

Idothea ochotensis (Brandt) is very much larger than the two species just described. It is dark in color and very large for an isopod, reaching a length of 42 mm. with a width of 16 mm. The abdomen ends in a short, but well defined, point. The range of this species is from the northeast coast of Asia to San Francisco.

Pentidotea resecata (Stimpson) is often found clinging to kelp and is frequently washed ashore on it. It is just the color of the yellowish plant to which it holds. The body is long (fig. 244) and narrow, 39 mm. by 8.5 mm., the abdomen occupying ⅓ of the total length or 13 mm. Two sharp points at the postero-lateral angles of the abdomen separate this from other species of *Pentidotea*. It may be found from British Columbia to the San Diego region.

Pentidotea aculeata Stafford occurs on the southern California beaches (fig. 245). The male is a red-brown or mahogany color, darker through the mid-dorsal region. The female is pink with white spots along the median line of the back and two rows of markings half way between the mid-dorsal line and the lateral margins. Dimensions; 23.5 mm. by 5.5 mm.

Pentidotea wosnesenskii (Brandt) is about as large as *Pentidotea resecata* but differs from it in that the terminal segment of the body is rounded and has a small median point (fig. 246). The male is larger than the female and lighter colored. It may be found at low tide in the sand and under rocks from Alaska to the vicinity of San Francisco. Length 32 mm., width 11 mm.

Synidotea harfordi Benedict is found in the San Diego region. All the epimera are united to the segments in the members of the genus *Synidotea*, but they superficially resemble the species of *Idothea*. There is a notch at the tip of the abdomen (fig. 247). The color is a pale golden

PHYLUM—ARTHROPODA 291

brown with lighter markings. Dimensions: 16 mm. by 6 mm.

Synidotea laticauda Benedict differs from *Synidotea harfordi* in being much broader and in having a broader,

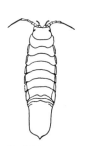

FIG. 246.—*Pentidotea wosnesenskii;* x1.

FIG. 247.—*Synidotea harfordi;* x2.

shallow notch at the posterior extremity. The flagellum, or whip, of the second pair of antennae is composed of 17 instead of 31 joints. Dimensions: 17.5 mm. by 7.5 mm. It occurs in San Francisco Bay.

Colidotea rostrata (Benedict) lives among the spines of sea urchins, and its coloring resembles that of the sea urchin. It reaches a length of about 12 mm. and a width of 5 mm.

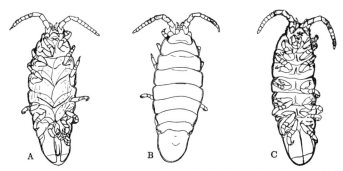

FIG. 248.—*Colidotea rostrata.* A. Ventral view of female showing incubatory pouch of overlapping scales and the infolded uropoda: x5. B. Dorsal view of female; x5. C. Ventral view of male; x5.

The head has a short rostrum and prominent lateral angles. The epimera are absent on the first four segments, narrow on the fifth, and wider on the sixth and seventh segments (fig. 248). It is at home on the coast of southern California.

Superfamily—ONISCOIDEA

This group is characterized by styliform, or bristle-like, uropoda attached at the end of the abdomen. The pleopoda contain air cavities and can be used for air breathing. The first pair of antennae are small and inconspicuous. The terrestrial isopods belong to this superfamily. A number of forms inhabit the beaches.

Tylos punctatus Holmes and Gay has the uropoda turned back to form a covering for the pleopoda, of which there are four pairs. The body is strongly convex (fig. 249) and

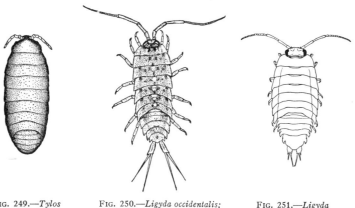

Fig. 249.—*Tylos punctatus;* x2.5.

Fig. 250.—*Ligyda occidentalis;* x1.

Fig. 251.—*Ligyda pallasii;* x1.

covered with microscopic spines. The head has triangular projections in front of the eyes. Epimera are distinct on all segments except the first where there is merely a thickened margin. Specimens may be found in the sand from San Pedro to Lower California. They roll up like a ball when alarmed, and remain so until the danger is past.

Ligyda occidentalis (Dana) is the most conspicuous isopod along the southern California coast (fig. 250). It may be found scurrying about upon the rocks at high-tide mark or hiding itself among them. It often escapes notice by remaining motionless in depressions or crevices in the rocks, and is aided in this by its dull gray and brown colors. The pattern is variable, as is the ground color, some individuals being much darker than others. The legs are tipped with orange. The uropoda are about one-half as long as the body and divided into two stiff projections, the inner one tipped with a little spine. *Ligyda occidentalis* is reported from the Sacramento River to San Francisco Bay and along the coast of California, south of San Francisco. Length of body 23 mm., width 9 mm., length of uropoda 12 mm.

Ligyda pallasii (Brandt) is found from the San Francisco region to Alaska, replacing *L. occidentalis* in the north. It is a flat, broad form with a granular surface and short uropoda (fig. 251). Body not quite twice as long as broad; 20 mm. by 11 mm.; uropoda 3 mm. long.

Order—EUPHAUSIACEA

The animals in this group are small, more or less transparent, phosphorescent crustaceans, much resembling the shrimps in form though they are often not more than 25 mm. long. As in the shrimps, the thorax is nearly covered by a carapace which may be produced into a rostrum. Below the stalked eyes are the antennules, or first antennae, which have two flagella. The antennae are below, outside of the antennules and have only one flagellum and are provided with a broad scale.

Attached to the thorax are the eight pairs of biramous, or two-branched, legs which give rise to the name *Schizopoda*, as this order is sometimes called, for these are the cleft-footed crustaceans. The segments of the abdomen bear pleopoda and the last segment of the abdomen, termed the telson, has an enlarged pair known as the uropoda on each side of it. The uropoda are articulated with the next preceding segment.

294 SEASHORE ANIMALS OF THE PACIFIC COAST

The phosphorescent organs are set like little lanterns along the sides of the body. Their exact use is not known, but they seem to light up the dark depths in which the animals spend all or a part of their days. The organs consist of lens, reflector, and light-producing tissue and are under the control of the nervous system. In the following species they are found on the eye stalks, the bases of the second and seventh legs, and under some of the segments of the abdomen.

Euphausids are taken with towing nets near the surface, or at some depth, and are sometimes found among the eelgrass along the shore.

FAMILY—EUPHAUSIIDAE

Euphausia pacifica Hansen is 20–25 mm. in length. The gills are attached to the legs from the second to the eighth pair, and are not covered by the carapace. The last two pairs of legs are much reduced and nearly hidden by the gills. The pleopoda are well developed and aid in swimming.

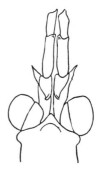

FIG. 252.—*Euphausia pacifica*, female; head, eyes and anterior part of thorax from above; x6.5. (After Esterly.)

FIG. 253.—*Nyctiphanes simplex*, female; anterior part of head, eyes and antennular peduncles from above; x6.5. (After Esterly.)

It may be distinguished from closely related species by the single denticle on the lateral anterior margin of the carapace (fig. 252) and the absence of any keel or process on the third to the fifth abdominal segments. In southern California waters it is abundant near the surface in June and July, and is reported as a common species in the Japanese region.

Nyctiphanes simplex Hansen may be distinguished from the preceding species by the presence of a backward-pointing scale or "leaflet" at the base of the antennules in front of the eyes (fig. 253). Dr. Esterly* gives the following measurements: length of adult males, 11-12 mm.; of egg-bearing females, 14-15 mm. In the San Diego region they are abundant at the surface in November and December. Hansen says they occur in the East Pacific.

Order—STOMATOPODA

(*The Mantis-shrimps*)

This order is marked by several well defined features. There is a very short carapace which does not cover the last three or four segments of the thorax. The eyes are borne on stalks and the part of the head which carries them is separated from the remainder by a joint. The first five pairs of thoracic appendages, or maxillipeds, are turned forward and end in claws. They are not flattened like the maxillipeds of the crabs and are not used as chewing organs but are adapted for grasping. The first pair are small and insignificant while the second pair are greatly developed with the last segment turned back and fitting into the one above it, much as a knife blade folds into its handle. This arrangement, which somewhat resembles the chela of a crab without the thumb (or pollex), is termed subchelate. The walking legs, six in number, are located on the last three thoracic segments. They are slender and weak and divided into two branches, one much smaller than the other. The broad, flattened abdomen bears a telson and large uropoda in addition to six pairs of swimming limbs, each consisting

*C. O. Esterly—The *Schizopoda* of the San Diego Region; Univeristy of California Publications in Zoology, Vol. 13, p. 9.

of two parts, one, flattened and oar-like for propulsion, the other, a branching gill for respiration. It is said that the female does not use her abdominal appendages for the attachment of eggs, as many crustaceans do, but that these are carried by the maxillipeds very much as if they were to be eaten. The development of the young is complicated, there being several larval stages. The adults live in burrows or among the crevices of the rocks off shore and are seldom found at tide level. The colors are brilliant.

Three species belonging to this order occur in our region. They are said to be edible but are not captured in sufficient quantities to be commercially important.

Family—SQUILLIDAE

(The Mantis-shrimps)

Chloridella (Squilla) polita (Bigelow) (fig. 254) is found south of Point Conception. The specific name is suggested by the polished upper surface of the animal. It reaches a length of 60 mm.

Pseudosquilla bigelowi Rathbun (fig. 255) is found as far north as Point Mendocino. The second maxillipeds are proportionally smaller than in the polished squilla, and the body is broader. The difference in the number and arrangement of the spines on the telson, or terminal segment of the abdomen, may be seen in the photograph, and serves to further distinguish the two species. It reaches 8 inches or more in length.

The third species is *Pseudosquilla lessonii* (Guèrin). Though it closely resembles the preceding form, it differs in a number of minor respects. There is a terminal spine on the rostrum and there are two spines at the base of the dactyl of each of the large chelipeds, that are not found in *P. bigelowi*. Moreover, the uropoda of *P. lessonii* are more spiny than those of the other species. It is found along the southern part of the California coast.

FIG. 254.—The polished squilla, *Chloridella polita;* approximately x5/6.

FIG. 255.—*Pseudosquilla bigelowi*, a mantis shrimp; less than ½ natural size.

FIG. 256.—*Pandalus gurneyi*, a shrimp; x5/4.

Order—DECAPODA

(The ten-footed crustaceans)

The most highly organized crustaceans, such as shrimps, lobsters, and crabs, belong to this group. The head and thorax are united into a cephalothorax which is covered by a carapace. The gills are situated in special chambers at the sides of the thorax under the carapace. There are five pairs of thoracic legs (periopoda) of which the first, at least, is usually developed into chelipeds, or claw-bearing legs, while those not so modified are used as walking legs.

There is considerable diversity of opinion about the systematic arrangement of the different groups in classification. According to Borradaile there are two suborders, *Natantia* and *Reptantia*, distinguished by a number of characters, of which some of the more apparent are listed below. In the *Natantia* the body is generally compressed or flattened from side to side and there is usually a forward-pointing extension of the carapace called the rostrum (fig. 257). The second antennal scale is typically large. The first abdominal segment is not much smaller than the rest and the first five abdominal limbs (pleopoda) are always present, well developed, and used for swimming.

On the other hand, in the *Reptantia* the body is not compressed, in fact is more often depressed as in the crabs. The rostrum is often small or entirely absent and is somewhat flattened if present. The second antennal scale is small or absent, never large. The first abdominal segment is distinctly smaller than the rest and the first five abdominal limbs are often reduced in size, or absent, and are rarely used in swimming.

Suborder—NATANTIA

The shrimp industry of California centers around San Francisco, where 909,844 pounds of shrimps were marketed in 1921. Those in the markets are mostly of one species,

PHYLUM—ARTHROPODA

Crago franciscorum, with a small percentage of *C. nigricauda,* and *C. nigromaculata.*

The following account* of the shrimp industry is taken largely from *California Fish and Game.* The shrimps drift with the tides back and forth along the bottom of San Francisco Bay but are able to select their environment to some extent. In the winter, when more fresh water is coming into the bay, they move down into deeper water, while in the summer they move farther up toward the mouths of the rivers. It is thought that while they are carrying their eggs they go into shallower water. The larger individuals are usually caught in deep water where the current is swift. They can live in water of varying degrees of salinity, even sometimes in fresh water. Little is known of the life-history. Females carrying eggs on the swimmerets are found at all seasons of the year. It is thought that they hatch in about two months and that it is two years before the shrimps are old enough to spawn. They feed on tiny animal and plant life near the surface and at the bottom and can swim through the water rather rapidly, moving with the head first.

The shrimps of the San Francisco markets are mostly caught in nets by the Chinese. The use of these nets was prohibited from 1911 to 1915 because a great many young of valuable kinds of fish were being destroyed in them and the shrimps were being overfished. Before the law was passed, the annual catch by the Chinese junks had been over 10,000,000 pounds per year of shrimps and fish combined, of which only about 800,000 pounds were used fresh, the rest being marketed as dried shrimp meat and fertilizer. The dried product was exported to the Orient.

The Chinese nets are funnel shaped and set side by side at the bottom of the bay with the opening toward the current. They are held in place by a series of stakes or spreaders which keep them open to their full extent, the

---

*N. B. Scofield—Shrimp Fisheries of California; California Fish and Game, Vol. 5, pp. 1–12.

opening being rectangular and about twenty-four feet across by about four and a half feet deep. An average day's catch for boats using forty nets was 6,000 pounds. At the time the use of the Chinese nets was prohibited, the numbers of shrimps in the bay had been so depleted that the other kinds of nets could not yield a large enough catch to make the business profitable, so the markets were without shrimps until 1915, when the use of the Chinese nets in southern San Francisco Bay was permitted. There are fewer young fish of the more valuable species at this point so the shrimp market can be supplied without serious injury to other fisheries. Drying of shrimp is now prohibited so that only enough to supply the fresh shrimp market is being caught. Gradually the demand for shrimps has returned so that now about 750,000 pounds are marketed each year. It is gratifying to know that the numbers of shrimps in all parts of the bay have increased as have also the numbers of young fish.

The shrimps have become so common an article of diet that they are recognized readily by everyone. In most localities where rocks are uncovered by the tides, and pools formed among them, shrimps may be found. They are free-swimming crustaceans that are abundant, not only in shallow water and in the tide pools, but also in the depths of the sea. The naturalists of the British *Challenger* Expedition found a few species at a depth of four miles, and many forms seem to pass their lives swimming in mid-water. They are agile creatures, darting backward by means of a sudden flip of the abdomen and making use of every bit of seaweed or crevice among the rocks to hide themselves from the intruder. Aided by their protective coloring and the smoothness of the elongated bodies they are surprisingly successful in their attempts to escape. There is need for quickness of movement, for they are the prey of countless fish and other carnivorous animals. In turn, they devour both animal and vegetable matter.

It is evident that such an alert animal is well served by its sense organs. The delicate antennules and whip-like antennae are probably used as organs of touch, and their great length, which frequently equals or exceeds that of the body, enables them to be useful over an extended area. Possibly the antennules also act as organs of smell or convey other sensations to the animal.

Many of the decapods are noted for their ability to regenerate lost limbs. A curious instance of this is to be found in certain members of the genus *Crangon* (*Alpheus*), in which one of the chelae is much larger than the other (fig. 264). Przibram,* a European scientist, found that if the larger claw was broken off, the smaller one would grow larger and take its place while the injured one would in time develop into a small claw.

Figure 258 shows the external structure of a shrimp. The head and thorax are united to form a single region, the cephalothorax, which is covered by the carapace, a protective shield. The gills are hidden in chambers at the sides of the thorax beneath the carapace. The thoracic legs (periopoda), are ten in number, the first two or three of them bearing claws or chelae. The abdominal segments bear pleopoda, or swimmerets. They are used in swimming and for the attachment of eggs. The swimmerets are branched, and so are the legs in a few species, but in most adult shrimps there is only one branch. The abdomen ends in a tail-fan composed of the last segment, the telson, and the sixth pair of swimmerets, the uropoda, which are greatly enlarged and paddle-shaped.

Some shrimps pass through a complicated metamorphosis. Members of one tribe, the *Peneides*, are hatched as nauplii, little larvae with three pairs of biramous appendages, then become protozoea and zoea, in turn, the latter with the rudiments of walking legs and uropoda. The last larval

---

*G. Smith and W. F. R. Weldon—*Crustacea;* p. 156, The Cambridge Natural History, The Macmillan Company.

stage greatly resembles the adult schizopod in form, with branched legs and swimmerets. This is the mysis stage, and corresponds to the megalops of crabs. In most of the *Carides* the nauplius and earlier zoea stages are passed in the egg, and the creature hatches as a zoea, passing through the megalops stage to the adult condition. Some of the *Crangonidae* hatch as megalops and have a very short larval existence. Little is known of the earlier stages of most of our west coast species.

There are three tribes. The *Stenopides* have the first three pairs of legs chelate, and one or both of the third pair longer and stouter than the first two pairs. The *Peneides* have the third legs chelate, too, but not stouter than those of the first two pairs. In the *Carides*, the tribe to which our common shore species belong, the third pair are without chelae. The sharp bend in the abdomen that gives a broken-backed appearance is typical of this tribe.

Tribe—CARIDES

Family—PANDALIDAE

The shrimps of the genus *Pandalus* have a long, slender, spiny rostrum. The dorsal spines extend back on the carapace forming a crest. The second pair of legs are of unequal length and very slender, the segment just above the hand being divided into many ring-like joints. The first pair of legs, if chelate, is only minutely so. The genus *Pandalus*, like most genera of shrimps, contains many species, but most of them are inhabitants of deep water so that they are not often found by the ordinary collector.

Pandalus danae Stimpson, the "coon-striped" shrimp, is found from the San Francisco region north to Alaska, usually at a considerable depth but occasionally in the tide pools. It may frequently be found among the shrimps in the markets. It reaches a length of 110 mm., the carapace and rostrum together being 50 mm. long. The rostrum (fig.

257-A) is never as much as one and a half times the length of the carapace. The dorsal spines extend a little more than half way back on the carapace. The third segment of the abdomen is not compressed or keeled and is without a median lobe or spine in front of the margin. The antennal scale, or acicle, which projects out in front of the eyes, has a blade which is no narrower toward the end than the thickened part of the acicle (fig. 257-C), in contrast with a less common species *P. stenolepis* Rathbun, the acicle of which is very narrow at the end, the outer margin being markedly concave. *P. stenolepis* is like *P. danae* in the other points that have been mentioned, and is found in the same region.

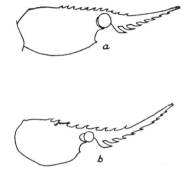

FIG. 257.—A. Rostrum of *Pandalus danae*. B. Rostrum of *Pandalus gurneyi*. C. Antennal scale of *Pandalus danae* (left) and *Pandalus stenolepis* (right).

Pandalus gurneyi Stimpson (fig. 256) generally seems to frequent deep water. The one photographed was dredged off Point Loma. It is a handsome species, being spotted with red on a lighter ground color. The carapace and rostrum together measure 42 mm., the whole length is 85 mm. The long, curved rostrum (fig. 257-B) distinguishes it from the other species. Moreover the rostrum is one and a half, or more than one and a half, times as long as the carapace.

FAMILY—HIPPOLYTIDAE

The members of this family of shrimps have the segment above the hand in the second pair of legs divided into a number of ring-like joints or annulations, seven, in the case of the forms listed below, except *Hippolyte californiensis* which has but three, and *Hippolysmata californica* which has about 32 annulations.

Hippolyte californiensis Holmes is a long, slender species. The rostrum has 3–5 teeth on both the upper and the lower margin. The base of the rostrum is rounded and not continued upon the carapace. The peduncle of the antennules is about half as long as the rostrum, the outer flagellum much shorter than the slender inner one. The first pair of chelipeds is short, the hand broad and thick at the base, fitting into a depression in the carpus. The second pair of chelipeds is more slender but much longer than the first pair; carpus three jointed, the first joint longest. The abdomen is not crested or keeled. The telson is truncated and spinulous at the tip. The length is 38 mm. It is reported from Bodega Bay among the eelgrass, also from San Pedro, San Diego, Puget Sound, and Alaska.

Hippolysmata californica Stimpson (fig. 258) is a beautiful species and one that is common in the tide pools along the coast of southern California. At very low tide one often finds

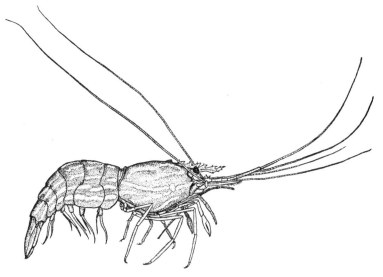

FIG. 258.—*Hippolysmata californica*; x1.2.

them under rocks from which the water has receded entirely, and this is perhaps the easiest location in which to capture them, for if they have enough water to enable them to swim, their quick backward darts are hard to follow, and they are very apt to elude the collector. The broken longitudinal stripes of red or red-brown make it quite inconspicuous among the rocks that are covered with red and brown algae. The ground color between the stripes is transparent and white or bluish. The telson is bluish, edged with pink, the legs red near the claws, and the long antennae and antennules red. The antennal scale is longer than the peduncle of the antennules and is truncated or broadly rounded at the end. The first pair of legs are chelate and stouter than the others, the second pair are long and slender with minute chelae, and the merus and carpus together are divided into about fifty short rings. Length 42 mm., carpace and rostrum 17 mm.

There are a number of shrimps of the genus *Spirontocaris* that are more or less abundant in the tide pools as well as at a greater depth. They are smaller than the species just described, so they are not found among the shrimps in the markets. They show a great variety of coloring. After seeing them against a white background one would conclude that they are easy to find because of their brilliant colors, but against the bright colored algae and corallines of the tide pools, the shrimps are almost indistinguishable so long as they remain quiet, their striking colors serving well for camouflage.

Spirontocaris carinata (Holmes) is rather less common than the other members of this genus. It has been found in Monterey Bay and south of this point. The rostrum (fig. 259-A) is about as long or longer than the rest of the carapace, the terminal half of it being without spines. The third abdominal segment has a hump or angle near the posterior end. The third, fourth, and fifth segments are not all keeled nor do they end in a sharp spine. The sixth abdominal segment is less than twice as long as wide. The color is said to vary with the surroundings, those among the green seaweeds being bright green, while other specimens match the red algae in which they live.

Spirontocaris paludicola (Holmes) (fig. 259-F) has a rostrum about as long as, or longer than, the rest of the

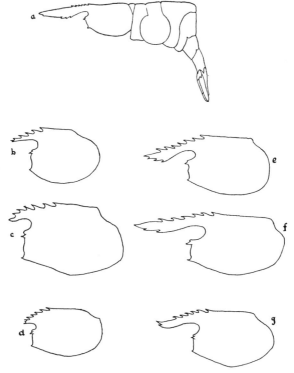

FIG. 259.—Outline of rostrum and carapace of the species of *Spirontocaris*.
A. *S. carinata.*   C. *S. brevirostris.*   E. *S. picta.*   G. *S. cristata.*
B. *S. palpator.*   D. *S. taylori.*   F. *S. paludicola.*   (After Holmes.)

carapace. The sixth abdominal segment is shorter than the seventh, and the third, fourth, and fifth are not keeled nor do they end in a sharp spine. It has been found in Humboldt and Bodega Bays. The color is green like the eelgrass in which it lives. This species may reach 32 mm. in length.

Spirontocaris palpator (Owen) (fig. 259-B) has the rostrum shorter than the rest of the carapace, and not reaching beyond the middle of the antennal scale, or as far as the second segment of the antennular peduncle, but reaching as far as, or beyond, the eye. It may be distinguished from *S. brevirostris*, a more northern species, by its smaller size and by the fact that the antennal scale is longer than the telson. It is found from San Francisco southward as a shallow water form and also lives at greater depths. Our specimens are 25 mm. long, the carapace and rostrum being 7 mm. long.

Spirontocaris brevirostris (Dana) (fig. 259-C) has the rostrum shorter than the carapace, and not reaching beyond the middle of the antennal scale or the second segment of the antennular peduncle, but reaching as far as, or beyond, the cornea of the eye. The antennal scale is the same length as, or shorter than, the telson. The length is 49 mm., and the length of carapace and rostrum 16 mm. It has been reported from Alaska as far south as San Francisco Bay, being taken along shore at low tide as well as in deep water.

Spirontocaris taylori (Stimpson) (fig. 259-D) is rather common from San Francisco Bay south to Lower California. The rostrum is shorter than the rest of the carapace, not reaching the cornea of the eye, and is armed above with five or six teeth, the last three or four being on the carapace. The length is 31 mm., carapace 8 mm. long. Specimens show a wide range of color, the smaller ones being nearly transparent and somewhat mottled in the anterior parts, but darker and shaded with green and brown posteriorly with transverse bands of white. The larger ones are variable also. One (fig. 263) was seal-brown with an orange stripe nearly one-third the width of the body from the tip of the abdomen to the tip of the antennules. Another (fig. 263) was brown with greenish shading in the telson and a pink blotch, edged with white, over the middle of the carapace.

The legs are banded with red, or brown and white and the antennae are brownish.

Spirontocaris picta (Stimpson) is abundant in the tide pools from San Francisco to San Diego. It is small, our largest specimen being 23.5 mm. long, the carapace and rostrum measuring 9.5 mm. It is semitransparent and greenish, with oblique red bands, and the legs barred with crimson. The rostrum (fig. 259-E) is shorter than the rest of the carapace but reaches beyond the middle of the antennal scale.

Spirontocaris cristata (Stimpson) is found from Alaska to San Diego. The rostrum (fig. 259-G) is shorter than the rest of the carapace, reaching the cornea of the eye and the second segment of the antennal peduncle, but not beyond the middle of the antennal scale. It can be distinguished by the slender dactyls of the walking legs which are longer than those of the nearly related species. The posterior lateral angles of the fifth abdominal segment are acute and bear a large spine. The species may reach 32 mm. in length.

Family—CRANGONIDAE

(This is the Alpheidae of many writers)

In all the California representatives of this group of shrimps, the eyes are hidden under the carapace. The legs of the first pair usually bear unequally developed chelae. The chelae in some cases are carried in an inverted position, with the dactyl or movable finger beneath, instead of uppermost as is usually the case.

Crangon (*Alpheus*) *dentipes* (Guerin) (figs. 260 and 264), the snapping shrimp, makes its presence known to the collector by snapping its large chela, but the task of following up the sounds and locating the shrimp is not always an easy one. It is very active and like its relatives, is protectively colored. The body is greenish, sometimes with a tinge of blue, and a border of orange on the telson. The chelae are mottled with brown, or green and orange with the tip of the palm darker and the rounded tip of the finger white. The antennae and antennules are yellow-brown. The legs are whitish except the first pair which are yellowish on the last few joints. The snapping is done with the finger

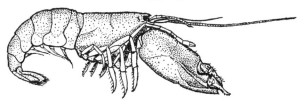

FIG. 260.—The snapping shrimp, *Crangon dentipes;* x1.5.

of the big claw as it strikes against the propodus in closing. When a shrimp snaps in the aquarium jars, it sounds as though a glass dish has cracked. If the shrimp snaps in one's hand the impact is distinctly felt and is somewhat painful.

This species has a rostrum with three frontal spines nearly equal in size. The eye stalks are short and concealed under the projecting margin of the carapace. The chelae of the first pair of legs are very unequal in size and the finger of the large one works horizontally. The second pair of legs is slender and bears small chelae. There is a spine on the lower side of the merus of the third and fourth pairs of legs. The body length is 31 mm., the carapace 11 mm., the large hand 18 mm., and the small hand 11 mm. It may be found in tide pools from the Farallones to Lower California.

Crangon bellimanus (Lockington) is much like *C. dentipes* but has no spine on the lower side of the merus of the walking legs. The fingers of the large hand are not longitudinal, the dactyl is short, curved, and works horizontally. The length is 33 mm., carapace 12 mm., large hand 17 mm., and small hand 13 mm. It has been collected from Monterey to San Diego.

Crangon equidactylus (Lockington), may be distinguished

from the two species above by the absence of the spine on the lower side of the merus of the legs, and by the horizontal, nearly equal fingers of the large chela. The dactyl is slightly the longer, and works nearly vertically. The length is 18 mm., carapace 6.5 mm., larger hand 7 mm., and smaller hand 5 mm. The species has been taken from Monterey and Santa Barbara.

Betaeus harfordi (Kingsley) is a small species in which there seems to be much color variation. Dr. Holmes* writes of finding purple ones on some sea urchins from deep water, while another specimen found under a rock at low tide was nearly white. Dr. Hilton† describes pale green specimens

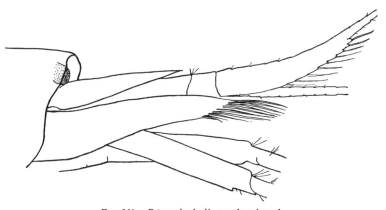

FIG. 261.—*Betaeus harfordi*; greatly enlarged.

found in kelp holdfasts. The range so far as reported is from Point Arena to Laguna Beach. The front of the carapace is without spines and projects over the eyes (fig. 261). The hands are oval and the dactyl is joined to the lower side of the hand. Length of body 19 mm., carapace 6 mm., hand 6 mm.

Betaeus longidactylus Lockington (fig. 262) is also an inhabitant of tide pools among and under rocks and in

---

*S. J. Holmes—Synopsis of the California Stalk-Eyed *Crustacea*; Occasional Papers on the California Academy of Sciences, No. 7, p. 190.

†W. A. Hilton—*Crustacea* from Laguna Beach; Journal of Entomology and Zoology, Pomona College, California, Vol. 8, p. 67.

FIG. 262.—A tide-pool shrimp, *Betaeus longidactylus;* x5/4.

FIG. 263.—*Spirontocaris taylori*, about natural size.

264   265

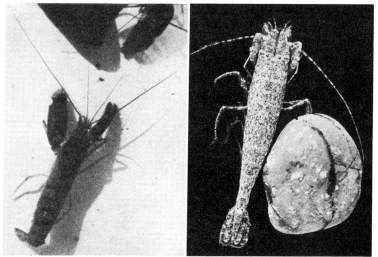

FIG. 264.—The snapping shrimp, *Crangon dentipes;* photographed from life; about natural size. FIG. 265.—The black-tailed shrimp, *Crago nigricauda:* photographed from life; about natural size.

eelgrass. It is usually abundant, and found from San Pedro to San Diego. The front has a smooth edge and does not quite cover the eyes as it does in *B. harfordi*. The finger is joined to the lower side of the hand in the large, nearly equal claws. The large claws are very long and narrow, largest in adult males. They vary from olive-green to olive-brown and purplish blue. The legs are red and there is a fringe of yellow hairs on the telson. One specimen, after molting in the laboratory, was almost transparent and pale greenish blue with claws lighter than the body. In a few hours the color was darker and less transparent. Large specimens are 40 mm. long with hands 23 mm. long.

FAMILY—CRAGONIDAE

(*The Crangonidae of many writers*)

The shrimps of this family may be recognized by the hands of the first pair of chelipeds which are subchelate; the thumb ends in a spine and the finger folds against the margin of the hand (fig. 267).

Crago nigricauda (Stimpson), the black-tailed shrimp, (figs. 265 and 266) according to Dr. Holmes,* occurs all the

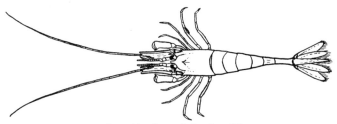

FIG. 266.—*Crago nigricauda;* x3/5.

way from Alaska to Lower California. It may often be found in the San Francisco markets. When alive it is dark grey with a blackish tail. The photograph illustrates the

*S. J. Holmes—Synopsis of the California Stalk-Eyed *Crustacea;* Occasional Papers of the California Academy of Sciences, No. 7, p. 171.

PHYLUM—ARTHROPODA 313

peculiar sandy effect of the coloration and shows how hard it would be to see the animal with the bottom of a sandy pool for a background. Its distinguishing characters are one median spine on the rostrum and a small spine just back of it; the dorsal profile of the carapace is nearly straight, not depressed; the hand is a little more than twice as long as wide and reaches the tip of the antennal scale; the antennal scale is about two thirds the length of the carapace; and the fifth abdominal segment has a keel on the dorsal side. The length is 68 mm., the carapace 17 mm., and the antennal scale 10.5 mm.

Crago nigromaculata (Lockington) is often found with *C. nigricauda* along the coast of California and is much like it except for a large circular spot on each side of the sixth segment toward the back margin, and the absence of the dorsal median keel on the fifth abdominal segment. The spot has a bluish center with a black ring around it, outside of which is a ring of yellowish. Also the hand is a little longer, the antennal scale is three-fourths the length of the carapace and the chelipeds reach only to about the middle of the antennal scale.

Crago alaskensis elongata (Rathbun) is taken in abundance off the coast from British Columbia to southern California. It resembles *C. nigricauda* but the hand is 2½ to 3 times as long as wide; the antennal scale is as long as the carapace without the rostrum, with a spine much longer than the inner angle of the blade. The length is 55.7 mm., carapace length 13.5 mm., and antennal scale 11.6 mm. long.

Crago franciscorum (Stimpson) is the common shrimp of the San Francisco markets, and is found from Alaska to San Diego. It can be easily distinguished by the narrow chelipeds in which the finger (fig. 267) when bent is almost parallel with the hand and the length of the hand is about 4½ times the width. The color is a mot-

FIG. 267.—Chela of *Crago franciscorum* (male); x2.5. (After Rathbun.)

tled yellowish gray. The length of a typical male specimen is 77 mm., carapace length 20.5 mm., and length of antennal scale 15 mm.

Crago munita (Dana) differs markedly from the other Cragos, here given, in its spiny carapace which has a depressed area in the middle and toward the front. The length is 36.2 mm., the length of carapace 10 mm., and length of antennal scale 4 mm. It is found from Alaska to southern California.

Crago munitella (Walker) is like *C. munita* except for an extra spine on the second ridge from the middle of the carapace. This spine is just back of the upper lateral spine on the same ridge. It is taken from Puget Sound to southern California.

Paracrangon echinata Dana. The carapace and upturned rostrum bear a number of spines. The second pair of walking legs are absent, the third pair slender and short. The length taken from the tip of the spine at the anterior base of the rostrum is 47.4 mm.; carapace from same point, 16 mm.; length of carapace from posterior line of orbit 12 mm. The species has been found in Alaska and Puget Sound.

Suborder—REPTANTIA

On the Pacific coast this suborder is represented by three tribes: *Palinura*, including the rock lobster; *Anomura*, with a variety of forms ranging from burrowing shrimp-like species to the hermit crabs; and *Brachyura* which includes all of the true crabs. There is also a fourth tribe, the *Astacura*, to which the true lobsters and the fresh water crayfish belong, but this is not represented by salt water species on our coast. As the different tribes show a large degree of specialization and of variation in structure, these features will be dealt with in connection with the separate groups.

PHYLUM—ARTHROPODA

Tribe—PALINURA

The California spiny lobster, so prominently displayed in the fish markets of the west coast, is the only species of importance in this group. It has several relatives but they occur in deep water. The first and third legs are alike in members of this tribe, either chelate or not, and the abdomen is straight and symmetrical and ends in a broad tail-fan composed of the last segment, or telson, and the two large biramous uropoda at the sides of it. The carapace is fused at the sides to the epistome and the rostrum is small or absent, the last two characters distinguishing the spiny lobsters and their relatives from the true lobster and the fresh water crayfish that belong to the tribe *Astacura*.

Family—PALINURIDAE

Panulirus interruptus (Randall) (fig. 268). According to *California Fish and Game** the catch of spiny or rock lobsters for the year 1921 was 334,271 pounds in California, augmented by 943,547 pounds from Mexico. It is caught in shallow pools left at low tide or at any depth up to 18 or 20 fathoms, more rarely at 35 fathoms, along the coast of southern California and Mexico from Point Conception, on the north, to the region around Manzanillo on the south. The lobsters live among the rocks, hiding in the crevices, and are especially numerous in the kelp which seems to offer them additional protection and food.

The lobster fishermen use traps which are commonly large boxes made of laths (fig. 269). There is a hole in the top which is surrounded by laths sharpened and pointing inward to prevent the escape of the lobsters after they have entered. Often a coarse wire netting is stretched over a framework and provided at the top or sides with funnel-shaped entrances guarded by sharp wire points where the netting is cut to form the opening. In the summer months

---
*27th Biennial Report, California Fish and Game Commission, p. 135.

FIG. 268.—The California spiny lobster, *Panulirus interruptus;* photographed from life.

FIG. 269.—Lobster traps ready for use.

the traps, baited with decayed fish or abalone are set in water from 2-7 fathoms deep and in winter at a depth of 8-18 fathoms. The trap is attached by a rope to a buoy so that it can be readily located by the fishermen. The fisherman who is efficient will visit his traps every day and can operate about twenty of them.

Like all wild animals subject to commercial exploitation, the spiny lobster is in danger of serious depletion of numbers, if not complete extermination, and the state of California has enacted laws designed to conserve the species. In this case the aim has been to protect females during the breeding season, hence it has been made unlawful to fish for spiny lobsters except from October 15th to March 1st, and to take females under $10\frac{1}{2}$ inches in length or over 16 inches at any time. As there is a more ready sale for individuals weighing 2 or 3 pounds, those over 16 inches in length are not in great demand in the markets, and since the large females produce a much greater number of eggs than those of smaller size it is desirable to protect them.

Most crustaceans pass through a series of developmental stages, and *Panulirus* has a curious larval existence. The coral-colored eggs which appear in May and June are attached to the inner branches of the swimmerets. A large female may produce as many as 250,000 to 500,000 at one time. They hatch in nine or ten weeks and during this period the females seek sheltered rocky places near shore where the water is shallow. The larvae are very fragile at the time of hatching and without yolk material (fig. 270). From the flattened condition of the body they have been named phyllosoma ("leaf-body") larvae. Only the appendages of the head, and some belonging to the thorax, are present, and the abdomen is much reduced in size. Owing to the great transparency of the body the relatively large pigmented eyes are the most striking characteristic. Comparatively little is known about the life-history. Efforts to carry on artificial propagation as is done with the true lobster of the eastern coast, *Homarus*, have not, up to the present time, been successful. The California Fish and

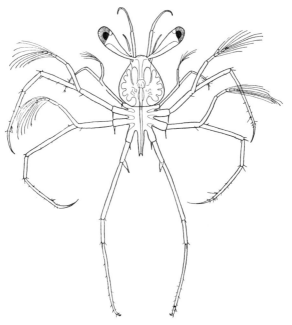

FIG. 270.—Phyllosoma larva of the California spiny lobster, *Panulirus interruptus:* drawn from life; x20.

Game Commission boat *Albacore* has taken phyllosoma larvae of advanced type from a depth of 75 fathoms and as far as 150 miles off shore. Following the phyllosoma stage is the puerulus or intermediate stage. The length of time necessary to complete development is not known.

After reaching adult form the lobster continues to grow by repeated moults, increasing in size after the moult and before the hardening of the "shell" or body covering. We have found some spiny lobsters about to shed and the freshly cast exoskeletons of others in July and August, and there is considerable evidence that the process may take place at any time of the year. As is the case with others of the group, the body shrinks away from the exoskeleton before it is withdrawn, and the animal escapes through an opening formed on the upper side between the carapace and the first abdominal segment, often leaving such a perfect counterpart of itself that only the colorless eye coverings, from which the pigmented eyes have been withdrawn, indicate that the skeleton is empty. After moulting the covering is soft for a time, and, until it hardens, the lobster is an easy victim for its enemies.

Among the animals which prey upon the spiny lobster are the sheephead, the jewfish, and the octopus. The stomach of a jewfish has been found to contain as many as ten lobsters. From these and other aggressive creatures it is protected by the hard exoskeleton, made firm by deposition of lime, and bearing numerous sharp-pointed spines that are capable of inflicting wounds upon a would-be captor. The rostrum, so commonly found on the anterior border of the crustacean carapace, is absent; but the stalked eyes are amply guarded by a pair of sharp spines above the orbits. The spiny lobster secures additional protection by hiding among rocks and beneath the kelp, and by its nocturnal habits, for it appears to feed at night since that is the time when it is trapped in largest numbers. It is claimed that fewer are caught on moonlight nights than on dark nights.

There is considerable variation in size and color. The first is correlated with age and food supply and depends also upon sex, the males reaching the larger size. Allen* reports that he saw a male specimen at Catalina Island weighing seventeen pounds, but ten pound lobsters are rare. The rate of growth and the length of life are both undetermined, but a large individual must be several years old. Spiny lobsters of several shades of color may appear in the same trap, and it is not apparent that color variation depends upon any one condition. It probably represents individual variation only.

---

*B. M. Allen—Notes on the Spring Lobster (*Panulirus interruptus*) of the California Coast. Univ. Calif. Publ. Zool., Vol. 16, pp. 139–152.

The sexes may be distinguished by the fact that the swimmerets of the male are shorter than those of the female, the latter having a leaf-like inner division upon which the eggs are fastened. The female bears broader paddles (uropoda) on the telson, and has small pincers on the fifth pair of walking legs instead of an ordinary claw such as the male possesses.

The spiny lobster moves by walking slowly forward, backward, or sideways, by the use of the ten feet, or by swimming backward, propelled by quick and powerful strokes of the abdomen made more effective by expanding the tail-fan. They thrive in aquaria provided with running water, and make an interesting exhibit.

The figure shows the main external characteristics. Un-

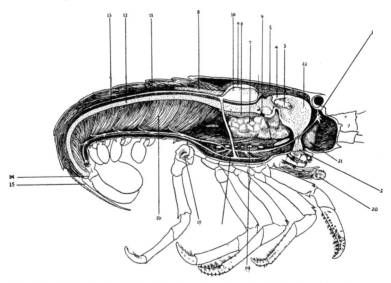

FIG. 271.—California spiny lobster, *Panulirus interruptus*, dissected to show internal organs.

1. Brain.
2. Mouth.
3. Lateral tooth of stomach.
4. Median tooth of stomach.
5. Muscles to stomach.
6. Opening of duct from liver.
7. Liver.
8. Male gonad.
9. Pericardium.
10. Heart.
11. Extensor muscles of abdomen.
12. Intestine.
13. Abdominal artery.
14. Anus.
15. Telson.
16. Flexor muscles of abdomen.
17. External opening of sperm duct.
18. Sternal artery.
19. Skeleton.
20. Maxillipeds.
21. Green gland.
22. Stomach.

doubtedly, there is a certain shrimp-like appearance when seen from above, but this disappears when the sub-cylindrical carapace and depressed abdomen are taken into account. The ponderous, crushing claw of the true lobster is absent in this species, indeed the first pair of legs are without chelae and are used in walking. The limbs and antennae are easily broken off but regenerate in course of time (compare fig. 306).

They feed upon a wide range of material, both of plant and animal origin, and do not discriminate against decaying

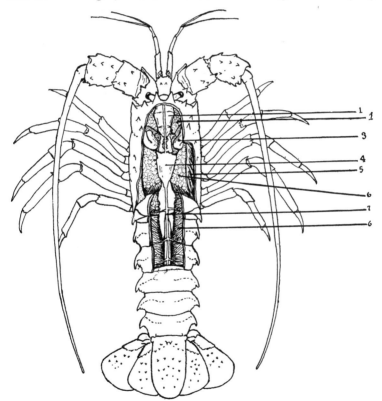

FIG. 272.—California spiny lobster, *Panulirus interruptus*, dorsal view, dissected to show internal organs.

    1. Stomach.   4. Heart.   6. Muscles.
    2. Liver.   5. Gills.   7. Dorsal blood vessel.
    3. Reproductive organs.

substances. The lobster has six pairs of mouth parts which enable it to take in food. The largest of these appendages resemble the walking legs except for the fact that they turn forward. Hence they are called maxillipeds. They help to hold the food and pass it on to the two pairs of maxillae and the pair of mandibles which break it up so that it may be taken by way of the short, wide esophagus into the stomach. The esophagus and stomach (figs. 271 and 272) are both lined with chitin which is a continuation of the chitinous skeleton on the outside of the body and is shed with it when the animal moults. This lining is thickened and hardened at certain points in the fore part of the stomach to form the "gastric mill." This consists of three opposed teeth which serve to grind the food. The digestive fluid from the liver joins the food in the small, hinder portion of the stomach. Absorption of digested food takes place from the intestine, which extends straight back to the telson on the under side of which the anus opens to the exterior.

At each side of the body of the lobster, between the carapace and the thorax wall proper, is a gill chamber, open at the anterior and posterior ends, and containing twenty-one gills. The gills bear enormous numbers of tiny filaments which give them a plume-like appearance and enable them to present a large surface to the water. Some are attached at the point where the limbs join the body wall, some above this point, and others at the basal joint of the appendages. Broad, flat, leaf-like structures are also attached to the basal joints of the appendages in such a way that movements of the legs or maxillipeds cause the expanded parts to act like paddles and force the water forward through the chamber. A live spiny lobster will be seen to keep its maxillipeds constantly in motion even when it is not taking food. This keeps the current of water flowing over the gills.

The heart lies above and back of the stomach and is a white, thick-walled sac with three pairs of openings leading into it from the pericardial sinus and a number of arteries leading from it (fig 271). When the heart contracts, the colorless blood is forced into the arteries, which lead to all parts of the body and finally end in fine capillaries, which open into sinuses or spaces among the muscles and organs. These spaces all communicate with a space in the ventral part of the body cavity where the ventral nerve cord and ventral arteries lie; from here the blood passes by blood vessels to the gills. In the gills the blood flows through tiny vessels out into the numerous filaments (fig. 273). Each fila-

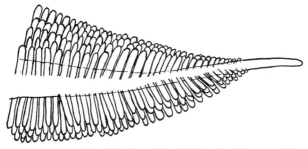

FIG. 273.—A small portion of the tip of one of the gills of a spiny lobster; greatly magnified.

ment is bathed with water so that the blood, as it flows along, is separated from the water in the gill chamber by a membrane so thin that oxygen can pass in by osmosis and carbon dioxide can pass out. The oxygenated blood flows back into the body and up into the pericardial sinus and thence into the heart through the three pairs of openings, or ostia, already mentioned. Valves in the arteries and ostia prevent the backward flow of the blood.

The muscles, which form the edible part of the lobster are found mainly in the abdomen and in the legs since these are the parts concerned with locomotion. The strands of the large abdominal muscles are arranged in a curious intricate fashion.

The green glands or kidneys are excretory organs which lie at the base of each antenna. The external opening of this gland is on a small tubercle at the outer posterior edge of the first joint of the antenna.

The brain is small and lies just below the eyes. It sends off nerves to eyes, antennae, and antennules. A ventral nerve cord extends from it back throughout the length of the body. This ventral cord is made up of two strands which separate in the mouth region, one running on each side of the esophagus, after which they come close together. The sternal artery also runs down between the two parts of the cord. Ganglia, enlarged points along the nerve, appear in each segment of the abdomen. In the thorax, however, the number of ganglia seems not to correspond with the number of segments, several adjacent ones having been fused in certain cases. Branch nerves connect each ganglion with all parts of the corresponding body segment.

The eyes are compound. The outer surface, or cornea, shows a large number of microscopic squares. For each tiny square there is a separate eye-like structure set off from its neighbors by a black layer of pigment. Each of these radially arranged "eyes" contains a structure which corresponds to a lens, and, below it, a rod-like sensory portion which connects at its lower end with nerve fibres from the optic nerve. Each separate part of the compound eye receives only the light coming directly into it and gives what has been called "mosaic vision." Creatures with this sort of eye are quick to observe movement in anything near them.

The auditory organ is located in the dorsal side of the first joint of the antennule near the base of the joint, and is formed by an infolding of the surface layer. The sac is lined with chitin, the surface of which is produced into setae with tiny branches. The sac is in communication with the outside and the water with which it is filled contains tiny pieces of sand. The setae on different parts of the body are thought to be useful as tactile organs, and the row of fine setae on the distal half of the outer branch of the antennule are thought to be olfactory organs.

The organs of reproduction in both male and female lie

in the thorax on both sides of the heart, often extending some distance posterior to it. The sperm duct opens on the basal segment of the fifth leg. Before the eggs are laid the male has placed the sperm upon the under side of the posterior part of the thorax of the female. This is a putty-like mass which is white and soft at first and later becomes hard and turns almost black. It is a tough substance in which are tubules containing the spermatozoa. The oviducts open on the basal segments of the third pair of legs, near the sperm patch, so that the eggs, as they are laid, come in contact with the spermatozoa and are fertilized.

Tribe—ANOMURA

Sand crabs, ghost shrimps, and hermit crabs belong to this group. They are greatly diversified in structure and often show remarkable adaptations that are correlated with strange habits and places of abode. For instance, the common sand crab of the California coast, *Emerita analoga* (fig. 290) which lives down in the sand of the beach where it is exposed to the action of the waves, is fitted by nature to burrow with extreme rapidity and to breathe and eat while buried in the sand. The flat crabs of the family *Porcellanidae* (fig. 299) with both carapace and chelipeds markedly depressed, are common inhabitants of the crevices between the rocks wherein they can dwell secure from molestation by larger animals. The hermit crabs (figs. 278-283) represent another correlation of habit and structure within the *Anomura*. With them the abdomen is soft and twisted spirally, but the crabs secure protection by living in empty mollusk shells. Sometimes these shells form a base for corals as is illustrated by the photograph (fig. 283) which shows a hermit crab brought up from deep water off La Jolla. The shell was covered with a handsome group of white coral polyps. Barnacles, sponges, and other encrusting forms, are also carried about in this way. In the tropics, certain close relatives of the hermit crabs (the *Cenobitidae*) are land forms capable of climbing shrubs and trees. The shrimp-like, translucent members of the family *Callianassidae* (fig. 276) burrow deeply under rocks and in the mud of bays and lagoons. They are remarkable for the transparency of their body covering which often permits the beating of the heart and the motions of the gills to show distinctly through the delicate carapace.

With such variety of habits and diversity of structure, it is difficult at first to think of these forms as members of the same tribe. To be sure, they are all decapods for they have ten thoracic legs, although in some of the groups, as, for example, the *Lithodidae* (figs. 287–288) this fact becomes evident only after close observation. Borradaile,* whose classification of the *Decapoda* we have followed in this volume, has placed in the *Anomura* those members of the suborder *Reptantia* in which the carapace is not fused with the epistome (the region between the eye and the antennules), the third pair of legs are not chelate, though the first pair are apt to be, and the last pair of thoracic legs are weak and clearly distinct in size and position from the others. Other characters help to distinguish the group, but the three given above, taken together (for some of them are shared with other tribes) will suffice.

The form of the abdomen varies widely. It may be straight and shrimp-like, soft and unsymmetrical, or bent under the thorax in a manner resembling the true crabs. Indeed, many kinds are commonly called crabs (as king crabs, sand crabs, etc.), in recognition of this resemblance, but it should be remembered that all true crabs are members of the next group, the *Brachyura*.

Family—CALLIANASSIDAE

The members of this family are sometimes called "ghost shrimps." The dealers in live bait commonly call them

---

*L. A. Borradaile—On the Classification of the Decapod Crustaceans; Annals and Magazine of Natural History, Vol. 19, pp. 457–486.

"crawfish," a name that more properly applies to the crayfish of inland waters. They are long, thin-bodied, and white, contrasting markedly with the black mud in which they live. "Ghost shrimps" burrow in the sand and mud, often a foot or more below the surface. The hole they live in may be only as large around as your finger and the shrimp may be nearly three inches long with a cheliped almost as long as itself, but it has no trouble turning around in the hole. It "rolls up" or rolls the abdomen under and in a moment it is headed in the opposite direction. One can get a good idea of their quickness and flexibility by observing several of them placed together in a dish of sea water. A small fish, the "blind goby" is often found in the holes with the *Callianassa*. Members of this family have no side plates on the abdomen, the first pair of legs, and sometimes the second, are chelate, one of the chelae of the male often being greatly enlarged.

Upogebia pugettensis (Dana) (fig. 274) is found from Alaska to Lower California on muddy beaches, in holes a foot or more in depth; occasionally it is taken in water several fathoms deep. The specimen photographed was from the mud of one of the tidal channels near Moss Landing, north of Monterey. Specimens measuring 112 mm. have been reported. They are a muddy color when alive, contrasting with the glistening white *Callianassa*. The rostrum is narrow and roughened anteriorly. The first pair of thoracic legs only is chelate, the chelipeds being equal in size. The hand has two parallel hairy lines on the upper edge and a line of setae on the outer surface which is continued obliquely across the carpus (fig. 277-D). The thumb is bent downward, has a tooth near the middle and is shorter than the sharp, curved dactyl. A parasitic crustcean, *Phyllodurus abdominalis* Stimpson, is often found attached to the abdominal appendages of this species, also a small bivalve mollusk *Pseudopythina rugifera*.

Professor Kincaid says that this species is a menace to

FIG. 274.—*Upogebia pugettensis*. Female specimen carrying eggs. Taken in February near Moss Landing, in the Monterey region; x¾.

FIG. 275.—*Calianassa californiensis*, the California ghost shrimp, two males (right) and two females (left) photographed from life; about ⅓ natural size.

FIG. 276.—*Calianassa longimana*, the long-handed ghost shrimp; males; slightly reduced.

the oyster industry in the Puget Sound region. In making its burrows, the creature often covers the young oysters with mud and debris so that numbers of them are smothered. The numerous burrows also allow leakage from the dykes which have been built to keep the oysters on the mud flats covered with water at low tide.

Callianassa californiensis Dana (fig. 275) is found all the way from Alaska to San Diego. It burrows in the mud and may be obtained by digging in the mud flats when the tide is out.

The species can be distinguished from its relatives by the shape of the chelipeds (fig. 277-B). In this one the hand and carpus are broader than in the others. The carpus has a sharp over hanging upper margin on the inner side and a projecting lobe at the posterior end, and is very little longer than broad. The female is much like the female of *C. longimana* but it can be distinguished by the rounded rostrum and by the fact that the cornea is in the middle of the eye stalk.

This species is often dug for bait. It is an interesting animal to study because it is so transparent that the contractions of the heart may be readily seen, also the gills and the movements of the gill bailers. The coral colored reproductive organs also show through the body covering. The appendages under the abdomen are used for swimming and are paddle-shaped. The huge chelae would seem to be too large to be very useful but specimens in captivity will use them to seize worms or the soft abdomens of other ghost shrimps. The length of our largest male specimen is 65 mm., the large cheliped being 58 mm. long and the small one 35 mm., the width of the carpus 16 mm., and length of carpus 17 mm.

Callianassa longimana Stimpson (fig. 276) is found in much the same localities as the foregoing species and is also reported from under rocks in tide pools. The large cheliped is proportionately longer and narrower than that of *C.*

californiensis, the carpus being nearly twice as long as broad. The female can be distinguished by the location of the cornea just behind the middle of the eye stalk, and by the subacute rostrum. Length of a male specimen 40 mm., length of large cheliped 40 mm., width of carpus 6.5 mm., length of carpus 12 mm.

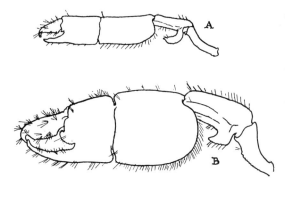

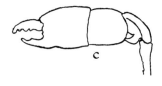

FIG. 277.—Large chelae (A) of *Callianassa longimana*, (B) *C. californiensis*, (C) *C. affinis* (after Holmes), (D) *Upogebia pugettensis* (small specimen); not drawn to the same scale.

Callianassa affinis Holmes (fig. 277-c) differs but little from the two species just given. The eye peduncles are oblong rather than acute, with the cornea in front of the middle of the peduncle. The "carpus of large cheliped of male very little longer than broad, and very little longer, sometimes even shorter, than palm" (Rathbun),* "palm oblong, both inner and outer faces convex" (Holmes).†

---

*Mary J. Rathbun—Decapod Crustaceans of the Northwest Coast of North America; Harriman Alaska Series, Vol. 10, Smithsonian Publication 1997, p. 154.

†S. J. Holmes—Synopsis of the California Stalk-Eyed *Crustacea*; Occasional Papers of the California Academy of Sciences, No. 7, p. 162.

Family—PAGURIDAE
(*The hermit crabs*)

Hermit crabs are common in the tide pools where they may be seen crawling about with shells on their backs. These are usually old gastropod shells which the crabs have taken for their own use after the original inhabitants have died. Because of this habit of living in a hard, protecting domicile the body of the crab has become greatly changed in form. Only the fore part of it protrudes and this is covered with a firm exoskeleton, but the abdomen which fits up into the spire of the shell is long, slender, and soft, without a hard covering. Some of the appendages are small, and some have entirely disappeared.

The large claws are usually well developed and serve to block the opening when the crab has retreated within the shell. There are two pairs of walking legs which end in long pointed claws. The fourth and fifth pairs of thoracic legs are small. The appendages of the abdomen are small or absent, except the sixth pair which are hook-like and near the end of the abdomen. With these, the crab clings to the columella of the shell and thus fastens itself securely in its abode.

As the hermit crab grows, it is obliged now and then to move into larger quarters. Empty shells of all sizes are common along the shore so that it probably is not hard to find one of the larger size when the old one has become too small. The crab merely lets go and crawls out of the old shell, thrusts its abdomen into the new spiral and fastens to it with the hooks. This moving operation is an interesting thing to watch. Let the sea water in which the hermit crabs are kept get stale so that they crawl out of their shells, and then give them fresh sea water; usually they will appropriate the empty shells without delay.

The shells in which the hermit crabs live are frequently covered with other animals (fig. 283) such as hydroids, bryozoans, sea anemones, sponges, and barnacles. These animals do the crabs no injury and enjoy the benefits of a wandering life. The hermit crabs, like most of their relatives, are fond of either fresh or old meat. They are great fighters and will even fight and devour one another.

The coral colored eggs are carried on the appendages of the lower side of the abdomen. They commonly appear in January, February, and March and are about the size of mustard seed. The young hermit crabs in the zoea stage have a very short frontal spine and the segments of the abdomen develop early. Following the zoea is the glaucothoe, which corresponds to the megalops of the crabs. In this stage, strange to say, the young hermit is perfectly symmetrical. Its thorax resembles that of an adult, but the abdomen has five pairs of symmetrical, two-branched appendages. This stage lasts four or five days, at the end of which time a shifting of the internal organs takes place. The liver, reproductive organs, and green gland move down into the abdomen and the posterior lobes of the liver move over from the right to the left side of the intestine. The right abdominal appendages then degenerate, more completely in the male than in the female. By this time the little crab is living in a shell, usually one of the right handed spiral type, so that his unsymmetrical shape fits the curve of the shell. Experiments show the interesting fact that if the crab can not find a shell the degeneration of the appendages takes place more slowly than it does if it gets a shell at the usual time.†

Genus *Paguristes*. In this genus the sexual appendages are borne on the first two abdominal segments in the male and the first segment in the female.

Paguristes turgidus (Stimpson) is reported from Puget Sound to Catalina. The anterior part of the carapace measures 5 mm. in length in a female specimen, 10.5 mm.

---

†M. T. Thompson—The Metamorphoses of the Hermit Crab; Proceedings of the Boston Society of Natural History, Vol. 31, pp. 147-209.

in a male, and is a little longer than wide. The swollen chelae are characteristic. Specimens reach a length of three inches.

Paguristes bakeri Holmes (fig. 282) is often taken along the southern coast. The shape of the chelae distinguishes the species from its nearest relatives. They are nearly equal in size, spiny, and with a straight inner margin. The anterior part of the carapace is longer than wide. The carapace of a male may be 9–35 mm. long.

Paguristes ulreyi Schmitt, taken on the southern California coast, resembles the preceding species but the eye stalks are longer and very slender and the hands are nearly oblong. The eye stalks are as long as the anterior part of the carapace is wide and more than three-fourths of its length. The length of the carapace of a typical male is 12.5 mm. and that of a female a little over 5 mm.

Genus *Pagurus*. The species included in this genus have no genital appendages on the first and second segments of the abdomen.

Pagurus ochotensis Brandt is a large species found occasionally off the California and Washington coast. The anterior part of the carapace is wider than long. The acicle is trigonal, naked, glossy, iridescent, and extends far beyond the eye stalks. The chelipeds are spiny. The sides of the carpus of the large cheliped are flat and granulated. A row of longer spines lies along the inner margin of the upper surface of the carpus. The hand is oblong, rounded above, the dactyl with a row of spines on the outer edge and another near the middle of the outer surface. The dactyls are spinous, twisted, grooved on each side and nearly as long as the two preceding joints.

The small hermit crabs found so commonly in the tide pools usually belong to one of four species and can be identified with the help of the following key.

    A. Antennal scale very nearly the length of the eye stalk. Peduncle of antenna exceeds length of eye stalk by at least

one-half the length of the last joint of the peduncle (fig. 284). Walking legs banded with blue or white.
> a. Anterior part of carapace as wide as or slightly wider than long. Claws of walking legs almost as long as the next joint. Tips of segments of walking legs banded with white or pinkish and upper extremeties of segments of walking legs sometimes banded with blue. Dactyls with four longitudinal stripes, one on each side, one above and one below. Antennae the same color as the body with fine white rings. *P. hirsutiusculus.*
> b. Anterior part of carapace longer than wide. Claws of walking legs markedly shorter than the next joint. Tips of last joint of walking legs banded with bluish, claws blue with red longitudinal stripe on each side, red spots on legs and chelipeds. Antennae and antennules red. *P. samuelis.*

B. Antennal scale markedly shorter than eye stalk. Peduncle of antennae exceeds the length of eye stalk by less than half the length of the last joint of the peduncle. Walking legs not banded with blue or white.
> a. Anterior median tooth of carapace broad and rounded (fig. 284).
> Large hand a little less than twice as long as wide. *P. granosimanus.*
> b. Anterior median tooth of carapace sharply pointed. Large hand twice as long as wide........*P. hemphilli.*

Pagurus hirsutiusculus (Dana) (figs. 279 and 280) is one of the most abundant hermit crabs in the tide pools from Alaska to Lower California. Large specimens from Pacific Grove and northward are often much more hairy than the small specimens found in the south. There are distinct white or pinkish bands on the walking legs, the joints of the second and third pairs being banded at the outer ends. Our large specimens show the white banding less markedly and have blue bands at the joints or just below them. The dactyls have longitudinal stripes of red on a light blue ground. The sides of the walking legs often have a longitudinal median stripe of blue or whitish. The antennae are the color of the body with white bands or rings. Our largest specimen is 30 mm. long; the hand is 13 mm. long.

Pagurus samuelis (Stimpson) (figs. 281 and 284) is also

FIG. 278.—*Pagurus granosimanus;* slightly reduced. FIG. 279.—A "hairy" specimen of *Pagurus hirsutiusculus,* removed from its shell; slightly reduced.

FIG. 280.—*Pagurus hirsutiusculus;* slightly reduced. FIG. 281.—*Pagurus samuelis;* slightly reduced.

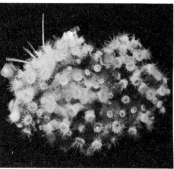

FIG. 282.—*Paguristes bakeri;* approximately x½. FIG. 283.—Gastropod shell inhabited by a hermit crab and covered with coral polyps; about ½ natural size.

found from Alaska to Lower California. The antennae and antennules are red and the tips of the last joint of the walking legs are banded with blue. The claws are bluish with red longitudinal stripes. Legs and chelipeds are spotted with red. Schmitt gives the measurements of his largest specimen as 42 mm. from the tip of the rostrum to the end of the telson, and 18 mm. as the length of the carapace.*

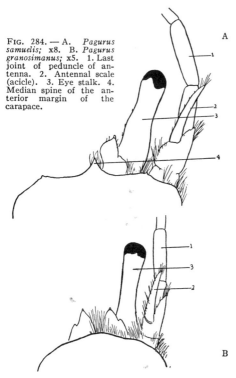

FIG. 284. — A. *Pagurus samuelis;* x8. B. *Pagurus granosimanus;* x5. 1. Last joint of peduncle of antenna. 2. Antennal scale (acicle). 3. Eye stalk. 4. Median spine of the anterior margin of the carapace.

Pagurus granosimanus (Stimpson) (figs. 278 and 284) occurs all along the coast from Alaska to Lower California. Teh antennae are red and there are no bands of color on the legs, but the tubercles of the chelipeds are whitish or bluish and the walking legs are spotted white or bluish. The granulations on the hands are very distinct. In large specimens the body is 29 mm. long and the large hand 8 mm.

Pagurus hemphilli (Benedict) is found on the northern California coast and as far south as Monterey. It grows to be larger than the preceding species and resembles it, but may be recognized by the sharp median tooth on the anterior margin of the carapace, and the greater length of

---

*Waldo L. Schmitt—The Marine Decapod *Crustacea* of California; University of California Publications in Zoology, Vol. 23, p. 140.

the large hand. The inner part of the hand is flat and the fixed finger has a depression in it. Length of a large specimen is 42 mm.

Family—LITHODIDAE

This family includes a number of crab-like deep-water forms and some that are occasionally found near shore when the tide is out. All of them have the fifth pair of legs reduced in size and folded within the gill chambers, and the antennae external to the eyes. They also have a prominent rostrum and a broad carapace, usually with many spines or other projections.

Hapalogaster mertensii Brandt (fig. 285) is a brownish or brownish red form. It has a soft sac-like abdomen that does not fold completely under the thorax. The cephalothorax is flattened and bears tufts of stiff bristles, as does the abdomen. There is also a fringe of soft hair along the posterior margin. The lateral margins of the heart-shaped carapace bear spines which are largest anteriorly. The chelipeds are flattened, as are the walking legs. The latter are red on the ventral surface. This species is distributed from Puget Sound to the Aleutian Islands and may be taken along the shore at low tide or in shallow water. An average specimen has a carapace 25 mm. long.

Hapalogaster cavicauda Stimpson ranges along the California coast from Mendocino County to Monterey and is common under rocks at low tide. It closely resembles the preceding species but is thickly covered with short hair and has a shorter rostrum. The length of the carapace is 17 mm., breadth 19 mm. This crab so nearly resembles the brown granite rocks that one is apt to pass it by for a pebble or a piece of moss. It remains motionless after it has been disturbed which helps to make it inconspicuous (fig. 286).

Oedignathus inermis (Stimpson) is another form with a soft thick abdomen, but the basal segment is strengthened by limy plates as, to a lesser extent, are the two terminal

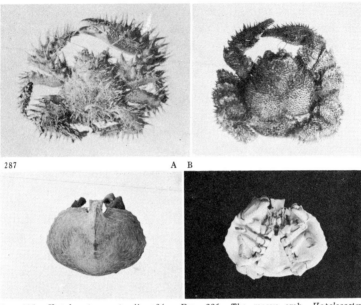

FIG. 285.—*Hapalogaster mertensii*; x⅔. FIG. 286.—The mossy crab, *Hapalogaster cavicauda*; x8/9. FIG. 287.—The Sitka crab, *Cryptolithodes sitchensis*, (A) dorsal view, and (B) ventral view; x5/6.

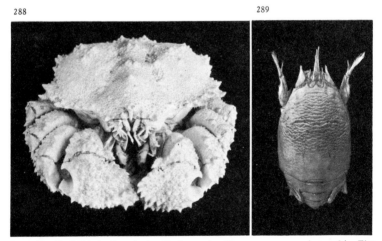

FIG. 288.—*Lopholithodes foraminatus*, the box crab. From a museum specimen; x⅓. The round opening between the cheliped and the first pair of walking legs permits the water to enter the gill cavity when the legs are tightly folded against the carapace. FIG. 289.—*Emerita analoga*, the sand crab; x5/6.

ones. The carapace is convex and without spines but covered with flat, scale-like plates. The chelipeds are very unequal and beset with wart-like tubercles. The length and breadth of the carapace are equal, 15 mm. in a large specimen. Like the species of *Hapalogaster*, this is often found on the rocks. It is said to range from Unalaska to Monterey.

Phyliolithodes papillosus Brandt has a triangular carapace, the apex of which is formed by the prominent rostrum, which ends in two prongs, the base being the posterior margin. The chelipeds and walking legs are covered with numerous long, blunt, slightly flattened spines. Dorsally, the carapace is strongly tuberculated and it is furnished with four prominent spines at the sides. The broad abdomen covers the entire under side of the body. It is protected by hard plates with membranous central areas which are the chief distinguishing character of this genus. Holmes gives a length of 52 mm. for the carapace. This curious spiny creature occurs from Dutch Harbor, Alaska, as far south as the Bay of Monterey.

The butterfly crab, *Cryptolithodes typicus* Brandt, is so named because the carapace is very wide at the sides and extends like wings over the walking legs, hiding them from view. It is also called turtle crab because the lateral extensions turn downward like the "shell" of a turtle. Miss Way* gives the color as "blackish-brown on dorsal side, light gray on ventral side." This species is not very common but has been reported from Alaska to Monterey and is sometimes seen on the rocks at low-water mark. The length of the carapace is about 34 mm.

Cryptolithodes sitchensis Brandt, the Sitka crab (fig. 287), is as peculiar as the preceding species. The carapace is high in the middle and produced into two shield-like extensions that project sideways, turning down to cover the

---

*Evelyn Way—*Brachyura* and Crab-like *Anomura* of Friday Harbor, Washngton; Puget Sound Marine Station Publications, Vol. 1, p. 352.

legs. They project further forward than in *C. typicus*. The rostrum is narrow where it meets the body and wider at the tip, leaving two deep notches between it and the lateral expansions. The eyes and antennae can be seen through these notches. The surface of the carapace is smooth, with occasional small rounded tubercles, and uneven, with a prominent ridge running lengthwise through the anterior region. The color of the males is bright red. A female, taken at Pacific Grove is light gray with streaks of darker gray and the central part of the dorsal region is rose pink. Holmes gives the measurements of a large specimen as follows: length of carapace 47 mm., width 64 mm. This species may often be found among the rocks when the tide recedes and ranges from Sitka to Monterey.

Figures 288 and 294 show the dorsal and ventral sides of the box crab, *Lopholithodes foraminatus* (Stimpson). The carapace is broad and covered with wart-like tubercles above and at the sides. The chelipeds and walking legs are tuberculated, some of the tubercles becoming short spines. The eye stalks are spiny, too. The abdomen is covered with calcareous plates, thickly beset with tubercles. Peculiar semicircular emarginations may be seen on the second joints of the chelipeds and similar ones on the first walking legs. When the legs are folded tightly together and against the carapace, a characteristic attitude, the two notches form a circular opening through which the current passes in breathing. Such an opening may be termed a foramen and this has given the specific name to the creature. The specimen photographed is in the museum of the Scripps Institution at La Jolla, California, and measures 131 mm. in width and 110 mm. in length of carapace. This species occurs from the Farallon Islands to the Puget Sound region.

Lopholithodes mandtii Brandt is called the king crab. It is a large species when fully grown, reaching a length of 220 mm. and a width of 260 mm. (Way), or $10\frac{1}{4}$ inches. It may be distinguished from its relative by the absence of the foramen. Furthermore, the carapace is more convex and bears four prominent cone-like elevations, of which the anterior one is the largest. Like *L. foraminatus*, this form is covered with tubercles and spines. When the legs and claws are folded up against the body, it resembles a very rough, stout box and in this manner secures protection from injury. Miss Way describes it as follows: "color very brilliant, scarlet or orange, with bright purple markings particularly on ventral part of the body and the spines on the legs." *L. mandtii* ranges from Monterey northward to Sitka, Alaska.

FAMILY—HIPPIDAE

Emerita analoga (Stimpson), the sand crab (figs. 289 and 290), is interesting from the standpoint of the fisherman because it is favored as bait for several kinds of fish and is readily obtained along sandy beaches. The carapace is ovate, somewhat elongated, and very convex dorsally. It is marked with ripple-like transverse lines which are less evident toward the sides and the posterior portion. Two of the lines are very definite, one across the back just behind the anterior margin and another near the middle of the thorax. Two small depressions exist, one on each side of the mid-line, about half way between the transverse lines. The eye stalks are long and slender, jointed near the base. The pigmented eyes are small. The antennae are plume-like and are carried coiled beneath the outer maxillipeds most of the time. The antennules may easily be confused with the antennae as they are longer than in any of the true crabs and are not reflexed under the cephalothorax (fig. 290). Each antennule has the typical two whips or flagella.

The postero-lateral part of the carapace extends downward and shields the legs and underparts laterally. When at rest, the abdomen and all the appendages except the antennules, eye stalks, and the last abdominal legs are folded compactly below the body, the antennae, in particular, being so neatly stowed away that only a careful observer

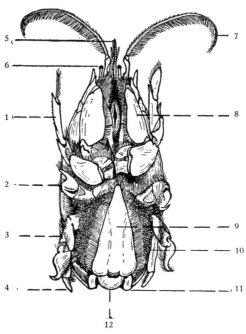

FIG. 290.—*Emerita analoga:* a female with eggs (Ventral view). 1, 2, 3, 4. First four pairs of thoracic legs. 5. Antennules. 6. Eye. 7. Uncoiled antenna. 8. 3rd maxilliped, the merus is large, the terminal joint narrow. 9. Telson. 10. Egg mass. 11. Uropod. 12. 6th segment of the abdomen; x1.

would notice them. This is accomplished by coiling them closely about the mouth and folding the broad, valve-like outer maxillipeds over them like a pair of trap doors, effectually concealing all but very small portions of their basal joints.

While named sand crabs, their resemblance to a true crab is not marked. When we consider habits, the two forms vary even more widely. The sand crabs live in the shifting, wave-washed sand. When the water sweeps in they may often swim about momentarily, but as the wave recedes, they burrow into the sand and bury themselves with astonishing rapidity, the shifting of the sand erasing the marks of their hurried digging. They always move backward, whether swimming, or burrowing, or even crawling (though this form of locomotion is less frequent). In swimming, progress is made by rapid movements of the oarlike uropoda, or last pair of abdominal appendages, which bend dorsally over the margins of the abdomen and extend outward at the sides. They burrow into the wet sand with the aid of the uropoda and the first four pairs of thoracic legs. As a rule, the animal lies just below the surface with the anterior part of the body almost at the top, and thrusts the slender eye-stalks and antennules into the water above. When the antennae lie parallel and close together, their long, fine bristles intermesh and form a sort of tube through which water is drawn into the gill chambers for breathing purposes (fig. 291).

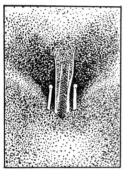

FIG. 291.—The sand crab, *Emerita analoga*, buried in sand with eyes and antennules above the surface; x2.4.

It would seem that a crustacean equipped to use its antennules as a respiratory tube, its maxillipeds as trap doors and its legs as trenching tools had about reached the limit of adaptation, but the observations of Weymouth and Richardson[*] indicate another remarkable coördination between the structure of this animal and its life in the ever moving sand. Lacking hard mandibles for chewing, such as most of the larger crustaceans possess, sand crabs feed on minute plants and animals. The small forms are found in the sandy water and are caught by means of the antennae. These appendages are stretched outward over the surface of the sand, forming a brush-like strainer through which

---

[*]F. W. Weymouth and C. H. Richardson, Jr.—Observations on the Habits of the Crustacean *Emerita analoga;* Smithsonian Miscellaneous Collections, Vol. 59, No. 7 (Publication 2082).

the current runs, leaving a residue of tiny organisms and sand grains behind it. When the wave goes out the antennae are whipped under and tucked in above the large maxillipeds, the food materials they bear, together with more or less sand, passing into the body. It is the extended antennae which produce the characteristic little V-shaped ripple marks on the beach that indicate the presence of *Emerita.*

Mead* has found that they show a strong tendency to move down hill when crawling on the surface. This would help to keep them from moving far away from water. We have occasionally found them buried under the sand at some distance from the edge of the water, but not beyond the reach of the flood tide.

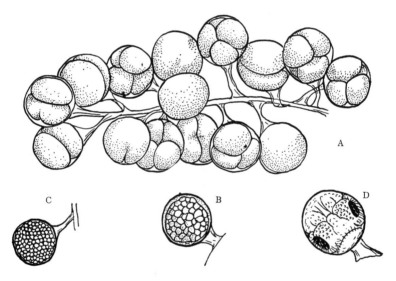

FIG. 292.—Early stages in the development of the sand crab, *Emerita analoga.* A. Eggs undergoing segmentation. B. and C. Advanced stages in the segmentation of the egg. D. larva shortly before hatching, the eyes are well developed and visible through the egg membrane; x35.

The female is much larger than the male, which it resembles except that there are three pair of abdominal

---

*H. T. Mead—Notes on the Natural History and Behavior of *Emerita analoga* (Stimpson); University of California Publications in Zoology, Vol. 16, pp. 431–438.

appendages that are lacking in the male. Measurements of the carapace, 29 mm. for a large female, 12 mm. for a male, show the disparity in size. The eggs are held between the ventral part of the thorax and the reflexed abdomen and may be found at any time between the first part of the year and early autumn. At first they are a bright coral color, changing with advancing development to a dull grayish brown. A pinkish white larva is produced which has black compound eyes of huge size for such a small creature. Figure 292 shows several stages in development. (a) shows the early divisions of the eggs and the manner in which they are held together by strands of colorless material. Figures 292-b and 292-c illustrate more advanced stages, the original egg cell having divided into many smaller cells. Figure 292-d shows the larva on the point of bursting through the covering membrane and beginning an independent existence. It is possible at this stage to see the eyes without the aid of a lens. The larva, as it looks immediately after hatching is pictured in figure 293. The creature is so transparent that the beating of the heart can be seen distinctly with the aid of a microscope.

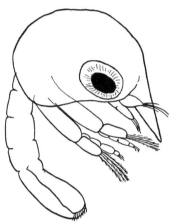

FIG. 293.—Larva of the sand crab, *Emerita analoga*, shortly after hatching; x60.

Soft-shelled sand crabs are often found which appear like the others except that their exoskeletons are less rigid. These individuals have just moulted (see page 322) and their skeletons have not yet hardened. The adults are steel-gray above, with transverse flecks of lighter color, the general effect harmonizing with the sand. The ventral surface and legs are white with a pinkish tinge.

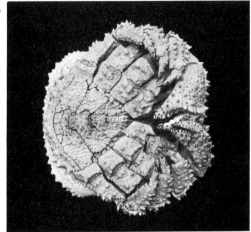

FIG. 294.—Ventral side of the box crab, *Lopholithodes foraminatus*. From a dried specimen; x⅛. FIG. 295.—The spiny sand crab, *Blepharipoda occidentalis*; x⅔. FIG. 296.—*Lepidopa myops*; x⅔.

PHYLUM—ARTHROPODA

Our common sand crab ranges from the Oregon shore down along the coast of South America where sandy beaches are exposed to wave action. Miss Rathbun* reports that a closely allied species is an article of diet on the coast of Peru. In this country its presence is the lure which brings certain kinds of fish into the breakers. Many fall prey to curlews and other birds along the shore. Another closely related species, *E. talpoida* Say, lives on the Atlantic coast.

FAMILY—ALBUNEIDAE

Blepharipoda occidentalis Randall, the spiny sand crab (fig. 295) is the largest of the sand crabs. The anterior legs terminate in sharp, toothed claws, flattened from side to side and bearing curved spines on the margins and outer surfaces with a crest of light colored hairs. The walking legs are similarly crested with the terminal segments blade-like and more or less sickle-shaped, those of the second pair being very broad and deeply notched. As in the case of the common sand crab, the fifth pair of thoracic legs is insignificant and hidden by the fourth. The maxillae are not flattened as they are in the smaller species and bear a series of sharp points on their inner surfaces. An outer branch is present that is lacking in *E. analoga*.

The oblong carapace is roughened in front and marked with transverse grooves. This might well be called the spiny sand crab for the anterior and lateral margins are armed with long, sharp pointed spines that curve forward. Possibly, they have some relation to the fact that the anterior region is necessarily the more exposed when the animal is burrowing in its native sands. One spine in the anterior region is located a little back of the front margin on the mid-dorsal line and forms the beginning of a low ridge from which the carapace slopes gently with a cornice-like

---

*M. J. Rathbun—The Stalk-Eyed *Crustacea* of Peru and the Adjacent Coast; Proceedings of the U. S. National Museum, Vol. 38, p. 554.

348 SEASHORE ANIMALS OF THE PACIFIC COAST

effect at the sides. In the main, the antennae and antennules are like those of *E. analoga* but the former are shorter in comparison to the size of the animal. The little eyes are borne on slender stalks which are jointed about midway to the tip. As in other *Anomura*, the eye stalks lie nearer the median line of the body than the antennae. The abdomen is much reduced and the telson, or terminal segment, is

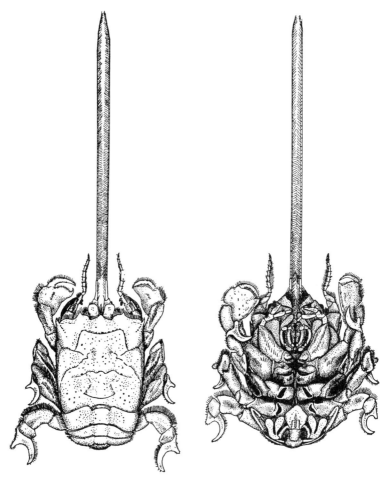

FIG. 297.—*Lepidopa myops:* dorsal and ventral views; x1.2.

small and bears thin leaf-like expansions at the sides. The egg masses are a brilliant carmine.

This animal is confined to the southern part of our coast from Monterey southward and is nowhere common. The habits have not been studied as carefully as those of *Emerita* and little is known about them. Specimens may be detected, when the tide is out, by a flattened dome of sand covering the carapace. One of our specimens measures 50 mm. along the mid-dorsal line of the carapace, with a breadth of 39 mm.

Lepidopa myops Stimpson (figs. 296 and 297) is another southern California species. The illustration shows most of the structural details. The length of the antennules, the slight growth of hairs upon the antennae (which are shorter) and the situation of the eyes on flat plates instead of stalks, will differentiate this species from either of the other sand crabs. The carapace of an adult specimen is about 20 mm. wide. *Lepidopa myops* is white with steel-blue and iridescent pink tones above. It inhabits sandy beaches near low-water mark, the same situations that shelter the other two forms. Occasionally, when the tide is out, an individual may be tracked to its temporary burrow by the marks left by the feet and the long antennules which have been allowed to drag along the sand. When placed in water this sand crab swims backward by movements of the abdomen. It can burrow in the sand very quickly and in practically the same manner as the common sand crab, keeping the long antennules extended directly outward and parallel to each other most of the time in order to form a breathing tube.

Family—PORCELLANIDAE

The thick-clawed crab, *Pachycheles rudis* Stimpson (fig. 298), has massive chelipeds. The rough tuberculated upper surface of the chelipeds will serve to distinguish this form from its relatives, the porcelain crabs. The carapace is convex longitudinally, quite smooth, slightly broader than long (length 14.5 mm., breadth 15.25 mm.), and finely striated in the gill regions. It is a light brown with some small transverse white streaks which resemble those of *Petrolisthes*. The little prominences on the claws give

FIG. 298.—The thick clawed crab, *Pachycheles rudis;* x1.

FIG. 299.—The flat-topped crab, *Petrolisthes eriomerus;* x4/3.

them a rough, sandy appearance. The walking legs are covered with light colored hair. This species lives in crevices and small holes in the rocks. It may be found at low tide under stones and among the *Chama* shells or rock oysters encrusting them. The range is from British Columbia to Lower California.

Pachycheles pubescens Holmes, the hairy *Pachycheles*, is found from the San Francisco region to Puget Sound. The carapace is flatter than in *rudis* and not so smooth, being marked anteriorly by small transverse punctures, and has a much shallower indentation at the rear. The front, also, projects a little more than in the preceding species. The claws are quite uniformly granulated, rather than tuberculated, the granules being completely or nearly hidden by the dense, thick pubescence which covers the upper surface of the hands; scattered through this pubescence are numerous tufts of longer, stiffer hairs which arise from the bases of many of the granules. The two dimensions of the carapace are the same; length and width 15 mm.

The porcelain crab, *Petrolisthes cinctipes* (Randall) is characterized by extreme flatness. Both the carapace and the abdomen are compressed and the latter is folded under the thorax much as in the true crabs, but more loosely. The eye stalks do not fit into orbits as they do in the *Brachyura*, and the antennae, which are long and whip-like, join the body outside of the eye region. Another difference is in the fifth pair of thoracic legs which are very small and fold over the base of the carapace. The maxillipeds on the under side of the body are flattened as are those of the true crabs.

The porcelain crab is an inhabitant of rocky coasts and secures food and protection by living in the crevices of the rocks, an existence for which it is well adapted by the flatness of the body and claws. Specimens may occasionally be seen in tide pools and under seaweed but never far from the rocks. When forced to swim, they commonly turn on their backs and progress by moving the abdomen, which is furnished with two lateral swimming appendages, corresponding to the uropods of the *Natantia*. The general color is brownish with transverse flecks of lighter color. The chelipeds are brownish, or bluish, and the walking legs are marked with grayish bands. The merus of the walking legs is hairy. The length is 20 mm. *P. cinctipes* is found from Vancouver Island to the Gulf of California.

Petrolisthes eriomerus Stimpson, the flat-topped crab (fig. 299) differs but slightly from *P. cinctipes*. The following details will distinguish the two forms. The merus (or first long segment) of the leg is more narrow in *eriomerus* and the antennae lack the setae, or small hairs, found on *cinctipes*, but the merus joints of the walking legs are not hairy. The last pair of sutures on the telson is transverse in the former and oblique in the latter. It exceeds *cinctipes* also in the length of the antennae and claws. Aside from these slight structural differences, the two species agree closely and they have the same general habits and live in the same kind of places. *P. eriomerus* has been reported from Puget Sound to Lower California.

Tribe—BRACHYURA

(The crabs)

Literally, the *Brachyura* are the short-tailed decapods. The group comprises the true crabs only. A number of our species are edible, but the only edible one that occurs in large enough numbers to be marketed is *Cancer magister*, the big crab. In 1921, 33,373 dozen crabs were caught in California.

The following account of fishing for crabs, is taken from a report by F. W. Weymouth* in *California Fish and Game* (1916). San Francisco has the largest fishery on the coast, the fishing being done on a sandy bar in about 5–10 fathoms of water. *Cancer magister* shows a distinct preference for sandy bottoms. Hoop nets are commonly used, the net being shaped like a deep dish two and one half to three feet in diameter at the top, fifteen to eighteen inches in diameter at the bottom, and about one and one-half feet high. Coarse netting covers the sides and bottom, the mesh being large enough to allow the small crabs to escape. Small fish, used for bait, are placed at the bottom of the net and protected by wire netting. The traps are lowered into the water and made fast to floats. A fisherman, after lowering about 20 of the nets, goes back to the first one, which will have been down about half an hour, and hauls each net up in turn. He endeavors to draw the nets up rapidly so that the crabs will not have time to escape. The crabs are sorted, the undersized ones being thrown overboard and the others kept in "live boxes" until they are sent to the market. In California, *Cancer magister* is not often taken in water less than six feet deep, but on Vancouver Island and northward

---

*F. W. Weymouth—Contributions to the Life History of the Pacific Coast Edible Crab; California Fish and Game, Vol. 2, pp. 22–27 and p. 38.

it is found in shallower water where it may be taken with a dip net at low tide.

Weymouth says, concerning the habits of the crab, that they are buoyed up by the water so that they move lightly over the bottom on the tips of their legs, and look much less clumsy than they do when moving about on land. When frightened, they show surprising speed, and they must be quick, to get the fish which they capture for food. Much of the time, they lie almost buried in the sand, with only eyes, antennae, and antennules visible. When the crab is buried in the sand, in this fashion, the stream of water that flows through the gill chamber is freed from sediment by a peculiar strainer. The large pincers, when folded, accurately fit the contour of the sides of the body which is here covered with a dense plush-like coat of hair. When the stream of water passes into this crevice between the

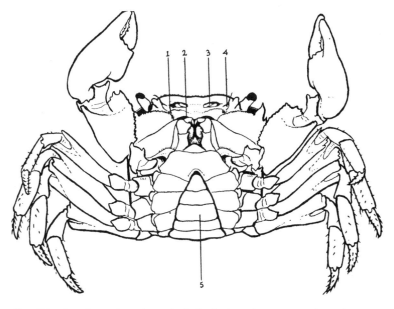

FIG. 300.—A male specimen (ventral view) of the striped shore crab, *Pachygrapsus crassipes;* x1.2. 1. Opening of auditory sac. 2. Gill slit. 3. Antennule (folded). 4. Antenna. 5. Abdomen, showing segments 3-7.

body and the pincers, the teeth on the edge of the shell exclude the large sand grains, while the hairs exclude the fine particles, so that only clear water reaches the gills. From time to time the water current is reversed.

Crabs are characterized by a wide carapace, frequently broader than long, either without any rostrum or with a small one, and by a much reduced abdomen, which is folded on to the ventral surface of the thorax in such a manner as to make it practically invisible from above (figs. 300 and 301). When fully mature, the sexes may be distinguished

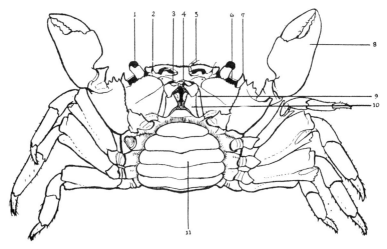

FIG. 301.—Female specimen (ventral view) of *Pachygrapsus crassipes:* x1.2.
1. Eye.
2. Antennae.
3. Gill slit.
4. Terminal joints of 3rd maxilliped.
5. Antennule (extended).
6. Orbit.
7. Post-orbitae spine.
8. Chela.
9. Exopodite of 3rd maxilliped.
10. Proximal portion of 3rd maxilliped.
11. Abdomen, showing segments 2-7.
The first walking leg on the right side is broken off.

by the shape of the abdomen. In the males it is narrow and more or less Λ-shaped, but broad and ∩-shaped in the females. The abdominal appendages are much reduced, the male having only two pairs, which are used in reproduction, the female possessing four pairs modified for carrying the eggs (fig. 303). The antennae are short and the antennules fold back into depressions in the carapace. They are situated between the eye and the median line of the body. The eyes are stalked and often extend from sockets in the carapace into which they can be retracted. The third maxillipeds are broad and flat and fold neatly against the ventral surface of the cephalothorax. The first pair of legs form chelipeds, often of great size and, in the case of a large individual, capable of pinching severely. The sixth leg-segment, the propodus, with few exceptions, has the outer angle produced to form the pollex or thumb; opposed to this is the movable dactyl or seventh segment (fig. 302). In

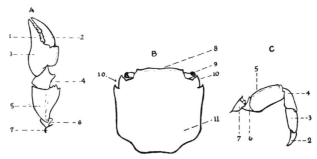

FIG. 302.—Diagrams to illustrate the terms used in connection with the carapace and limbs of the crabs (drawn from *Pachygrapsus crassipes*). A. External view of a cheliped (left). B. Dorsal view of a carapace. C. Dorsal view of a fifth leg (right).

1. Pollex.
2. Dactyl.
3. Propodus.
4. Carpus.
5. Merus.
6. Ischium.
7. Coxa.
8. Front.
9. Eye on stalk.
10. Post-orbital spine in antero-lateral region.
11. Branchial region.

many species the chelipeds, especially of the male, are unequal in size, the male fiddler crab being an extreme example of this kind. As a rule the walking legs are much alike (usually not folded up over the carapace or tucked under the abdomen as in the crab-like *Anomura*) but in one group the posterior pair is flattened and adapted for swimming (fig. 339).

Crabs move sideways, the legs on one side pushing while those on the other pull the body along. As there are eight of them used in locomotion the movement is continuous and uninterrupted. Many species, particularly certain active shore forms, will scuttle away in a very lively manner if disturbed.

The eggs are carried between the thorax and abdomen, attached to the modified abdominal legs (fig. 303). There they form a mass of such size that the abdomen is prevented from closing upon the carapace. When hatched, the young larva appears as a zoea (fig. 304), a minute, transparent form so different from the adult that it was originally described as another species of crustacean —something quite apart from its parents. It is ornamented with one or often two long spines, one pointing forward, the other backward, and is furnished with two relatively huge compound eyes located upon short stalks. After several moults the larva appears as a megalops (fig. 305) which in turn transforms into a miniature of the adult.

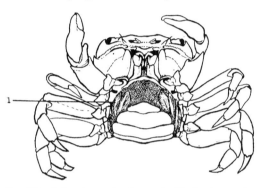

FIG. 303.—A female purple shore-crab, *Hemigrapsus nudus*, showing the egg mass held in place by the pleopoda and the abdomen; x½. 1. One of the pleopoda.

After this metamorphosis, growth continues at intervals succeeding each moult. The body covering is chitinous, but it is stiffened with a deposit of calcium carbonate (lime) to form a hard shell. Were it not for the little spaces at the joints which are free from the deposit of lime the animal would be perfectly rigid. As any increase in size is impossible within such an unyielding structure the covering is periodically shed and replaced. As this exoskeleton is not truly a part of the body but is formed by a secretion from the underlying membrane, the animal is able to loosen it while preparing a substitute beneath. Finally, the old covering is cast off and the crab appears soft and tender, without the

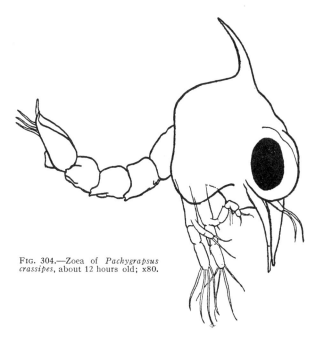

Fig. 304.—Zoea of *Pachygrapsus crassipes*, about 12 hours old; x80.

protection normally afforded by the hard parts, but capable of considerable growth in the few days before the new covering hardens (figs. 306 and 307). Individuals in this condition are commonly called "soft-shelled" crabs. Some of the decapods, in moulting, develop a crack along the dorsal part of the cephalothorax through which the body and appendages are drawn, but the true crabs, when shedding, emerge through a slit-like opening on the back between the carapace and abdomen. Cast exoskeletons, slit in this manner, are often found on the beach.

The width of the carapace, which exceeds the length in many forms, is largely due to the broad gill chambers at the sides (fig. 308). Properly, these chambers are not within the thorax, but are formed by an extension of the upper part of the carapace which projects outward and then turns downward and inward some distance toward

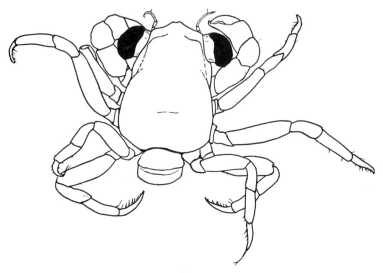

FIG. 305.—Megalops of *Pachygrapsus crassipes*; x9.

the mid-ventral line, forming the side and roof of a space large enough to inclose the gills and to allow considerable room for the circulation of water. The floor of this cavity is formed by the side of the thorax. The gills attach at the base of the legs and maxillipeds and have a feathery appearance. Water passes into the cavity through the slit which separates the side wall of the chamber from the thorax proper, and comes out through a narrow aperture just anterior to the mouth.

The heart is located back of the mid-dorsal region and immediately below the carapace. The almost colorless blood is forced out into arteries by the rhythmic beating of the heart and passes into a series of spaces known as sinuses and thence, by blood vessels, to the gills. Here it obtains a supply of oxygen and releases carbon dioxide, returning to the heart by veins and sinuses. As in the spiny lobster, the blood does not pass directly into the heart proper, but lies

Fig. 306.—Young specimens of *Cancer productus*. The small specimen shows the striped pattern often seen in young individuals of this species. Below are dorsal views of a specimen that has just moulted (left), together with the cast-off exoskeleton (right). Note the increase in size which followed the moulting process; x4/5.

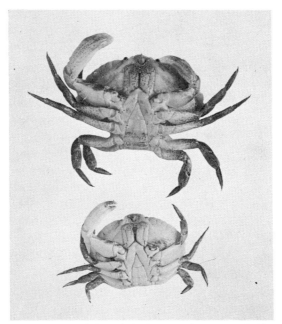

Fig. 307.—Ventral view of the larger specimen shown in Fig. 306 and its cast exoskeleton (below). It will be noted that the crab before moulting had but one claw and after moulting had two, the new one smaller than the other. The crab emerged through an opening at the posterior margin of the carapace, leaving the cast-off exoskeleton otherwise intact; x4/5.

in a thin-walled sac surrounding that organ, called the pericardial sinus (peri, around; cardium, heart). Passage to the heart is effected by means of openings supplied with valves to prevent a reversal of the current.

Ordinarily, crabs are scavengers, though there is considerable variation in their food habits, some even being vegetarians. The food is held by the maxillipeds and nibbled off by the mandibles, two hard structures at the sides of the mouth, passing thence into the stomach through a short gullet or esophagus. The stomach is not properly a digestive organ but is provided with hard ridges and tooth-like structures controlled by muscles that grind the food until it is very fine. It is then strained through the posterior part which is lined with a growth of setae, or hair-like projections, on the inner surface. At the sides of the stomach are two glands, yellowish in color and composed of a mass of tubes ending blindly like the fingers of a glove. These digestive glands are sometimes called livers but differ in function from the liver of vertebrate animals. The fluid they secrete enters the intestine where it makes the food soluble. The digestive tract is surrounded by a cavity filled with blood which absorbs the products of digestion.

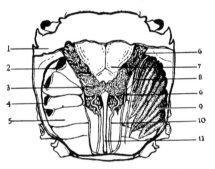

FIG. 308.—*Pachygrapsus crassipes* (male) with top of carapace and heart removed to expose the structures below. On the left side the gills have also been removed, leaving the gill cavity empty; x1.
1. Anterior part of carapace left intact.
2. Gill slit.
3. The gonads.
4. Apertures through which the gills attach to the legs.
5. Inner wall of the carapace.
6. The liver.
7. Top of stomach.
8. Whip-like process connected with maxilla.
9. Gills
10. Vas deferens.
11. Intestine.

Above the liver and posterior to the stomach lie the gonads in which the reproductive cells are produced (fig. 308). The part of the thorax which lies below the gill chambers is occupied principally by the muscles which move the legs. This mass of white muscular tissue is the edible part of the market crab. Excretion is carried on by small glands situated at the base of the antennae and opening to the exterior by minute pores. The antennules each have a small sac-like structure within their basal portions that is thought to function as an auditory organ.

The nervous system in its essentials resembles that of the lobster, consisting of an anterior ganglion or "brain" and a ventral nerve cord composed of two parts which encircle the gullet (fig. 309). The large size of the thoracic ganglion, however, and the nerves radiating from it as a center, is in marked contrast with the linear nervous system of the *Natantia*.

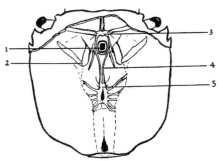

FIG. 309.—*Pachygrapsus crassipes* dissected to show principal parts of the nervous system; x1. 1. The oesophagus. 2. A supporting framework connected with the mandibles. 3. Brain. 4. Ventral nerve cords. 5. Thoracic ganglion.

The *Brachyura* include the largest of the crustacea, the great spider crabs, which are brought up from the depths of the Pacific Ocean. These crabs measure 12 feet or more across, including the extended legs. The following account deals briefly with our more common shallow-water and shore forms.

Systematists divide the *Brachyura* into three subtribes based on the structure of the mouth parts and legs and the number of gills. They are the *Oxystomata, Dromiacea,* and *Brachygnatha*. The following key (adapted from Miss Rathbun),[*] with the help of the drawings accompanying it, will serve to distinguish them. The last group is the largest

---

[*] M. J. Rathbun—The Grapsoid Crabs of America; U. S. National Museum Bulletin No. 97, p. 14.

and, with a single exception, all the crabs described in this volume belong to it. There are two superfamilies.

    A. Mouth field prolonged forward to form a gutter. Last pair of legs either normal or abnormal. Gills few.
        Subtribe *Oxystomata* (narrow-mouthed crabs) (fig. 310).

FIG. 310.—An oxystomatous crab, *Randallia ornata* (female); x⅔.

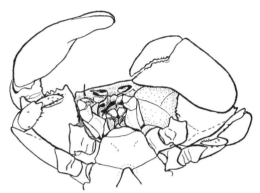

FIG. 311.—Mouth parts of a brachygnathous crab, *Pachygrapsus crassipes* (male); x1.

    B. Mouth field roughly square.
        a. Last pair of legs abnormal, dorsal. Gills usually many. Subtribe *Dromiacea*.
        (The members of this group on our coast are uncommon forms and therefore not treated in the text).
        b. Last pair of legs normal, rarely reduced, not dorsal (except in some forms outside our territory). Gills few. Subtribe *Brachygnatha* (short-jawed crabs) (fig. 311).
            1. Forepart of body narrow, usually forming a distinct

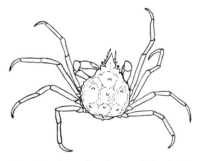

FIG. 312.—An oxyrhynchous crab, *Inachoides tuberculatus* (female); x1.

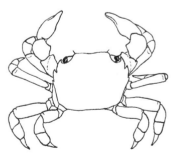

FIG. 313. — *Hemigrapsus oregononsis* (young male). An example of the *Brachyrhyncha*; x⅔.

rostrum. Body more or less triangular. Orbits generally incomplete.
Superfamily *Oxyrhyncha*
(narrow-nosed or spider crabs) (fig. 312).
2. Forepart of body broad. Rostrum usually reduced or wanting. Body oval, round, or square. Orbits nearly always well inclosed.
Superfamily *Brachyrhyncha*
(short-nosed crabs) (fig. 313).

Subtribe—OXYSTOMATA

Family—LEUCOSIIDAE

Randallia ornata (Randall), the purple crab (fig. 314), lives along the entire coast of California and southward. The carapace is nearly hemispherical and convex dorsally, giving the animal an inflated appearance. There are two prominent teeth on the posterior margins of the carapace, two others a little farther forward and a number of denticles along the sides. The antennae are minute. The color is light, mottled with blotches of red. Specimens are frequently brought in by fishermen. At Balboa we have found them on the shore of the bay, at exceptionally low tide (length and breadth, 48 mm.).

Subtribe—BRACHYGNATHA

Superfamily—OXYRHYNCHA

Family—PARTHENOPIDAE

Heterocrypta occidentalis (Dana) (fig. 316) is easily recognized by its triangular carapace, the margins of which extend over the walking legs, partly hiding them from view, and by the long chelipeds with short "thumb and finger." The carapace is marked with prominent ridges of a purplish color, bearing numerous white tubercles. There are also tubercles on the margins of the carapace and the ridges of the chelipeds. The general color is pinkish, somewhat mottled with lighter spots. *H. occidentalis* is reported

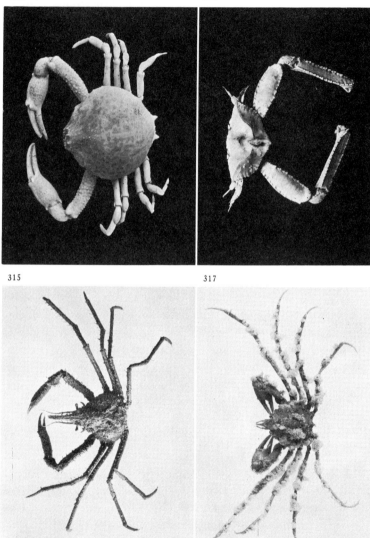

FIG. 314.—The purple crab, *Randallia ornata;* x5/6. FIG. 315.—*Oregonia gracilis.* The usual covering of algae, sponges, Bryozoa and other fixed forms has been cleared away. The long rostrum with the two spines almost parallel is a prominent character; x½.
FIG. 316.—*Heterocrypta occidentalis;* about x⅔. FIG. 317.—*Inachoides tuberculatus,* decorated with sponges; x5/6.

from Monterey, Avalon, and La Jolla. We have found it in the bay at Balboa when the tide was very low. Dimensions: carapace 26 mm. wide, 17 mm. long.

Family—INACHIDAE

(*The spider crabs*)

The graceful decorator crab, *Oregonia gracilis* Dana (fig. 315) is an abundant species on our northwest coast, where it may be found from low-tide mark to deep water all the way from Monterey Bay to Bering Sea. The carapace is triangular with the apex formed by the two spines of the slender rostrum, and is covered with stiff, recurved hairs and minute tubercles. There is a sharp spine curving forward a short distance behind each orbit and a spine is formed by the lengthening of the partition between the two pits into which the antennules fold. The chelipeds are long, especially in old males, roughened with fine tubercles and little hairs, and end in long, smooth, inward-curving fingers. The walking legs are slender and bear a few small hairs. The terminal segments are strongly curved. The grayish or tan body color is relieved with small spots of red; this color, however, and often the entire shape of the animal, is completely hidden by a mass of attached seaweed, sponges, and *Bryozoa*. The length from rostrum to posterior edge of carapace is 45 mm. in the specimen figured.

Inachoides tuberculatus (Lockington) (fig. 317) is found from Monterey Bay southward to Panama and is often abundant on the piling in harbors, as at San Pedro, where it may be seen crawling slowly about among the tunicates (*Ciona*) and the seaweeds growing on the wharf piles and floats. Like most of the spider crabs it covers the body with foreign growths, making it difficult to distinguish from its surroundings. The specimen photographed was covered with sponges. The anterior half of the rostrum is slender. The length of a male specimen is 11 mm., width 8 mm.

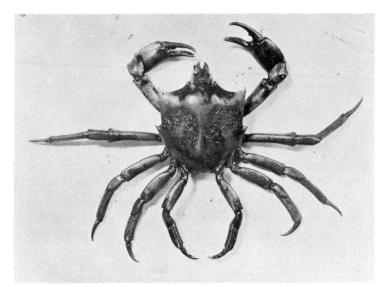

FIG. 318.—The northern kelp crab, *Epialtus productus*. The rough patches on the carapace are algae; x2/5. (Photographed by Ivan C. Hall.)

FIG. 319.—The southern kelp crab, *Epialtis nutallii;* x½.

Epialtus productus Randall, the kelp crab, (fig. 318) occurs abundantly from Alaska to Lower California. Small individuals are common on shore and in tide pools and adults are occasionally found in the same situations but they are usually hidden among the kelp offshore. The color is a reddish, or olive-brown, mottled with small round spots of darker shade. The ventral surface is lighter, strongly marked with red. Young or recently moulted crabs are not conspicuously spotted and are frequently light in color. Ordinarily, the carapace does not exceed 100 mm. in length; but Dr. Weymouth* reports a large male taken in Monterey Bay to have a length of 107 mm., width 93 mm., and chelipeds 195 mm. in length. The characteristic shape of the carapace will distinguish this species from *Epialtus nuttallii* Randall (fig. 319), its nearest relative on the Pacific coast.

The latter species may be called the southern kelp crab, as it is found south of Point Conception. It differs from *E.*

FIG. 320.—*Mimulus foliatus;* x5/4.   FIG. 321.—*Pelia tumida*, the dwarf crab; x4/3.

productus in having a more convex carapace, a more prominent rostrum with a small triangular notch at the tip. Furthermore, a small tooth found just in front of the eye

*F. W. Weymouth—Synopsis of the True Crabs (*Brachyura*) of Monterey Bay, California; Stanford University Publications, University Series, No. 4, p. 29. Dr. Weymouth tells us that the figure 170 mm. in the original is a misprint, 107 mm. being the correct figure.

in *E. productus* is missing in *E. nuttallii*. In color it is a dark red-brown.

Mimulus foliatus Stimpson (fig. 320) is a small crab found among the rocks at low tide. It has a bifid rostrum the horns of which are short, never exceeding one-fourth the total length of the carapace. The surface of the carapace is smooth and undulated, with a thin margin which is produced in a leaf-like expansion. The color is variable, sometimes being light red, tan, or purplish. The legs are often crossed by light bands. Adults may have a covering of bryozoan or sponge growth. It is found from Alaska to Mexico. The length of the carapace of a large specimen is 30 mm. and the width 32 mm.

Three small crabs of the genus *Pugettia* live along the coast. They may be found at low tide crawling among the seaweed or in clumps of eelgrass. The body is covered with seaweed, hydroids, and bryozoans in such profusion that it is difficult to distinguish the crabs from their surroundings. Stiff, recurved "hairs" upon the carapace serve to attach the algae and other decorations and the crabs are said to pinch off bits of seaweed and fasten them upon their backs. They are deliberate in all their movements. Stripped of their covering, one finds that they have a rostrum divided into two diverging prongs, also several prominent lateral processes. The three species differ only slightly and the following points together with the outline of the carapace (fig. 322) will be enough to distinguish them.

FIG. 322.—Carapace outlines of three species of *Pugettia*. *P. richii* (left). *P. dalli* (center). *P. gracilis* (right) (all from male specimens); x ⅔.

In *Pugettia gracilis* Dana (fig. 322), the second tooth

behind the eye is broad and triangular. The merus of the cheliped has a prominent irregularly toothed crest on its upper side. The species is found from Alaska to southern California. The length is 28 mm.

In *Pugettia richii* Dana (figs. 324 and 322), the carapace in the anterolateral region is dilated into two flattened horizontal spines. The merus of the cheliped has a few tubercles on the upper side but no crest. The color is reddish. It is found on the Pacific coast of the United States and southern Canada. The length is 37 mm.

Pugettia dalli Rathbun (fig. 322) is much smaller than the other two species. The anterolateral region of the carapace has one slender spine and a flattened, obtuse, oval tooth, not horizontal. The merus of the cheliped has an irregular crest on its upper and inner margins, and the hand is wider than that of *P. richii*. This species is found at San Pedro, Catalina Island, and southward. The length is 11 mm., length of cheliped 13 mm., and width of hand 3.3 mm.

Pelia tumida (Lockington), the dwarf crab, is a tiny alga-covered crab with a bifid rostrum (figs. 321 and 323). It has a pear-shaped carapace without spines but covered with pubescence which helps to hold the algae and debris in place. The walking legs are compressed and the margins have stiff hair-like setae. When turned on its back, one of these crabs will remain motionless with its legs upcurved and is almost indistinguishable from the surroundings. The length is 12 mm. and the width 8 mm. Southern California is the northern limit of the species. We once had a specimen which was decorated with red seaweed. We removed the decorations and put the crab in a small aquarium in which there were some brown seaweeds. The next day the crab was again decorated, this time with brown seaweed.

FIG. 323.—*Pelia tumida;* x1.4.

FIG. 324.—*Pugettia richii;* x5/4.

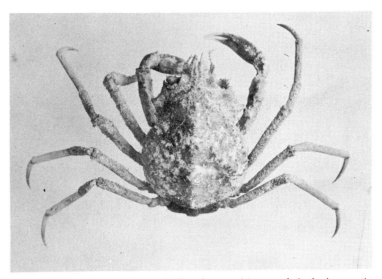

FIG. 325.—The toad crab or lyre crab, *Hyas lyratus*, with most of the foreign growths removed. The living crabs are usually completely hidden under a covering of attached animals and plants; x3/4.

FIG. 326.—*Loxorhynchus grandis*, the sheep crab, male; x¼.

FIG. 327.—*Loxorhynchus grandis*, the sheep crab, female; x½.

The range of the lyre crab or toad crab, *Hyas lyratus* Dana (fig. 325) extends from Puget Sound to Bering Sea. It is occasionally found along the shore at low tide in Alaska and is fairly abundant offshore. The carapace of a large male may reach 105 mm. in length. The rostrum is short with two prongs that curve inward like hooks. The contour of the carapace describes the outline of a lyre. A peculiar feature is the presence of small calcareous knobs on the ventral side, below and at the side of the base of the antennae, and the second joints of the antennae are expanded somewhat. There is a leaf-like projection just behind the eyes formed by the union of two spines. The carapace bears blunt tubercles. Like a majority of the species in this family, the toad crab is covered with algae, sponges, and hydroids. When the decorations are scraped away the color is seen to be reddish gray or tan, sometimes more or less distinctly marked with red spots on the carapace or red or gray bands upon the legs. The specimen photographed was an egg-bearing female taken by dredge among the San Juan Islands where it is common in deep water.

Loxorhynchus grandis Stimpson (figs. 326 and 327) popularly called the sheep crab, lives in deep water but is occasionally taken inshore or brought up by fishermen. The rostrum is armed with two divergent spines, and spines occur both in front of the eyes and behind them. The walking legs are progressively longer anteriorly. The carapace is inflated, ovate in outline, and sparsely covered with tubercles. The chelipeds are tuberculated and larger in the male than in the female, especially in older specimens. Owing to the habit of carrying living barnacles, bryozoans, and algae upon the carapace and appendages, and to its peculiar shape and deliberate movements, this crab presents a ludicrous appearance. It is a large species, reaching a length of 165 mm., and a width of 123 mm. It is found along the coast of California as far north as the Farallon Islands. A fosssil specimen has been discovered in the Miocene deposits of

FIG. 328—*Loxorhynchus crispatus*, the moss-covered crab. The specimen shown above, has been stripped of its moss covering while the one below is shown just as it was found, covered with seaweeds; x¾.

Fresno County, California, showing that it was resident there when the San Joaquin Valley was an arm of the sea.

The moss crab, *Loxorhynchus crispatus* Stimpson, (fig. 328) is often so completely covered up with hydroids, bryozoans, sponges, seaweed, and other growths as to lose all resemblance to a crab. All its movements are made with great deliberation. It is smaller than *L. grandis*, has fewer tubercles (9–12 usually), and has a short, thick, plush-like coat of hairs. The carapace is somewhat flattened and narrowly triangular. The walking legs are short for a spider crab. The slender chelipeds are much longer in the males than in the females, becoming remarkably long in old individuals. This crab ranges from the vicinity of San Francisco to San Diego and may be found from the tide line to a depth of 50 feet. Length of carapace of a large specimen (as given by Weymouth) 115 mm., width 84 mm.

Scyra acutifrons Dana (fig. 329), the sharp-nosed crab, is reported from Alaska to San Diego but seems to be more abundant north of Point Conception. The flattened rostrum is produced into two divergent horns. The carapace is pear-shaped with a very rough surface and marked with several elevations and depressions. The length considerably exceeds the breadth as the following measurements (from Holmes) indicate; length of carapace 35 mm., width 26 mm. The sharp-nosed crab is usually well covered with sponges, bryozoans, and hydroids.

Superfamily—BRACHYRHYNCHA

Family—CANCRIDAE

The genus *Cancer* includes the large edible crab that is found in the markets (fig. 335) as well as several smaller species that resemble it in form. Some of these smaller kinds are edible but are not taken in sufficient quantities to make them important from that standpoint. As there are nine species of *Cancer* found on the coast the following key

Fig. 329.—*Scyra acutifrons*, the sharp-nosed crab; x1.

Fig. 330.—The red crab, *Cancer productus;* x4/5.

(adapted from Weymouth) has been included to simplify the task of identifying them.

Dr. Weymouth* says that only a few of the nine species of *Cancer* are common. The edible *C. magister* is the largest and is the one seen in the markets. *C. productus* and *C. antennarius* are next in size and abundance. *C. gracilis* may be fairly abundant but all the others are either small or rare or both small and rare.

- A. Carapace nearly circular in outline without prominent angles at the sides; 12-13 teeth on the anterolateral margin.............................*C. oregonensis*.
- AA. Carapace markedly wider than long with a marked angle where the front and back margins meet; 9-12 teeth.
  - B. The front portion between the eyes extends forward and is made up of 5 nearly equal teeth; the finger is black on the inner margin...................*C. productus*.
  - BB. The front portion does not extend forward markedly and is made up of 5 teeth of unequal size and spaced at unequal distances; the fingers tipped with either black or white.
    - C. Fingers of the cheliped dark tipped.
      - D. The carpus of the cheliped has two spines, one above at the distal end and the other below on the inner angle.
        - E. The carapace is not hairy........*C. amphioetus*.
        - EE. The carapace is hairy.
          - F. The 10th tooth back of the eye is conspicuous and an 11th one is present....*C. gibbosulus*.
          - FF. The 10th tooth back of the eye is not conspicuous and an 11th one is not present..*C. jordani*.
      - DD. The carpus has a spine above at the distal end but none below on the inner angle.
        - E. Under parts are blotched with red..*C. antennarius*.
        - EE. Under parts are a uniform light color..*C. anthonyi*.
    - CC. Fingers of the cheliped white tipped.
      - D. Carapace with fine granulations; of the ten teeth on the margin back of the eye, the 10th is markedly the largest....................*C. magister*.

---

*F. W. Weymouth—Synopsis of the True Crabs (*Brachyura*) of Monterey Bay, California; Stanford University Publications, University Series, No. 4, p. 37.

FIG. 331.—*Cancer antennarius*, the rock crab; x2/5.

FIG. 332.—Ventral view of *Cancer antennarius*, showing characteristic spotting; x2/5.

DD. Carapace without granulations and the posterior tooth is not noticeably the largest....*C. gracilis.*

The red crab, *Cancer productus* Randall (figs. 306, 307 and 330) is used for food but is not abundant enough to be important commercially. It may be found from Lower California to Kadiak, Alaska. The surface of the carapace is smooth, without marked granulations. The anterolateral teeth are large and close together, giving a serrate appearance to the margin. The margin of the carapace between the eyes is produced forward and is made up of five nearly equal teeth that are equally spaced. The upper surface is dark red and the lower parts much lighter. Young specimens vary greatly in color and are often mottled or streaked. Length up to 103 mm., width 173.5 mm.

Cancer amphioetus Rathbun has prominent broadly triangular marginal teeth, nearly equal in size and the carapace is not hairy. It has two spines at the distal edge of the carpus of the cheliped. Found from San Diego southward. Length 24.2, width 33.4 mm.

The rock crab, *Cancer antennarius* Stimpson, (figs. 331 and 332) is dark red-brown, sometimes mottled with a yellowish tinge and the under surface is marked with red blotches. The walking feet and the margins of the abdomen are fringed with hair and the lower surface bears many hairs. The younger specimens are typically covered with a thick pubescence but this disappears from the upper surface with age. The large marginal teeth are bent forward slightly. The rock crab is an edible form but is seldom seen in the markets, probably because of the difficulty of obtaining sufficient numbers from the rocky shores it inhabits. Nevertheless, it is common along the coast of California in shallow water and is frequently obtained at low tide. The range extends from British Columbia to Lower California. Length (according to Holmes) 76 mm., breadth 113.5 mm.

Cancer gibbosulus (de Haan) has sharp anterolateral teeth, alternately large and small. The tenth tooth is conspicuous and an eleventh spine is present. The carapace is hairy. The carpus of the cheliped has a conspicuous spine near the hinge, another inside of this one at the end of a slight ridge, and a third below this on the inner side. There are spines on the finger of the cheliped. Found from Alaska to Lower California though it is not common on the shore. Length of a large specimen 46 by 32.5 mm.

Cancer anthonyi Rathbun (fig. 334) has but one spine on the distal end of the carpus of the cheliped and no spines on the hand. The under parts are not spotted. Color brownish red. Found south of San Pedro. Length 42.4 mm., width 65 mm.

Cancer jordani Rathbun (fig. 333) is a smaller species with strong, sharp, anterolateral teeth, alternately large and small (in small and medium-sized specimens only) but the tenth tooth is inconspicuous and there is no eleventh tooth. The carapace is hairy but there are no spines on the finger of the cheliped.

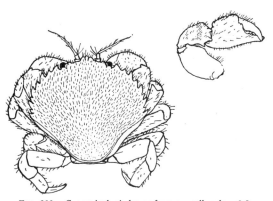

FIG. 333.—*Cancer jordani*, drawn from a small male; x2.8.

There are two spines at the distal end of the carpus of the cheliped. Found from Monterey Bay southward. Length of a large specimen 25.4 mm., width 33.4 mm.

Cancer magister Dana (fig. 335), the big crab, is one of the largest edible crabs on the Pacific coast and the one commonly found in the markets. It reaches a length of 120.7 mm. and a width of 177.8 mm. (Schmitt). The carapace is covered with small granulations and the anterolateral margins each bear 10 small teeth or denticles set

FIG. 334.—*Cancer anthonyi;* x1.

FIG. 335.—The edible crab, *Cancer magister*, young specimen. Ordinary specimens seen in the markets frequently measure more than seven inches (width of carapace).

at intervals along the edge, the posterior one being the largest. The color of the upper surface is a light reddish brown while the legs and ventral surface are more yellowish. While this species is said to occur from the Aleutian Archipelago to Magdalena Bay, Lower California, it is unimportant commercially south of Monterey Bay. It lives upon sandy bottoms and is not found upon the shore, at least in California. The ability to live in the sand, which it shares with several other kinds of crabs, is due to an arrangement whereby the sand particles are strained out of the current of water entering the gill chamber. The water passes through a crevice between the shell and the chelae, the latter being folded tightly against the carapace. The edge of the shell is provided with a number of teeth which keep out the larger grains of sand while below the edge of the carapace there is a patch of fine hair which acts as a strainer for the smaller particles. Many sand-dwelling species of crustacea show similar adaptations.

As the animal is carnivorous, the crab traps are usually baited with fish. Most of the commercial crab-fisheries use nets or traps (page 352). Spears and dip nets are useful to some extent in sheltered places. At San Francisco a large fishery exists on the bar beyond the Golden Gate in water ranging from 5 to 10 fathoms in depth. A similar fishery exists at Eureka. Altogether, the crab fisheries of California produced 33,373 dozen crabs in 1921, and Washington and Oregon added considerably to this total.

Unfortunately, the edible crab, like the Atlantic lobster, *Homarus americanus*, and the Pacific coast spiny lobster, *Panulirus*, is becoming scarcer from year to year and many of the old fishing grounds have been depleted. Laws have been enacted to protect the species and it is to be hoped that this valuable food resource will be conserved.

Cancer gracilis Dana (fig. 336) has much the same distribution as the preceding. In the north it may often be found along the shore at low tide but in the southern part

Fig. 336.—*Cancer gracilis*, male; x1.

Fig. 337.—*Cancer oregonensis*. This small *Cancer* may be distinguished from others of the same genus by the nearly circular outline of the carapace; x1.

of the range it usually lives in deeper water. The ventral surface is less hairy than in other representatives of the genus and the carapace, while finely granulated, is not tubercled. The walking legs are slender and graceful, giving the specific name to the form. It some respects much like *Cancer magister*, it may be distinguished by its smaller size (about 60 mm., instead of 100 mm. or more), the greater convexity of the carapace and the absence of tubercles upon its surface. The color is grayish or tan, dotted with small spots of red. Weymouth has found the megalops of this crab together with young ones in the adult form clinging to the underparts of jellyfish in Monterey Bay.*

Cancer oregonensis (Dana) is a small crab (fig. 337). It is reported to range from the Aleutian Islands to Lower California but is not common in the southern part of this area, though it becomes rather plentiful in the Puget Sound region. It is quite distinct in appearance for the antero-lateral and posterolateral margins do not form an angle as in other *Cancers*, giving a more circular shape to the carapace. There are 12 or 13 distinct, granulated teeth along the sides and the carapace reaches its greatest width at the 7th or 8th tooth. The walking legs are quite hairy and the chelipeds are large with dark colored claws. The general color is dark red above and lighter below, but the species is somewhat variable in this respect. Holmes gives the following dimensions for a male, length of carapace 17 mm., breadth, 21 mm. At very low tides this crab may be found buried in the sand or under rocks but it is more abundant in fairly deep water.

The horse crab *Telmessus cheiragonus* (Tilesius) is a peculiar form from the northwest coast (fig. 338). The carapace and walking legs are covered with bristle-like scales and stiff hairs arranged in irregular rows. Six large teeth along each side and three anterior ones give a saw-toothed effect to the margins. The color varies from a

---

*F. W. Weymouth—Synopsis of the True Crabs (*Brachyura*) of Monterey Bay, California; Stanford University Publications, University Series, No. 4, p. 42.

yellow-brown to a distinct red and the dactyls of the chelipeds are black for more than half their length. Dimensions:

FIG. 338.—*Telmessus cheiragonus*, male; x5/4.

length of carapace 50 mm., breadth 64 mm., according to Holmes. This crab may be found on the shore at low tide around Puget Sound but is not abundant.

Family—PORTUNIDAE

The swimming crab, *Portunus xantusii* (Stimpson), (fig. 339) is occasionally found swimming at the surface of the water. We have taken specimens from the mud of Balboa and San Diego Bays at low tide, also from Mission Bay, near San Diego. They are often seen swimming in the water or walking along on the bottom but when they try to evade a pursuer they flatten down and partly bury themselves in the soft mud until it is almost impossible to see them. The carapace is broad, flat, and finely pubescent and

FIG. 339.—The swimming crab, *Portunus xantusii;* x1. FIG. 340.—*Cycloxanthops novemdentatus;* x5/4.

each lateral angle is armed with a long spine. The last pair of thoracic legs are paddle-like and adapted for swimming. This species differs from the other forms treated in this volume in its manner of locomotion, being able to swim freely in the open sea. Though closely related to the blue crab of the Atlantic coast, the most important edible species of the eastern part of the United States, it is not sufficiently numerous to be exploited commercially. *P. xantusii* is recorded from San Pedro, Santa Catalina Island, Balboa Bay, La Jolla, San Diego, and thence south to Chile.

Family—XANTHIDAE

Cycloxanthops novemdentatus (Lockington) (fig. 340) is common in pools and under rocks at low tide. It inhabits the southern California shore and has been found as far north as Monterey. The carapace is wide and flattened, the surface finely granular and rugose in front, smoother posteriorly. The anterolateral border bears a row of small teeth. A large male measures 53.4 mm. long by 94 mm. wide. The color is dark, with a tinge of purple, and occasionally of red.

The members of the genus *Lophopanopeus* are smaller than *C. novemdentatus* and are rather plentiful on rocky shores. Figure 341 illustrates an adult egg-bearing female of *L. leucomanus* (Lockington), a species found from Monterey to Lower California. Though rarely seen at the northern limit of its range, where it is replaced by *L. heathii*, it is more frequent in the south. The carpus and the upper and outer surfaces of the claw are pitted with irregular depressions separated by a network of ridges. A plainly marked lobe occurs at the base of the hand. A brown-

Fig. 341.—*Lophopanopeus leucomanus*, female; x2.

ish, often nearly black, band crosses the thumb and finger but does not extend back upon the palm. The carpal segments of the walking legs have two small tubercles on the dorsal crest. The front of the carapace is marked by a very narrow median notch, a generic character. Average specimens vary from 10 to 15 mm. in width and slightly less in length. Individuals 18 mm. wide are reported by Holmes. The color characters are variable.

Lophopanopeus diegensis Rathbun has been reported from San Diego and Monterey. Instead of a pitted surface, the carpus and the upper part of the hand are tuberculated. Width of carapace 10 to 12 mm.

In the Monterey Bay region *Lophopanopeus heathii* Rathbun (fig. 342) is the common species. This form can be distinguished from the preceding two by the smooth carpus. The little lobe on the upper margin of the claw is present in this form also.

The black-clawed crab (fig. 343), *Lophopanopeus bellus* (Stimpson) is the only member of the genus in Puget Sound and neighboring regions. It inhabits rocky shores as far south as Monterey. The lobe or tooth on the upper margin of the hand is absent and both the hand and the carpus are smoother than in the species we have described. *L. bellus* is the largest, too, Rathbun citing the following dimensions for a large male: length 22.5 mm., width 33.8 mm. The color is subject to considerable variation, ranging from red-brown or purplish through shades of gray almost to white. The pattern is variable, the color appearing in irregular spots, but the black band across the finger and thumb is constant.

Lophopanopeus frontalis (Rathbun) of the San Diego region resembles the northern species in having smooth chelipeds and carpus, but possess a prominent tooth on the dorsal surface of the hand. It is the only species of this genus on our coast with the dark band of the thumb and finger extending back on the hand. Size up to 23.7 mm.

FIG. 342.—*Lophopanopeus heathii;* x9/10.   FIG. 343.—*Lophopanopeus bellus;* x9/10.

FIG. 344.—The lumpy crab, *Xanthias taylori;* x5/4.   FIG. 345.—*Pilumnus spinohirsutus,* the "hairy" crab; x5/4.

Xanthias taylori (Stimpson), the lumpy crab, can readily be recognized by the tuberculated chelipeds and carapace (fig. 344). Young individuals, 5 or 6 mm. across the carapace, have sharp curved teeth instead of the blunt tubercles. The legs are covered with stiff "hairs" or setae. The animal is dull red above, the lower surface being much lighter. The black fingers are characteristic. The favorite haunts of this species are holes and crevices in the rocks to which it retires when alarmed. Specimens may be obtained by turning over stones exposed at low tide. *X. taylori* ranges from Monterey to Lower California.

The San Diego region is the home of a peculiar little hairy crab (fig. 345), *Pilumnus spinohirsutus* (Lockington). Like the two species last described, it is very retiring, hiding in the sand under and among the rocks where the prevailing light-brown, rather sandy color helps to render it inconspicuous. Though confined to the southernmost part of our shore, it is fairly abundant in rocky places within this range, and can be taken at low tide. The carapace, walking legs, and chelipeds are covered with stiff setae as shown by the photograph.

Family—GONEPLACIDAE

Speocarcinus californiensis (Lockington) (fig. 346), the burrowing crab, ranges from San Diego to San Pedro on the

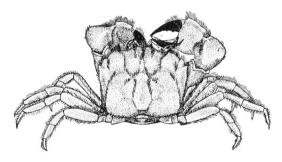

FIG. 346.—*Speocarcinus californiensis*, the burrowing crab; x1.

mud flats of shallow bays and inlets. Burrows are dug in the soft mud and the crabs may be found at the mouths of these with their claws upraised toward the intruder. Like the fiddler crabs, they commonly rear upward until the body is practically in a vertical position but are much less alert and active than their smaller relatives. The carapace is convex longitudinally, but almost straight transversely, and a prominent spine occurs at each anterolateral corner. Fringes of light-colored hair are found in front and along the sides of the carapace, crests of similar hairs mark the depressions on the upper surface, the eye stalks are similarly adorned and the walking legs are sparsely fringed. The animal is brownish or nearly white with the tips of the great claws black. A male specimen measures 16 mm. in length and 22.6 mm. in width.

Family—PINNOTHERIDAE

The crabs of this family are all of small size and the females inhabit the gill chambers of living mollusks and other marine animals. As they do no harm to the hosts that shelter them, at least in the sense of physical injury, and are not demonstrably helpful, they appear to have the status of commensals (literally, mess-mates). The males are free swimming and are rarely seen. They are apt to be considerably smaller than the females, a departure from the general rule among the *Brachyura*. The species are numerous and the group is widely distributed in shallow water. Some of the more common kinds are included here for the convenience of those of our readers who may be fortunate enough to find specimens.

Pinnotheres pugettensis Holmes is sometimes found in the gill cavities of tunicates from Puget Sound and neighboring regions. The convex carapace is soft and yielding, squarish in outline, with the front a little protruding and turned downward. The broad, nearly circular abdomen of the female covers practically all of the lower surface. The last pair of legs is the longest. The carapace may have a length of 10 mm. and a width of 10.5 mm.

Pinnotheres nudus Holmes, which lives in the mantle cavities of mussels, has a nearly square carapace with rounded corners. The front is not produced. This species is larger than *P. pugettensis*, specimens with a breadth of 24 mm., length 20 mm., being reported. It lives along the California coast.

Another crab that is found in mussels and clams is *Fabia subquadrata* (Dana) (fig. 347). The male is unknown but the

FIG. 347.—*Fabia subquadrata* (female) dorsal and ventral surfaces; x1.4.

female has been taken from Alaska to Laguna Beach. As in the genus *Pinnotheres*, the carapace is smooth, rather membranaceous, squarish, with the angles rounded off and the front turned downward. The abdomen is larger than the carapace. The distinguishing feature of the genus is the presence of two longitudinal grooves which extend from the orbits nearly half way back on the carapace. The palm of the chela has two rows of fine hairs on the lower margin, the outer extending to the base of the immovable finger, the inner reaching the tip. This crab is whitish, marked with orange spots. It reaches a width of from 12.5 to 13 mm. and a length of 10 mm. to 11.5 mm.

A similar crab, *Fabia lowei* Rathbun, occurs from Santa Monica to San Diego in the shells of living piddocks and mussels. It differs from *F. subquadrata* chiefly in two respects: first, a shallow transverse groove, which occurs in *subquadrata* and is covered with fine hair, is absent in *lowei;* second, the palm of the latter does not widen distally as it does in *subquadrata*. There are other minor differences, but the two cited are sufficient to distinguish the species. In color and dimensions, they resemble each other closely. The male is unknown.

About a dozen species of the genus *Pinnixa* exist along our coast. They are characterized by a short, wide carapace, and the relatively large size and great length of the third pair of walking legs. Owing to their small size and the fact that they inhabit the tubes of annelid worms and the cavities of bivalve mollusks and holothurians, they are easily overlooked. Specimens are not very commonly taken except by skilled collectors. The four most common kinds can be distinguished by the help of the following key adapted from Rathbun.*

    A. Dactyl of the third pair of walking legs straight, or slightly curved.
        B. Carapace nearly three times as wide as long...*P. longipes.*
        BB. Carapace not exceeding two and one half times as wide as long..............................*P. tubicola.*
    AA. Dactyl of the third pair of walking legs markedly sickle-shaped.
        B. Carapace oblong, about one and one-half times as wide as long................................*P. faba.*
        BB. Carapace pointed at sides, about twice as wide as long.
                                                                        P. littoralis.

Pinnixa longipes (Lockington) has the third pair of walking legs enormously developed (fig. 348). The carapace is about three times as wide as long. The last pair of legs are weak and small. It will be noted that the direction of elongation is the same as that of locomotion, namely lateral. This is of service in moving along the narrow worm tubes which form its habitat. Holmes† says of it, "This species lives in the tube of an annelid worm (*Clymenella*). It forms the extreme point of modification of this peculiar genus. There

---

    *M. J. Rathbun—The Grapsoid Crabs of America; U. S. National Museum Bulletin No. 97, p. 129.

    †S. J. Holmes—California Stalk-Eyed Crustacea; Occasional Papers of the California Academy of Sciences, No. 7, p. 92.

is probably no other crab which has such great width relatively to its length, there is certainly no known species

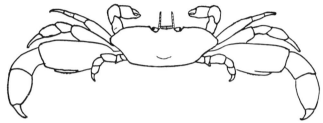

FIG. 348.—*Pinnixa longipes;* x5. (After Holmes.)

in which the fourth pair of pereopods is so enormously enlarged; and I believe there is no Brachyuran which exceeds it in smallness of size." Rathbun gives 2.2 mm. as the length and 5.7 mm. the width of an egg-bearing female. This strange little crab may be found along the coast of California from the southern part northward to Tomales Bay.

Pinnixa faba (Dana) is an orange-spotted form that ranges from Prince of Wales Island, Alaska, to northern California. It occurs in the shells of bivalve mollusks. Rathbun gives the following measurements: "Female (17468), length of carapace 15.2, width of same 22.8 mm." The male is smaller.

Pinnixa littoralis Holmes (figs. 349 and 356) is also found in the mantle cavities of clams and other bivalves. It is gray or greenish white with brown bands on the walking legs and red spots on the anterior part of the carapace. The size is about the same as that of the foregoing species. The range is from Sitka, Alaska, south to San Diego, California. Large specimens commonly live in pairs

FIG. 349.—*Pinnixa littoralis.* Female at left, male at right; x1.

within the mantle cavity of the giant gaper, *Schizothaerus*. In the Puget Sound region a large percentage of the gapers are so inhabited and smaller specimens occur in *Macoma* and *Mya*.

Pinnixa tubicola Holmes inhabits the burrows of sand worms from Puget Sound to San Diego. We have taken it under rocks on Brown Island, near Friday Harbor, Washington. The color is light brown with gray spots on the dorsal surface. The dimensions of the female are as follows: width 10 mm., length 4 mm., length of third walking leg 10 mm. The carapace of the male may be a trifle wider, and one of our specimens measures 4 mm. by 8 mm.

Two other commensal crabs of this group should be mentioned. *Scleroplax granulata* Rathbun has a hard, strongly convex carapace. This is smooth in the center but finely granulated toward the margins. The third pair of walking legs are the longest, but markedly so, and the fourth pair are the smallest, yet not much reduced. The color is gray-white. The range extends from British Columbia to Lower California. The females make their homes in the mantle cavities of mussels and other bivalve mollusks.

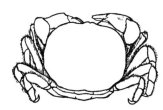

FIG. 350.—*Opisthopus transversus* (female); x1.6.

Opisthopus transversus Rathbun (fig. 350) has a hard, slightly oblong carapace without any granulation or tubercles. It is highest in the middle portion and slopes downward on all sides from this elevation. The polished surface is covered with a network of reddish lines and spots on a light background. The walking legs are nearly equal, the second pair being the longest. All have hairy edges. The female of this handsome species reaches a length of carapace of 14 mm. and a width of 18 mm. The male is smaller but may exceed 11 mm. in width. Specimens have been found in the sea cucumber (*Stichopus*), in the siphons of piddocks (*Pholas*), and in the mantle cavities of clams and keyhole limpets (*Megathura*). The species ranges as far north as the region about Monterey.

Family—GRAPSIDAE

The striped shore crab, *Pachygrapsus crassipes* Randall (figs. 354, 355 and frontispiece) is the most common representative of the *Brachyura* along the California coast. Scurrying among the rocks, running sideways over the sand, or congregating about a dead fish (for they excel as scavengers), these crabs are both numerous and lively. On mud flats exposed by the tide and in the channels which wind among them they are commonly found in company with species of the genus *Hemigrapsus* but are seldom seen on long stretches of sandy beach. They can remain out of water for a considerable time and may frequently be found at a distance of several feet from the water. Some tropical species of *Grapsidae* have come to live inland as land crabs and others have adopted a fresh-water habitat.

The carapace is rather square, the sides converging slightly posteriorly, and the upper surface is transversely striated anteriorly. The stripes and markings vary through red, green, and dark purple. The chelae are marked with a network of fine purple lines. A large male will attain a length of 34.5 mm. and a breadth of 40 mm.

Many call this the rock crab, a name that seems very suitable considering the favorite habitat, but it should not be confused with the edible rock crab, *Cancer antennarius*, a larger species which prefers deeper water. The striped shore crab occurs from Oregon southward to the Gulf of California on our side of the Pacific and is reported from Japan and Korea on the other side.

Two other shore crabs are found upon the western coast of the United States and Canada. The first of these, the purple shore crab, *Hemigrapsus nudus* (Dana) is found in the same situations as the species described above. The carapace is about equal as to length and breadth, smooth and convex forward (fig. 351). The corners are rounded, the anterior ones with conspicuous, forward-pointing spines. The males bear a patch of soft hair on the inner sides of the chelae. The walking legs are without hair. Color and markings vary greatly, greenish yellow, reddish brown, and purplish specimens appearing in the same vicinity, though the purple tinge is prevalent. The red spots on the claws seem to be constant and serve to distinguish it from the yellow shore crab. Specimens are usually not over 40 mm. across the carapace. *H. nudus* appears to be rather scarce on the southern California shore but is a more common species to the northward.

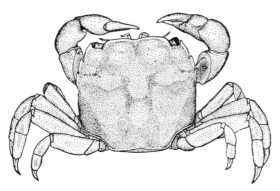

FIG. 351.—*Hemigrapsus nudus*, the purple shore-crab (female); x⅔.

Figure 352 shows *Hemigrapsus oregonensis* (Dana), variously called the yellow shore-crab, the hairy shore-crab, and the mud crab. There is a close resemblance to *H. nudus* in size and form but not in color. The round, red spots which characterize the chelae of the latter are absent on this species. Moreover, the walking legs of this form are noticeably hairy, whereas they are naked in *H.*

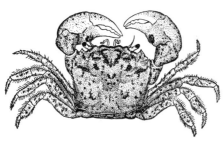

FIG. 352.—The yellow shore-crab, *Hemigrapsus oregonensis* (male); x1.

nudus. The mud crab is yellow (with a tendency toward buff) or gray, the carapace and legs mottled with brownish purple, or black spots and the tips of the claws light yellow or white. The under surface is white. A large male might measure as much as 30 mm. in breadth. The females are smaller. The species is found inhabitating mud flats from Alaska to Mexico. Great numbers, literally swarms, may be seen in such locations, scampering about or snapping their chelae at the intruder. Like the related forms, the mud crab is a scavenger.

While not a common crab upon the shore, where, indeed, its occurrence is exceptional, *Planes minutus* (Linnaeus) (fig. 353), merits consideration for several reasons. In the first place, it is more widely distributed than any other form we have here, living in all temperate and tropical seas. Moreover, it is a pelagic species, at home on the surface of the open sea where it finds shelter among the floating masses of gulfweed or *Sargassum*. It has also been found on drifting logs, on jellyfish, and turtles. The members of the famous *Challenger* Expedition found it attached to *Janthina*, a pelagic gastropod mollusk with a frail purple shell.

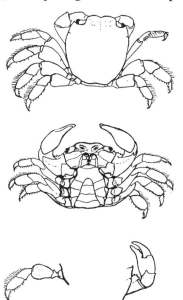

FIG. 353.—The pelagic crab, *Planes minutus*. Male. Dorsal view (above). Ventral view (center). Posterior left leg. Right cheliped. One of the walking legs is broken off in this specimen; x1.3.

In keeping with its drifting environment *Planes minutus* is a good swimmer, differing markedly in this respect from the shore crabs of the genera *Hemigrapsus* and *Pachygrapsus*

FIG. 354.—The striped shore-crab assumes a threatening attitude when he has no chance for retreat; x⅓. FIG. 355.—The striped shore-crab backs into a crevice when danger threatens; x⅓. FIG. 356.—*Pinnixa littoralis*, from the mantle cavity of the gaper. Dorsal and ventral views of the female. Dorsal view of the male in the center; x5/4. FIG. 357.—The fiddler crab, *Uca crenulata*. A. Anterior view of a male; x1. B. Anterior view of female; x1. C. Two male fiddler crabs fighting.

to which it is closely related. The latter, like most crabs, are creeping forms in the sense that they require a substratum to which they cling or upon which they walk in their peculiar sideways manner. *Planes*, though walking in the usual way, can swim straight ahead. This is done by using the four pairs of walking legs as oars, sweeping backward and forward in unison. For this purpose they are much flattened and bear crests of thickly set white hairs. Owing to their lack of color the fringes are inconspicuous but they probably afford considerable aid in swimming. We have noticed that a specimen kept in the laboratory held its chelae pressed tightly against the ventral surface of the body while swimming, the attitude closely resembling that shown in the upper figure.

The color is variable, ranging from a pale greenish yellow to an olive-brown and the carapace may be mottled with irregular whitish blotches or with darker areas. In a male specimen taken by us on the beach near La Jolla, the body color was a kelp-brown with indistinct patches of a slightly darker shade occupying a central position on the upper surface of the carapace. There were tiny flecks of white on the part above the gill chambers. The ventral surface was white. The legs and chelipeds resembled the carapace in color but were somewhat lighter, the hands tending toward yellow. The eyes were light green. This specimen, now in the museum of the Scripps Institution at La Jolla, measured as follows: length of carapace 14 mm., width the same, length of hands along the dorsal crest 9 mm. each. Some specimens reach a length and breadth of 18 mm.

Family—OCYPODIDAE

The fiddler crabs are easily recognized by the one huge claw of the male (figs. 357 and 358). The female has two small chelae of equal size. These crabs dig burrows in the sand far up where only the high tides reach them.

The following account is written from our observations of colonies of *Uca crenulata*, on the shores of Mission Bay.

As one approaches a fiddler crab colony on the beach, the animals may be seen in large numbers brandishing the big claws in a peculiar fashion but when one has reached a spot which a few moments before was well supplied with crabs, not one is to be seen for every crab has retreated into his hole. If the observer will seat himself close to a promising colony and remain quiet for five or ten minutes, the little fellows will begin to come out, hesitatingly at first, but soon more freely, until they are apparently continuing their regular program, uninfluenced by the presence of the intruder so long as he remains motionless.

We have watched colonies thus and have counted as many as thirty-five crabs in an area about six feet square. Some are digging their burrows, bringing out balls of sand and piling them around the mouth of the burrow. The sand is loosened by the claws of the anterior legs and brought up by them. If the crab emerges from the burrow with the left side of the body leading, the ball of sand is carried by the first three walking legs of the right side. The males eat with the small hand only, while the females can eat with both hands. They scoop up the sandy mud and the mouth parts seem to sort it over and reject the hard particles which collect in a mass just below the mouth. Later these little balls either drop down to the ground or are handed down by the cheliped. In this way they feed on the tiny animals and plants that are in the sand. The males are frequently seen to brandish the large claw in a peculiar way. First, they reach out with it as far as it will go, then they bring it in toward the body with a sudden movement. This motion which has probably suggested the name of "fiddler" crab is carried on during the breeding season and is prosecuted more vigorously when a female crab is nearby.

Figure 357-c shows two males fighting. They lock the two large chelae together as if they were shaking hands and then each apparently tries to break off his opponent's claw by a sudden twist. Sometimes there is a tearing sound as though the claw had been broken off. We have heard the sound a number of times but we have never seen a claw really broken off. However, we have found severed claws on the beach which would indicate that the struggle sometimes ends disastrously. Frequently we have seen the combat terminated by one of the fighters losing his hold and being thrown back over his opponent for a distance of a foot or more.

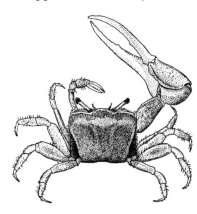

FIG. 358.—The fiddler crab, *Uca crenulata* (male); x1.

If the watcher moves suddenly, the crabs dart to their holes and are out of sight in a few seconds. If the movement is slow or slight, they merely stand at attention, remaining motionless until they decide whether to flee or to resume their usual activities.

We have been successful in keeping fiddler crabs alive for several weeks in captivity. Sand enough was brought from the bay shore where the crab colony lived to make a bank several inches deep in the bottom of a wash tub. Salt water was added every few days and was two or three inches deep in one side of the tub. During the day time the crabs made burrows, ate, fought, and carried on their normal activities except for the brandishing of the great claw; this we did not see them do after they left the bay shore.

The carapace of a fiddler crab is rectangular, wider than long, markedly convex, and smooth.

Uca crenulata (Lockington) (figs. 357, 358). A large male is 12.6 mm. long and 19.7 mm. wide with the large chela 36.3 mm. long. The outer surface of the palm rounds gradually into the upper surface forming a keel-like ridge. San Diego is the northern limit of the species.

Uca musica Rathbun ranges from Mexico northward, having been reported from as far north as Vancouver Island. A recent report states that they have been found at Balboa. The third to sixth abdominal segments in the male are almost completely fused. There is an oblique stridulating ridge on the inside of the large claw near the lower angle. When the cheliped is bent, this ridge plays against a line of granules on the surface of the first walking leg. The carapace of a male specimen is reported to be 8 mm. long and 12.9 mm. wide.

Class—ARACHNOIDEA

Order—PYCNOGONIDA

(The "sea spiders")

Pycnogonids remind one of "daddy long legs" or spiders. In a few species the legs grow to be an inch long but most of our common kinds are very small. Their size, their habit of feigning death, and their color help to make them almost invisible and accounts for the fact that many people never notice them at all. They live among seaweeds, bryozoans, and hydroids, in the tide pools, and among clumps of mussels and sea anemones on the piling of old piers. They are also dredged at greater depths.

This order of arthropods is variously classified by different writers. Owing to a certain resemblance between the larvae and the nauplii of crustaceans some have placed them in that class. Others have assigned them to an independent class, the *Pantapoda*, of equal rank with the *Crustacea* on the one hand and the *Arachnoidea*, which includes the spiders, mites, and other related forms, on the other. That they resemble both groups is evident, but their vernacular name leaves no doubt that the likeness which appeals to the public, at least, is the resemblance to the spiders. Many scientific workers agree, also, that the structure of the pycnogonids is virtually that of the spiders and classify them as an order under the subclass *Arachnida*. According to this view, they are marine arachnids with a large segmented cephalothorax and a minute, rudimentary abdomen. There are about 200 species known at present, most of them of small dimensions and spindling form.

The body of the pycnogonid is short and thin with four short processes on each side to which the eight long, thin legs are attached. In some species the legs are six or seven times as long as the body. They have four simple eyes which are situated on a tubercle on the first segment of the body. This tubercle is in some cases high and pointed and in others low and blunt. The anterior end of the body forms a proboscis. The animal takes its food with the suctorial mouth which is at the anterior end of the proboscis. The masticating apparatus is in the alimentary tract just beyond the sucking mouth. On each side of the proboscis, in some species, are the chelifori which bear pincers and just behind and below the chelifori, palps are sometimes found. The abdomen is narrow, unsegmented, without appendages, and sometimes projects upward at an angle. The stomach cavity has long branches out into the palps and legs. These branches can be seen through the body wall in many of the species. There are no special organs for breathing.

A pycnogonid living upon *Tubularia* has been seen to grasp the hydroid with its claws and suck the animal in bit by bit. Some of them swim by treading water or by kicking themselves along but most of them move by crawling slowly over the rocks and weeds. When disturbed they draw their legs up over the back and remain quiet and usually escape detection. Their legs seem to fold either way equally well. They resemble their surroundings very closely, the red forms being ordinarily found among red seaweeds and the white ones on a light colored background. They instinctively cling to things and a number in a dish together will soon form a ball by clinging to each other. Their legs are easily injured and quickly regenerated.

The male often carries the eggs about on the under side of the body until they hatch. In front of the first pair of legs are the "ovigerous legs," small appendages which are used for carrying the eggs. These appendages are present on the males and often on the females also. The males usually have cement glands in the fourth joints of the walking legs. The secretion from these glands cements the eggs together into masses and fastens them to the walking legs. The reproductive organs extend out into the legs and the openings are usually on the second joint.

Dr. Leon J. Cole* watched the egg laying of *Anoplodactylus* and says that the male climbed upon the female and crawled over her head to lie beneath her, head to tail, and then fastened his egg carrying legs into the mass of eggs as they were being laid. The whole operation required only about five minutes.

The egg masses look like little tufts of cotton and one male may carry several balls. Some larval pycnogonids from the first have the full number of legs while others have five or three pairs and still others have long tendril-like organs by which they attach themselves to the hydroids on which they live. Some pycnogonids do not go through a well marked larval stage.

Family—AMMOTHEIDAE

Chelifori are chelate only in young stages. Palpi and ten-jointed ovigerous legs are present in both sexes.

Ammothella bi-ungulata Dohrn, variety *californica* Hall (fig. 359), a straw colored pycnogonid, is found under stones

---

*L. J. Cole—Notes on the Habits of Pycnogonids; Biological Bulletin, Vol. 2, p. 205.

PHYLUM—ARTHROPODA

at low tide. The body length is nearly 3 mm. and the legs are about 4.5 mm. long. The palps have nine joints and the chelifori three. The eye tubercle is low and rounded. The legs are long but stout and have very few hairs except

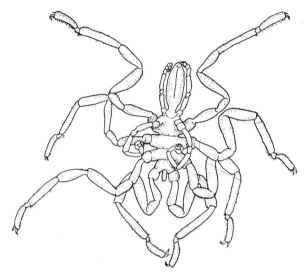

FIG. 359.—*Ammothella bi-ungulata* var. *californica*, under side showing ovigerous legs; x10.

those which are in a double row on the sole of the last segment. The branches of the stomach are easily traced in the legs when the animal is seen under the microscope. The males among our specimens were bearing egg masses in July (1918).

Tanystylum intermedium Cole (fig. 360) is one of the smallest pycnogonids. The body is disciform with the lateral processes close together and the body segments not well marked. Each of the first three pairs of lateral processes has a spine on the upper margin. The first joint of each leg has two spines, one at each side of the distal margin. The chelifori have two joints and pincers in immature specimens, but mature specimens show only a groove in the second joint where the chelae have disappeared. The palps are seven jointed and reach a little beyond the pear-shaped proboscis. Developing ova of various sizes may frequently be seen in the third and fourth joints of the legs of the female. This species is whitish and so tiny that it is difficult to find among the light-colored bryozoans upon which it lives. It is reported from the San Diego region and Laguna. Body 1 mm., leg 3 m m., extent 6.5 mm.

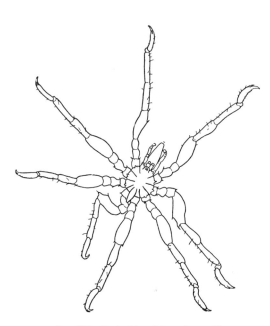

FIG. 360.—*Tantystylum intermedium;* x12.

FAMILY—PHOXICHILIDIIDAE

Chelifori are present and bear pincers. There are no palps and the five or six jointed ovigerous legs are present only in the male.

Halosoma viridintestinalis Cole (fig. 361) is a tiny species, white or light brown, which lives among colonies of bryozoans along the California coast. The first three pairs of lateral processes are close together, but there is a space between the third and fourth pairs. The first segment of the trunk projects forward beyond the base of the proboscis a short distance, making a thick neck of moderate length. The legs are smooth except for microscopic hair-like setae and the three setae on each leg as shown in the figure. The body covering is thin and transparent so that the light-green intestine with its prolongations into the legs shows through very distinctly. Length about 1 mm., first leg 3 mm., extent 7 mm.

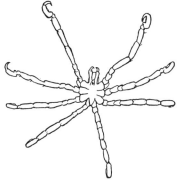

FIG. 361.—*Halosoma viridintestinalis;* x10.

Anoplodactylus erectus Cole (fig. 362) is much larger than the two preceding species, the body measuring 2.5 mm. and the legs having an extent of 19 mm. It is red with tints of green and yellow like the seaweeds upon which we find it. The body is long

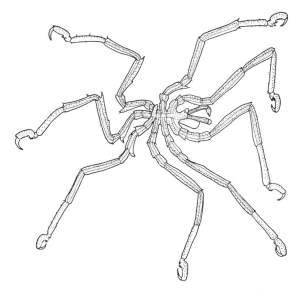

FIG. 362.—*Anoplodactylus erectus;* x5.

and slender with the lateral processes well separated at the bases. The abdomen is erect and the first trunk segment projects well beyond the base of the proboscis, forming a slender neck. The second leg joint of the male has a projection on the ventral side at the distal end. The female lacks this projection but the end of the joint is distinctly swollen.

This species is found in the San Diego region and at Laguna. It is reported as living in great numbers upon hydroids found on the piles at Balboa. Dr. Hilton* of Pomona College describes the life history as he has worked it out from specimens found on these hydroids. From his observations he concludes that the eggs are produced in the summer and early fall and the long armed larvae (fig. 363-b) are the

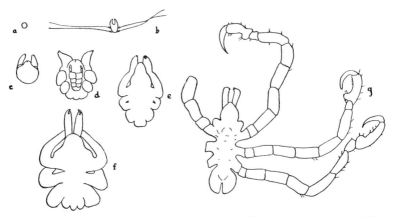

FIG. 363.—Stages in the development of the pycnogonid, *Anoplodactylus erectus*. The egg is shown in A. The form B takes up its abode within a hydroid and loses its long appendages. Forms C, D, E and F are found within the hydroid and G is found clinging to the tentacles of the hydroid; x23. (After Hilton.)

first stage after the egg. These take up their abode within the hydroids, at the same time losing their long appendages. Within the hydroids are found the forms shown in figure 363-*c*, *d*, *e* and *f*. Cast skins indicate that there are two moults within the hydroid. The next form with three pairs of well-developed legs, as shown in *g*, is much larger and is found clinging to the tentacles of the hydroids. By No-

---

*W. A. Hilton—The Life History of *Anoplodactylus erectus* Cole; Pomona College Journal of Entomology and Zoology, Vol. 8, pp. 25–34.

PHYLUM—ARTHROPODA

vember these had probably become adults for no immature stages were to be found.

Family—PYCNOGONIDAE

Chelifori and palpi are both absent and ovigera are present only in the male.

Pycnogonum stearnsi Ives (fig. 364) is a small species which is often found clinging to the base of the common

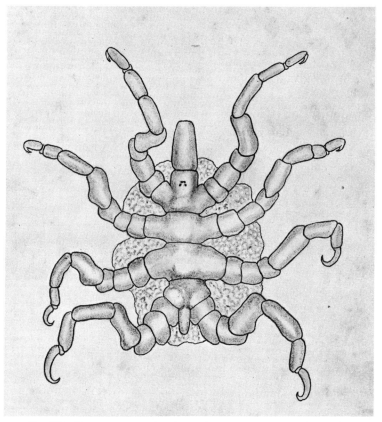

Fig. 364.—*Pycnogonum stearnsi*, male carrying eggs. (Drawn from life; x10.)

large sea anemone. It is readily recognized on account of its thick, broad trunk and stout legs with powerful claws.

Each of the trunk segments has a prominent tubercle in the mid-region and a smaller one at the outer edge of each lateral process. The eggs are carried on the ovigera in one or two large masses. The females are from 4 to 8 mm. long and the males a little smaller. It has been reported from Humboldt County to San Diego.

CHAPTER XI.

Phylum—MOLLUSCA*

(The chitons, snails, clams, and devil fish)

A mollusk is known by its soft, unsegmented body, a prominent part of which is the muscular foot used for locomotion. The body is, in most cases, protected by a limy shell which has been produced by the mollusk itself. In exceptional cases this shell is internal and a few adult mollusks have no shells at all though they may have had them in early life.

The beauty and variety of form of the shells of mollusks early attracted the attention of students, so that thousands of species have been described and named. The first writers based their descriptions on the shells alone, but later work showed that the shell itself could not be depended upon to show the true relationships. Accordingly the present day descriptions of groups include anatomical features of the soft parts which were often disregarded by earlier students.

---

*Authorities quoted throughout this chapter:

R. Arnold—The Paleontology and Stratigraphy of San Pedro; Memoirs of California Academy of Sciences, Vol. 3.

W. H. Dall—Summary of the Marine Shell-bearing Mollusks of the Northwest Coast of America, etc.; U. S. National Museum Bulletin 112.

Josiah Keep—West Coast Shells (1911 Edition); Whitaker and Ray-Wiggin Co.

Ida S. Oldroyd—Marine Shells of Puget Sound and Vicinity; Publications of Puget Sound Biological Station, Vol. 4, pp. 1–272.

E. L. Packard—Molluscan Fauna from San Francisco Bay; University of California Publications in Zoology, Vol. 14, pp. 199–452.

Julia Ellen Rogers—The Shell Book; Doubleday Page and Co.

F. W. Weymouth—The Edible Clams, Mussels and Scallops of California; California Fish and Game Commission, Fish Bulletin No. 4.

Dr. Fred Baker, Mr. F. W. Kelsey, and Mrs. Kate Stephens, conchologists of San Diego have given us data from their unpublished records.

The shell is secreted by the mantle, a flap-like fold from the dorsal side, which more or less completely envelopes the animal. A shell is bivalve if it consists of two halves and univalve when it is in one piece. One group, to which the chitons belong, has a shell made up of eight similar plates. The shell, if univalve, may be cylindrical with a hole at each end as in the scaphopods (fig. 478), or spiral as in the ordinary snail shells (fig. 484), or cap shaped as in the limpets (fig. 608).

Some forms, such as the slugs and snails, have a distinct head with mouth, tentacles, and eyes, while others, such as

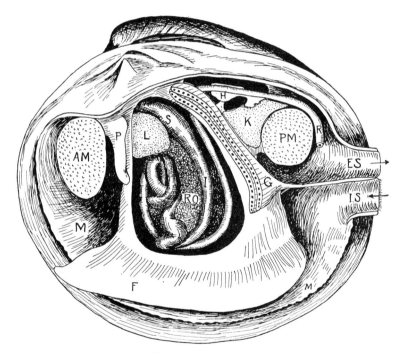

FIG. 365.—Diagram of a clam, *Chione undatella*, dissected to show the internal organs.

AM. Anterior adductor muscle.
ES. Excurrent siphon.
F. Foot.
G. Gills.
H. Heart.
I. Intestine.
IS. Incurrent siphon.
K. Kidney.
L. Liver.
M. Mantle.
P. Palps.
PM. Posterior adductor muscle.
R. Rectum.
RO. Reproductive organs.
S. Stomach.

the oyster and clam have no distinct head, the mouth being at the anterior end of the visceral mass and the tentacles and sense organs usually at the margin of the shell. The muscular foot forms a flat, sole-like creeping organ in the *Gastropoda* and *Amphineura*, and an axe-shaped digging organ in *Pelecypoda*, but its form in the *Cephalopoda* is quite different a portion of the foot being modified to form the long tentacle-like arms (fig. 676).

The classes of mollusks may be distinguished as follows:

Pelecypoda: With bivalve shell (clams and oysters).
Scaphopoda: With cylindrical shell (tooth shells).
Gastropoda: With a shell that is either conical or spirally coiled, a distinct head and one or two pairs of tentacles (snails and limpets).
Amphineura: Either with a naked body or with a shell made up of eight plates (chitons).
Cephalopoda: With a head which has long arms bearing suction discs (octopus and squid).

Class—PELECYPODA (LAMELLIBRANCHIATA)

(The clams and oysters)

The pelecypods are symmetrical mollusks that form bivalve shells, that is, shells made up of two similar parts. Clams, oysters, mussels, and the destructive shipworms are familiar examples of the class. They have no head, and the mantle, divided into a right and left lobe, secretes the shell which encloses the animal.

Pelecypods may be found in the shallow water all along the coast. Clams live in the sand of the smooth, sandy beaches, mussels and certain of the clams are attached to piling of the wharves or fastened to the rocks, while other clams live in the mud of bays and inlets, or even in burrows in rocks. Teredos bore into the wharf piles, and certain pectens attach themselves to the seaweed or kelp while others swim about by clapping their two shells together.

Many of the members of the group are commercially important in that they furnish food for man, in fact most kinds of clams, mussels, and oysters are edible. Pearls which are frequently found in them are cysts of mother of pearl, which have been formed around foreign objects that have washed in between the mantle and the shell. Larval forms of parasitic worms are often the nucleus about which the mother of pearl cysts are built. The material from which the pearl is made is the same as that which lines the shell of the mollusk.

When the animal moves about, the axe-shaped foot is thrust out between the two halves of the shell and is used as a very active and effective digging tool. Certain parts of the mantle edge are modified to control the incoming and outgoing currents of water, most species having two well-developed tubes or siphons (fig. 365). Through the lower one of the two siphons the water flows in to the mantle cavity bringing with it microscopic organisms which serve as food. Water is carried into the gills by the beating of numerous cilia, and it flows on out through the other siphon carrying with it the waste products. Siphons in some species are long and well developed, while in others they are scarcely more than a slight fusion of the mantle edges. In the oyster there are no siphons, the water merely flowing out in the region of the anus and in at other points. The current is kept up by cilia, tiny whip-like processes which continually beat in one direction and keep the water in motion.

The two halves of the shell are not always symmetrical and the thickness is variable in the different species. A thin, horny, outer layer or periostracum is present on some shells. It is a hairy layer in some instances, and in others looks like a thin coat of varnish. The inner, mother of pearl layer of the shell frequently is beautifully iridescent. This effect is due to refraction of light by the delicate plates of which it is made up. These fine lines on the iridescent surface may usually be seen with a hand lens. This pearly layer is

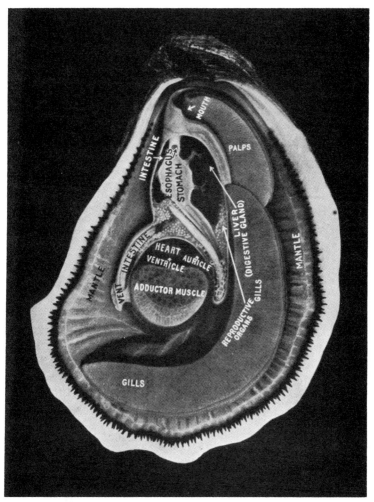

Fig. 366.—Model of an oyster. (From a photograph furnished by courtesy of the American Museum of Natural History, New York.)

secreted by the whole surface of the mantle while the middle and outer layers are secreted by the thick outer edge of the mantle. The umbo, or beak, of the shell is the oldest part of it, as will be seen if one studies the lines of growth which mark the successive additions to the shell (fig. 367).

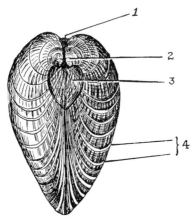

FIG. 367.—*Chione succincta* (anterior end); x1. 1. External ligament of hinge. 2. Umbo. 3. Lunule. 4. Lines of growth.

A ligament, made up of the external ligament and internal cartilage, is usually behind the umbones, and unites the two halves of the shell (fig. 368). It may be external, fitting into excavated portions of the edge of the valves, or it may be internal, in which case, it is contained in a groove or spoon-shaped pit. Interlocking teeth below the umbo, make up the hinge. These teeth and the impressions of the muscle and mantle attachments, which are features used in identifying shells, are shown in the diagrams (figs. 367 and 368). The adductor muscles are the ones which close the two valves of

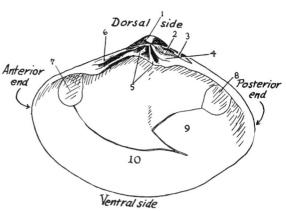

FIG. 368.—Inside of right valve of *Tivela stultorum* (small specimen; x⅔). The length is the antero-posterior measurement; the height or altitude is the dorso-ventral measurement.
1. Umbo.
2. Cartilage.
3. Ligament.
4. Posterior lateral tooth.
5. Cardinal teeth.
6. Anterior lateral tooth.
7. Anterior muscle scar.
8. Posterior muscle scar.
9. Pallial sinus.
10. Pallial line.

the shell. Some species have one adductor and some have two. When these muscles relax, the hinge ligament causes the shells to open and other muscles extend the siphons and the foot.

The thin leaf-like gills lie in the mantle cavity where they are washed by the water currents. The diagram (fig. 369)

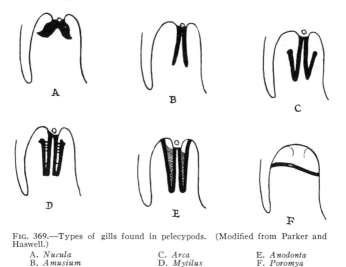

FIG. 369.—Types of gills found in pelecypods. (Modified from Parker and Haswell.)
    A. *Nucula*   C. *Arca*   E. *Anodonta*
    B. *Amusium*   D. *Mytilus*   F. *Poromya*

shows the different types of gills found among the pelecypods. The other organs are shown in figures 365 and 366. There are no horny jaws nor rasping tongues such as we find in the *Gastropoda*.

The group is poorly provided with sense organs, and has little need for them. The mantle edges and siphons are sensitive. Ocelli are often absent, but, if present, they usually are found around the edge of the mantle. These "eyes" probably serve only to make the mollusk sensitive to any change in the amount of light at the mantle edge. A faint shadow falling on a scallop is enough to cause it to close its valves quickly. The ocelli of this free-swimming form shine like tiny jewels (fig. 383) and may be seen clearly when the shells gape. Lithocysts or otocysts, organs for maintaining equilibrium, are present in the foot. Osphradia, which are thought to be olfactory organs, are patches of sensory epithelium which lie near the gills.

The reproductive organs usually open into the mantle cavity. The sexes are separate in most cases, but there are hermaphroditic species. Fertilization takes place in the water, sometimes within the mantle cavity. Egg cases so commonly seen among the gastropods are never formed by pelecypods. The eggs may remain protected in the mantle cavity during part of their growth period.

Order—PRIONODESMACEA

Family—NUCULIDAE

Nucula (Acila) castrensis Hinds, the camp nut-shell (fig. 370). The valves of this mollusk are convex and of medium thickness with the outside sculptured in such a way as to represent rows of tents. The hinge has a prominent internal cartilage-pit with many sharp saw-like teeth on each side of it. The altitude reaches 14 mm., the length 15 mm. The range is from Bering Sea to San Diego.*

Family—LEDIDAE

Leda taphria Dall, the grooved *Leda* (fig. 371), has oblong valves rounded in front but elongated and pointed behind. The surface is sculptured with sharp, concentric, raised lines. The hinge has a prominent internal cartilage-pit and about twenty sharp teeth on each side (R. Arnold). The altitude reaches 20 mm. The range is from Bodega Bay to Lower California.

Leda hamata Carpenter, the hooked *Leda* (fig. 372), has thin valves, the surfaces sculptured with strong concentric raised lines and a raised band passes from the umbo around the escutcheon to the posterior end. The escutcheon, the

---

*The ranges given for mollusks are taken from the publications of Dr. W. H. Dall unless otherwise noted in the text.

PHYLUM—MOLLUSCA 419

depression behind the beaks, is deep-set and smooth. Altitude 5 mm., length 8 mm. (R. Arnold). The range is from Puget Sound to Panama Bay.

Yoldia cooperi Gabb, Cooper's *Yoldia* (fig. 375), has thin and compressed valves, the surfaces sculptured with small

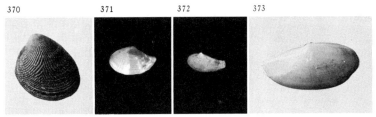

FIG. 370.—*Nucula castrensis*, the camp nut-shell; x5/4. (small specimen). FIG. 371.—*Leda taphria*; x1 FIG. 372.—*Leda hamata*; x1 (small specimen). FIG. 373.—*Yoldia limatula*; x5/4.

FIG. 374.—*Yoldia ensifera*; x5/4. FIG. 375.—*Yoldia cooperi*; x5/4. FIG. 376.—*Ostrea lurida*, the native oyster; x1.

concentric ribs. The epidermis is glistening olivaceous. The muscle scars are large and the hinge teeth numerous. Altitude 32 mm., length 64 mm. The range is from San Francisco to San Diego.

Yoldia ensifera Dall (fig. 374) has thin valves covered with a shiny olivaceous epidermis in which lighter and darker zones can usually be distinguished. There are about 30 teeth in front of the beaks and about 24 behind them. The beaks are less pointed than those of *Y. cooperi* and the umbones are more nearly central. It greatly resembles *Y. limatula* (fig. 373) but the latter is proportionately longer. A specimen of *Y. ensifera* 26 mm. long is 12 mm. high and 5 mm. in diameter. The length may reach 30 mm. The species has been found from Alaska to the San Luis Obispo coast.

Family—OSTREIDAE

(*Oysters*)

Many people think that there are no native oysters on this coast because many of those sold in the markets have come from the Atlantic. We have native oysters in abundance but they are too small to be of much commercial value. Huge oysters thrived here long ago when the ocean covered a large portion of what is now California. Le Conte gives the measurements of a fossil specimen of *Ostrea titan* as 13 inches high, 8 inches wide, and 6 inches thick. Fossil remains of large oysters have been found in the San Joaquin Valley, and on Coyote Mountain in the Colorado desert.

The eastern oyster, which is intermediate in size between these giants and our present day natives, has been introduced on our coast repeatedly, but does not seem to reproduce to any great extent. It is raised in considerable numbers in San Francisco and Tomales Bays from "spat" (as the small oysters are called) imported from the eastern coast.

The history up to the "spat" stage is interesting and a knowledge of it will make one understand the difficulties that attend the introduction of oysters into new locations.

It has been estimated that a large Maryland oyster may produce 60,000,000 eggs per year, while an oyster of average size produces 16,000,000 per year. The sexes are separate and the males produce spermatozoa in even greater numbers. Fertilization takes place in the water and the eggs may

PHYLUM—MOLLUSCA

easily fail of fertilization if the number of oysters is small or the genital products greatly scattered. The fertilized eggs develop into the ciliated larvae which swim at the surface for a time and are known as "fry." In this stage they are easily killed by sudden changes of temperature. Storms or unfavorable currents may cause them to drift off to neighboring beds. The oyster-grower is anxious about his charges until the free creatures have safely settled and become "spat." If they settle upon a muddy bottom they may easily become suffocated. It is customary to strew the bottom of a bed with old shells upon which the young oysters can safely settle and attach. The "fry" usually lie upon the left side and soon attach firmly to the support. At this time they are about $1/80$–$1/90$ of an inch in diameter. Oysters grow most rapidly in warm temperatures and where there is a good food supply. Under such favorable conditions they may reach $2\frac{1}{2}$ inches in length in 6–7 months. The density of the water is an important factor in the environment since the eastern oyster must live in brackish water. If there is a good seasonal catch, the "spat" will be too crowded so they are thinned out and the surplus, shells covered with the little oysters, may be sold by the bushel to be used in stocking new beds or replenishing old ones.

In regard to oyster culture in San Francisco Bay, *California Fish and Game* reports the following note by Mr. McKnew.[*] "The oysters which are brought from the Atlantic coast as seed oysters are shipped out in cold storage, taking about fifteen days for the trip. They are, at the time of planting, from $\frac{1}{2}$ inch to 1 inch in length, and on their arrival are planted in beds where they are left for three to four years. They are then tonged up, boxed and shipped to the markets." As to growth, Mr. McKnew says that oysters near Burlingame Point, planted in May, 1915, averaged 8,000 to 10,000 to the bushel and in August, 1916 averaged 1,100 to 1,200 to the bushel.

---

[*]California Fish and Game, Vol. 2, p. 208.

Edmondson* in a U. S. Bureau of Fisheries Report (1922) says, "That there are prospects of inducing the eastern oyster to propagate on the west coast is indicated by the spawning of the species in certain localities in Willapa Bay during the season of 1917."

California Fish and Game reports† that considerable damage is done to the clam and oyster beds, particularly in Tomales Bay, by the sting rays which during March and April move into the shallow water of the bay and remain through the summer.

Ostrea lurida Carpenter (fig. 376), the native oyster, has valves irregular in shape being sometimes circular and sometimes ellipsoidal and elongated. The surface is often made up of thin plates and may be irregularly plaited. The altitude reaches 60 mm. The range is from Sitka to Cape San Lucas.

Ostrea virginica Gmelin, the eastern oyster, widens gradually from the narrow pointed beak at the apex to the rounded end. The upper valve is the smallest and flattest with leaf-like scales and a leaden color. The altitude reaches 100 mm. and the species is found in San Francisco Bay, Tomales Bay, and Puget Sound where it has been introduced from the eastern coast (fig. 366).

Family—PECTINIDAE

The *Pectens*, or scallop shells, with their rounded valves and wing-like "ears" are found on nearly every beach. They have a small foot, and a single muscle with which to close the valves. They swim through the water with a jerky motion by clapping the two valves together.

Around the edge of the mantle of the *Pecten* are the ocelli, or tiny eyes (figs. 379 and 383), which often shine quite brightly. A shadow passing over is readily detected by the

---

*C. H. Edmondson—Shellfish Resources of the Northwest Coast of the United States; Dept. of Commerce Bur. Fish. Doc. 920, p. 17.

†California Fish and Game, Vol. 2, p. 157.

FIG. 377.—*Pecten hericius*; x⅔.   FIG. 378.—*Pecten hindsii navarchus*; x⅔.

FIG. 379.—*Pecten circularis*, the circular scallop, a living specimen showing the ocelli or eyes and the fringe of the mantle edge. (Photographed by Frank W. Peirson.)   FIG. 380. —*Pecten circularis aequisulcatus*, the speckled scallop; x⅔.

FIG. 381.—*Pecten latiauritus monotimeris*; x3/2.   FIG. 382.—*Hinnites giganteus*, the purple-hinged or rock scallop, a young, nearly symmetrical specimen; x1.

animal and the shells come together with a snap. Pectens make good food and are often seen in the northern markets. Among the largest *Pectens* are *P. caurinus* Gould from the north coast, a good sized specimen of which measures seven and a half inches in height and length, and *P. subnodosus* Sowerby from Lower California which frequently exceeds six inches in height. Present day *Pectens* of southern California are much smaller than this but large fossil shells which are numerous in this region often measure six inches across.

Pecten hericius Gould, the pink pecten (fig. 377), has long been known as a variety of *Pecten hastatus*. It is a northern form, found as far south as San Diego but dredged in large numbers at Puget Sound. Both valves have numerous, strongly spinose, radiating ribs. The color is pink to greenish and the left valve is the darker. An average specimen is 58 mm. high, 53 mm. long, and 16 mm. in diameter.

Pecten hindsii navarchus (Dall) has also been known as a variety of *Pecten hastatus*. The left valve is pink, the right one lighter colored. The radiating ribs are more numerous than those of *P. hericius* and the spines are much smaller, the right valve having few spines (fig. 378). The species is found from Bering Sea to San Diego and is dredged in considerable numbers in Puget Sound. One specimen measures 62 mm. high, 61 mm. long, and is 21 mm. in diameter.

Pecten caurinus Gould, the weather-vane scallop, has thin, flat, broad valves with about twenty squarish ribs. The ears are small and the shell is white inside. Outside the upper valve is purplish red and the lower one white with reddish margins (Oldroyd). A large specimen measures 190 mm. high and wide. The range is from Alaska to Oregon.

Pecten latiauritus Conrad, the broad-eared scallop (fig. 383), has thin, delicate shells with the two valves unequal and not exactly symmetrical. There are 12–16 low, squarish ribs separated by channeled interspaces. The ears are wide and pointed above. The color is mottled brown and white. The altitude reaches 23 mm., the length

25 mm., hinge line 20 mm. (R. Arnold). The range is from Monterey to Lower California. The variety *monotimeris*

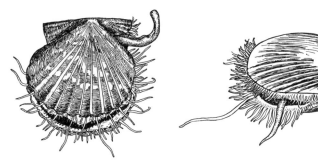

FIG. 383.—*Pecten latiauritus*, the broad-eared scallop. Drawn from a living specimen. The tentacles and foot are extended and the ocelli or "eyes" around the edge of the mantle are visible; x1.

FIG. 384.—*Lima dehiscens*, the file shell. Drawn from a living specimen; x2.6.

Conrad (fig. 381) is "more oblique, inflated, and markedly shorter, with smaller ears" (R. Arnold). The altitude reaches 20 mm., the length 19 mm., the hinge line 16 mm. The range is from Monterey to Lower California. Great numbers of these mollusks, alive and attached to seaweed are sometimes washed on shore.

Pecten circularis Sowerby, the circular scallop (fig. 379), is about as wide as high and very convex, the diameter being much more than half the altitude. The left valve is darker colored than the right and the ribs are narrower. The altitude reaches 46 mm., the length 47 mm., the diameter 33 mm., and the range is from Monterey to Peru.

The variety *aequisulcatus* Carpenter, the speckled scallop (fig. 380), is flatter, wider, and reaches a larger size. The convexity, the prominence of the ribs, and the color are variable features. A large specimen measured 72 mm. in height, 83 mm. long, and 36 mm. in diameter. The range is from Santa Barbara to Lower California.

Hinnites giganteus Gray, the purple-hinged or rock scallop (fig. 382), is free-swimming when it is young and has a shell that is nearly symmetrical. When it reaches a length of about 20-30 mm., the animal attaches itself to a fixed object such as a rock or another shell. The object that is chosen often has an irregular surface so that the shell grows to be more and more irregular. The left valve is less convex than the right. On the inner side of the valves the hinge area is tinged with purple. Our largest specimen is 85 mm. high. The range is from Alaska to Lower California.

Family—LIMIDAE

Lima dehiscens Conrad, the file shell (figs. 384 and 391), looks like a small and very oblique *Pecten*. It has a thin, frail, white shell, with fine radial striations. The length of a large specimen is 15 mm. and the height 22 mm. (Stephens). The range is from Monterey to Mexico.

Family—ANOMIIDAE

Anomia peruviana d'Orbigny, the rock oyster or jingle (fig. 385), has a white or copper-brown upper valve which is

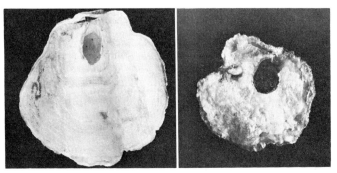

Fig. 385.—*Anomia peruviana*, the rock oyster or jingle; x5/6. Fig. 386.—*Pododesmus macroschisma*, a rock oyster or jingle; 5/6.

bluish green inside. It is attached to rocks or shells by a byssus which passes through a large notch in the lower valve. The muscle scars inside the upper valve form a nearly even straight row, radiating from the direction of the hinge (Dall). Height and length reach about 75 mm. The range is from San Pedro to Peru.

Pododesmus (Monia) macroschisma (Deshayes), also called rock oyster or jingle (fig. 386), like the last named species adheres to fixed objects and has a small and nearly flat, deeply notched lower valve. The upper valve has only two muscle impressions and the larger scar is radiately striated (R. Arnold). The height reaches 80 mm. (Packard) and the range is from Alaska to Lower California.

Family—MYTILIDAE

The mussels have brown or black shells and are attached to firm objects by a byssus of stout, tough threads. If one detaches a small mussel and puts it into a dish of sea water he can usually see the way in which these byssus threads are laid down. If the dish is undisturbed for a few minutes, the mussel thrusts out its foot far enough to show the byssus gland. The tip of the foot is bent up until it touches the gland and then bent down and extended until it touches the dish. When the foot is drawn back, a fine brownish thread will be seen extending from the mollusk to the dish along the pathway over which the foot just traveled. The foot tip repeats the performance at intervals, each time making the attachment on the dish at a different spot so that the byssus finally consists of a group of strands attached to the mollusk and radiating outward to the supporting structure like so many guy-ropes. The glandular secretion is glue-like when secreted, but hardens when drawn out into fine threads in the water. The hinge is near the edge of the shell. The mantle edges are fused at only one point to form an exhalent siphon.

Mussels are excellent food. The reader will find several good recipes for preparing mussels for the table in Bulletin 4 of the California Fish and Game Commission.

Mytilus californianus Conrad, the California sea mussel (fig. 387), forms great "beds" on the surf beaten rocks all along the coast. The glossy black covering is sometimes worn away on the older parts of the shell. Specimens measuring more than eight inches in altitude have been reported but individuals half that size are more common. The range of the species is from Alaska to Mexico.

These mussels are used extensively for food and may often be found in the markets. Cases of poisoning have been reported from their use but investigation* has shown these to have been due to the lack of care or knowledge in selecting the mollusks. One should always discard mussels with broken shells and those that are not capable of closing their shells.**

It is estimated that this mussel deposits fully 100,000 eggs annually and Dr. Heath found that it may grow from the egg to a length of three and a half inches in one year.†

Mytilus edulis Linnaeus, the bay mussel (fig. 388), is wedge shaped, and the shell is proportionately wider than the last named species. The height reaches 65 mm. and the length 32mm. The range is from the Arctic Ocean to Lower California. This mussel is often to be found in the markets of San Francisco. The same species is also found on the Atlantic coast and is extensively cultivated in Europe.

Under the title, "Fish-catching Mussels," the following note appears in *California Fish and Game*.‡ "After lying at anchor in San Francisco Bay for the past winter (1914-1915), the *Phelps* was taken across the bay and put in dry-dock, being rasied from the water on the day following the trip across the bay. The bottom was densely covered with mussels (*Mytilus edulis*), and other organisms. The ship is 312 feet in length and 45 feet in width, and on the bottom area about 50 or 60 fish were found trapped by the mussels. The fish were all in the same condition and comparatively fresh. * * * It is probable that, on the trip across the bay, the ship ran through a school of anchovies (*Engraulis mordax* Girard). Attracted by the edges * * * of the mantle of the

---

*E. P. Rankin—The Mussels of the Pacific Coast; California Fish and Game, Vol. 4, p. 117. **Information above incorrect and dangerous. See correction p. 581.
†California Fish and Game, Vol. 3, p. 35.
‡California Fish and Game, Vol. 1, p. 194.

mussels, many of the anchovies tried to secure a bite, with fatal results. The valves of the mussels were snapped shut with sufficient speed and strength to enclose and hold the more or less pointed head of the fish."

Septifer bifurcatus (Conrad) Reeve, the branch-ribbed mussel (fig. 389), is small and wedge-shaped, with pointed, terminal beaks. The surface of the shell has numerous radiating ridges which divide near the margins. A small shelly deck stretches across the interior of each valve near the hinge. The height reaches 45 mm., the width 18 mm. and the range is from northern California to the Gulf of California.

The horse-mussels are usually solitary and found partly buried in mud or gravel. They are anchored by the byssus. The hinge is not at the extreme end and the color is brownish rather than black.

Modiolus modiolus (Linnaeus), the giant horse-mussel (fig. 390), is an edible mollusk with a large, coarse, oblong shell. The basal margin is concave with a fissure for the byssus. The epidermis is thick, blackish or brown and roughly bearded near the margin of the shell. The animal is orange or reddish. Specimens nine inches high and four inches in diameter have been found (Keep). The range is from Alaska to San Pedro.

Modiolus rectus Conrad, the straight horse-mussel (fig. 393), is narrow, oblong, convex, the lower margin is nearly a straight line, and the shell is often bearded. The color is brown, the length reaches 150 mm., the height 50 mm., and the diameter 38 mm. The range is from Bolinas to Lower California.

Modiolus capax Conrad (fig. 394), the fat horse-mussel, is found from Santa Barbara to Peru. Like *M. modiolus*, this species is bearded, and covered with a brown, glossy epidermis, but the diameter of the shell is proportionately greater. A specimen 39 mm. high and 23 mm. wide is 21

FIG. 387.—*Mytilus californianus*, the California sea mussel; ×2/5. FIG. 388.—*Mytilus edulis*, the bay mussel; ×4/5. FIG. 389.—*Septifer bifurcatus*, the branch ribbed mussel; ×4/5. FIG. 390.—*Modiolus modiolus*, the giant horse mussel; ×5/4. FIG. 391.—*Lima dehiscens*, the file shell; ×5/4. FIG. 392.—*Modiolus fornicatus*, the arched horse mussel; ×5/4. FIG. 393.—*Modiolus rectus*, the straight horse mussel; ×⅔. FIG. 394.—*Modiolus capax*; ×⅔. FIG. 395.—*Botula falcata*, the pea-pod rock-borer; ×⅔. FIG. 396.—*Lithophaga plumula*, the "rock-eating" mussel; ×½.

mm. in diameter, while a specimen of *M. modiolus* 46 mm. high and 23 mm. wide, is only 16 mm. in diameter.

Modiolus fornicatus Carpenter, the arched horse-mussel (fig. 392), is short and swollen, "somewhat wedge shaped, having a breadth more than half of its length" (Keep). The beaks are not quite terminal, but are marginal and bent forward. The shell is white with a light brown epidermis. One specimen measures 54 mm. from the beaks to the ventral margin, 31 mm. long, and 30 mm. in diameter. It is found all along the California coast.

Botula (Adula) falcata (Gould), the pea-pod rock-borer (fig. 395), has a long, slender shell with dull, chestnut epidermis, and many transverse wrinkles. The adult mollusks live in burrows in the hard rocks. Within the burrow they are attached to the rock by the byssus. Lloyd* (1897) investigated the habits of the rock-boring mollusks and came to the conclusion that *B. falcata* occupies burrows made by pholads. He calls attention to the fact that the mollusk does not always fill the burrow and is attached by the byssus near the open end of the burrow out of which the posterior end of the valves protrude. A large specimen measures 14 mm. high and 82 mm. long. The range is from Oregon to San Diego.

Lithophaga plumula Hanley, the "rock-eating" mussel (figs. 396 and 397), also lives in burrows in the rocks. It has a slender cylindrical shell rounded in front and tapering behind. Two radial grooves extend back from the beaks and the space between these is frequently filled with a plume-like encrustation which projects in a symmetrical fashion beyond the ends of the valves. From the beaks to the ventral mar-

FIG. 397.—*Lithophaga plumula*, the rock-eating mussel. Drawn from a living specimen, with foot and siphons extended; x1.

---

*F. E. Lloyd—On the Mechanisms in Certain Lamellibranch Boring Molluscs; Transactions of the New York Academy of Sciences, Vol. 16, p. 312.

gin a large specimen measures 83 mm., the length is 25 mm. and the diameter 23 mm. It is found from Monterey to Patagonia as well as on the Atlantic coast.

Family—PERIPLOMATIDAE

Periploma planiuscula (Sowerby), the silver lantern-shell (fig. 400), has a white, smooth shell with a silvery lining. There is a spoon-like hinge tooth in each valve. The length of one specimen is 46 mm., the altitude 33 mm., the diameter 18 mm. From the beak to the anterior end is 13 mm., to the posterior end is 33 mm. The range is from San Pedro to Mexico.

Family PANDORIDAE

Pandora filosa (Carpenter) has a small, thin shell. The right valve is flat, the left convex. The minute beaks are about one-fourth the length from the anterior end (fig. 398). The dorsal margins are straight. Both valves have fine lines of growth and the right valve has fine radiating grooves. The left valve has one thin hinge ossicle and the right one has two ossicles. The length of one specimen is 21 mm., the height 6 mm., the diameter 3 mm. The range is from Bering Sea to San Pedro.

Pandora punctata Conrad, the rare pandora (fig. 401), has an unusual shape. The shell is much compressed. The interior is pearly and marked with fine pits. The length reaches 40 mm. The range is from Vancouver Island to the Gulf of California.

Family LYONSIIDAE

Lyonsia californica Conrad, is bulged at the anterior end (fig. 399) while the posterior end is narrow, thin, and crooked. The valves are equal and the epidermis has radiating striae. The epidermis is often worn off leaving the pearly layer exposed. The length of a large specimen is 31 mm., the

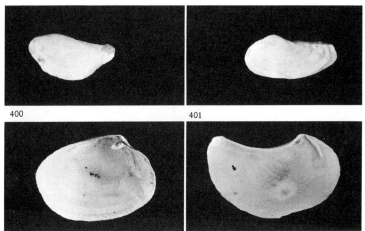

FIG. 398.—*Pandora filosa;* x3/2. FIG. 399.—*Lyonsia californica;* x⅔. FIG. 400.—*Periploma planiuscula,* the silver lantern-shell; x5/6. FIG. 401.—*Pandora punctata;* x1.

FIG. 402.—*Astarte alaskensis;* x1. FIG. 403.—*Entodesma saxicola;* x1. FIG. 404.—*Venericardia ventricosa;* x3/2.

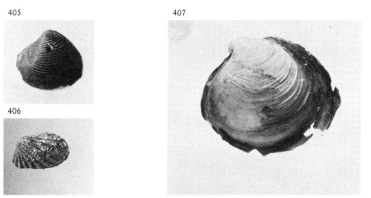

FIG. 405.—*Astarte esquimalti;* x1. FIG. 406.—*Cardita subquadrata,* the little heart-shell; x3/2. FIG. 407.—*Mytilimeria nuttallii;* x1.

height 14 mm., and diameter 11 mm. (R. Arnold). The range is from Puget Sound to Lower California.

Entodesma saxicola (Baird) is somewhat oblong and bulges in the hinge region (fig. 403). The inside of the shell is white and the outside covered with an olive-brown epidermis. The siphon end gapes and is prolonged by a thin, paper-like epidermis. According to Dr. Dall, the horny periostracum in drying always cracks the calcareous part of the valves. Internally, the hinge is covered by an oblong white ossicle. A large specimen measures 100 mm. in length and is 50 mm. high (Oldroyd). Dr. Dall* says that it sometimes reaches six inches in length. The species is found from the Aleutian Islands to Lower California.

Mytilimeria nuttallii Conrad, the sea bottle shell, may be found imbedded in masses of compound ascidians. The shell is thin and fragile, convex, with central beaks and equal valves (fig. 407). The surface is sculptured with fine concentric and radiating lines. The shell is white with a brown epidermis and has a pearly luster. A large specimen measures 36 mm. long and 36 mm. high (Baker). The range is from Vancouver Island to San Diego.

FAMILY—ASTARTIDAE

Astarte alaskensis Dall (fig. 402) is often dredged in Puget Sound. It has a strong shell which is white within and covered outside with a dark brown periostracum. There are 12-14 concentric ribs with wide interspaces. The inner margin is not toothed but is smooth and entire. Dr. Dall† gives the measurements of a large specimen as follows: height 26 mm.; length 31.5 mm.; diameter 14 mm. The species has been found from Bering Sea south to Puget Sound.

---

*W. H. Dall—A Review of some Bivalve Shells of the group *Anatinacea* from the West Coast of America; Proceedings of the United States National Museum, Vol. 49, p. 456.

†W. H. Dall—Synopsis of the Family *Astartidae*, etc.; Proceedings U. S. National Museum, Vol. 26, p. 946.

Astarte esquimalti Baird (fig. 405) is also taken by the dredge in Puget Sound. The shell is heavy, nearly equilateral, and covered with a dark brown periostracum. The concentric ribs are not quite uniform but may be interrupted in some places and coalesce in others. Dr. Dall* gives the measurements as follows: height 21 mm.; length 23 mm.; and diameter 11 mm. The range is from the Aleutian Islands to Puget Sound.

FAMILY—CARDITIDAE

Cardita subquadrata (Carpenter), the little heart-mollusk (fig. 406), has a strong shell with about fourteen prominent radiating ribs. The color is brownish white. The altitude of an ordinary specimen is 10 mm. and the range is from the Queen Charlotte Islands to Lower California.

Venericardia ventricosa (Gould) (fig. 404) has well marked radiating ribs and the inner margin of the shell is toothed. The white shell is overlaid with an olive-brown periostracum. It is often dredged in Puget Sound and is found from Alaska to the Coronado Islands. One specimen measures 18 mm. long, 16 mm. high, and 11 mm. in diameter.

FIG. 408.—*Chama* sp. drawn from a living specimen with siphons extended; x1.

Milneria minima (Dall) is tiny, delicate, and markedly convex (fig. 410). The color is light brown. It is found on abalones from Monterey to Lower California. The height is only about 5 mm.

FAMILY—CHAMIDAE

The chama mollusks (fig. 408) are always sessile, usually attached to the underside of rocks. The two valves are unequal in size and shape.

*W. H. Dall—Synopsis of the Family *Astartidae*, etc.; Proceedings U. S. National Museum, Vol. 26, p. 945.

Chama pellucida Sowerby, the agate *Chama*, lives attached to rocks or other support. The shell is firm and strong and the surface is made very rough by the concentric frills which are somewhat translucent and remind one of agate (fig. 408 and 409). It is usually white, sometimes tinged with red. The length and height reach 20 and 25 mm. respectively (R. Arnold). The range is from Oregon to Chile and the Galapagos Islands.

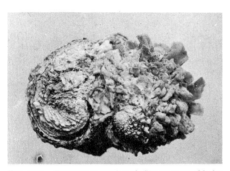

FIG. 409.—These two species of *Chama* grew side by side, *Pseudochama exogyra* on the right and *Chama pellucida* on the left; x1. (Specimen in the Kelsey-Baker collection, at the Scripps Institution.)

Pseudochama exogyra (Conrad), the reversed *Chama* (fig. 409), greatly resembles the last named species but is larger, and has fewer frills which are somewhat less spiny. It is generally attached by the left valve while in the case of *C. pellucida* the right valve is the one that is usually attached. Keep * gave the following method of distinguishing the species. "If you stand a specimen of this species on its edge, with the beaks uppermost and curving towards you, the side which was attached to the rock will be towards your left hand. But if you place a specimen of *pellucida* in the same position, the rocky side will be towards your right hand." The range is from Oregon to Panama.

FAMILY DIPLODONTIDAE

Diplodonta orbella (Gould), the round *Diplodonta* (fig. 412), has a white shell very much inflated and marked by lines of growth, some of which are more prominent than others so that the surface is somewhat uneven. The liga-

---

*Josiah Keep—West Coast Shells (1911 Edition); Whitaker & Ray-Wiggin Co., p. 71.

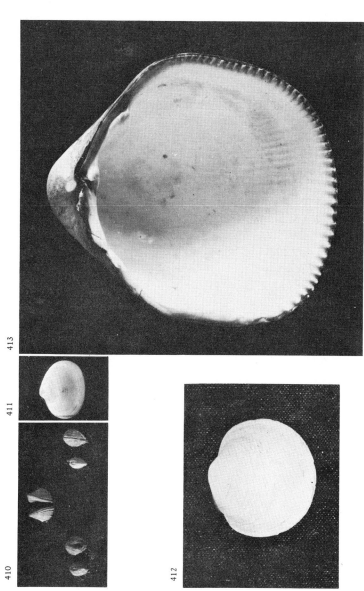

FIG. 410.—*Milneria minima;* x3/2.   FIG. 411.—*Kellia laperousi;* x3/2.   FIG. 412.—*Diplodonta orbella;* x5/4.   FIG. 413.—*Cardium elatum,* the giant cockle; x3/4.

ment is conspicuous and there are two cardinal hinge teeth in each valve. A large specimen measures 26 mm. in height and length and 20 mm. in diameter. The range is from Bering Sea to the Gulf of California. This species forms a protecting covering of sand cemented by mucus. The covering has long tube-like extensions in which the siphons lie so the mollusk is quite hidden (fig. 416).

FAMILY LUCINIDAE

Phacoides nuttallii (Conrad) has small, thin, circular valves with central unbones (fig. 414). Prominent concentric lines and radiating grooves give a cancellated appearance to the surface. The lunule is small and nearly all in the left valve. The height reaches 35 mm. The range is from Santa Barbara to Mexico.

Phacoides californicus (Conrad) (fig. 415), has a circular shell, of medium size and rather thick. The umbones are central and there are fine, sharp, close-set concentric lines and very faint radiating lines. The lunule is small, deep-set, and wholly in the right valve. The length reaches 37 mm. and the height 35 mm. (Baker). The range is from northern California to Lower California.

FAMILY LEPTONIDAE

Kellia laperousi (Deshayes), the smooth Kelly-shell (fig. 411), is of medium size, convex, and thin, with fine lines of growth. There is no lunule. Each valve has one very prominent cardinal tooth, the right valve has two laterals and the left but one. The hinge area between the cardinal and lateral teeth is lacking, giving the impression that it has been broken. The measurements of one specimen are: height 19 mm., length 23 mm., diameter 11mm. (R. Arnold). The range is from Bering Sea to San Diego.

FIG. 414.—*Phacoides nuttallii;* x1.   FIG. 415.—*Phacoides californicus;* x1.

FIG. 416.—*Diplodonta orbella* in its covering of sand cemented with mucus; x1.

FIG. 417.—*Cardium substriatum*, the egg shell cockle; x½. FIG. 418.—*Cardium quadragenarium*, the forty-ribbed cockle or spiny cockle; x½. FIG. 419.—*Cardium corbis*, the basket cockle; x½.

Family CARDIIDAE

Mollusks of this family are known as cockles or "heart shells." If the shell is viewed from the end (fig. 420) one sees the heart-shaped outline best.

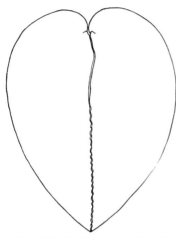

FIG. 420.—*Cardium elatum*, view of anterior end; x½.

The cockles are edible but are not found in large enough numbers to make them of much importance commercially. *Cardium corbis* (fig. 419) is the only one that is abundant enough to appear in the markets. It is found on the tide flats at many points along the California shore but according to Weymouth* (1920) it is not as abundant here as it is on the tide flats of Puget Sound and the Gulf of Georgia. The cockles burrow only a short distance down into the sand though they are active mollusks. The mantle margins are fused at two points to form the siphon openings.

Cardium quadragenarium Conrad, the forty-ribbed cockle or spiny cockle, is large, inflated, and thick, with central, prominent umbones (fig. 418). There are about forty squarish radiating ridges which are smooth except for conspicuous pointed tubercles toward the margin of the shell. The color is brownish, the altitude reaches 120 mm., the length 115 mm., and the diameter 98 mm. The range is from Santa Barbara to Lower California.

Cardium corbis (Martyn), the basket cockle, has a large, inflated shell with about 37 squarish, close-set, radiating ridges which are marked only by lines of growth (fig. 419).

---

*F. W. Weymouth—The Edible Clams, Mussels and Scallops of California; California Fish and Game Commission, Fish Bulletin No. 4, p. 28.

A large specimen measures 71 mm. in height and length and has a diameter of 57 mm. (R. Arnold). The range is from Bering Sea to San Diego where it may be found on tide flats in bays and sounds or even on exposed beaches where there is coarse, loose sand.

Cardium elatum Sowerby, the giant cockle, is the largest of the cockles and can be distinguished by its great size and by the fact that the ribs are faint and the lines of growth are faint, wavy lines (figs. 413 and 420). The length reaches 144 mm., the height 167 mm., and the diameter 136 mm. (R. Arnold). The range is from San Pedro to Panama.

Cardium substriatum Conrad, the egg-shell cockle, is small, inflated, and thin (fig. 417). The surface of the shell is smooth except for fine radiating lines which usually show only near the margin. The shell is mottled with reddish brown especially on the inside. The width of an ordinary specimen is 17 mm., the height 18 mm., and the diameter 13 mm. (R. Arnold). The range is from San Pedro to Mexico.

FAMILY VENERIDAE

Tivela stultorum (Mawe), the Pismo clam, is perhaps our best known clam. Its habits and distribution are well described by Weymouth* (1920 and 1923). From these excellent accounts we extract the following data.

The shell is thick and heavy (fig. 421 and 368). One large specimen weighed four pounds, three ounces. The color is grayish but some specimens are marked with chocolate-brown in radiating lines. Young specimens are more apt to show the brown coloration. A glossy covering over the shell gives it a varnished appearance. The range is from Halfmoon Bay to Socorro Island, off Mexico (Weymouth).

The home of the Pismo clam is on exposed sandy beaches

---

*F. W. Weymouth—The Edible Clams, Mussels and Scallops of California; California Fish and Game Commission, Fish Bulletin No. 4, p. 29, and The Life-History and Growth of the Pismo Clam; California Fish and Game Commission, Fish Bulletin No. 7.

where it is pounded by the heavy surf. In the early days it was said to be more abundant but even in recent years (1916-1919) Morro, Pismo, and Oceano have furnished an output of 150,000 individuals or 200 tons as a yearly average.

FIG. 421.—*Tivela stultorum*, the Pismo clam; x⅓. Drawn from a living specimen with siphons and foot extended.

At low tide the clam digger, in old clothes and armed with a potato fork, wades out into the sand bars. He "feels" for the clams, thrusting the fork into the sand and when a shell is struck he lifts it out. The clams are carried in a long net "drag" or sack fastened to the clam digger's belt. A light wooden hoop holds the mouth of the bag open. The water is usually waist deep but it is not unusual to see the surf break over the heads of the men.

Weymouth says, "At times this may partake of the exhilaration of surf-riding with an anchor of clams to prevent being swept too far, but at sunrise of a foggy day with a cold wind whipping the spray from the breakers it is a life of exposure."

It has been estimated that a Pismo clam about five inches long lays 75,000,000 eggs in one spawning season. In spite of this enormous reproduction, the clams are so much less abundant now than they were in former years that laws designed to protect the species have been passed. At the present time it is not lawful to gather in one day or have in your possession more than 15 of the clams or to gather, destroy, or have in your possession, any Pismo clams measuring less than 4¾ inches across. According to Weymouth, a clam of the legal size is about seven years old. In the Monterey Bay district it is unlawful to gather any of the clams between May 1st and August 31st. Since these laws are changed from time to time, it is well for the collector to inform himself on this point before he begins to collect.

Amiantis callosa (Conrad), the sea cockle (fig. 422), has a heavy, pure white shell. It has no radiating ribs but numerous concentric rounded ridges. Large shells are covered with a smooth, white, shining covering. It is found from Santa Monica (Weymouth) to Mexico but is not abundant. A large specimen is 100 mm. long and 78 mm. high.

FIG. 422.—*Amiantis callosa*, the sea cockle; x½.

Both *Saxidomus nuttallii* Conrad (fig. 423) and *Saxidomus giganteus* (Deshayes) are known as the Washington clams, butter clams, or money shells. They are often marketed together being fairly abundant in bays from Humboldt to Bolinas. The valves of *S. nuttallii* are white within but have slight purple markings near the siphon end. They reach nearly 5 inches in length. The pallial sinus is deep. The range is from Humboldt Bay (Weymouth) to Lower California (Baker). *S. giganteus* is similar but is a little more nearly circular in its outline and smaller. The outer surface is not so rough, the growth lines are finer and less marked and the interior is white without the purple tinge. They average about 3 inches long. They are found from Alaska to Monterey.

FIG. 423.—*Saxidomus nuttallii*, the Washington clam, butter clam, or money shell; x½.
FIG. 424.—*Chione undatella*; x1.

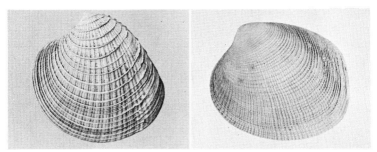

FIG. 425.—*Chione succincta*; x⅔.   FIG. 426.—*Paphia staminea laciniata*; x½.

FIG. 427.—*Tellina bodegensis*; x1.   FIG. 428.—*Metis alta*, the yellow *Metis*; x¾.

Three species of *Chione* may be found among the hardshell cockles which are sold in the markets of Los Angeles and San Diego. The clams have rounded, heavy shells which rarely exceed 2½ inches in length. The siphons are short which indicates that the clam must live near the surface of the mud. They are all southern forms being found from San Pedro to Mexico. The following descriptions of *Chione* are taken largely from Weymouth (1920).

Chione fluctifraga (Sowerby) (fig. 429) is the largest of the three species. It has no distinct lunule (fig. 367) and its radiating ribs are more prominent than its concentric lines of growth. The ribs are more prominent at the posterior end of the shell than elsewhere. The pallial sinus is sharply triangular. The length of a large specimen is 60 mm.

Chione undatella (Sowerby) (figs. 365 and 424) has a conspicuous lunule. The concentric ribs are numerous and more prominent than the radiating ribs which are everywhere about equally conspicuous. A large specimen measures 46 mm. and has 33 concentric ridges.

Chione succincta (Valenciennes) (fig. 367 and 425) also has a conspicuous lunule. The concentric ribs are few and more prominent than the radiating ribs. The latter are more conspicuous at the posterior end of the shell. A large specimen 47 mm. long has 16 concentric ridges, while another one 45 mm. long has 21 ridges.

Paphia staminea (Conrad), the rock cockle (fig. 430), is also known as the little-neck clam, hard-shell clam, Tomales Bay cockle, rock clam, and the ribbed carpet-shell. It seldom grows to be more than 3 inches long. The radiating ribs are more clearly marked than are the lines of growth and the length is distinctly greater than the height. The inner margins of the valves are rough. The color is yellowish white, often with brown markings. A number of varieties have been described. *P. staminea* var. *petiti* (Deshayes) is a form abundant north of the Columbia River. According to Dall, it is larger than the southern variety, yellowish, chalky white, or dull gray without the color spots. *P. staminea* var. *laciniata* (Carpenter) has radiating ribs (fig. 426) and concentric ridges that are about equally prominent and small spines are present at the intersections. A worn specimen cannot be distinguished from the variety *petiti*. *P. stamina* var. *ruderata* (Deshayes) is a northern form with concentric ridges more prominent than the radiating ribs. *P. staminea* var. *orbella* (Carpenter) includes the specimens that have nestled in the borings of pholads and so have become abnormal in shape.

According to Weymouth*(1920) most of these clams found in the San Francisco markets are gathered from Tomales Bay, about 7,000 pounds per month having been obtained in 1911. The species and the varieties named are found on the California coast and northward on the coast of Alaska.

Venerupis lamellifera (Conrad), the rock Venus (fig. 431), has an irregular white shell with prominent concentric ridges which are like thin folds of shell. Radiating ribs are absent or show as a mere trace. These clams nestle among the rocks. One specimen measures 26 mm. long, 21.5 mm. high, and has a diameter of 16 mm. The species is found from Monterey to San Diego.

Gemma gemma (Totten), the gem shell, has a tiny shell 2-5 mm. long, equilateral, nearly round, and somewhat inflated. It is said to have been introduced here with "seed" oysters from the Atlantic coast. The shell is white, sometimes tinged with purple, and is marked with delicate, transverse lines. It has been reported from the San Juan Islands to San Francisco Bay.

Family PETRICOLIDAE

Petricola carditoides (Conrad), the rock-dweller (fig. 436), often nestles in the holes made by the boring clams. The oval shell often becomes misshapen to fit the hole in which

---

*F. W. Weymouth—The Edible Clams, Mussels and Scallops of California; California Fish and Game Commission, Fish Bulletin No. 4, p. 40.

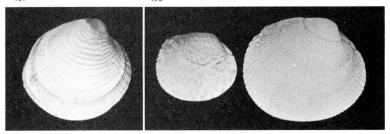

FIG. 429.—*Chione fluctifraga*; x3/5. FIG. 430.—*Paphia staminea*, the rock cockle; x1.

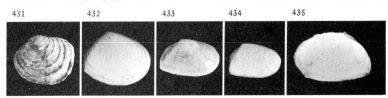

FIG. 431.—*Venerupis lamellifera*, the rock Venus; x5/4. FIG. 432.—*Tellina meropsis;* x1.
FIG. 433.—*Tellina carpenteri;* x1. FIG. 434.—*Tellina modestus;* x3/2. FIG. 435.—*Tellina buttoni;* x3/2.

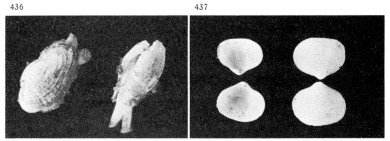

FIG. 436.—*Petricola carditoides*, the rock dweller; x1. Living specimens with siphons extended. FIG. 437.—*Cooperella subdiaphana;* x1.

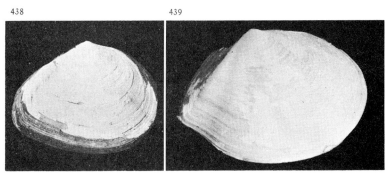

FIG. 438.—*Macoma nasuta*, the bent nosed clam; x1. FIG. 439.—*Macoma indentata;* x4/3.

the mollusk lives. The surface of the shell is marked with fine radiating lines and with fine concentric lines which sometimes form ridges. The margin is smooth. The length is 14-45 mm. (Packard). It is found from Vancouver Island to Lower California.

FAMILY—COOPERELLIDAE

Cooperella subdiaphana (Carpenter) has a fragile, white shell about one-half an inch long (fig. 437). The beaks are slightly raised, sharp, prominent, there are two narrow, bifid teeth in the left valve and three in the right, and the pallial sinus is very large (R. Arnold). It is found from the Queen Charlotte Islands to the Gulf of California.

FAMILY—TELLINIDAE

The tellens have flat shells with a prominent external ligament. The siphons are long and slender and the pallial sinus is large. *Tellina meropsis* Dall has a thin white shell sometimes tinged with yellow. Posteriorly the shell is bent and angled as shown in the figure (fig. 432). One specimen measures 19 mm. long, 15 mm. high, and 6 mm. in diameter. It is found from San Diego to the Gulf of California.

Tellina carpenteri Dall has a tiny, thin, pink and white, glossy shell (fig. 433 and 441). A large specimen is 14 mm. long, 8 mm. high, and 3 mm. in diameter. It is found from Forrester Island to Panama.

Tellina modestus Carpenter has a thin, white, glossy shell with fine lines of growth (fig. 434). It may reach 13 mm. in length. It is found from Vancouver Island to Lower California.

Tellina buttoni Dall also has a thin, white, glossy shell (fig. 435) distinguished by an oblique rib across the inside of each valve at the anterior end. It reaches 19 mm. long, 7 mm. high, and 2.5 mm. in diameter. It is found from Alaska to the Gulf of California.

Tellina bodegensis Hinds is long and flat (fig. 427). The shells are pure white with a smooth, polished surface. The lines of growth are fine and regular. A large specimen is 50 mm. long and 23 mm. high. The range is from British Columbia to the Gulf of California.

Metis alta (Conrad), the yellow *Metis*, is large, nearly circular (fig. 428) and unsymmetrical. The posterior side of the deeper valve shows two angles and the opposite valve has an angular groove. The lines of growth are prominent and irregularly spaced, and the radiating lines are faint. The external ligament is nearly concealed by the shell. The polished white interior is often tinged with bright yellow. It reaches 75 mm. in length. The range is from Santa Barbara to San Diego.

Macoma nasuta (Conrad), the bent-nosed clam (fig. 438), is not a large species as it seldom grows to be more than 57 mm. long and 42 mm. high. The siphon end is elongated and bent to the right (fig. 440-a). The shells are white but

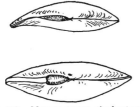
FIG. 440.—*Macoma nasuta*, the bent nosed clam (above). *Macoma secta*, the white sand clam (below). (Dorsal views, after a photograph by Weymouth); x2/3.

FIG. 441.—*Tellina carpenteri*, drawn from life, showing foot and siphons extended; x4/3.

are often covered with a thin gray layer or periostracum. The siphons are long and separate. It can be distinguished from *M. secta* by the fact that its ligament is less conspicuous and the posterior end of the shell shows a more oblique truncation. This clam inhabits muddy bays and is abundant all along the coast, from Alaska to Lower California. According to Weymouth* (1920) it used to be marketed in San

---

*F. W. Weymouth—The Edible Clams, Mussels and Scallops of California; California Fish and Game Commission, Fish Bulletin No. 4, p. 43.

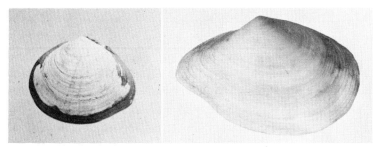

FIG. 442.—*Macoma inquinata;* x1.   FIG. 443.—*Macoma indentata tenuirostris;* x1.

FIG. 444.—*Semele decisa,* the flat clam; x¾.

FIG. 445.—*Semele rupicola;* x1.   FIG. 446.—*Semele pulchra;* x1.

Francisco by the Chinese. Though the species does not seem to be used for food now to any extent, the abundance of the shells in the kitchen middens of the Indians indicates that this was the species most frequently eaten by them.

Macoma inquinata (Deshayes) is medium sized, convex, and thin. The shell is whitish with a brown periostracum (fig. 442). There are two small cardinal teeth in each valve and the pallial sinus does not reach the anterior muscle impression in the left valve (R. Arnold). The length reaches 55 mm. The range is from Alaska to Japan and to Monterey. The variety *arnheimi* Dall is more plump, the basal margin is somewhat twisted, the beaks are 15 mm. behind the anterior end. Length 38 mm., height 30 mm., and diameter 15 mm.; range Alaska to San Francisco.

Macoma balthica (Linnaeus) is a small shell 3-35 mm. long. The original description, as quoted by Packard, says, "Shell of the size of the seed of the white lupine, somewhat delicate, very fragile, interior white, exterior flesh-colored, of a rounded triangular shape." It is found from Point Barrow to North Japan and San Diego.

Macoma indentata Carpenter (fig. 439) is small and resembles *M. nasuta*, but may be distinguished by its smaller shell, its umbones posterior to the center, and the indentation in its ventral line near the posterior end. It is found from Puget Sound to Lower California. A large specimen measures 47 mm. long and 32 mm. high.

Macoma indentata tenuirostris Dall (fig. 443) is longer than *M. indentata* in proportion to its height, with its posterior portion long and narrow. A specimen measures 55 mm. long and 32 mm. high. The range is from the Santa Barbara Islands to San Diego (Kelsey).

Macoma secta (Conrad), the white sand clam or giant *Macoma*, is the largest of the genus since the shells reach 4 inches in length. The ligament is strong and broad, the umbones nearly central (fig. 447). The left valve is flatter

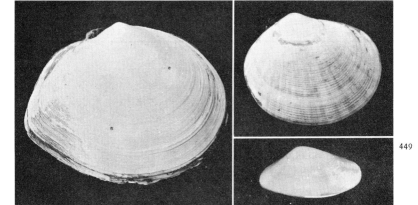

FIG. 447.—*Macoma secta*, the giant *Macoma*, or white sand clam; x1. FIG. 448.—*Semele rubropicta*; x1. FIG. 449.—*Donax californica*, the wedge shell; x1.

FIG. 450.—*Cumingia lamellosa;* x¾.

FIG. 451.—*Donax gouldii*, the common wedge shell or bean clam. Photographed from life; a number of individuals show foot and siphons extended; x⅔.

PHYLUM—MOLLUSCA 453

than the right (fig. 440-b). According to Thompson,* in British Columbia it is found at a depth of a foot and a half in pure sand on the exposed portion of the larger flats. The clams are said to make excellent food though the alimentary canal is likely to be full of sand. Weymouth suggests that this difficulty might be overcome if the clams are left for a time in tanks of salt water, a method which used to be employed with the bent-nosed clam when it was marketed by the Chinese. The range is from Vancouver Island to the Gulf of California.

FAMILY—SEMELIDAE

Semele decisa (Conrad), the flat clam (fig. 444), is a southern form, found from San Pedro to Lower California (Baker). A large specimen measures 80 mm. long and 76 mm. high. The valves are heavy and only slightly arched. Outside the shell is rough and granular, but inside, it is smooth, white, and tinged with purple on the margins. The cardinal teeth are obsolete. Although it is listed among the edible clams, it is not an abundant species.

Semele rupicola Dall (fig. 445) has a rough and somewhat uneven exterior with conspicuous concentric ridges. The interior of the shell is pure white to deep rose or almost purple, the color being especially marked around the margins. The lateral teeth are prominent, the cardinal ones small. Old wave worn shells with their bright rose-pink color are conspicuous among the broken shells washed up on the beach. The range is from Santa Cruz to the Gulf of California. A large specimen measures 50 mm. in length (Kelsey). They are often found in the holes of boring mollusks in the rocks at La Jolla and their shape is determined, more or less, by the shape of the holes in which they have lived.

Semele rubropicta Dall is heavier than the next species and has a more convex shell (fig. 448). Fine, regular con-

---

*F. W. Weymouth—The Edible Clams, Mussels and Scallops of California; California Fish and Game Commission, Fish Bulletin No. 4, p. 45.

centric ridges are crossed by radiating ribs. The latter are obsolete near the hinge. The color is white inside, but outside the shell is marked with conspicuous rays of rose color. It is found from Alaska to Lower California. One specimen measures 38 mm. long, 30 mm. high, and is 16 mm. in diameter.

Semele pulchra (Sowerby) has a small, oval shell sculptured with fine, close, radiating and concentric lines (fig. 446). The cardinal tooth is prominent. A large specimen is 20 mm. long, 16 mm. high, and 8 mm. in diameter. The range is from Monterey to Ecuador. The shell is yellowish outside, tinged with rose. The pink color inside is in the center of the shell, while outside it is near the hinges.

Cumingia lamellosa Sowerby has a thin shell much like *Macoma nasuta* in general outline but with numerous sharp concentric ridges (fig. 450). The hinge has a triangular, spoon-shaped cartilage pit and a small anterior cardinal tooth in each valve. There are two lateral teeth in the right valve, which are less developed in the left. The pallial sinus is wide, expanded interiorly, and deep (R. Arnold). The length is 24.5 mm., altitude 18 mm., and diameter 4.9 mm. They are often found in the holes of boring mollusks in the rocks at La Jolla. In such cases the shape is distorted, to conform to the holes in which they have lived. The range is from Crescent City to Peru.

Family—DONACIDAE

Donax californica Conrad (fig. 449), the wedge mollusk, has a thin light shell with very faint radiating furrows. The interior of the margin is finely crenulated. One specimen measures 28 mm. long, and 14 mm. high. The range is from San Pedro to Panama.

Donax gouldii Dall (fig. 451), the common wedge shell or bean clam, is often beautifully marked with radiating or concentric patterns of purple, brown, or gray green. Both color and pattern show great variety. The shells are strong, marked with radiating lines and lines of growth. The umbones are close to the posterior end instead of being nearly central as in *D. californica*. The siphons are short and the clam lies near the surface of the sand. Its favorite habitat seems to be a hard smooth sandy beach where it may occupy the strip of shore from mid-tide to low water.

There are many interesting unsolved problems in connection with the appearance and disappearance of these clams. In 1910 the beach in front of the Scripps Institution at La Jolla was thickly populated with *Donax*. Each low tide would expose a crowded strip that averaged 10-20 feet wide and extended the entire length of the sandy beach—about a mile. The clams over most of this area lay closer together than they do in the photograph so that one could scoop up a dozen of them in a handful of sand. People used them for making clam bouillon but the few hundreds of pounds per day that were taken made no impression on their numbers.

After one high tide period in the fall of 1910 few were to be seen and they have not been found in any great numbers here since that time. Search along the whole beach would perhaps yield a hundred clams now (1925) where at one time millions could have been secured.

In 1895 the *Donax* were so abundant in the region between Long Beach and Wilmington that canneries were established for putting up clam extract. The accepted method for gathering clams was to scoop up the sand and clams into a box with a screen bottom and allow the surf to wash away the sand and leave the clams. A sack of clams too heavy to carry could be obtained in a few minutes by this method. Recently we have heard of no places along the coast where the clams have been so abundant as they formerly were or even abundant enough to make them available as a food supply. Great numbers of fossils found near False Bay indicate that this species was abundant in Pleistocene times. They have been found from Oceano (Weymouth) to Mexico. A large specimen may reach 27 mm. in length.

Family—PSAMMOBIIDAE

Psammobia californica Conrad, the sunset shell (fig. 453), is white with red rays from umbo to margin and fainter concentric lines of red. The umbo is nearly central and the valves fit loosely. The margins are often covered with a brown periostracum. A large specimen measures 105 mm. long, 63 mm. high, and 33 mm. in diameter. It is found in Japan and from Alaska to San Diego.

Sanguinolaria nuttallii Conrad, the purple clam (fig. 452), has a thin, flat, shell and a prominent external ligament. The right valve is flattened more than the left and the siphon end is bent to the left. The shells are tinged with purple

Fig. 452.—*Sanguinolaria nuttallii*, the purple clam; x⅔. Fig. 453.—*Psammobia californica*, the sunset shell; x⅔. Fig. 454.—*Heterodonax bimaculata*; x1.

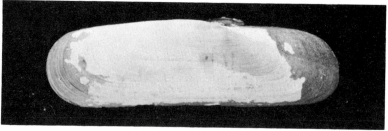

Fig. 455.—*Tagelus californianus*, the jackknife clam; x1.

and the periostracum is a brown, varnish-like covering. A large specimen is 87 mm. long, 63 mm. high, and 25 mm. in diameter. The range is from San Pedro to lower California.

Heterodonax bimaculata (Linnaeus) (fig. 454) has an oval shell that is flat and thin. The color is usually white or purple. It is found from Monterey to Panama and also on the Atlantic coast. The length of a large specimen is 24 mm., the altitude 19 mm., diameter 10 mm.

Tagelus californianus (Conrad), the jackknife clam (fig. 455), is sometimes incorrectly called a razor clam. The length of the shell is three or more times the altitude, the long margins are nearly parallel and the umbo is central. The white or grayish shell is covered with a dull brownish periostracum. The shell reaches a length of 3-4 inches. According to Weymouth* (1920) the clam digs a smooth-lined permanent burrow about 15-16 inches deep within which it moves readily

FIG. 456.—*Solen rosaceus*, drawn from a living specimen showing foot and siphons extended; x1.

up and down. When it is at the top of the burrow the clam is about its own length from the surface and the two siphons reach up to the water through two small holes in the mud. When disturbed, it pulls in its siphons and rapidly retreats to the bottom of the burrow. Beyond that point it can dig but progress is necessarily slower. It is not much used as food though the flavor is said to be fairly good. A great many are used for bait. The range is from Santa Barbara to Mexico.

Tagelus subteres (Conrad) (fig. 459) also has a central umbo and is much like the last named species but the shell is smaller and the periostracum is a darker brown. Violet colored radial markings are just visible from the outside, and the

---

*F. W. Weymouth—The Edible Clams, Mussels and Scallops of California; California Fish and Game Commission, Fish Bulletin No. 4, p. 49.

inside is also tinged with violet. A specimen measures 45 mm. long, 15 mm. high, and 9 mm. in diameter. The range is from Santa Barbara to Panama.

FAMILY—SOLENIDAE

Solen rosaceus Carpenter is also one of the jackknife clams (figs. 456 and 457). The umbos are at the extreme anterior end and the siphons are united. The shell is small and about five times as long as wide, with a thin transparent periostracum through which the rosy color of the shell shows. The foot has little pigment (Weymouth). It burrows in the mud of bay shores from Santa Barbara to the Gulf of California. A large specimen measures 75 mm. long, 14 mm. wide, and 19 mm. in diameter.

Ensis californicus Dall is much like *Solen rosaceus* in its proportions, but is slightly curved. It is found from Monterey to the Gulf of California.

Solen sicarius Gould, also one of the jackknife clams (fig. 460), has a slightly curved shell which is about four times as long as wide. The foot shows dark pigment (Weymouth). The range is from Vancouver Island to Lower California. The length is 82 mm., the height 19 mm., and the diameter 12 mm. It is said to live in burrows similar to those of *Tagelus*. According to Weymouth it is sometimes mistaken for the razor clam which only lives on sandy ocean beaches.

There are two razor clams found commonly on the coast, *Siliqua lucida* (Conrad) (fig. 458) and *Siliqua patula* (Dixon) (fig. 461). The shells are thin, fragile, and smooth with a polished epidermis. A prominent rib extends from the umbo to the margin across the inside of the valve. According to Weymouth,[*] they are found on sandy beaches exposed to the ocean, especially on those that are broad and level. Weymouth says that small razor clams placed on wet sand have completely buried themselves by 8–10 movements of the foot

[*] F. W. Weymouth—The Edible Clams, Mussels and Scallops of California; California Fish and Game Commission, Fish Bulletin No. 4, p. 51.

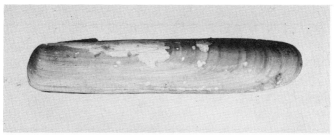

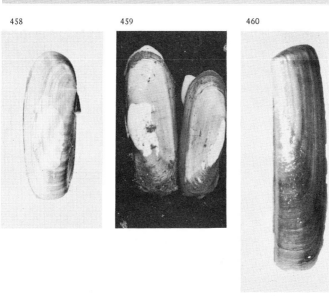

FIG. 457.—*Solen rosaceus*, one of the jackknife clams; x1. FIG. 458.—*Siliqua lucida*, a razor clam; x1. FIG. 459.—*Tagelus subteres;* x4/5. FIG. 460.—*Solen sicarius*, one of the jackknife clams; x5/6. FIG. 461.—*Siliqua patula*, a razor clam; x5/6.

within 7 seconds. The foot is extended until it projects half the length of the shell. The tip of the foot is pointed so it penetrates the sand readily. Having reached its full extent the tip swells up until its cross section is greater than that of the shell. Hence, when the foot contracts, the shell moves more readily through the sand than does the foot and is drawn down. The Alaska and Washington forms are canned under the name of sea clams. Although these are among the best of the edible clams, the living ones do not market well because the shells close so incompletely that they do not hold moisture and are thin and easily broken.

S. lucida is the smaller species, with a thin, fragile shell, tinged with violet. The umbones are a little less than one-fourth the distance from the posterior end. The interior rib is narrow and perpendicular to the dorsal margin. The range is from Monterey to Lower California. The length is 39 mm., height 15 mm.

S. patula is the species that is commercially valuable. It has a larger shell than *S. lucida* with a polished, horn-colored epidermis. The rib is oblique and broad. The umbones are a little more than one-fourth the length from the posterior end. The range is from Alaska to the San Luis Obispo coast, though they are not so abundant in California as they are in Washington and Alaska (Weymouth). Dr. Baker found them in northern Japan. The length of a large specimen is 138 mm., and the height is 57 mm.

Family—MACTRIDAE

Mactra californica Conrad has valves that are nearly equilateral, and white, with a brownish epidermis. The unbones are nearly central (fig. 462) and the pallial sinus is deep and rounded. The lines of growth are fine, irregular ridges. The length is 55 mm., the height 41 mm., and the diameter 22 mm. The range is from the Straits of Juan de Fuca to San Diego.

Mactra nasuta Gould (fig. 463) shows only traces of the lines of growth and has a grayish epidermis. The umbones are nearer the posterior end than the anterior, the pallial sinus is deep and rounded. The length is 50 mm., height 35 mm., diameter 15 mm. This species is found in the San Diego region.

Fig. 462.—*Mactra californica:* x5/4.

Schizothaerus nuttallii (Conrad), the gaper (fig. 464), is also known in some localities as the summer clam, the horse clam, or as the otter shell. This is one of the largest clams to be found on our coast. Shells reach 6–8 inches in length and the weight of the whole clam may reach four pounds. The cartilage pits are large and of equal size in the two valves. Even when retracted, the siphons project far beyond the shell, and are covered with a heavy, brown epidermis. The pallial sinus is deep and its lower edge is united with the pallial line. The collector is made aware of the presence of these clams on a mud flat by the jets of water they squirt out periodically from their siphons. Most clams have the habit to some extent but we have seen no other species send a jet two or three feet into the air as we have seen these do on the mud flats at Tomales Bay. It requires good digging to secure the clams as they lie from a foot and a half to three feet deep in the mud. The huge siphons reach to the surface and are closed at will by two valves of a horny material. The clams are found from Alaska to San Diego and often live on the outer beaches as well as in muddy bays. Dr. Fred Baker also reports specimens from Japan. As to their value as food, Weymouth* (1920) reports that the clams are of excellent quality, siphon and all being used, though the latter has to be "skinned" before it is cooked. On account

---

*F. W. Weymouth—The Edible Clams, Mussels and Scallops of California; California Fish and Game Commission, Fish Bulletin No. 4, p. 56.

FIG. 463.—*Mactra nasuta;* x4/3.

FIG. 464.—*Schizothaerus nuttallii*, the gaper or summer clam; x2/3.

of the comparatively thin shells and the fact that they can not close tightly and hold their moisture the clams can not be marketed at any distance so they can be used only by local residents and campers.

Family—MYACIDAE

Mya truncata (Linnaeus), like the next species, has a peculiar spoon-like hinge tooth in the left valve. The posterior end is markedly truncate and a brownish, papery epidermis often extends beyond the shell (fig. 465). The lines of growth are distinct but radiating lines are absent. One specimen measures 65 mm. long and 44 mm. high. The species is circumboreal and found as far south as Bering Island and Puget Sound.

Mya arenaria Linnaeus, the soft shell (fig. 467), is also known as the soft clam, long clam, and mud clam. The umbo is central, the valves gape slightly at the posterior end, the cartilage pits are unequal, the one on the left forming a conspicuous spoon-like projecting tooth. The pallial sinus is deep, not united with the pallial line. The color is whitish with a gray periostracum which is often a rust color. Specimens five inches long have been gathered but most of them are smaller.

It is not a native species but seems to have been introduced with seed oysters from the Atlantic coast about fifty years ago. It is now found from British Columbia to Monterey though it is most abundant in San Francisco Bay where it obtained its first foothold. In fact it was the only form reported as being marketed from San Francisco Bay in 1919 (Weymouth 1920*) having displaced the native species. At some points, fences of stakes have been put in around large areas of the mud flats in order to keep out the skates, sting rays, and sharks which devour numbers of the clams. According to Weymouth, similar methods might make profit-

---

*F. W. Weymouth—The Edible Clams, Mussels and Scallops of California; California Fish and Game Commission, Fish Bulletin No. 4, p. 56.

FIG. 465.—*Mya truncata;* x¾.   FIG. 466.—*Cryptomya californica*, the false *Mya;* x4/5.

FIG. 467.—*Mya arenaria*, the soft shell, soft clam, long clam, or mud clam; x4/5.   FIG. 468.—*Saxicava arctica;* x1.

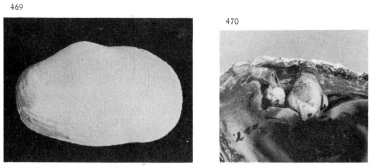

FIG. 469.—*Platyodon cancellatus;* x1.   FIG. 470.—*Pholadidea parva*, the little piddock. Two specimens are shown in their burrows in an abalone shell; x4.3.

able clam farms on some of the tide flats of Humboldt, Tomales, Morro, and False Bays. On the Atlantic coast, *Mya* is the shellfish most used, next to the oyster, and in California the Pismo clam is the only one sold more extensively. In 1919, 324,824 pounds of *Mya* were marketed.

Cryptomya californica (Conrad), the false *Mya* (fig. 466), has a small, elliptical shell, smoother than that of *M. arenaria*, showing only faint lines of growth. The radiating ribs show only at the posterior end. The right valve has a prominent lamellar tooth in which is a shallow cartilage pit. The pallial sinus is obsolete. The color is white with a grayish epidermis. One specimen measures 33 mm. long, 25 mm. high, and 13 mm. in diameter. The range is from Alaska to Mexico.

Platyodon cancellatus (Conrad) (fig. 469), like *Mya*, has unequal cartilage pits, the one on the left forming a conspicuous spoon-like tooth, but unlike *Mya*, it has the umbo nearer the posterior end of the shell. The valves are markedly truncated and gape widely at the posterior end. The lines of growth are distinct, even, and regular. The species is found from Bolinas Bay to San Diego. The length is 55 mm., height 34 mm., and diameter 28 mm.

Family—CORBULIDAE

Corbula luteola Carpenter, the yellow basket shell (fig. 472), is small and tinged with yellow or bluish. The surface of the shell is marked by fine, distinct, concentric ridges. The margins turn inward, forming a sub-marginal ridge. The length is 10.5 mm., height 7 mm., and diameter 5 mm. The range is from Monterey to Lower California.

Family—SAXICAVIDAE

Panope generosa Gould (fig. 471), is called the geoduck by the Indians of the north, the "eo" having the sound of "oi" in oil. The geoduck is the largest of the burrowing clams, reaching a length of 7 inches and a weight of 6.5 pounds (Weymouth, 1920). The huge siphons cannot be drawn within the shell nor can the valves be completely closed. The animal seems to be entirely too large for its shell. The shells are white with clearly marked lines of growth. There is a prominent tooth in each valve and the pallial sinus is wide and shallow. The range is from Puget Sound to San Diego.

Saxicava arctica (Linnaeus) is small with a thin shell which is wrinkled and irregular in shape (fig. 468). An ashy periostracum covers the white under layers. The hinge is small and near the anterior end. The length reaches 25 mm. The species is found from the Arctic Ocean to Panama and also in the Atlantic. They are often found attached to kelp holdfasts.

FAMILY—PHOLADIDAE

The mollusks included in this family, burrow into the rocks along shore. The shell is decidedly roughened so that it provides a rasping surface for digging. The following account is based on data given by Lloyd* (1897). The pholads' boring tools consist of the foot and four adductor muscles for motive power and the anterior part of the shell for the cutting instrument. The first movement is the extension of the foot at the anterior end of the shell. The foot is applied to the base of the burrow where it clings as a sucking disc. The contraction of the foot muscles brings the anterior part of the shell close to the wall of the burrow. Before or while this is done the anterior part of the anterior adductor and the ventral adductor have brought the front margins of the shell close together. The cutting now begins when the posterior adductor and the posterior part of the anterior adductor muscles contract. This pulls the anterior part of the shell open, thereby rasping the wall of the burrow, the motion of

---

*F. E. Lloyd—On the Mechanisms in certain Lamellibranch Boring Molluscs; Transactions of the New York Academy of Science, Vol. 16, pp. 307–316.

FIG. 471.—*Panope generosa*, the geoduck; x⅔.

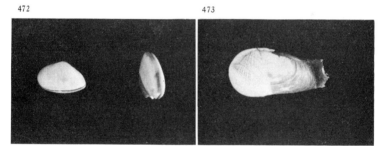

FIG. 472.—*Corbula luteola*, the yellow basket shell; x3/2. FIG. 473.—*Pholadidea penita*, the common piddock; x½.

FIG. 474.—*Barnea pacifica*, the western piddock; x1. FIG. 475.—*Zirfaea gabbi*, the rough piddock; x3/5.

the shell being upward and backward. The axis of motion passes through the umbo, swinging with the latter as a center from the ventral adductor to the posterior and back again with one round of attack. In most clams the axis of motion is constant and the hinge ligament causes the gaping of the valves in opposition to the adductor muscles which close them. A hinge ligament is absent in the piddocks. On account of its passive character it would be useless in boring.

Lloyd found that *Pholadidea* and *Parapholas* in early life resemble *Zirphaea* when they are in, what he calls, the working stage. The mantle is fused except where there is an opening for the muscular foot. After *Pholadidea* and *Parapholas* have provided themselves with burrows, the foot degenerates, and the opening in the mantle becomes only a pore through which water may pass. *Zirphaea* remains active throughout its life. The foot of the working forms has strong muscles whose points of attachment are the spoon shaped processes in the umbo of each valve.

Barnea pacifica (Stearns), the western piddock (fig. 474), is a mud borer. The shell is cylindrical, delicate, and marked with clear lines of growth and faint radiating lines. The anterior and posterior regions are not sharply marked off from each other as they are in others of the family, though the ridges are much less prominent at the posterior end. Anteriorly at the intersections of the radiating and concentric lines, there are sharp points which become more or less worn off. There is a dorsal accessory plate which is bent downward back of the umbones. The internal rib is short, curved, and flattened. The length is 20–60 mm. The range is from San Francisco to Lower California.

Zirfaea gabbi Tryon, the rough piddock (fig. 475), is a large borer and may grow to be 4.5 inches long. The huge siphon projects a distance equal to the length of the shell even when it is retracted as far as possible. The shell gapes widely at both ends. There is no accessory plate over the

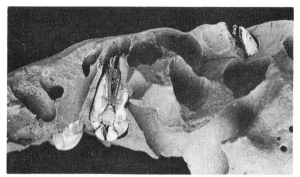

Fig. 476.—*Parapholas californica* in its rock burrow; x3/5.

Fig. 477.—A piece of piling showing burrows of shipworms; x1/3. The pile from which this section was taken had been in the water only three months. (This specimen is on exhibition in the Scripps Institution museum.)

hinge area, though there is a pad-like membrane, and the valves are reflexed. It is found from Alaska to San Diego, either in hard clay of the bay shores or in reefs on the ocean shore. A specimen 73 mm. long is 46 mm. high and 50 mm. in diameter.

Parapholas californica Conrad (fig. 476) is smaller than the last named species but reaches a length of two or three inches. The anterior end of the shell is nearly round and much roughened but the siphon end is smaller and covered with brown epidermis in the form of scales. There are two long, narrow plates over the hinge region. The range is from Coos Bay, Oregon to San Diego.

Pholadidea penita (Conrad), the common piddock (fig. 473), is globular at the anterior end and tapers to the posterior end where the shells gape somewhat and are covered with brown epidermal scales. The accessory plate over the hinge area is triangular. The length is 2–3 inches. The range is from Alaska to Guayaquil.

Pholadidea parva (Tryon), the little piddock, is a very small species. It burrows into the shells of the abalone and is frequently responsible for the formation of blister pearls. The range is from Monterey (Berry) to Lower California in *Haliotis* shells (fig. 470).

Family—TEREDIDAE

To this family belong the shipworms of universal ill repute. They are wood-boring mollusks, worm-like in shape. The valves of the shell are small, deeply notched, and borne at the anterior end of the body. Posteriorly at the sides of the slender siphons there are two hard structures, called pallets, which can be used to close the burrow against intruders while permitting enough water to enter to serve the needs of the animal.

For their food the teredos depend upon microscopic organisms that enter through the incurrent siphon with the water used for respiration. Like the rock-dwelling piddocks and the razor clams that burrow in mud, these mollusks enter wood for protection. The mouth of the burrow is very small because it is made by the young animal and serves as an opening from which the slender siphons, guarded by the pallets, may be extended. At first the burrow crosses the grain of the wood but it soon turns and runs with the grain (fig. 477). The piling of wharves may be completely riddled with the burrows in a few months and in the present day they do more damage in that way than by drilling into the hulls of wooden ships. The cost of their ravages in terms of replacements and repairs to waterfront structures amounts to many million dollars annually. For example, engineers estimate that in two years in an area including only the upper part of San Pablo Bay and the region up to Antioch on the lower San Joaquin River the damage done by marine borers, the teredos, and the isopod crustacean *Limnoria*, amounted to more than $15,000,000.

The boring is done by a peculiar motion of the two valves of the shell. They are not attached by a continuous hinge of the sort usually found in bivalves but meet in two separate places where knob-like projections occur on each half of the shell. With these as pivots a rocking motion is brought about by the action of the two adductor muscles. The anterior muscle pulls the valves toward each other in front. Then the stronger posterior adductor, with added leverage due to its attachment to an extension of the shell, spreads these anterior edges of the shells apart forcibly. Thus the two valves, rocking on the dorsal and ventral articulations as fulcra are able to rasp the wood. The cutting is done by the outer anterior part of the valve when the shell is opened outward. Toothed ridges parallel to the margin provide the roughened surface for cutting. As these ridges are worn down, others are secreted at the edge by the mantle, keeping the boring mechanism in constant repair. During the boring the animal is held in position by the foot. The sawdust is removed by way of the mouth, passing through the digestive tract apparently without being digested. The elongated body is covered with the mantle, which not only secretes the shell and the pallets and lays down a thin deposit of nacre as a lining for the burrow, but is thickened behind the shell to act as a washer, filling the burrow too tightly to permit particles of wood to get in between the animal and the wall of its burrow.

Three species of shipworms are found on the Pacific coast. The largest is *Bankia setacea* (Tryon), which Professor Kofoid* calls the giant or plumed pileworm. Burrows belonging to this species reach a depth of three feet and a diameter of seven-eighths of an inch. It may be distinguished by the feather-like pallets which are made up of a series of fifteen to thirty similar parts, reaching in large specimens a length of two inches and a width of about one-fourth of an inch. Dr. Dall gives the range as extending from Bering Sea to the Gulf of California. It has caused trouble at San Pedro and also in San Francisco Bay where it is responsible for much of the destruction of timber that has taken place below the Straits of Carquinez.

Teredo navalis Linnaeus is the species most commonly figured in text books. It is widespread in distribution but seems to have been recently introduced in San Francisco Bay, the first report being from Mare Island (Barrows 1914), and at present it is not known elsewhere on the Pacific coast. As it is capable of living in brackish water, it is very destructive to piling above the Straits of Carquinez, particularly when the rivers are carrying less than the usual amount of water (Kofoid).† The larvae are free-swimming and the rapidity with which infection can spread among the piling can be realized when it is known that the Dutch naturalist Sellius estimated the number of eggs produced by one individual at 1,874,000. This is considered a conservative

---

*C. A. Kofoid—The Marine Borers of the San Francisco Bay Region; Report on the San Francisco Bay Marine Piling Survey (1921) p. 25.

†C. A. Kofoid—The Marine Borers of the San Francisco Bay Region; Report on the San Francisco Bay Marine Piling Survey (1921), p. 46.

estimate.† *T. navalis* can be distinguished from the plumed pileworm by the shape of the pallets which are each composed of a single paddle-like element with the ends concave and the sides extended backward in the form of two sharp points. The excurrent siphon is about half the length and diameter of the incurrent one and both are marked with spots of reddish brown.

Teredo diegensis Bartsch is a smaller, relatively stouter form which adheres very firmly to the lining of the tube. The pallets have two backward projecting prongs of greater proportionate length than the points found on the pallets of *T. navalis*. The siphons are shorter than in the preceding species and differ also in being nearly equal in length and unpigmented. The perforations reach a depth of about six inches and the mollusk is from one to four inches long. It is known from San Diego to San Francisco Bay but seems to be scarcer and less destructive than the other two borers.

The economic importance of these mollusks has led to an extended study of their habits and depredations and many interesting details which can not be included here may be obtained from the papers of Kofoid, Miller, Sigerfoos, and others listed in the bibliography.*

Class—SCAPHOPODA

(*The tooth shells*)

Among shells collected at the beach one frequently finds the tooth shells, small, white, cylindrical, and shaped like miniature elephant tusks. They are open at both ends and

---

†C. A. Kofoid—The Marine Borers of the San Francisco Bay Region; Report on the San Francisco Bay Marine Piling Survey (1921), p. 39.

G. Sellius (1733)—Historia naturalis Teredinis, seu Xylophagi marini tubuloconchoidis speciatim Belgici (Trajecti ad Rhenum, Besseling).

*Articles by C. A. Kofoid and R. C. Miller in annual reports of the San Francisco Bay Marine Piling Survey, beginning with the year 1921.

R. C. Miller—The Boring Mechanism of *Teredo;* University of California Publications in Zoology, Vol. 26, pp. 21–80.

C. P. Sigerfoos—Natural History, Organization, and Late Development of the *Teredinidae*, or Shipworms; Bulletin U. S. Bureau of Fisheries, Vol. 27, pp. 191–231.

unless they are taken by the dredge are not likely to contain the living animal, for they live buried in the sand beyond low tide limit, sometimes at great depths.

Miss Rogers* gives an interesting account of the way the Indians of the northwest used these shells. They strung

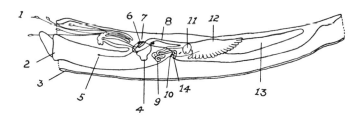

FIG. 478.—*Dentalium*, left side view. (After Pilseneer in Lankester's Treatise on Zoology.)
1. Captacula.
2. Foot.
3. Mantle.
4. Radular sac.
5. Pedal ganglion and otocyst.
6. Cerebral ganglion.
7. Pleural ganglion.
8. Esophagus.
9. Intestine.
10. Visceral ganglion.
11. Kidney.
12. Liver.
13. Gonad.
14. Anus.

them on fine threads of deer sinew and used them both for money and for ornament. They valued the perfect shells more highly than the worn ones and obtained them by dredging in shallow water along shore. They dredged from canoes and used long fine toothed rakes to bring up the mollusks. A fathom string of 25 shells was valued at about £50 and a string the same length, made up of 40 smaller, and for this reason, less valuable shells, would buy a slave.

In many ways, the scaphopods are unlike the rest of the mollusks, having no heart, gills, eyes, or tentacles, but they are like them in having a radula, a rasping, tongue-like structure, a foot, and a mantle which secretes the shell. The arrangement of the organs is shown by the diagram. The animal digs into the sand with the foot and lies buried obliquely with the posterior opening projecting up into the water and serving both as an opening for the incoming and outgoing currents of water. Cephalic filaments or captacula (fig. 478) extend out beyond the mouth, radiating in all

---

*Julia Ellen Rogers—The Shell Book; Doubleday, Page and Co., p. 300.

directions. They are long and thread-like, with club-shaped ends and are thought to help in capturing the minute plants and animals that make up the food. There are no blood vessels, the only structure that in any way suggests a heart, being a blood space near the anus which is somewhat contractile. The sexes are separate and the ova laid singly.

Family—DENTALIIDAE

Dentalium neohexagonum Pilsbry and Sharp, the hexagonal tooth shell (fig. 479), is thin and white. The cross section is a

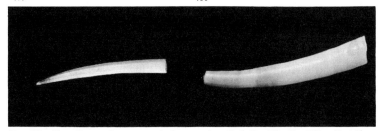

FIG. 479.—*Dentalium neohexagonum*, the hexagonal tooth shell; x3/2. FIG. 480.—*Dentalium pretiosum*, the Indian's tooth shell; x3/2.

hexagon. The length is 30–50 mm. The range is from Monterey to Central America.

Dentalium pretiosum Sowerby, the Indian's tooth shell, is curved, tapering posteriorly, heavy, with fine incremental rings on the surface, and striated posteriorly (fig. 480). The aperture and cross section are circular (R. Arnold). The length is 10–25 mm. The range is from Alaska to San Diego.

Cadulus californicus Pilsbry and Sharp (fig. 481). The surface is smooth and glossy, bluish white and slightly translucent. The length is 4–10 mm. The range is from Alaska to Ecuador.

Class—GASTROPODA
(*The snails and sea slugs*)

The gastropods include the familiar snails, sea slugs, limpets, and abalones, as well as the less known pteropods. They are unsymmetrical, usually form a spiral shell which is more or less of a protection to them, have a distinct head which usually bears tentacles, and a broad flat foot for locomotion.

The shell is perhaps the most conspicuous feature of the group as a whole, though it is absent in the nudibranchs and rudimentary in the sea hare.

FIG. 481.—*Cadulus californicus;* x3/2.

Mollusks which have no shell when they are adults, usually in the young stages possess a shell which disappears later. The shell is formed by certain cells of the mantle, particularly those at the margin which build on shell material at the edge of the aperture. The shell can be thickened through the addition of successive pearly layers by all parts of the mantle. This addition of material often

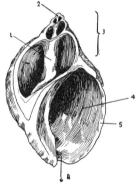

FIG. 482.—Shell of *Thais* with part of the outer wall removed to show the columella and whorls; x1.5. 1. Columella. 2. Apex. 3. Spire. 4. Aperture. 5. Outer lip. 6. Anterior canal.

FIG. 483.—*Olivella biplicata* with a part of the outer wall removed to show the extent to which the inner structures have been absorbed; x2.

takes place when the mollusk is being attacked by boring mollusks, worms, or sponges. The drawing (fig. 482) shows the interior of the shell and the columella to which the mollusk is attached by muscles. It is this columellar muscle which, by contracting, pulls the mollusk back into its shell.

The shell building process is not continuous but is interrupted by rest periods which leave their mark on the shell as lines of growth. Some shells such as *Bursa* (fig. 484) and *Murex* (fig. 546) form a thick lip at the aperture before each rest period, which remains as a prominent longitudinal rib or varix to mark the end of each period of growth. Some species have the power when they are adults of absorbing parts of the shell that have been previously formed. The drawing of *Olivella biplicata* (fig. 483) shows the condition within the shell where the axis and the internal portions of the whorls have been absorbed. Some species on the other hand, *Caecum* for example (fig. 583), cease to occupy the tip of the spire and close off a portion of the tip with a shelly septum, after which the unoccupied portion frequently breaks off. The marginal teeth or ribs frequently seen within the outer lip of the shell may be dissolved away as the animal grows or they may be left to become permanent features of the shell. Names of the various parts of the shell that are used in species descriptions are shown on figure 484.

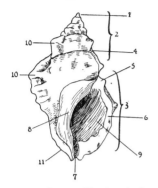

FIG. 484.—*Bursa californica*, the frog shell; x⅓.
1. Apex. 6. Outer lip.
2. Spire. 7. Anterior canal.
3. Body whorl. 8. Columella.
4. Suture. 9. Aperture.
5. Posterior canal. 10. Varix.
11. Umbilicus.

When the mollusk retreats within its shell, the aperture is usually closed by an operculum, a plate of cuticle which is situated on the dorsal side of the foot just over the ends of the fibres of the columellar muscle. When the animal is crawling, the operculum may be out near the tip of the foot (fig. 485), but when the animal is contracted within its shell, it just closes the aperture (fig. 571). The head of the mollusk bears one or two pairs of tentacles, and usually a pair of eyes. The latter may be near the base of the tentacles or half way out on them, but marine gastropods never have them at the tip of the tentacles as the land snails do. These eyes show great differences in complexity of structure all the way from a simple depression in the skin lined with pigmented and retinal cells to those of

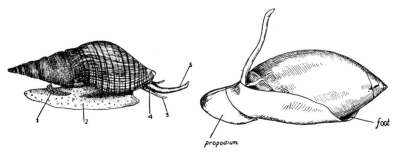

FIG. 485.—*Cryptoconus carpenterianus* as it appears when fully extended; x1.
1. Operculum.
2. Foot.
3. Tentacle.
4. Eye.
5. Siphon.

FIG. 486.—*Olivella biplicata*, as it appears when fully extended; x2.

Murex in which the "eye ball" encloses a sort of lens. At the anterior end of the head is the mouth, which is in some cases in a long proboscis, but more often is on the under side of a short snout. The gastropods with a long proboscis are often carnivorous and the others vegetable feeders, but there are many exceptions to this statement. Just inside the lips is the pharynx or muscular throat, which contains one or two jaws and the radula or lingual ribbon. The jaws are used to bite off the food and the radula tears it into fine pieces. Some carnivorous species have no jaws, the radula serving to tear up the food sufficiently. The radula is a chitinous band which lies in the floor of the mouth curving upward a little on each side. The upper surface is covered with teeth of various numbers and shapes, arranged in rows. The teeth can be raised and depressed by means of muscles and, together with movements of the radula as a whole, serve to rasp or card the substances that pass over them. The pattern formed by the teeth of the radula is usually characteristic for each species so that it frequently serves as a test in careful species distinction. The foot forms a broad, flat sole, which can creep over flat surfaces (fig. 485). Mucus secreted by a gland which opens on the under side aids in locomotion. In the pelagic pteropods and heteropods the foot is modified to form swimming organs. The main bulk of the body is taken up by the visceral mass. The mantle covers this and extends beyond it, leaving only the head and foot projecting on the under side. Either in front, at the back, or at the side, is a space between the body and the mantle, called the mantle cavity. In this cavity are the gills or lungs, the openings of the anus and the ducts from the kidneys, the osphradium, and a mucous gland. On account of the spiral twisting of the body and the displacement of certain organs, one member of the usual pair is lost in many instances, so that there is usually only one gill, kidney, auricle, and osphradium.

Three sorts of respiratory organs are found among the gastropods, ctenidia or gills, "lungs," and adaptive gills. Ctenidia, also known as branchiae, are the usual form of breathing organs. The name ctenidia (little combs) suggests the feathery character of the organs. A few species are air breathers having a lung cavity, while others are provided with both a gill and lung so that they can breathe equally well in the water and out of it. The lungs consist of a sac with its walls richly supplied with blood vessels so that the blood receives its oxygen supply from the air that enters the cavity. Adaptive gills are the projections from the dorsal body wall of the nudibranchs, but many of the group are without these, and breathe through the surface of the body wall, which may be smooth or covered with papillae.

The sense of touch is well developed over the whole body surface, but is especially delicate in the tentacles. The osphradia lie near the ctenidia and seem to be sense organs for testing the respiratory fluid. The rhinophores (fig. 500) are thought also to contain organs of smell or of some chemical sense. Experiments show that mollusks are sensitive to odors, many species being attracted by bait which has a strong odor. Otocysts, tiny sacs lined with sensory cilia and filled with fluid secretion in which are suspended

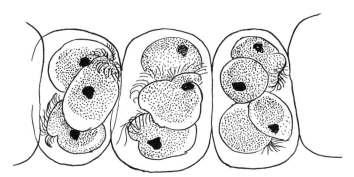

FIG. 487.—Egg chain of *Navanax inermis*, greatly magnified, to show the developing larvae within the capsules.

calcareous particles, are usually present in mollusks in the foot, and enable them to preserve their equilibrium and sense of direction.

The opisthobranchs and pulmonates are usually hermaphroditic, but the other gastropods are as a rule unisexual. The eggs in most cases are laid in protective chitinous or gelatinous capsules (figs. 487, 489, 493). The larvae leave the capsules when they are in the veliger stage (fig. 488). They swim by cilia and lead a pelagic life until the adult form is reached.

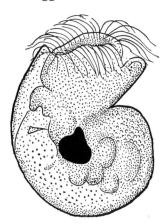

FIG. 488.—The veliger larva of *Navanax inermis* swims by means of cilia, hair-like structures which serve as oars. This drawing was made one week from the time the eggs were laid. (Greatly enlarged.)

ORDER—PTEROPODA

The pteropods or "wing-feet" are pelagic animals. They live in the open ocean, many of them rising to the surface only at night and sinking to greater depths in the day time. Others remain close to the surface for a part of the day at least, seeming to prefer the warmer water near the surface to the cold of the great depths.

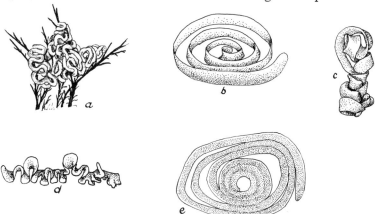

FIG. 489.—Egg masses of mollusks that were laid in laboratory aquaria. With the exception of A which were deposited on seaweed, these ribbon-like masses were attached to the glass of the aquarium near the surface of the water.
  A. Egg mass of *Flabelina iodinea*.
  B. Egg mass of *Chromodoris macfarlandi*.
  C. Egg mass of *Chromodoris californiensis*.
  D. Egg mass of *Acanthodoris rhodoceras*.
  E. Egg mass of a brown dorid.

Pteropods have wing-like appendages on each side of the head, which are modifications of the lateral lobes of the foot. The body is, in some cases, enclosed in a thin, glassy, transparent shell. They are extremely active in their movements and swim rapidly by means of the wing-like "fins" so that one thinks of them as aquatic butterflies.

Pteropods are usually associated together in enormous numbers. One species, *Clione limacina*, which is especially abundant in northern waters even among the ice floes, is known as

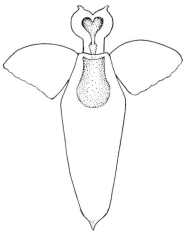

FIG. 490.—*Clione kincaidi*. A pteropod, or wing-footed mollusk which swims the high seas; x4.

"whales' food." Deposits at the bottom of the ocean contain great numbers of pteropod shells in some regions. These pteropod ooze deposits, according to Murray and Hjort,* are in the shallower waters usually far from continental land on oceanic ridges, especially within the coral reef regions where the surface water is comparatively warm throughout the year. The deposits cover a relatively small extent of ocean floor, but great numbers of pteropods, particularly of the polar species, are without shells and so would leave no trace.

The *Pteropoda* are classified in two suborders, the *Thecosomata* which have a shell and are mainly carnivorous creatures, and the *Gymnosomata* which have a distinct head, no shell or mantle and are limited to a vegetable diet.

Suborder—GYMNOSOMATA

Family—CLIONIDAE

Clione kincaidi Kjerschow-Agersborg (fig. 490) is small and translucent like a bit of frosted glass. The organs show through the body wall as bright orange colored masses. The length is about 15 mm. We saw large numbers of them at Friday Harbor, Puget Sound, in July.

Order—OPISTHOBRANCHIATA

(*Sea slugs*)

The order *Opisthobranchiata* includes the sea slugs. They are all marine snails which, though they have shells in the embryo form, frequently have lost them by the time they have reached adult life. They usually have two pairs of tentacles, sometimes only one pair, and the eyes are near the base of the posterior pair. The auricle and gills are back of the ventricle and the blood flows from the gills into the

---

*Sir John Murray and J. Hjort—The Depths of the Ocean; The Macmillan Company, p. 163.

auricle. The members of one group, the nudibranchs, have no true ctenidia, respiration being carried on through the branchiae or through the body wall.

SUBORDER—TECTIBRANCHIATA

The suborder *Tectibranchiata* includes the opisthobranchs that have the gill more or less enclosed in the mantle cavity, instead of uncovered, as in the *Nudibranchiata* or naked gilled opisthobranchs. The tectibranchs usually have a shell although sometimes, as in the case of the sea hare, it is a mere rudiment.

TRIBE—CEPHALASPIDEA

FAMILY—ACTAEONIDAE

Actaeon (*Rictaxis*) *punctocoelatus* (Carpenter), the barrel shell, is white with two series of narrow black bands (fig. 491). There is a fold on the columella and the aperture is long and narrow. The body of the mollusk is pure white. Unlike most of the *Opisthobranchs* it has an operculum. The specimen shown was found in January in a tide pool near the caves at La Jolla. Keep* says they come to the tide pools to lay their eggs at certain times of the year. The height is 10 mm., diameter 5 mm. The range is from Monterey to Lower California.

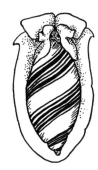

FIG. 491.—*Actaeon punctocoelatus*, drawn from life; x3.

FAMILY—BULLARIIDAE

Bullaria gouldiana (Pilsbry), the cloudy bubble-shell, has a large, globular, moderately thin shell. The ground color is gray or brown, often mottled with a lighter color, which gives a cloudy effect over the largest part of the body whorl. There is a pit at the apex (fig. 495) and the aperture extends the full length of the shell. The

---

*Josiah Keep—West Coast Shells (1911); Whitaker, Ray-Wiggin and Co., p. 120.

photographs (figs. 492 and 493) of the living specimens show how a large part of the shell is covered and how the

Fig. 492.—*Bullaria gouldiana*, photographed from life; x½.

Fig. 493.—*Bullaria gouldiana*, with egg mass, photographed from life; x½.

foot extends posteriorly. The mantle, foot, and cephalic shield are pale orange or yellow with white, irregular spots, and the eyes are small black pigmented spots. The cloudy bubble-shell is found on mud flats. The eggs are laid in the summer time, in long, gelatinous strings (fig. 493) which may be found caught on the eel-grass. Miss Rogers* states that the food of *Bulla* (*Bullaria*) consists of small bivalves and snails which are swallowed whole and ground to pieces between the strong walls of the gizzard, also that the mantle flaps are sometimes used in swimming. The length of the shell is 50 mm. The range is from Santa Barbara to Mexico.

Family—AKERATIDAE

Haminoea virescens (Sowerby) and *H. vesicula* (Gould), the white bubbles, are small with thin shells. They are found on mossy rocks and in tide pools or on the mud flats of the bay shore where they plow along on or just under the surface of the mud. When in motion, the yellowish brown body extends far beyond the shell (fig. 496). One individual with a shell 11 mm. long measured 28 mm. when fully extended.

*Julia Ellen Rogers—The Shell Book; Doubleday, Page and Co., p. 244.

PHYLUM—MOLLUSCA

Often in light colored specimens, the pulsations of the heart may be seen through the shell. During the summer the eggs are usually abundant on the mud or adhering to the eel-grass. They are arranged in large spirals within a soft, ribbon-like roll of gelatine. The aperture of *H. virescens* is wider below than is that of *H. vesicula* (figs. 497 and 498).

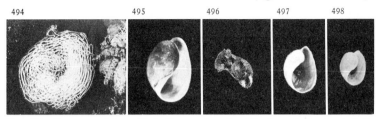

FIG. 494.—Egg mass of the mollusk, *Navanax inermis*. (About one-half natural size.)
FIG. 495.—*Bullaria gouldiana;* x⅛. FIG. 496.—*Haminoea* sp., photographed from life; x½.
FIG. 497.—*Haminoea virescens;* x4/5. FIG. 498.—*Haminoea vesicula;* about one-half natural size.

H. virescens is 15 mm. long. The range of this species is from Santa Barbara to Mexico. *H. vesicula* is 20 mm. long. Its range is from Vancouver Island to the Gulf of California.

FAMILY—AGLAJIDAE

Navanax inermis (Cooper), the striped mollusk (plate 8), has beautiful colors and is abundant on the mud flats at Mission Bay and Balboa as well as in the tide pools of rocky beaches. The range of the species is from Catalina to San Diego. The drawing indicates the color and pattern, but the markings are variable. The mollusk is nearly always the same shade of brown, dark above, lighter below, but the yellow color may be irregularly arranged or in narrow or wider lines, sometimes with dots and dashes between the lines. Often the lines are white instead of yellow. In some specimens the yellow border of the mantle is even in width and continuous while in others it is a series of dots and dashes. The blue is generally a series of dots or dashes but was very scattering in our smallest specimen. In most of our specimens, one posterior flap is longer than the other.

Almost without exception individuals of this species are infested with a parasitic copepod, *Pseudomolgus navanaci* Wilson. The copepods attach themselves to the gills so firmly that it is hard to dislodge them. In the laboratory *N. inermis* devoured *Haminoea* in considerable numbers, swallowing the creatures whole and later discarding the unbroken shell. In specimens taken on the mud flats at Mission Bay, one can frequently feel the *Haminoea* shells inside the digestive tract of the mollusk. We have found the eggs in the summer time and in January. Figure 494 shows an egg mass deposited in the laboratory. The eggs are looped and festooned on the eelgrass in a skein-like gelatinous mass. The mollusks may reach 7 inches in length.

TRIBE—ANASPIDEA

FAMILY—APLYSIIDAE

Tethys californica (Cooper) (formerly known as *Aplysia*) the sea hare (fig. 499), the largest of our sea slugs, sometimes reaches a length of fifteen inches. It is found from Monterey to San Diego. The shell is internal and is little more than a thin horny plate buried in the mantle (fig. 504). There are four tentacles on the head and the eyes are in front of the posterior pair. The parapodia form large flaps which extend up over the dorsal part of the animal. When disturbed, these creatures eject a purplish fluid into the water which serves to conceal them and thus make their capture difficult or frighten the enemy away. In fact, the fluid was once believed to be extremely poisonous. Miss Rogers states that one Mediterranean species was charged with causing baldness, but she also reports that natives of the Friendly

FIG. 499.—*Tethys californica*, the sea hare; about one-fourth natural size.

Julia Ellen Rogers—The Shell Book; Doubleday, Page and Co., p. 246.

and Society Islands eat allied species raw, so we can rest assured that we may at least handle the animals with safety. The sea hares are often very abundant in the rocky tide pools of the southern coast, while at other times they are scarce. It is thought that when they come together in large numbers it is their breeding time. We have found the eggs in November and also in March.

The color of the sea hare varies from greenish to redbrown, sometimes with irregular lines or blotches of a darker color. Its color makes the animal inconspicuous as it crawls about over the red-brown seaweeds which constitute its food.

This species is so large and so abundant that it makes an excellent subject for dissection. The mouth is a vertical slit at the anterior end of the body. Just inside the mouth are the horny mandibular plates and beyond them is the radula. With this radula, which bears many small teeth so that its surface looks like that of a file (figs. 501 and 502), the seaweed is torn or broken into bits small enough to be taken into the stomach. The first stomach is thin walled and large, but the second has thick muscular walls and inside it is provided with pyramid shaped teeth (fig. 503). These grind the food so that it is in small pieces when passed into the third stomach which is lined with finer teeth. The long intestine and the hepatic caecum are imbedded in the mass of liver which occupies most of the body cavity (fig. 500). The anus opens into a funnel made by the posterior part of the mantle.

The large cerebral ganglia above the pharynx show clearly, and nerves may be traced from them to tentacles, eyes, and the integument near by. Connectives from the cerebral ganglia lead down to the buccal ganglia, also laterally and downward to the pleural and pedal ganglia. From these ganglia nerves may be traced in different directions, the ones leading back into the foot, and those to the visceral ganglia being especially clear.

The heart lies in the pericardial chamber, just in front of the ctenidium, and has two chambers, the ventricle being anterior. The nephridia extend to the left of the heart, within the mantle, above the pericardial cavity. The organ of Bohadsch or mucous gland is a nodular body with an opening into the mantle cavity.

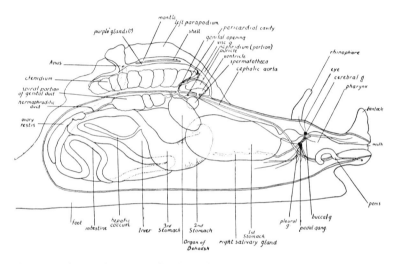

Fig. 500.—Diagram of the organs of *Tethys californica*, the sea hare. (Greatly reduced.)

The ovotestis is imbedded in the liver at the posterior end of the body, and from it leads the small hermaphroditic duct extending to the genital mass, which contains nidamental and albumen glands and the fertilization chamber. The spermatocyst lies next and the large hermaphroditic duct extends to the external opening. This duct is made up of two parts, the ovo-spermatic duct and the copulatory duct. Into the copulatory duct opens the spermatotheca near the genital opening. The ovo-spermatic duct, which acts both as an oviduct and as vas deferens, continues forward as an external ciliated groove, which leads down to and into the

PHYLUM—MOLLUSCA

opening of the penis sheath to the tip of the penis, forming a conduit for the spermatozoa in copulation. The sea hare

FIG. 501.—Radula of *Tethys californica*; x2.

FIG. 502.—Two types of teeth from the radula of *Tethys californica*; greatly enlarged.

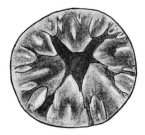

FIG. 503.—Teeth in the second stomach of *Tethys californica*; enlarged.

FIG. 504.—Shell of a small specimen of *Tethys californica*; x⅔.

thus belongs to the group of mollusks having two reproductive openings, so that one individual may perform the part of male to a second individual and female to a third.

SUBORDER NUDIBRANCHIATA

Anyone who has seen color plates of nudibranchs is eager to know more about them, to find some for himself and to see if the artist has not exaggerated the coloring and beauty of these relatives of our well known and unpopular garden slug. If the colors are not correct the difficulty lies in the inability of the artist to get paints that are clear and bright enough to do the creatures justice. Nudibranchs are found on the algae in the tide pools, or on the kelp at some distance out from the shore. Even when they are in the tide pools they are not easily seen for their bright colors often blend with the brightly colored seaweeds, and if they have been left stranded out of water on rocks, algae, or sponges, they are shapeless things, and even the expert finds it hard to recognize them. The nudibranchs have no shells, though a small coiled shell is always present in the embryo. They do not have true gills, respiration being carried on either through the body surface or through gill-like structures on the back. They have two pairs of tentacles. The *Aeolididea* often have nematocysts or sting cells in their cerata, the substitute gills which form a waving fringe on the back. It has been shown however that these nematocysts have been derived from the hydroids upon which the animals have fed, and may be present at one time and absent at another time. Professor Herdman experimented with various species and found that some very conspicuous nudibranchs are inedible and are let alone by larger animals, while other forms that are palatable harmonize so perfectly with their usual background that they are not readily seen by their enemies. Professor Herdman even went so far as to eat a live specimen of *Ancula cristata*, a conspicuous form, to find out whether it was palatable or not. He reported that "the taste was pleasant, distinctly like that of an oyster," but that the fish did not seem to agree with him since they did not attempt to taste the mollusks though they swam close and looked at them.†

Much of the following descriptive material is taken from the publications of Dr. MacFarland,* to which the reader is referred for more complete descriptions and lists of species.

---

†A. H. Cooke—Molluscs; Cambridge Natural History, Vol. III, p. 72.
*F. M. MacFarland—Opisthobranchiate *Mollusca* from Monterey Bay, California; Bulletin of the Bureau of Fisheries, Vol. 25, pp. 109–151.
—The Nudibranch family *Dironidae;* Zoologische Jahrbücher, Supplement XV, 1 Band, pp. 515–536.
—The Morphology of the Nudibranch Genus *Hancockia;* Journal of Morphology, Vol. 38, pp. 65–104.
—The *Acanthodorididae* of the California Coast; The Nautilus, Vol. 39, pp. 49–65 and 94–103.

Tribe—HOLOHEPATICA

Family—TRITONIIDAE

Tritonia festiva (Stearns) is translucent cream color or white, with a delicate pattern of chalky-white markings as shown in the color sketch (plate 7). The digitate veil on the head and the characteristic tentacles also show well in the figure. The gills are arborescent plumes arranged on a ridge on each side of the back. The original description was based on specimens from Point Pinos, near Pacific Grove. We have found it there and also at Corona del Mar near Balboa.

Family—DORIDIDAE CRYPTOBRANCHIATAE

This family has the branchial plumes in an arc or circle and usually joined together at their bases. These branchiae are retractile, being drawn down into a common cavity by the animal when it is disturbed. The rhinophores, or posterior pair of tentacles, are usually club-shaped and bear a number of folds, more or less spirally or obliquely arranged. The mantle is large and covers the head. The dorsal surface of the body is generally roughened.

Archidoris montereyensis (Cooper) (plate 8) is abundant in Monterey Bay, both in tide pools and on the piles of the wharf at Monterey. It is light yellow, sprinkled on the back with brown, greenish, or black dots. There are also patches of the darker color especially toward the middle of the dorsal side. These black patches of color are on the low, conical, closely set tubercles as well as between them. It is one of the giants of the group, for its length may reach 50 mm., width 25 mm., and height 12 mm. There are seven large, spreading branchial plumes which have a dusty appearance because of the scattered brown or black dots. The rhinophores are elongated and finger-like.

Anisodoris nobilis (MacFarland), is a large species resembling the preceding in some respects (plate 8) but larger, reaching a length of 20 cm. The color varies from rich orange-yellow to light yellow, mottled with patches of dark brown or black between the tubercles rather than on them, as is the case in the preceding species. The tubercles may be knobbed on the tips and small ones appear between the larger ones. The oral tentacles, unlike those of *Archidoris montereyensis*, are flattened and auriculate with an external groove. The branchial plumes are large and spreading, pinkish, and tipped with white.

Rostanga pulchra MacFarland (plate 8) is abundant at Pacific Grove and a number of them have been found in the tide pools at La Jolla and Laguna. They are especially numerous on a red sponge which encrusts the underside of overhanging rocks. They closely resemble this sponge in color and are covered with minute red or brown papillae, the ground color is irregularly mixed with white, yellowish, orange, red, and black. The body is elliptical and the ends of the body are equally rounded. The largest specimens that have been reported were 18 mm. long. The eggs are deposited within a narrow, flat ribbon of gelatinous substance which is attached by one edge in a closely wound coil. The ribbon is orange-red and matches the sponge upon which it is frequently laid.

Diaulula sandiegensis (Cooper) is pale yellow or brownish and easily distinguished by the row of dark brown or black rings (fig. 506) down each side of the dorsal surface. These rings may vary in size, number, and position. The head is entirely covered by the mantle but the tip of the foot shows beyond the posterior edge of the mantle when the animal is crawling. The surface is covered with minute processes giving it a velvety appearance. There are six branchiae which can be wholly retracted. Specimens 5–6 cm. long are frequently found. The species has a long range, being found from Sitka to San Diego, and a variety lighter colored than this is found off the coast of Patagonia.

Aldisa sanguinea (Cooper) is light to dark red, dotted with minute black spots, and thickly covered with small conical tubercles (plate 8). In the middle line just in front of the branchiae is a rounded or oval spot of black and another similar spot sometimes lengthened into two spots, appears just back

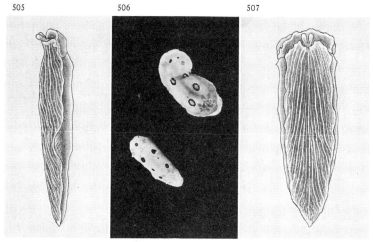

FIG. 505.—*Pleurophyllidia californica*, side view, showing the gills under the mantle; x1.
FIG. 506.—*Dialula sandiegensis*, photographed from life; x½. FIG. 507.—*Pleurophyllidia californica*, dorsal view; x1.

of the rhinophores. This species is only 17 mm. long and 8 mm. wide and is found in rocky tide pools near Monterey.

Cadlina marginata MacFarland is translucent yellowish white (plate 9) with low tubercles, which are each tipped with a spot of lemon yellow surrounded by a ring of white. The margins of mantle and foot have a narrow band of lemon yellow. The tips of the branchiae and the rhinophores are also yellow. Large specimens are 45 mm. long. The range of the species is from British Columbia to the Monterey Bay region.

Cadlina flavomaculata MacFarland (plate 9) also has a yellowish white mantle with low rounded tubercles. Down each side from the rhinophores to the branchial plumes is a row of lemon-yellow spots on low tubercles. The rhinophores are darker than the mantle, sometimes being brown or black, and the branchial plumes are white. This species reaches a length of 20 mm. The breadth is 8 mm. They are found from Pacific Grove to San Diego.

Chromodoris californiensis Bergh (plate 9) is a beautiful ultramarine blue, a little lighter on the margins of the mantle and the foot, with numerous, bright, orange-colored, oblong spots in two lengthwise rows on the mantle, one row down each side of the foot, and a group of round spots on the mantle in front of the rhinophores. The body is narrow and the mantle is not extended at the sides. The mantle projects beyond the oral tentacles but the foot reaches out nearly 20 mm. behind the mantle when the animal is crawling. There are about twelve plumes in the branchiae. This species is fairly abundant in the tide pools from Monterey to San Diego. The length is about 6 cm. Figure 489-c shows the appearance of an egg mass laid in the laboratory.

Chromodoris porterae Cockerell is smaller than *C. californiensis* being only 11–12 mm. long. The color is deep ultramarine blue, with two broad, orange stripes running lengthwise on the mantle, almost coming together back of the branchiae, and ending at the rhinophores in front. Anterior to the rhinophores is a transverse band of orange which is apparently a continuation of the side stripes of the mantle. Down the median line of the mantle is a light blue stripe and the mantle margin is edged with white. The species is found from Monterey to San Diego.

Chromodoris macfarlandi Cockerell (plate 11) has a ground color of reddish purple which is pale in contrast to the rich colors of *C. porterae* and *C. californiensis.* The mantle has a margin of orange, bordered with white and there are three longitudinal stripes of yellow, the lateral ones beginning at the rhinophores and meeting back of the branchiae and the median one beginning in front of the rhinophores and running between them. Large specimens are 5–6 cm. long.

FAMILY—DORIDIDAE PHANEROBRANCHIATAE

Laila cockerelli MacFarland (plate 9) is a small slug-like species only 20 mm. long and 7 mm. wide. The ground color is white with the rhinophores, tail, and the numerous club-shaped papillae of the body tipped with bright orange-red. These papillae, which are 1–6 mm. long, are closely set around the mantle margin in short oblique rows with the largest ones toward the middle line. Tiny spicules support the papillae, and may also be seen through the dorsal surface giving the appearance of a network of fine lines. There are small, low tubercles on the dorsal surface of the body between the papillae. The rhinophores may be drawn down into sheaths, which have smooth margins. There are five non-retractile branchial plumes, which are often tipped with orange-red. It is found from Monterey to San Diego in tide pools.

Triopha carpenteri (Stearns) is large, measuring 60 mm. in length, 15 mm. in width, and 29 mm. in height. The ground color is white, sometimes yellowish, often with white spots on very small tubercles (plate 10). The frontal margin is wide in the region of the rhinophores, is continued behind into a dorso-lateral ridge, and bears along its whole length a number of papillae which are tuberculated and irregularly lobed. These papillae, the tips of the branchiae, the rhinophores, and the scattered tubercles of the back are a bright orange color. There are also irregularly arranged spots of orange along the sides of the animal. This species lives on the brown kelp and in the tide pools at Monterey, and on account of its bright coloring is very conspicuous. Dr. MacFarland* says, "It is avoided by the tide pool fishes as apparently inedible, its bright colors seemingly serving a warning purpose."

Triopha maculata MacFarland (plate 10) is slug-like with a broad flattened frontal margin, which bears a number of short branched processes. The series of processes is extended along the side of the animal at the dorso-lateral angle of the body. The ground color is yellowish brown

---

*F. M. MacFarland—Opisthobranchiate *Mollusca* from Monterey Bay, California; Bulletin of the Bureau of Fisheries, Vol. 25, p. 137.

varying in depth of shade in different individuals. There are bluish white, round, or oval spots scattered over the whole surface of the body but in young specimens these spots are inconspicuous (plate 10). The branching processes and the branchial plumes are bright orange or vermilion, shading into dark brown. The rhinophores have a yellowish stalk, but the leaves and middle line of the club-shaped portion, and the border of the sheath are bright orange-red. This nudibranch is abundant in the rocky tide pools at Monterey and is fairly common at La Jolla. The two figures show the coloring of young and older individuals.

Polycera atra MacFarland (plate 10) is a small species, our specimens from La Jolla being 15 mm. long. Dr. MacFarland reports specimens at Monterey reaching 23 mm. The body is slug-like and highest just in front of the branchiae. Blue-black longitudinal lines are separated by lighter bands which are almost white and have numerous orange-yellow spots. The frontal margin is adorned with six slender, pointed, yellow processes. The rhinophores are not retractile and are furnished with a yellow band near the tip. There are eight branchiae, the anterior ones being longest. They are tipped and spotted with orange. Back of the gills and at the side of them are four sharp thorn-like processes, also orange color, which in some cases are only small tubercles. The specimens we had in the laboratory in July were very active. Eggs were laid in a short, broad ribbon of gelatine one centimeter long. These were attached to algae, as were others seen in the tide pools, where the specimens were found. Range, Monterey to San Diego.

Acanthodoris rhodoceras Cockerell and Eliot (plate 11) is light brown with dark spots. Some of the spots are at the tips of conical tubercles which are scattered over the whole dorsal surface of the body, while others are between the tubercles. The rhinophores and branchiae are tipped with dark red-brown. The specimen figured was small, about

10 mm. long. The species has been reported from San Pedro, the specimen figured was found at La Jolla.

Ancula pacifica MacFarland (plate 11) is a small yellowish white species with orange markings. There is hardly any frontal margin, and the anterior tentacles are short and blunt. The rhinophores are non-retractile and at the base of each are two long, slender, forward pointing processes tipped with orange. The rhinophores are also tipped with orange. There are three nearly equal branchiae, not retractile, and tipped with orange. On each side of the branchial plumes are three or four blunt club shaped processes, tipped with yellow. A narrow, orange stripe runs down the median line from rhinophores to branchiae and from there along the crest of the tail, and an indistinct one from each rhinophore along the dorso-lateral margin. These are found rarely in the tide pools at Pacific Grove, Laguna, and La Jolla. We found that this nudibranch frequently swam upside down at the surface of the water and was very active in captivity.

Hopkinsia rosacea MacFarland, is a beautiful rose pink form (plate 11), covered with numerous long papillae. The length of a large specimen is 29 mm., width 16 mm., height 5 mm. The longest dorsal papillae are 18 mm. in length. The body is flattened, firm, and fragile, many spicules making it almost calcareous. The dorsal portion slopes gradually to the margin of the foot so that there is no ridge marking the boundary between back and sides. The foot, has a broad, short tail and has a deep triangular notch in front. The anterior tentacles are broad, thin, and united in front, forming a veil-like expansion. The rhinophores are long, slender, and non-retractile. The branchial plumes number 7–14, their bases forming a wide arc. This species is reported from Monterey and San Pedro under shelving rocks between tide marks. "The eggs are laid in the usual spiral form, the band being narrow and the same color as the animal" (MacFarland).

Family—DORIDOPSIDAE

In this family the body is nearly always soft, the gill plumes are completely retractile into a pocket, and the oral opening is pore-like, suctorial, and there is no radula. *Doriopsis fulva* MacFarland (plate 11) is a rich yellow with yellowish white branchial plumes and dark rhinophores. The surface of the mantle bears papilla-like elevations tipped with white. The foot and under side of the mantle are light yellow, the mantle having a mesh work of fine white lines. There are five wide spreading, retractile branchial plumes. Large specimens are 65 mm. long, 30 mm. wide, and 12-13 mm. high. Dr. MacFarland reports this as one of the most common nudibranchs at Pacific Grove. He also reports the egg bands to be yellow, about 7 mm. wide, closely coiled, and fastened to the sloping sides of rocks or to brown algae. The egg laying takes place at any time during the year, but mainly in the summer time.

Tribe—CLADOHEPATICA
Family—PLEUROPHYLLIDIIDAE

Pleurophyllidia californica Cooper (figs. 505 and 507). We have found this rather sluggish mollusk along with sea pansies at Mission Bay. It always devours sea pansies if it is confined in an aquarium with them. The foot is white, tinged with salmon pink and edged with pure white, while the mantle is dark red-brown with fine longitudinal lines of white. There is a prominent head shield in front, half-moon shaped and dark brown above, edged with white. The two rhinophores have their bases between the mantle edge and the head shield, and disappear under the mantle when retracted. The eyes are tiny black spots, one at the outer side of the base of each tentacle. The breathing organs are little folds on the under side of the anterior margin of the mantle. When the animal is quiet the mantle covers the foot, but when it is crawling the mantle appears narrower and the foot extends out half the width of the mantle on each side of it. They are 4-5 cm. long.

Family—DIRONIDAE

Dirona picta MacFarland sometimes appears in large numbers in the tide pools at La Jolla and has been found as far north as Monterey. It lives upon the brown algae but often floats at the surface of a tide pool with the sole of the foot uppermost. The color is variable being reddish yellow to brown, the dorsal parts sprinkled with yellowish white, olive-green, and pink spots. In the mid-dorsal region and toward the sides the dark green liver shows through. The cerata are somewhat inflated, pointed at the tip, and slightly contracted at the base. They readily become detached so that a specimen that has been in the laboratory for a few days is likely to have few cerata or only small ones. The length reaches 40 mm., the width 7 mm.

Family—DENDRONOTIDAE

Dendronotus giganteus O'Donoghue (fig. 508) is one of the largest American nudibranchs, in some instances exceeding 200 mm. in length. All the *Dendronotidae* are remarkable for the long branching, dorsal appendages and in this species they are well developed. The rhinophores can be withdrawn into branching sheaths which much resemble the cerata in appearance. The latter are very thick at the base, the anterior three or four pairs dividing almost immediately into two main trunks. The anterior end of the creature is formed into an oral veil, which is provided with a number of elongated papillae. The color varies from grayish to brownish. The individual photographed, a specimen taken at Friday Harbor, Washington, was tan with the side branches of the dorsal appendages dark brown and their tips yellow or opaque white. A narrow white stripe on the margin of the foot seems to be characteristic of the species. The animal swims by a writhing, sideways movement of the body (shown in fig. 508-b). We do not know the

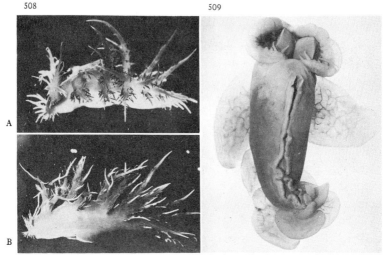

FIG. 508.—Two views of a living specimen of *Dendronotus giganteus*; x⅔. FIG. 509.—*Melibe leonina*, a preserved specimen, seen from the ventral side. (Photographed by H. P. K. Agersborg.)

range of this nudibranch, but it occurs at Vancouver Island and in the San Juan Archipelago.

The presence of 3–5 small papillae on the posterior edge of the rhinophore sheath serves to distinguish this species from *D. arborescens* (O. F. Muller), a circumpolar form found on both sides of the Atlantic and as far south as Vancouver Island on our coast. Another external character that will help to separate them is the more slender, cylindrical shape of the basal parts of the cerata in the latter species. Agersborg* found that the dorsal tentacles or rhinophores are the parts of the body most sensitive to touch.

FAMILY—JANIDAE

Antiopella aureocincta MacFarland mss. is a handsome species with perfoliate tentacles, and cylindrical gills. The

---

*H. P. K. Agersborg—Some Observations on Qualitative Chemical and Physical Stimulations in Nudibranchiate Mollusks with Special Reference to the Rôle of the "Rhinophores;" Journal of Experimental Zoology, Vol. 36, p. 439.

latter are simple, crowded, and arranged along the sides of the back. As the name indicates, the gills are circled near their tips by a band of gold color. This band may be bordered with brown. We took a specimen about an inch long on the kelp off La Jolla. The species is found as far north as Monterey Bay.

FAMILY—TETHYMELIBIDAE

Melibe leonina (Gould) (fig. 509) is a pelagic nudibranch of curious shape. It has a most unusual veil or hood surrounding the head and opening on the lower side. When filled with air, this may help to float the animal or it may aid in obtaining food. It is composed of two lobes and provided with two rows of tentacles located just within the margin, the outer row larger and less numerous than the inner. On the upper side of the hood are two ear-like flaps, the rhinophores, and where it joins the body there is a constriction giving the appearance of a neck. The dorsal appendages look like inflated leaves, the large anterior ones placed opposite each other, and the others alternating in position and diminishing in size until the last is very small. They are easily broken off from the living animal or may be shed, making it difficult to secure or keep a perfect specimen. While the foot, though narrow and poorly developed, can be used for crawling as well as for clinging to the eelgrass upon which it lays its eggs, the animal is more frequently observed floating or swimming in the water. Agersborg,* who has made a study of the habits of *M. leonina,* states, "When the hood is open it is tossed sideways and held in a direct position for the capture of small horizontally swimming crustacea. *Melibe* is actively predaceous. Its stomach has been found so completely filled with minute crustacea, such as copepoda, amphipoda and isopoda that it bulged out in almost a perfect sphere." The color may be yellowish, brown, or gray.

---
*H. P. K. Agersborg—Notes on *Melibe leonina* (Gould); Publications Puget Sound Biological Station, Vol. 2, p. 272.

The internal organs can be seen clearly in the living animal. Most of the body and all of the cerata are marked with a network of fine branched lines of a darker shade. This gives to the cerata an appearance reminiscent of a moss agate. The range of the species, so far as recorded, is from Nanaimo, British Columbia to La Jolla.

FAMILY—AEOLIDIADAE

Galvina olivacea O'Donoghue (frontispiece), is a small nudibranch that occurs plentifully upon masses of the hydroid, *Obelia longissima*, at East Sound, Washington. They vary in size from 6 to 9 mm. The egg masses are white and attached to the stems of the hydroid. They are oval or crescentic in outline, rather than ribbon-like or linear as in many nudibranchs, and between 1 and 2 mm. in length. Our specimens were taken late in July.

Hermissenda crassicornis (Eschscholtz) is a beautiful opalescent species the colors of which are well shown in the sketch (frontispiece). There is some variation in the colors, as the cerata range from light yellow to red-brown. We have found this species to be abundant all along the coast. Hydroids form, at least, a part of its food. A favorite habitat is the mud flats of the southern bays. In February and March, the species was abundant in the tide pools at Pacific Grove. At this time numbers of egg masses similar in appearance to those laid by *H. crassicornis* in the laboratory, were seen in the tide pools. The length is usually about one to two inches though much larger specimens, probably of the same species, have been seen. Agersborg, in experimenting with one species of *Hermissenda*, concluded that the oral tentacles have the power of discriminating between certain substances, such as food and odorous oils, while the dorsal tentacles lack this power.*

---

*H. P. K. Agersborg—Some Observations on Qualitative Chemical and Physical Stimulations in Nudibranchiate Mollusks with Special Reference to the Rôle of the "Rhinophores;" Journal of Experimental Zoology, Vol. 36, p. 443.

Flabellina iodinea (Cooper) (frontispiece) is a beautiful violet colored mollusk that sometimes comes in on the pieces of kelp that wash on shore. The orange-red rhinophores and orange colored cerata contrast strikingly with the violet purple of the body. The form is long and narrow with small head and long tentacles. The margin of the foot is produced at the anterior angles into long tentacle-like processes. This species lives on the colonies of hydroids growing on the kelp and is found in such situations from Puget Sound to San Diego, being washed on shore only rarely. Dr. MacFarland reports it as occurring on hydroids on the piles of wharves at Monterey. Specimens in the aquarium deposited many salmon-pink, gelatinous egg masses on the kelp (fig. 489-A) in December. The mollusks swim by bringing together the two margins of the narrow foot and then doubling and twisting the body. Specimens live well in the aquarium. The length of the specimen figured was 9 cm., width 7 mm., height 8 mm.

Order—PULMONATA

The order *Pulmonata* is made up largely of the snails found on land and in fresh water but there are a few marine mollusks in the group. They are air-breathers having a "lung" or respiratory sac instead of gills. Some members of the group must come to the surface of the water in order to take in air but some can get oxygen from the water that is taken into the mantle cavity. They have no operculum. Each animal possesses both male and female organs. The marine forms haunt the salt marshes near high-water mark.

Family—ELLOBIIDAE

Pedipes unisulcatus Cooper, the furrowed *Pedipes*, has a light brown, smooth shell with a short spire and a large body whorl (fig. 510). The columella has prominent white folds. The length is 6 mm. The range is from San Pedro to the Gulf of California.

Melampus olivaceus Carpenter, the olive ear-shell, is pear shaped with a short, flattened spire (fig. 511). The aperture is narrow and there are two folds on the columella, one near the center and one near the posterior end. The ground color is brown or greenish, frequently banded with a lighter color. It is abundant on the salt mud flats from Monterey Bay (Salinas River) to Mexico, being found near the high tide line where it is seldom under water. It reaches 10–12 mm. in length.

FAMILY—GADINIIDAE

Gadinia reticulata (Sowerby), the netted button-shell, is nearly circular, pure white, low arched, with a blunt, nearly central apex (fig. 521). Fine ridges radiate from the apex to the margin and are crossed by the lines of growth. It reaches 13 mm. in diameter. The range is from the Farallon Islands to Cape San Lucas.

ORDER—CTENOBRANCHIATA

This order includes mollusks in which the breathing organs, the ctenidia or branchiae, are in front of the heart and the auricle in front of the ventricle. The species are all unsymmetrical and have spiral shells which show great diversity in color, form, and size. An operculum, either horny or calcareous is usually present.

SUBORDER—ORTHODONTA

FAMILY—TEREBRIDAE

Terebra pedroana philippiana Dall (formerly known as *Terebra simplex*), the simple augur-shell, has a sharp spire, and a small aperture with a short, recurved canal (fig. 512). A spiral, beaded band encircles the shell just below the suture. The color is gray or brown, the length 35 mm. The range is from San Pedro to the Gulf of California.

FIG. 510.—*Pedipes unisulcatus*; x3/2. FIG. 511.—*Melampus olivaceus*, the olive ear shell; x4/3. FIG. 512.—*Terebra pedroana philippiana*, the simple augur shell; x1. FIG. 513.—*Conus californicus*, the California cone; x1. FIG. 514.—*Pseudomelatoma moesta*; x4/3. FIG. 515.—*Fusinus kobelti*, the Kobelt spindle shell; x1. FIG. 516.—*Macron lividus*; x5/4. FIG. 517.—*Cryptoconus carpenterianus*, Carpenter's turret shell; x1. FIG. 518.—*Clathrodrillia incisa*, the incised drill shell; x5/4. FIG. 519.—*Strigatella idae*, the Ida miter shell; x1.

Family—CONIDAE

Conus californicus Hinds, the California cone, has a heavy porcelaneous shell with a short spire and a long, narrow aperture (figs. 513 and 520). When the mollusk is living, the shell is covered with a dull brown, hairy epidermis but the worn and empty shells are smooth and chestnut color. The drawing shows the animal with siphon extended. The foot and siphon are white, sprinkled and streaked with black. The eyes are on slight elevations at the outer side of the tentacles.

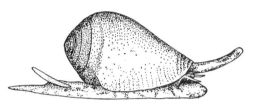

FIG. 520.—*Conus californicus*, the California cone, showing the animal fully extended on a smooth surface. The siphon is extended at the anterior end and the operculum is near the posterior end of the foot; x3.

Some of the species of *Conus* from other regions are large and beautiful. One reaches 9 inches in length, and another handsome one is so highly prized that single specimens have brought $200 at auction sales. According to the *Cambridge Natural History*[*] one species of *Conus* is accused of inflicting poisonous wounds. In these species each tooth of the radula bears a barb and poison duct. The California cone is small and can be safely handled. An ordinary specimen is 25 mm. long. The species may be found from the Farallones to Lower California.

Family—TURRITIDAE

Cryptoconus carpenterianus (Gabb), Carpenter's turret, has a strong, solid shell with a short canal, a thin lip, and a notch near the suture where the lip is recurved (figs. 517 and 522). The spire is red-brown, the body whorl often lighter with spiral lines of brown. The living mollusk is orange colored with fine white spots sprinkled over the mantle. A large

---

[*] A. H. Cooke—Molluscs; Cambridge Natural History, Vol. 3, p. 65, The Macmillan Company.

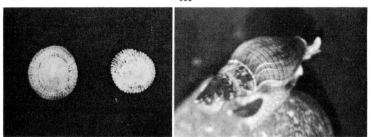

FIG. 521.—*Gadinia reticulata*, the netted button shell; x1. (Photographed by F. W. Kelsey.)
FIG. 522.—*Cryptoconus carpenterianus*, photographed from a living specimen; x¾.

FIG. 523.—*Kellettia kellettii*; x1.

specimen measures 75 mm. in length and the range is from Bodega Bay to San Diego (Kelsey).

Clathrodrillia incisa (Carpenter), the incised drill, has a slender, eight-whorled spire, a long, narrow aperture notched near the suture, and a short canal (fig. 518). The shell is brownish, with delicate spiral and cross lines. An ordinary specimen measures 30 mm. in length. The range is from Puget Sound to San Diego (Kelsey).

Pseudomelatoma moesta (Carpenter), the doleful drill, has about nine whorls, is brown and smooth except for prominent cross ribs upon all except the body whorl (fig. 514). There is a prominent notch at the suture and the canal is short and deep. The length is 25-30 mm. The range is from Monterey to Lower California.

Mangilia angulata Carpenter has a small shell with six angular whorls and ten prominent transverse ridges which are most prominent on the angle of the whorl (fig. 525). The canal is short and narrow. The length is 8 mm. and the body whorl is 5 mm. wide. The range is from Puget Sound to the Gulf of California.

FAMILY—OLIVIDAE

Olivella biplicata Sowerby, the purple olive, has a strong, solid shell, pearl-gray shaded with purple or pure white except for a narrow purple band winding on to the columella. There is a wide, white callus on the columella and there are two small folds at its base (fig. 526). The living mollusks may be found when the tide is low. They are often out of sight in the sand so that in hunting for them one is guided by the marks they have left as they crawled along just under the surface. A favorite haunt of the species seems to be where the sandy beaches merge into the rocky parts of the shore near low tide line.

Strings of these shells have been found in the old Indian graves and mounds. The Indians rubbed down the apex in order to thread the shells and used them for ornament or for

Fig. 524.—*Chrysodomus tabulatus*, the tabled Chrysodome; x5/4.

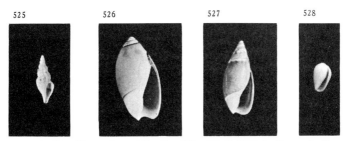

Fig. 525.—*Mangilia angulata*; x3/2. Fig. 526.—*Olivella biplicata*, the purple olive; x1. Fig. 527.—*Olivella pedroana*; x1. Fig. 528.—*Marginella jewettii*, the California rice shell; x3/2.

money. The fact that *Olivella* absorbs the internal portion of its whorls as it grows (fig. 483) made the stringing process an easy one. The length of a good sized shell is 25 mm. The range is from Vancouver to Lower California.

Olivella pedroana (Conrad) resembles *O. biplicata* but the shell is smaller, more slender, and has a proportionately longer spire (fig. 527). A large specimen measures 13 mm. The range is from Puget Sound to Cape San Lucas.

Family—MARGINELLIDAE

Marginella jewettii Carpenter, the California rice-mollusk, has a tiny, pure white shell, porcellaneous, and with a long aperture slightly less than the length of the shell (fig. 528). The spire is short and there are three prominent folds on the columella. It reaches 5 mm. in length. The range is from Monterey to Lower California.

Marginella californica (Tomlin), the colored *Marginella*, is often called the "wheat-shell" because the shell in color, shape, and size reminds one of a grain of wheat (fig. 534). It is vitreous and polished and often banded with lighter shades of brown. There are four well marked oblique folds on the columella. It reaches 9 mm. in length and is found from San Pedro to Mexico.

Family—MITRIDAE

Strigatella idae (Melvill) has a fusiform shell which is covered with a black epidermis (fig. 519) when the mollusk is alive. The worn shells are brown outside and smooth. An unworn shell has fine spiral lines crossed by faint lines of growth. The columella has three conspicuous oblique folds or ridges. A large specimen measures 50 mm. in length. The range is from the Farallon Islands to Cortez Bank, off San Diego.

Mitromorpha filosa (Carpenter), the threaded miter-shell (fig. 535), is dark brown with even, thread-like spiral lines.

A large specimen is 6 mm. long. The species is reported from Monterey to the Gulf of California.

FAMILY—FASCIOLARIIDAE

Fusinus kobelti (Dall), the Kobelt spindle-shell (fig. 515), has deep sutures and the convex whorls are crossed by 11–13 elevations. Dark spiral lines and indistinct spiral ridges encircle the larger whorls. A large specimen is 40 mm. long. the range is from Monterey to San Diego.

FAMILY—CHRYSODOMIDAE

Kellettia kellettii (Forbes), formerly known as *Siphonalia kellettii*, has a large shell with a high spire of seven or eight whorls (fig. 523). The whorls are ornamented with 8–10 prominent nodes and numerous fine, deep, spiral grooves. The canal is long, narrow, and curved back. The length reaches 114 mm. The range is from Santa Barbara to Lower California.

Macron lividus A. Adams is covered with a dark brown epidermis when it is alive (fig. 516) but worn shells are light brown, with a polished surface. The shell is thick and near the top of the aperture the columella has a strong fold. A good sized specimen measures 20 mm. The range is from the Farallon Islands to Lower California.

Chrysodomus tabulatus Baird, the tabled *Chrysodome*, has a large shell with a high spire of eight whorls (fig. 524). The whorls are sharply angulated and keeled above. Below this keel they are ornamented with ridges of unequal size. The suture is deep, the canal long, narrow, and curved back. The color is usually yellowish white. A large specimen measures 87 mm. long and 38 mm. in diameter. The range is from British Columbia to San Diego.

The shell of *Chrysodomus liratus* (Martyn), which is found from the Arctic Ocean to Puget Sound, is not so slender as *C. tabulatus*, is less angulated and has three

FIG. 529.—*Searlesia dira*; x1.　FIG. 530.—*Alectrion fossata*, the channeled basket shell; x5/4. (Photographed by F. W. Kelsey.)

FIG. 531.—*Alectrion mendica*, the lean basket shell; x5/4. (Photographed by F. W. Kelsey.)　FIG. 532.—*Alectrion perpinguis*, the fat basket-shell; x5/4. (Photographed by F. W. Kelsey.)　FIG. 533.—*Alectrion tegula*, the covered-lip basket-shell; x5/4. (Photographed by F. W. Kelsey.)

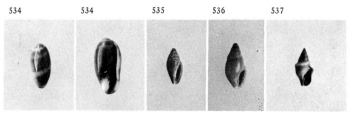

FIG. 534.—*Marginella californica*, the wheat shell; x3/2.　FIG. 535.—*Mitromorpha filosa*, the threaded mitre shell; x3/2.　FIG. 536.—*Columbella gausapata*, the common dove-shell; x3/2.　FIG. 537.—*Columbella carinata*, the keeled dove-shell; x3/2.

PHYLUM—MOLLUSCA

distinct ridges on the whorls of the spire. A good sized specimen measures 115 mm. long and 70 mm. in diameter. *Searlesia dira* (Reeve) is spindle shaped and marked with deep spiral channels (fig. 529). The whorls of the spire are crossed by longitudinal ridges, nine to a whorl. The color is red-brown overlaid with an ashy covering. There is a clearly marked umbilicus. The length may reach 50 mm. The species is found from Alaska to Monterey.

FAMILY—ALECTRIONIDAE

Alectrion fossata (Gould), the channeled basket-shell, is the largest of the basket-mollusks (fig. 530). The shell has prominent spiral grooves which show plainly within the aperture and longitudinal elevations cross these grooves. There is a callus on the columella and above the reflexed canal is a deep fossa or furrow. The color is brown or ashy white and the callus often has an orange tinge. A large specimen measures 45 mm. The range is from Vancouver to Lower California.

Alectrion mendica (Gould), the lean basket-shell, is a comparatively slender form with fine spiral grooves and prominent longitudinal ridges or varices (fig. 531). The color is light brown and the interior is white. The length is 20 mm. The range is from Alaska to Magdalena Bay, Lower California.

Alectrion perpinguis (Hinds) is called the fat basket-shell. The spiral and longitudinal lines of this shell are of about the same height and cut the surface into tiny and uniform squares (fig. 532). The color is whitish or brown, sometimes with one or two narrow bands of darker color. There is a deep groove at the base of the body whorl. The length reaches 20 mm. The range is from Puget Sound to Lower California.

Alectrion tegula (Reeve), the covered-lip basket-shell (fig. 533), is at home on the mud flats. The shell is small, and stout, with a recurved canal, a groove at the base of the body whorl, and a broad, smooth, white callus over the inner lip and front of the shell. The color is dark gray with lighter color on the longitudinal ridges that cross the whorls. The length of a large specimen is 20 mm. The range is from San Francisco to San Diego.

Family—COLUMBELLIDAE

Columbella gausapata Gould, the common dove-mollusk, has a tiny shell which is so abundant all along the coast that most people are familiar with it. The color is brown or reddish, sometimes almost white and marked with lines or spots (fig. 536). The surface often glistens as though the shell were polished. Empty shells are numerous among the broken bits along the shore and the live mollusks may be found under the alga-covered rocks. The usual length is 8 mm. The range is from Alaska to Lower California (Baker).

Columbella carinata Hinds, the keeled dove-shell, is similar to the last named species but there is a distinct whitish keel just below the suture of the body whorl (fig. 537). An ordinary specimen is 8 mm. long. The range is from San Francisco to Cape San Lucas.

Amphissa columbiana Dall, the wrinkled *Amphissa*, is a northern species which may occasionally be found along the California coast. The shell has seven whorls, each with 18–20 wavy-rounded ridges extending from suture to suture (fig. 538). Fine spiral lines encircle the whorls. The color is light yellowish-brown. The length may reach 20 mm. The range is from Alaska to San Pedro.

Amphissa versicolor Dall, the varicolored *Amphissa*, resembles the last named species but shows more varied coloring (fig. 539). Specimens may be brownish, red, or yellowish and striped or mottled with brown. The body whorl has a proportionately larger diameter than the last named species, the spire is shorter and the whorls more angulated. The transverse ridges are larger, and more

538
539
540
541

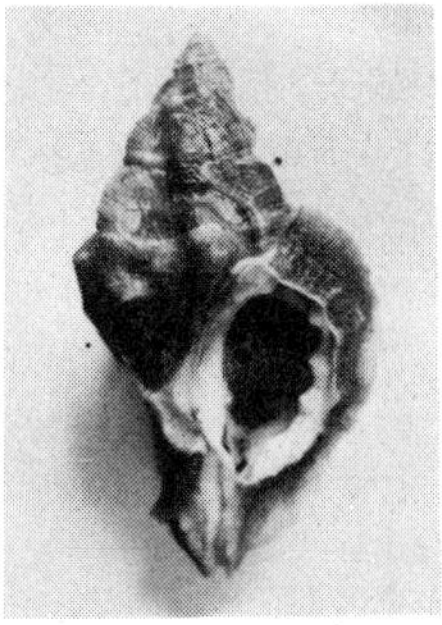

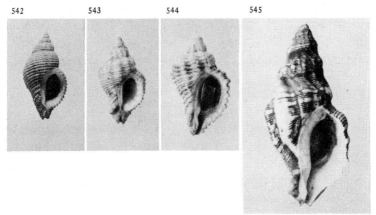

FIG. 538.—*Amphissa columbiana*, the wrinkled Amphissa; x3/2. FIG. 539.—*Amphissa versicolor*, the varicolored Amphissa; x3/2. FIG. 540.—*Murex gemma*, the incised *Murex*; x1. FIG. 541.—*Purpura nutallii*, Nuttall's hornmouth; x1.

542
543
544
545

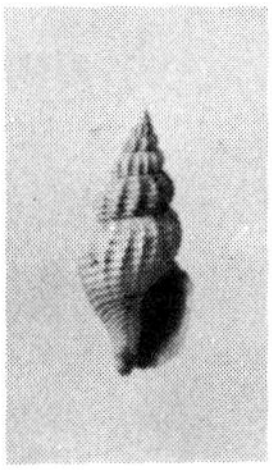

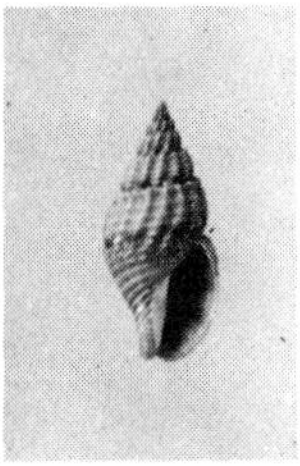

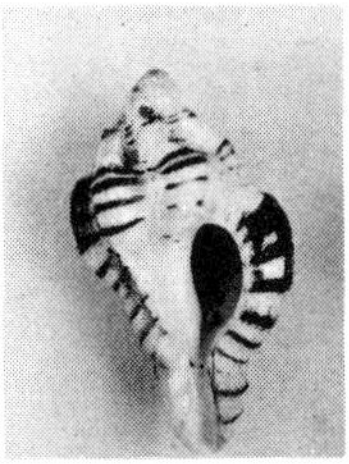

FIG. 542.—*Tritonalia gracillima*; x8/7. FIG. 543.—*Tritonalia lurida*; x3/2. FIG. 544.—*Tritonalia circumtexta*; x5/4. FIG. 545.—*Tritonalia poulsoni*; x10/9.

546
547
548

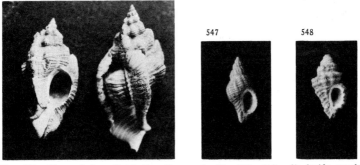

FIG. 546.—*Murex festivus*, the festive *Murex*; x1. FIG. 547.—*Tritonalia lurida munda*; x3/2. FIG. 548.—*Tritonalia interfossa*; x3/2.

oblique. The usual length is about 13 mm. The range is from Oregon to Lower California. A number of varieties of this species have been named.

Family—MURICIDAE

Murex gemma Sowerby, the incised *Murex*, is not a very common mollusk and as a rule only much worn shells are found. The shell has strong, rounded vertical ridges and is white with dark brown cross stripes (fig. 540). The length may reach 30 mm. The range is from Santa Barbara to Lower California.

Murex festivus Hinds, the festive *Murex*, has three prominent, reflexed, frill-like varices on each whorl and between each of the frills is a rounded node (fig. 546). The sutures are deep. The color is whitish, marked with numerous spiral lines in the space between the varices. The length reaches 40 mm. The range is from Santa Barbara to Lower California.

Purpura foliata (Martyn), the leafy hornmouth, has three broad, wing-like varices which seem to be made up of overlapping plates (fig. 549). The canal is long and curves backward and the old canals are left at the base of each varix when the animal builds on to the shell. Spiral ridges encircle the whorls and spread out fan-like upon the varices. The outer lip is crenulate with a sharp horn near the lower end. The specimen figured was found at Pacific Grove, measures 75 mm. in length and is white, stained with brown. The range is from Alaska to San Diego.

Purpura nuttallii (Conrad), Nuttall's hornmouth, like the two last named species, has three longitudinal varices but these are not reflexed (fig. 541). There are teeth inside of the outer lip corresponding to the spiral ribs on the body whorl. The canal is usually closed and there is a sharp tooth at the base of the outer lip. In young specimens this tooth may be absent and the canal may be open. The length reaches 50 mm. The range is from Monterey to Lower California.

Tritonalia gracillima Stearns is a small, heavy, brown shell, pure white within the aperture (fig. 542). An average specimen is 13 mm. long. The range is from Monterey to Lower California.

Tritonalia lurida (Middendorff) is spindle shaped with numerous rounded, raised spiral lines and a distinct suture (fig. 543). The upper whorls have transverse ridges which become less prominent on the body whorl and on the one next to it. The outer lip is thickened and bears a row of teeth within it. The canal is narrow and sometimes partly closed. The length reaches 25 mm. The range is from Alaska to Catalina Island.

The variety *munda* (Carpenter) may be distinguished by its less prominent transverse ribs which are of the same size on all the whorls and the spiral lines are regular (fig. 547). The length is 13 mm. and the range is from Alaska to San Diego.

Tritonalia poulsoni (Carpenter) has a slender spindle-shaped shell with six whorls (fig. 545). Rounded, lengthwise ridges are prominent at the widest part of each whorl but do not extend on to the concave upper portion of the whorl. There are spiral ridges on the lower portion of the whorl and fine spiral lines encircling all the whorls. The aperture is white within and has a row of 5–6 teeth inside the outer lip. The length reaches 32 mm. The range is from Santa Barbara to Lower California.

Tritonalia circumtexta (Stearns), the circled *Murex*, has a heavy shell with deep, distinct, and regular spiral grooves encircling it and with the whorls crossed by low but prominent ridges (fig. 544). The aperture is marked with brown and has about five teeth within the outer lip. The outside is grayish, marked with brown spots. It reaches 20 mm. or more in length. The range is from Trinidad to Lower California (Baker).

Tritonalia interfossa (Carpenter) is spindle-shaped with well marked spiral grooves, deep sutures, and sharp, transverse ridges which give it a very rough appearance (fig. 548).

The specimen figured measures 13 mm. The range is from Alaska to San Diego.

Urosalpinx cinereus (Say), the oyster drill (fig. 555), probably came with young oysters from the east for it is an Atlantic coast species which has recently appeared on our shore. The shell has spiral ridges crossed by transverse ridges and is yellowish gray. It reaches 30 mm. in length. It has been reported from San Francisco to San Pedro.

These mollusks are hunted by the oystermen for though they appear to be so insignificant and slow they multiply

FIG. 549.—*Purpura foliata*, the leafy hornmouth; x⅔.

FIG. 550.—*Trophon triangulatus*, the three-cornered Trophon; x⅔.

rapidly and have an enormous appetite for fresh oysters. A drill will crawl on to an oyster, bore a neat round hole through one of the valves of the shell near the hinge with its strong-toothed radula and devour the helpless mollusk.

Miss Rogers* says, "Each female lays during a period of several weeks a total of ten to one hundred egg cases. Each one is vase-shaped, vertically flattened and keeled, of clear,

---

*Julia Ellen Rogers—The Shell Book; Doubleday, Page and Co., p. 39.

parchment-like membrane, containing about a dozen eggs. The cases are attached by broad foot-like bases in regular rows, forming patches on the under sides of overhanging rocks, or other support, just above low water mark."

Trophon triangulatus Carpenter, the three-cornered *Trophon*, has a large shell with six prominent thin-edged varices on each whorl (fig. 550). The shell is light in weight, reddish brown and reaches 75 mm. in length. It has been found in the vicinity of Catalina and San Diego (Kelsey).

The species of *Thais* (*Nucella*) show great variation in form and marking. Dr. Dall† has examined a large series of specimens and says, "The sheltered rocky beaches of a well-protected harbor will afford slender elongated and lamellose specimens with small apertures. The outer rocks exposed to the ocean surf have short-spired, relatively smooth, wide-mouthed shells, which afford the least leverage to the waves. For, washed from his perch and carried to the muddy bottom off the shore by the undertow, an adult *Nucella* can hardly survive; and those offering the least friction and having the stronger hold on their situs are most likely to survive. There is also a connection between the situs and the shell which is less easily explained, and that is that, on rough surfaces such as an "oyster reef," or bar, the specimens of *lamellosa* are almost unanimously rough and laminate, while in undisturbed water on rocks with sandy surroundings the finest and most delicate development of lamellae and crenulations is to be found, according to the reports of collectors. In all cases *Nucella* seems to prefer a rocky habitat, especially if it affords young oysters or other sluggish species serving it as food."

Dr. Dall says further that the species are mostly carnivorous and he has seen them feed on small bivalves like *Anomia*, on ascidian colonies, and on the egg-capsules of their own

---

†W. H. Dall—Notes on the Species of the Molluscan Subgenus *Nucella* Inhabiting the Northwest coast of America and Adjacent Regions; Proceedings U. S. National Museum, Vol. 49, p. 559.

and related species. He has never found them attacking mollusks with very thick shells.

The egg-capsules, he says, are abundant in the early spring and are slender, vase-like objects, of parchment-like texture, with a flat, circular top and mounted on slender stalks in groups on rocks or old shells. The young, of which there are a number in each capsule, are cannibals, the weaker ones being eaten by the others. The survivors emerge from the top of the capsule leaving it unsealed.

It was from relatives of *Thais* that the famous Tyrian dye used to be made. The mollusks were ground in a stone mortar and the shade of color could be modified somewhat by the use of different species. Dr. Dall says, "Like the other species, these produce a purple dye, which I tested on an old handkerchief. It gave a dull purple color, which faded badly; but I afterward learned that it could be made permanent by the addition of lemon juice, which is used with the purple of the tropical species by the natives of Central America and Peru. However, I have never seen any article dyed with this substance which had any brilliancy or attractiveness of tint. If the classical descriptions of the Tyrian dye are correct, the American purple can not compete with it." We found the various species of *Thais* abundant at Pacific Grove, Dillon's Beach, and many of the islands of Puget Sound.

Thais lamellosa (Gmelin), the wrinkled purple, may be distinguished by 9–20 laminae or thin plates of shell, elevated at the edge and wrinkled and extended where they intersect a large spiral ridge (fig. 552). The shell is white to dark brown through various shades of yellow and either plain or banded. The length reaches 86 mm. The range is from Alaska to Santa Barbara. A number of varieties of this species have been named.

Thais lima (Martyn) has a fairly uniform spiral sculpture of alternated major and minor spiral cords (fig. 551). The aperture is large and the outer lip more or less crenulate.

PHYLUM—MOLLUSCA

The length reaches 60 mm. The range is from Alaska to Lower California.

Thais canaliculata (Duclos), the channeled purple, has uniform spiral ridges separated by distinct interspaces, rarely

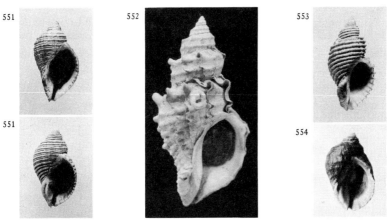

FIG. 551.—*Thais lima*: x4/5. FIG. 552.—*Thais lamellosa*; x⅔. FIG. 553.—*Thais canaliculata*; x4/5. FIG. 554.—*Thais emarginata*: x¾/5.

FIG. 555.—*Urosalpinx cinereus*, the oyster drill; x8/7. FIG. 556.—*Acanthina spirata*, the angled unicorn; x10/9. FIG. 557.—*Acanthina paucilirata*, the square-spotted unicorn; x3/2.

with minor spirals (fig. 553). There are about ten spirals on the last whorl. The height reaches 41 mm. The range is from Alaska to Monterey.

Thais emarginata (Deshayes), the short-spired purple, has a short spire of 3 or 3½ whorls without the nucleus and the body is rough with coarse spirals that are roughened with nodules which in some cases almost become spines (fig. 554). Many individuals are uniform in color, white, grayish or dark brown. The length reaches 40 mm. The range is from Alaska to Mexico.

Acanthina spirata (Blainville), the angled unicorn, is abundant and the shell nearly matches the algae-covered rocks to which it clings (fig. 556). There is a little horn or spine near the base of the outer lip and a row of teeth within the aperture. The whorls are marked with dark, more or less broken spiral bands which give it a pebbled appearance. The range is from Puget Sound to San Diego and Socorro Island. The height is 30 mm.

Acanthina paucilirata (Stearns), the square-spotted unicorn, has the horn and teeth and spiral bands of the other species and the whorls are distinctly angled. There are few spiral ridges, four on the body whorl, and between these ridges the color is in squarish dark blotches (fig. 557). The length is 20 mm. The range is from San Pedro to Lower California.

Suborder—STREPTODONTA

Family—EPITONIIDAE

Epitonium wroblewskii (Mörch), the northern opal-shell, has 7-8 whorls which are only slightly convex and bear eight blunt, transverse ribs which form radiating ridges from the apex of the shell (fig. 558). The color is white, the length reaching 28 mm. The range is from Alaska to San Diego.

Epitonium crenimarginatum Dall, the ladder shell, has a solid, strong, spiral shell having 7-8 whorls, each with 12-13 cross ribs (fig. 559). There is a prominent keel just below the suture which makes the whorl appear flattened above. The cross ridges often appear to be nearly worn away near the lower part of a whorl though they may be very distinct above. A large specimen is 17 mm. long. The range is from Monterey to Mexico.

PHYLUM—MOLLUSCA

Epitonium bellastriatum (Carpenter), the striped wentletrap, has a large body whorl and a short spire (fig. 560). The whorls number 6 and the cross ribs or varices 15. Between the varices are fine spiral ridges which may be traced well up toward the apex. A large specimen measures 19 mm. The range is from Monterey to San Diego.

Epitonium indianorum (Carpenter), the Indian wentletrap, has a thick shell with ten rounded whorls, each with 12–16 heavy, reflexed cross ridges. These ridges are striated with fine lines. The length may reach 26 mm. The range is from Alaska to Lower California.

Epitonium fallaciosum Dall, the white wentletrap, has a thin shell with 8 rounded whorls, 11–14 varices which are sharp and thin, sometimes reflexed and with a prominent keel near the suture (fig. 561). A large specimen measures 22mm. The range is from Monterey to the Gulf of California.

Family—JANTHINIDAE

Janthina exigua Lamark, the violet snail, is a pelagic mollusk with a thin, delicate shell (fig. 562) which is sometimes washed ashore on the coast of southern California and Mexico.

The foot of these mollusks exudes a viscid gelatinous secretion which forms a raft to which the animal clings in order to keep afloat. The egg capsules are attached to the underside of this raft. The mollusk is said to feed upon the small jellyfishes which also float at the surface of the sea. It has also been reported that *Velella* is devoured by the little mollusk. A large school of the animals floating under their rafts is not easily seen by the ocean traveler since the blue of the snail blends with the blue of the water. This is necessary if the animals are to continue their existence since the hungry seabirds find them good eating. The mollusks are more or less gregarious and great schools of them are sometimes washed on shore over night. There

FIG. 558.—*Epitonium wroblewskii*, the northern opal shell; x3/2. FIG. 559.—*Epitonium crenimarginatum*, the ladder shell; x3/2. FIG. 560.—*Epitonium bellastriatum*, the striped wentletrap; x3/2. FIG. 561.—*Epitonium fallaciosum*, the white wentletrap; x3/2.

FIG. 562.—*Janthina exigua*, the violet snail; x3/2. FIG. 563.—*Melanella micans*; x3/2. FIG. 564.—*Trivia solandri*, the large coffee-bean; x1. FIG. 565.—*Trivia californiana*, the small coffee-bean; x1.

FIG. 566.—*Erato vitellina*; x3/2. FIG. 567.—*Cypraea spadicea*, the nut-brown cowry; x1. FIG. 568.—*Erato columbella*; x3/2.

is a deep notch in the outer lip of the shell. The length is 10-18 mm.

Family—MELANELLIDAE

Melanella (*Eulima*) *micans* (Carpenter), the shining *Melanella*, has a slender, glossy shell with a straight, sharp spire made up of ten flat whorls (fig. 563). It is white with darker tints on the spire. A large specimen is 10 mm. long. The range is from Vancouver Island to Lower California and the species is also found in fossil form in Pleistocene deposits of southern California.

Family—PYRAMIDELLIDAE

The family *Pyramidellidae* contains many tiny mollusks. Over three hundred species are described by Dr. Dall and Dr. Bartsch in the U. S. National Museum Bulletin, No. 68. We mention only three of them here and refer the student to the above bulletin for detailed description.

Turbonilla tenuicula (Gould) is a variable species, the shell being either banded or uniformly colored and white to dark brown in color. Each whorl has 18-28 vertical ribs which are excurved, somewhat thickened and connected at their summits (fig. 575). There are 10-16 spiral lines on each whorl and the summits of the whorls are not strongly shouldered. The length is 7.5 mm. The range is from Monterey to Lower California.

Odostomia helga Dall and Bartsch has a white, conic shell marked with four broad, low, spiral bands on each whorl (fig. 578). The first few whorls have short, indistinct, vertical ribs. The length is 4.2 mm. The range is from San Pedro to the Coronado Islands.

Suborder—NUCLEOBRANCHIATA

The mollusks of this group swim the high seas. The foot is laterally compressed into a rounded fin or it may be rudimentary and provided with a float which enables the animal to keep near the surface. The *Pterotracheidae* have rudimentary shells or none at all though the body is large. The *Atlantidae* have a well developed spiral shell into which the animal can withdraw its whole body.

Family—PTEROTRACHEIDAE

The pterotracheid shown in figure 569 was taken in a tow net off La Jolla. The body is transparent and the

Fig. 569.—A pterotracheid mollusk which swims the high seas; x1.4.

organs can be seen through the body walls. About one fourth of the distance back from the anterior end are the tentacles and the large eyes. The visceral mass is at the posterior end of the body and there is a long, narrow caudal projection below and behind this mass. The nerves and ganglia show through the body wall with remarkable distinctness. One specimen measures 35 mm. long. The swimming organ is the large vertically flattened keel-like foot.

Family—ATLANTIDAE

A number of atlantids have been described from the North Pacific. Figure 570 gives a good idea of their appearance as they swim about through the water. They are sprightly animals and fascinating to watch under the microscope. The body is enclosed in a spiral shell the aperture of which is closed by an operculum when the mollusk contracts.

The shell of this individual was about 2 mm. in its largest diameter. The head forms a comparatively large cylindrical proboscis and has large eyes at the sides just back of the tentacles. The gills are in the mantle cavity. The foot is

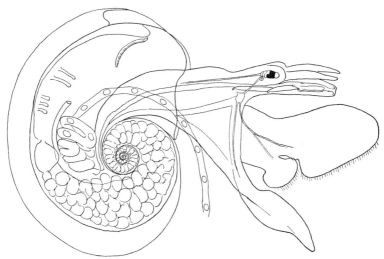

FIG. 570.—An atlantid, a pelagic mollusk. The chain of eggs appeared while the figure was being drawn; x30.

divided into a fore part which is vertically flattened and a hind part, horizontally flattened, bearing the operculum. The nervous and vascular systems show clearly through the body wall and shell. While we were looking at a living specimen, the animal began to deposit eggs as shown in the figure. The eggs were disposed at regular intervals within a transparent gelatinous tube which gradually advanced from the external genital opening, out of the aperture of the shell and into the water where it later broke off into short lengths. According to Adams,* the atlantids are distributed throughout the seas of warm latitudes, coming to the surface in calm weather, especially after nightfall and disporting themselves in vast numbers.

*H. Adams and A. Adams—The Genera of Recent *Mollusca* (3 volumes); Vol. II, p. 90. Van Voorst.

Suborder—PECTINIBRANCHIATA
Family—CYPRAEIDAE

Cypraea spadicea Swainson, the nut-brown cowry (fig. 567), has a beautiful shell, though perhaps rather ordinary in comparison with the handsome cowries found in tropical oceans. The living mollusk nearly envelops the shell with its red-brown mantle. The spire is not visible in a large specimen since it becomes covered as growth proceeds. Young shells are thin walled and the spire is still visible. It is thought that the outer, thickened portions of the shell are periodically dissolved as a preliminary to its enlargement. A large specimen measures 45 mm. The species is found from Monterey (Berry)† to Lower California.

Family—TRIVIIDAE

Trivia solandri Gray, the large coffee-bean shell (fig. 564), has a deep longitudinal groove on the upper side which does not extend to either end of the shell. About 13 cross ridges extend around the shell ending at the groove with rounded whitish nodes. The spire in this shell is wholly concealed. The shell is flesh color or ashen. The length reaches 17 mm. The range is from Catalina to Panama.

Trivia californiana Gray, the little coffee-bean shell (fig. 565), is crossed by 8–9 transverse ridges which are interrupted at the dorsal side by a whitish line. The color is red or red-brown. The length is 11 mm. The range is from Crescent City to Mexico.

Erato vitellina Hinds, the brown *Erato* (fig. 566), is pear-shaped, dark red-brown on the upper side, and white near the toothed lips. The length reaches 15 mm. The species may be found from Bodega Bay to San Diego though it is nowhere abundant.

Erato columbella Menke, the dove *Erato* (fig. 568), has a

---

†S. S. Berry—Miscellaneous Notes on Californian Mollusks; The Nautilus, Vol. 22, p. 37.

small shell with a short spire and long narrow aperture the lips of which are white and finely toothed. The upper side is olive-brown. The length reaches 7 mm. The range is from Monterey to Panama.

FAMILY—RANELLIDAE

Bursa californica (Hinds), the California frog-mollusk (fig. 571), formerly known as *Ranella californica*, has a large

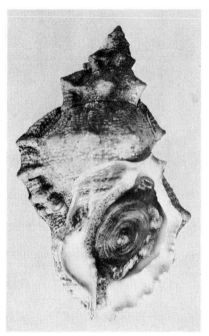 

FIG. 571.—*Bursa californica*, the California frog shell; x⅔. FIG. 572.—*Argobuccinium oregonensis*, the Oregon *Triton*; x⅔.

strong shell with five convex whorls, each having two prominent longitudinal ridges which form a more or less continuous ridge from the apex to the base. Between these are rows of rounded nodes and numerous revolving ridges. The length is 90 mm. The range is from Monterey to Lower California.

530 SEASHORE ANIMALS OF THE PACIFIC COAST

Family—CYMATIIDAE

Argobuccinum oregonensis (Redfield), the Oregon Triton, has numerous varices and a hairy, brown epidermis. The

FIG. 573.—*Bittium eschrichtii*; x3/2. FIG. 574.—*Cerithiopsis carpenteri*; x3/2. FIG. 575.—*Turbonilla tenuicula*; x3/2. FIG. 576.—*Cerithiopsis columna*; x3/2. FIG. 577.—*Bittium quadrifilatum*; x3/2.

FIG. 578.—*Odostomia helga*; x3/2. FIG. 579.—*Trichotropis cancellata*, the checked hairy-shell; x1. FIG. 580.—*Cerithidea californica*, the California horn-shell; x1. FIG. 581.—*Cerithidea californica hyporhysa* Berry; x1. FIG. 582.—*Turritella cooperi*, Cooper's tower-shell; x5/4.

inside of the shell is pure white (fig. 572). The length reaches 115 mm. The range is from Alaska to La Jolla (Baker).

Family—CERITHIOPSIDAE

Cerithiopsis carpenteri Bartsch is small, dark chocolate-brown in color, the rounded whorls marked with three spiral bands and numerous axial ribs (fig. 574). The length is 8 mm. and the range from San Pedro to the Coronado Islands.

Cerithiopsis columna Carpenter is chestnut-brown, with three spiral keels crossed by numerous vertical ribs (fig.

576). The suture of the adult shell shows an exposed keel. The length is 9 mm. The range is from Vancouver Island to San Diego (Kelsey).

Family—CERITHIIDAE

Bittium eschrichtii (Middendorff), the threaded *Bittium* (fig. 573), has nine whorls that are marked by four flattened spiral ridges separated by strong spiral grooves. Fine spiral lines and lines of growth are also present in unworn specimens. The color is white to brown, rarely spotted. The length is 14 mm., diameter 5 mm. It may be found from Alaska to Puget Sound. The variety *montereyense* Bartsch is smoother and more slender and is frequently variegated being whitish, mottled with rust brown. It has ten whorls and measures 13.8 mm. in length and 5 mm. in diameter and is found from Crescent City to Cape San Lucas.

Bittium quadrifilatum Carpenter, the four-lined *Bittium* (fig. 577), has a dull brown shell the whorls of which are marked with four prominent equal spiral ridges crossed by vertical ribs dividing the surface into numerous four sided pits. It has 10–11 whorls and a large specimen measures 11 mm. in length and 4.5 mm. in diameter. The range is from Monterey to Lower California.

Cerithidea californica (Haldeman), the California hornshell (fig. 580), is an inhabitant of the mud flats where it may often be seen in great numbers when the tide is out. The shell is dark brown or black and the 9–10 whorls are ornamented with spiral ridges crossed by vertical ribs. This sculpturing may be clearly marked or very faint. The length of a medium sized specimen is 27 mm. The range of the species is from Bolinas Bay to Lower California (Baker).

Family—TRICHOTROPIDAE

Trichotropis cancellata Hinds, the checked hairy-shell, has longitudinal ribs crossing the spiral ones (fig. 579). The hairy epidermis is light brown or grayish and the aperture is often pink. The length reaches 24 mm. The range is from Bering Sea to Oregon.

Family—CAECIDAE

Caecum californicum Dall, the California tube-mollusk (fig. 583), lives in a tiny white tube, slightly curved and

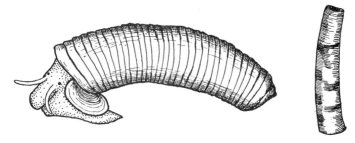

Fig. 583.—*Caecum californicum*, the California tube-mollusk; drawn from the living specimen; x24.

Fig. 584.—*Micranellum crebricinctum*, the close-ringed tube-mollusk; x7.

tapering. The surface is marked with 30–40 prominent rings. The length is 3 mm. The range is from Monterey to Lower California.

Micranellum crebricinctum (Carpenter) the close-ringed

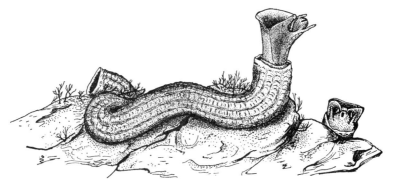

Fig. 585.—*Aletes squamigerus*, the scaly worm-mollusk, from life; x1.2.

tube-mollusk, has a shell (fig. 584), like that of the last named species but may be 5.5 mm. long and the rings are less distinct. The range is from Monterey to Lower California.

FAMILY—VERMETIDAE

Aletes squamigerus Carpenter, the scaly "worm-mollusk" (fig. 585), has a tubular, irregularly twisted shell fastened to rocks or to other shells. The masses of twisted tubes, often large, are among the conspicuous things seen on the beach. The surface may show longitudinal and circular markings but often only the lines of growth are evident. The color is dingy white and the circular aperture may be 15 mm. in diameter. The range is from Monterey to Peru.

Spiroglyphus lituellus (Mörch), the crooked "worm-mollusk" (fig. 586), has an irregular shell in the form of a

FIG. 586.—*Spiroglyphus lituellus*, the crooked worm-mollusk; x4.

spiral tube which is often less symmetrical and regular than the one shown in the figure. The color is dingy white and the surface is sculptured by fairly regular lines of growth. The aperture is circular and 1–2 mm. in diameter. The range is from Alaska to San Diego. The specimen shown in the figure was found on the under side of a large rock, which was left uncovered at an extremely low tide.

FAMILY—TURRITELLIDAE

Turritella cooperi Carpenter, Cooper's tower-shell (fig. 582), has a slender tapering spire, a distinct suture, faint spiral ridges, and a circular aperture. It is yellowish, sometimes with brown spots. The length reaches 46 mm. It is found from Monterey to Cerros Island (Baker), on sandy beaches.

Family—LITTORINIDAE

The littorines have small spiral shells. They live so far up on shore that even a moderately low tide leaves many of

Fig. 587.—A group of littorines clustered on a rock near high-tide line.

them uncovered. Groups of them are often seen on the rocks (fig. 587).

Littorina sitchana Philippi, the Sitka littorine (fig. 588), has a strong shell with about five spiral ridges traversing the whorls. The columella is rounded so that the aperture is

Fig. 588.—*Littorina sitchana*, the Sitka littorine; x5/4. (Photographed by F. W. Kelsey.) Fig. 589.—*Littorina planaxis*, the gray littorine; x5/4. (Photographed by F. W. Kelsey.) Fig. 590.—*Littorina scutulata*, the checkered littorine; x5/4. (Photographed by F. W. Kelsey.)

nearly circular. The height is 12 mm. The range is from Alaska to Puget Sound.

Littorina planaxis Philippi, the gray littorine (fig. 589), has a broad and conical shell with three whorls; the surface

FIG. 591.—*Lacuna porrecta*, the wide chink-shell; x3/2. FIG. 592.—*Iselica fenestrata*; x3/2. FIG. 593.—*Truncatella stimpsoni*, the California looping-snail; x3/2. FIG. 594.— *Lacuna unifasciata*, the one-banded chink-shell; x3/2.

FIG. 595.—*Hipponix antiquatus*, the ancient hoof-shell; x5/4. (Photographed by F. W. Kelsey.) FIG. 596.—*Hipponix tumens*, the sculptured hoof-shell; x5/4. (Photographed by F. W. Kelsey.)

is smooth except for the faint lines of growth and faint spiral striae. The outer lip is thin and the inner lip and columella are flattened as though that portion of the shell has been dissolved away. The length may reach 19 mm. The range is from Puget Sound to Magdalena Bay.

Littorina scutulata Gould, the checkered littorine (fig. 590), has a shell with four whorls, the surface is smooth except for the faint lines of growth. The color, as in the last named species, is brownish with various bands or checks of white. This species, however, is without the flattened area in the region of the columella. The length is 16 mm. The range is from Alaska to Lower California.

FAMILY—LACUNIDAE

Lacuna porrecta Carpenter, the wide chink-shell (fig. 591), has a large umbilical chink, a depressed spire, and a wide spreading aperture. The height reaches 4 mm. The range is from Bering Sea to San Diego.

Lacuna unifasciata Carpenter, the one-banded chink-shell (fig. 594), is brown and glossy with the color broken into dots on the keel of the body whorl. The flattened columella has a small umbilical fissure (Keep). The height reaches 3 mm. The range is from Santa Barbara to Lower California.

FAMILY—FOSSARIDAE

Iselica fenestrata (Carpenter), the windowed shell (fig. 592), has four whorls ornamented with squarish spiral ridges, there being 12–13 of the spirals on the body whorl. There are oblique riblets in the inter-spaces. The umbilical chink is small. The length reaches 8 mm. The range is from Puget Sound to the Gulf of California.

FAMILY—TRUNCATELLIDAE

Truncatella stimpsoni Stearns (fig. 593), is cylindrical, with even, regular longitudinal ribs. The color has been described as light reddish horn color or amber. The length is about 5 mm. The range is from Catalina to Lower California.

FAMILY—HIPPONICIDAE

Hipponix antiquatus Linnaeus, the ancient hoof-shell (fig. 595), is an oblique cone with a flattened apex and with the surface roughened by prominent leaf-like lines of growth. The name was given because of the hoof-like muscle scar inside the shell. The species is variable in appearance. The color is white, the length about 20 mm. The range is from Crescent City to Peru.

Hipponix tumens Carpenter, the sculptured hoof-shell (fig. 596), is more regular in shape than the last named species, the apex is small and distinct and curves over. Radial and concentric lines are present but the former are the most prominent in this species. The shell is white and frequently

PHYLUM—MOLLUSCA

is bearded with a light brown hair-like covering toward the margin. The length is 15 mm. The range is from Crescent City to Lower California (Baker).

FAMILY—CREPIDULIDAE

Crepidula onyx Sowerby, the onyx slipper (fig. 599), is glossy dark brown within with a white deck. The outside is

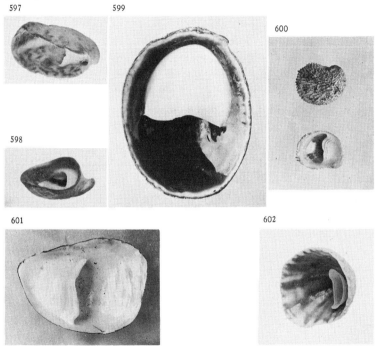

FIG. 597.—*Crepidula excavata*, the excavated slipper-shell; x4/3. FIG. 598.—*Crepidula adunca*, the hooked slipper-shell; x4/3. FIG. 599.—*Crepidula onyx*, the onyx slipper-shell; x1. FIG. 600.—*Crepidula aculeata*, the prickly slipper-shell; x1. FIG. 601.—*Crepidula nummaria*, the white slipper-shell; x4/3. FIG. 602.—*Crucibulum spinosum*, the cup-and-saucer limpet; x4/3.

light brown, often much roughened. The apex is at the margin of the shell and turned to one side. The shape depends somewhat upon the surface to which the animal clung while it was growing; some are almost symmetrical while others are much twisted. The length reaches 45 mm. The range is from Monterey to Chile.

Crepidula excavata Broderip, the excavated slipper (fig. 597), has a thin shell with almost parallel sides. The apex is slightly elevated and the cavity leading into it may be seen above the deck. The shell is whitish, mottled with brown except for the deck which is white. A medium sized specimen is 21 mm. long. The species ranges from Monterey to Peru.

Crepidula adunca Sowerby, the hooked slipper (fig. 598), has a prominent apex which curves out and upward away from the margin. The color is dark brown with a lighter deck. Individuals of this species are often found growing on turban shells as well as upon one another. A good sized specimen is 14 mm. long. The range is from Vancouver Island to Cape San Lucas.

Crepidula aculeata (Gmelin), the prickly slipper (fig. 600), has a low-arched shell with the apex curved far to one side and the upper side of the shell ornamented with rough, radiating ribs. The length of a fair-sized specimen is 25 mm. The species ranges from Santa Barbara to Chile.

Crepidula nummaria Gould, the whites lipper-shell (fig. 601), varies greatly in shape due to the contour of the larger shell to which it may cling. Some forms are nearly circular and flat while others are long and narrow, often fitting the curved surface within the aperture of a spiral shell. The deck is thin and delicate. The interior is white and glistening and the outside may be tinged with light brown. A large specimen is 16 mm. in length. The range is from Alaska to Lower California.

Family—CALYPTRAEIDAE

Crucibulum spinosum (Sowerby), the cup and saucer limpet (fig. 602), is a high arched shell with the apex nearly central and the outer surface near the margin usually bears many spire-like projections. The "cup" is attached on only one side. A large specimen measures 25 mm. in length. The range is from northern California to Chile.

PHYLUM—MOLLUSCA

Family—NATICIDAE

Polinices lewisii (Gould), the Lewis moon-snail (fig. 603), is a large, predaceous mollusk that is at home on the mud flats, where it plows along just under the surface of the mud. The foot is of enormous size and when fully extended seems to be far too large to ever be stowed away in the shell again (fig. 607). The shell is strong, brownish white, the interior chocolate color, the umbilicus narrow and deep with a callus extending partly over it from above. A medium sized specimen measures 105 mm. in length. The range is from British Columbia to San Diego.

H. P. Kjerschow-Agersborg* says, "It destroys oyster beds by its burrowing in them in search of clams, but it is not known whether it attacks oysters directly. On account of its burrowing habits, the oystermen, at the head waters of Puget Sound, destroy large numbers of them. * * At low tide, when rowing along the shores of Dyes Inlet near Chico, Washington, a large number of *Polynices* was found. As the tide was very low it was possible to pick them up by using a dip-net. Some of them, however, were not so easily removed from the bottom as others, holding to the same by means of the enormous foot, or having sucked down into the sand to the depth of about ten centimeters, leaving only part of the shell uncovered in the middle of a pit. It was soon found that there was a definite cause for their holding on to the bottom so firmly; these individuals of *Polynices* were feeding. * * In the case of *Mya*, the gasteropod sucks itself over the syphon down into the sand until its victim is dead from suffocation, and then when the clam has opened, *Polynices* simply sends its proboscis between the valves and devours the content. As for the hard-shelled clams, * * * * * the prey is held in the "sole" of the foot until the adductor muscles are relaxed or the victim is dead,

---

*H. P. K. Agersborg—The Utilization of Echinoderms and of Gasteropod Mollusks; American Naturalist, Vol. 54, pp. 420–425.

when the feeding begins. Several dead clams * * were found in possession of *Polynices,* but none of them were drilled. * * * * * As a natural enemy, *P. lewisii* seems to have none more dangerous than the twenty-rayed starfish (*Pycnopodia helianthoides*). * * I found in all instances, that when the slug came into contact with the star, it withdrew its foot at once. The monstrous foot, though it seemed impossible that it could be withdrawn within the shell, was very quickly covered thereby. Upon withdrawing the foot in a hurry, as it does when in contact with *Pycnopodia,* the periphery of the foot, which is perforated, throws a spray like a garden sprinkler with the holes in the spray-disk plugged except those around the periphery. No matter how much larger the animal is than its shell, when all the water is squeezed out of the foot, the former can be completely covered by the latter. In such a condition, however, *Polynices* can not live very long. * * * If it is not allowed to take in fresh water supply when it comes out to breathe it soon relaxes, an easy prey to the gluttonous *Pycnopodia.* In fact, when leaving *Polynices* with *Pycnopodia* in an aquarium, two of the former were killed and eaten by the latter within three days, leaving the shells and opercula."

Professor Kincaid says that these shells are numerous in the Indian kitchen middens which suggests that the mollusks must have been sought by the Indians for food.

Polinices draconis (Dall) is similar to *P. lewisii* but is shorter in proportion to its diameter (fig. 604), the angle where the lip meets the body is filled with a smooth white callus, and the anterior angle of the lip is thickened. The length is 60 mm. The range is from Alaska to Catalina.

Polinices recluziana (Deshayes), the southern moon-snail, is a flattened spiral (fig. 605), and has a thick covering of enamel on the columella which extends down and fills the umbilicus. Occasionally specimens are found with the umbilicus incompletely covered with the callus. In this and the two preceding species the eggs are laid in a gela-

FIG. 603.—*Polinices lewisii*, the Lewis moon-snail; x⅓.

FIG. 604.—*Polinices draconis*; x½.

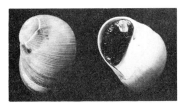

FIG. 605.—*Polinices recluziana*, the southern moon-snail; x2/5.

FIG. 606—Egg collars of *Polinices*; much reduced.

FIG. 607.—*Polinices lewisii*. Photographed from life, showing the mollusk crawling in a tide pool on the mudflats of Tomales Bay. The foot appears to be entirely too large to be wholly withdrawn into the shell.

tinous sheath which is apparently moulded over the foot of the animal for it is of about the same size and shaped like a collar (fig. 606). These "collars" are encrusted with sand and are familiar objects in places where the moon-shell is found. The length is 46 mm., diameter 48 mm. The range is from Crescent City to Chile.

FAMILY—ACMAEIDAE

Acmaea mitra Eschscholtz, the white cap (fig. 608), has a heavy white, conical shell with an erect, nearly central apex. The outer surface is often tinged with green from the corallines that encrust it. It is found from Bering Sea to Point of Rocks, Mexico (Kelsey). A large specimen measures 38 by 33 mm. and is 28 mm. high.

Acmaea cassis pelta (Eschscholtz), the shield limpet (fig. 611), has a large, strong shell, with low, coarse ribs, which are often almost obsolete. The dark central spot inside the shell is small or wanting. The outside is grayish white with radiating black stripes which are often broken into a tassellated pattern. There is a dark band around the inner margin. It is found from Alaska to Lower California. A large specimen measures 38 by 31 mm. and is 17 mm. high.

Acmaea scutum patina (Eschscholtz), the plate limpet (fig. 612), is less conical and pointed than the last species. Fine striations radiate from the apex. Small specimens are olive-gray and tassellated or striped with black. Inside there is an irregular central area of brown and a dark border. The length is 14–38 mm. A specimen 22 mm. long is 19 mm. wide and 7 mm. high. The range is from Alaska to Lower California.

Acmaea limatula Carpenter, (formerly known as *A. scabra*) the file limpet (fig. 613), has a low apex situated between the center and the anterior third, the outer surface has fine radiating scaly ribs with larger ones at regular intervals. The color is light yellowish brown, often indistinctly spotted or striped with brown. It is found from

FIG. 608.—*Acmaea mitra*, the white cap; x5/6. (Photographed by F. W. Kelsey.) FIG. 609.—*Acmaea scabra*, the ribbed limpet; x2/3. FIG. 610.—*Lottia gigantea*, the owl shell; x⅔.

FIG. 611.—*Acmaea cassis pelta*, the shield limpet; x5/6. FIG. 612.—*Acmaea scutum patina*, the plate limpet; x5/6. FIG. 613.—*Acmaea limatula*, the file limpet; x⅔.

FIG. 614.—*Acmaea persona*, the mask-limpet; x⅔. FIG. 615.—*Acmaea insessa*, the seaweed limpet; x1.

FIG. 616.—*Acmaea instabilis*; x3/2. FIG. 617.—*Acmaea asmi*, the black seaweed limpet; x1. FIG. 618.—*Acmaea depicta*, the painted limpet; x1. FIG. 619.—*Acmaea paleacea*, the chaffy limpet; x1.

northern California to Cerros Island. A large specimen measures 40 mm. in length.

Acmaea scabra Gould (formerly known as *A. spectrum*), the ribbed limpet (fig. 609), dwells high up on the rocks so that living specimens are easily found. The outside of the shell is gray and the inside is white, sometimes with a brown spot in the center. A large specimen measures 30 mm. in length. The species is found from Humboldt County, California (Stephens) to Socorro Island.

Acmaea persona Eschscholtz, the mask limpet (fig. 614), has a long, convex posterior slope and the apex points forward. The sculpturing is variable, the ribs sometimes being tuberculated and sometimes almost obsolete. The outside is whitish with dark markings. The inside border of the margin is dark and there is usually a large, brown area in the center. It is found from Alaska to Socorro Island. A good sized specimen measures 33 by 26 mm. and 11 mm. high.

Acmaea instabilis (Gould), the unstable seaweed limpet (fig. 616), is found from Alaska to San Pedro on kelp stalks. It is brown outside and light colored inside. It reaches 20 mm. in length. A specimen 12 mm. long is 9 mm. wide and 4 mm. high.

Acmaea insessa (Hinds), the seaweed limpet (fig. 615), is found on the flat ribbon-like portions of the brown seaweeds which are abundant on the rocky coasts. The shell is brown, the sides are flattened and the apex is high. It is found from northern to Lower California. A good sized specimen measures 16 by 11 mm. and is 10 mm. high.

Acmaea asmi (Middendorff), the black seaweed limpet (fig. 617), is smaller than the preceding species, rusty black outside, and black inside with a brown zone just outside the mussel scar. The surface is lusterless and often corroded. Its range is from Alaska to San Diego. A good sized specimen measures 8 by 6 mm. and 6 mm. high. At San Diego it has been taken in large numbers on *Tegula funebralis* by Mr. F. W. Kelsey.

Acmaea depicta (Hinds), the painted limpet (fig. 618), has a narrow shell with straight, flat sides. It is nearly white with fine, brown lines radiating from the apex. It is found from Santa Barbara to Lower California growing on eelgrass. A large specimen measures only 9 by 4 mm. and is 3 mm. high.

Acmaea paleacea Gould, the chaffy limpet (fig. 619), has a shell that is narrower than that of the last named species. It is white at the apex but the rest of the shell is brown, without any stripes. A large specimen is 10 mm. by 2 mm. and 2.5 mm. high. It is found from Trinidad to Lower California on eelgrass.

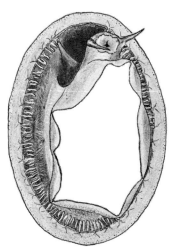

FIG. 620.—*Lottia gigantea*, the under side of living specimen showing the foot, mantle, mouth, and tentacles; x1.7.

Lottia gigantea Gray, the owl shell (figs. 610 and 620), is brown and often roughened on the outside but inside the shell is dark with a bluish white center which is often marked with brown. The apex is close to the anterior end and the shell is rather flat. It is found from Crescent City to Cerros Island. Large ones may be 90 mm. in length.

FAMILY—PHASIANELLIDAE

Phasianella compta Gould, the pheasant-shell (figs. 621 and 625), though tiny, is beautiful and shows great variety in its markings. Usually there are irregular longitudinal lines of white crossed by fine spiral lines of red, purple, or drab. The operculum is solid and white shaded with greenish gray, almost hemispherical, with the convex side facing outward.

FIG. 621.—*Phasianella compta*, drawn from a living specimen, showing foot and siphon extended; x6.

A large specimen may reach 6 mm. in length. The species is found from Monterey to the Gulf of California. Living specimens covered with a dull epidermis may often be found on the seaweed but it is the empty shells that are most often seen.

FAMILY—TURBINIDAE

Astraea undosa (Wood), the wavy top (fig. 623), is one of our largest mollusks. The whorls are crossed by wavy ridges

FIG. 622.—*Norrisia norrisii*, the smooth turban; x⅔. FIG. 623.—*Astraea undosa*, the wavy top-shell; x½. FIG. 624.—*Astraea inaequalis*, the red top-shell; x½.

FIG. 625.—*Phasianella compta*, the pheasant shell; x3/2. FIG. 626.—*Leptothyra carpenteri*, the red turban; x3/2. FIG. 627.—*Leptothyra bacula*, the berry turban; x3/2.

and just above the suture is a more prominent spiral ridge. The operculum is thick and heavy and ornamented on the outer side with curved ribs. The epidermis is fibrous and brown but under it the shell is pearly. It is found from Laguna Beach to Cerros Island. A large specimen may reach a diameter of six inches.

Astraea inaequalis (Martyn), the red top (fig. 624), also has regular oblique ridges crossing all the folds, but the ridge at the edge of the whorl is much less prominent and the concentric furrows on the base are deeply cut instead of being almost obsolete as in the preceding species. The operculum is thick and heavy but without the curved ribs. The color is brick-red when the animal is alive. It is found from Vancouver to San Diego (Baker) and reaches a diameter of 60 mm.

Leptothyra carpenteri Pilsbry, the red turban (fig. 626), is encircled by about fifteen fine spiral ridges. These ridges are distinct and are separated by furrows which are about as wide as the ridges. The operculum is solid and calcareous. The color may be red, ashen, or purple and sometimes specimens are marked with spiral bands or cross lines of darker color. A large specimen may be 8 mm. in diameter. The range is from Alaska to Coronado Islands (Baker).

Leptothyra bacula (Carpenter), the berry turban (fig. 627), is similar to the preceding species but is smaller and does not have the conspicuous spiral ridges. It is found from Puget Sound to Lower California and reaches 4.5 mm. in diameter.

Family—TROCHIDAE

Norrisia norrisii (Sowerby), the smooth turban (fig. 622), is red-brown becoming almost black near the umbilicus which is large and deep and tinged at the margin with bright green. The apex is blunt. The surface is smooth except for faint spiral ridges and lines of growth. The body of the mollusk is red and quite gorgeous when seen crawling upon the seaweed in the tide pools. It has been reported from Monterey to Lower California. A good sized specimen measures 40 mm. in diameter.

Tegula funebralis (A. Adams), the black turban (fig. 628), is abundant in some places. It is black, sometimes with a purple tinge. The apex is nearly always eroded, the teeth on the columella are white and there is no yellow coloring on the base. The body whorl appears to be gathered at the upper edge where it shows numerous lamellae next to the

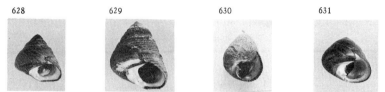

FIG. 628.—*Tegula funebralis*, the black turban; x⅔. FIG. 629.—*Tegula funebralis subaperta*; x⅔. FIG. 630.—*Tegula gallina*, the speckled turban; x⅔. FIG. 631.—*Tegula gallina tinctum*; x⅔.

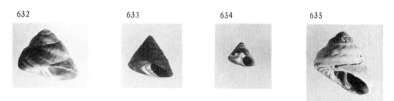

FIG. 632.—*Tegula brunnea*, the brown turban; x⅔. FIG. 633.—*Tegula montereyi*, the Monterey turban; x⅔. FIG. 634.—*Tegula pulligo*, the dusky turban; x⅔. FIG. 635.—*Tegula aureotincta*, the gilded turban; x⅔.

FIG. 636.—*Calliostoma annulatum*, the ringed top-shell; x1. FIG. 637.—*Tegula ligulata*, the banded turban; x⅔. FIG. 638.—*Calliostoma canaliculatum*, the channeled top-shell; x¾.

FIG. 639.—*Calliostoma costatum*, the blue top-shell; x1. FIG. 640.—*Margarites pupilla*: x3/2. FIG. 641.—*Calliostoma tricolor*, the three-colored top-shell; x1.

clearly marked suture. The umbilicus is nearly closed. It is found from Vancouver Island to Lower California and commonly reaches 23 mm. in diameter. The variety *subaperta* (Carpenter) has prominent spiral ridges and a deeper umbilical pit (fig. 629). Its range is the same and it often reaches 27 mm. in diameter.

Tegula gallina (Forbes), the speckled turban (fig. 630), has a black shell with white markings which give it a speckled appearance. It may be found from San Francisco to the Gulf of California. A large specimen measures 30 mm. The variety *tinctum* (Hemphill) Pilsbry shows almost none of the speckled pattern (fig. 631) but the base and larger whorls are tinged with yellow over the dark ground color. It is found from San Francisco Bay to the Gulf of California. A large specimen measures 23 mm.

Tegula brunnea (Philippi), the brown turban (fig. 632), is variously reported as russet-yellow, brown, orange, or deep crimson. The sutures are deep, the umbilicus is somewhat excavated, and there are one or two teeth on the columella. It is found from Mendocino county to Mexico (Kelsey). A good sized specimen measures 27 mm. in diameter.

Tegula montereyi (Kiener), the Monterey turban (fig. 633), is more distinct than the last few species named. The whorls and the circular base are almost flat. The umbilicus is open, funnel shaped and white inside. The color is light brown or olive and the diameter often reaches 28 mm. It is found from Bolinas Bay to Mexico (Kelsey).

Tegula pulligo (Martyn), the dusky turban (fig. 634), also has flat whorls and a flat base but the umbilicus is partly covered by a callus and has no spiral ridge within it as the last named species has. The color is dull brown sometimes tinged with purple. Specimens are said by Keep to reach 32 mm. in diameter. The range is from Alaska to San Diego (Kelsey-Baker collection).

Tegula aureotincta (Forbes), the gilded turban (fig. 635), has two or three prominent spiral ridges and numerous oblique cross ridges. The color is gray and the sutures are deep. An unmistakable character is the bright yellow stain in the large umbilicus. A large specimen measures 30 mm. The range of the species is from the Santa Barbara Islands to Magdalena Bay, Lower California.

Tegula ligulata (Menke), the banded turban (fig. 637), has a strong, solid shell, rusty brown and banded with raised spiral lines. These lines may be beaded or dotted with black. It is found from Monterey to Mexico. An average specimen measures 16 mm. in diameter.

Calliostoma costatum (Martyn), the blue top (fig. 639), has a chestnut-brown shell with enough of a blue tinge, especially in the worn specimens, to give it its common name. The shell is strong and the whorls, which number about seven, are rounded and encircled by smooth spiral ribs of which there are 7–9 on the next to the last whorl. The species ranges from Alaska to San Diego. A large specimen measures 25 mm. in diameter.

Calliostoma annulatum (Martyn), the ringed top (fig. 636), is one of the most beautiful of our shells. To be fully appreciated it must be seen alive since the salmon coloring of the body of the animal adds to the beautiful color effect. The whorls are flattened and encircled by small beaded spiral ridges. The shell is thin, straw colored, the sutures outlined with a purple band which extends along the angle of the body whorl. There is also a circle of the purple color in the region of the columella. The species is found from Alaska to San Diego (Baker). It reaches 23 mm. in diameter.

Calliostoma canaliculatum (Martyn), the channeled top (fig. 638), is light brown with prominent whitish spiral ridges. In some cases the upper ridges of the body whorl are slightly beaded. The shell is thin and rather delicate though stronger than the preceding one. It is found from Alaska to San Diego. A large specimen measures 28 mm. in diameter.

Calliostoma tricolor Gabb, the three-colored top (fig. 641), is light brown, encircled with fine beaded ridges. On each whorl three of the ridges are marked with alternate dashes of purple and white and the whorl is slightly angled at the lower one of these marked ridges. A large specimen is 17 mm. in diameter. The species is found from Santa Cruz to San Diego.

Margarites pupilla (Gould) is small, with a rather solid shell which has five convex whorls that are slightly flattened above, forming a narrow, tabulate band just below the suture (fig. 640). There are five small, flattened, equidistant ribs on each of the upper whorls. The base of the body whorl is nearly flat and has numerous fine revolving lines. The umbilicus is small and groove-like. It is found from Bering Sea to San Pedro. The height reaches 12 mm.

FIG. 642.—*Margarites lirulata*; x7.

Margarites lirulata Carpenter has also a solid little shell (fig. 642). The color is purplish or variegated. In some cases the surface is smooth and in others there are spiral ridges. The body whorl is convex beneath, the aperture brilliantly iridescent within, the umbilicus tubular and striated. The range is from Alaska to San Diego. The diameter is about 5 mm.

Family—HALIOTIDAE

The abalones have conspicuous and handsome shells. The spire is greatly flattened and the epipoda bordered with a fringe and tentacles, which project around the margin of the shell. Under the row of perforations near the left margin, one finds the respiratory chamber containing two gills. The mollusk holds to the rocks by means of the powerful muscle which forms the foot.

A large red abalone is said to produce annually one to two million eggs. In spite of this, the numbers of abalones have decreased in recent years, so that special legislation has been enacted in the effort to conserve them. In former years, great numbers of abalones were dried and exported to China. An old report for the year ending July 1912, states that there was shipped from Long Beach alone 14 tons of dried abalone meat. There is now a closed season during the breeding period from January 14 to March 16. During the open season, there is a limit to the size and number of abalones that may be taken in a day. Drying of the mollusks is now prohibited. The catch amounted in 1922 to over 1,500,000 pounds, taken chiefly from the Monterey region.

Abalone steak is excellent fare. To prepare it, remove the animal from the shell by heating, trim off the viscera and mantle from around the foot, and slice the latter into thin steaks. Put each slice between cotton cloth and pound two or three times with a wooden implement, then fry or broil, seasoning to taste.

Abalone pearls are often beautiful. The blister pearl is produced as a covering over a foreign body in the shell, or to reinforce the shell at a point which is being invaded by a boring animal. Culture experiments have been carried on to induce the formation of the pearls. Forms are fastened to the shell under the mantle and within a short time the mantle covers these with a layer of nacre similar to that with which it lines the shell. The experimenters say that the formation of a pearl takes 75–100 days. Many of the pearls may be produced in one individual. The industry might be a profitable one if laws were enacted for the protection of the culture beds.

Haliotis cracherodii Leach, the black abalone (fig. 643), is greenish black on the outside and smooth except for the lines of growth. The shell is markedly convex and the perforations usually number 5–8. It is not carinated at the line of perforations. Inside of the thick, black layer it is silvery with red and green reflections. It ranges from Oregon to Lower California. A large specimen measures 6 inches in length.

Haliotis rufescens Swainson, the red abalone (fig. 644), has

FIG. 643.—*Haliotis cracherodii*, the black abalone; x3/8. FIG. 644.—*Haliotis rufescens*, the red abalone; x3/10.

FIG. 645.—*Haliotis fulgens*, the green abalone x2/5. FIG. 646.—*Haliotis corrugata*, the corrugated abalone; x3/10.

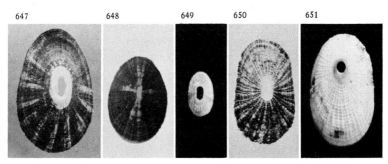

FIG. 647.—*Fissurella volcano*, the volcano-shell; x5/4. (Photographed by F. W. Kelsey.) FIG. 648.—*Fissurella volcano crucifera*; x4/3. FIG. 649.—*Megatebennus bimaculatus*, the spotted key-hole limpet; x3/2. FIG. 650.—*Lucapinella callomarginata*; x5/6. FIG. 651.—*Diadora murina*, the white key-hole limpet; x3/2.

a large heavy shell less convex than most of the other abalones, and the outside is sculptured with low irregular radiating waves. The red outside layer projects at the edge of the lip making a narrow coral-red edge. The perforations are tubular, large, and number 3–4. The species has been found from Bodega Bay to Lower California. It reaches 9 inches in length.

Haliotis fulgens Philippi, the green abalone (fig. 645), is dull reddish brown, sculptured on the upper surface and with 30–40 rounded spiral ridges which are nearly equal in size. There is an angle at the row of perforations. The holes are small, elevated and circular and number 5–6. The length is usually about 6 inches. The range is from the Farallon Islands to the Gulf of California.

Haliotis wallalensis Stearns, the northern green abalone, is found, according to Thompson,* along the coast between Westport and the Russian River. He says it resembles the green abalone of southern California in appearance, but is not greater than 5½ inches in length, has 5–6 open holes and the edges of these holes are not elevated.

Haliotis corrugata Gray, the corrugated abalone (fig. 646), is more nearly circular than the other species, with a high arched shell the outside of which is markedly corrugated. There are 3–4 perforations which are large and somewhat tubular. A large specimen measures 6 inches in length. The range is from Monterey to Lower California.

Family—FISSURELLIDAE

Fissurella volcano Reeve, the volcano-shell (fig. 647), has an oval shell which is somewhat narrower in front, and has an oblong orifice which is a little in front of the center of the shell. The color is ashy-pink with 13–16 purplish rays. Living specimens are very handsome with red-striped mantles

---

*W. F. Thompson—The Abalones of Northern California; California Fish and Game, Vol. 6, p. 50.

and yellow foot. The usual length is 25 mm. The range is from Crescent City to Panama. The variety *crucifera* Dall differs only in color, having four white, equidistant rays extending down from the apex (fig. 648). It is found from Monterey to Lower California.

Megathura crenulata (Sowerby), the giant keyhole limpet (figs. 652 and 653), has numerous fine radiating ridges and lines of growth. The margin of the shell is finely and evenly crenulated. The outline is oblong oval as is also the shape of the perforation which is a little in front of the center of the shell. The black mantle of the living animal nearly covers the shell and the huge yellow foot is much longer than the shell. Large shells measure 115 mm. The species extends from Monterey to Lower California.

Megatebennus bimaculatus (Dall), the two-spotted keyhole limpet (fig. 649), is oblong oval with a comparatively large "keyhole" of the same shape as the shell. There are numerous radiating ridges from the apex which widen slightly toward the margin of the shell. The anterior and posterior ends are somewhat concave so that when the shell is laid on a flat surface the two ends of the shell do not touch it. The color is white, brown, or slate, sometimes with darker rays on the sides. The species rarely exceeds 12 mm. in length. It is found from Alaska to Cape San Lucas.

Lucapinella callomarginata (Carpenter) has a low arched shell with rough radiating ribs, oblong perforation, and crenulate margin (fig. 650). The exterior is gray or brown sometimes marked with darker color. A large specimen is 23 mm. long. The range is from Bodega Bay to Valparaiso.

Diadora aspera (Eschscholtz), the rough keyhole limpet (fig. 654), has a high arched shell with a small, oval or nearly circular hole which distinguishes it from the preceding species. The margin is crenulate and the color is gray with darker radiating rays. A large specimen measures 50 mm. in length. The range is from Alaska to Lower California.

Diadora murina (Carpenter), the white keyhole limpet

FIG. 652.—*Megathura crenulata*, the giant key-hole limpet; x1. FIG. 653.—*Megathura crenulata*, the giant key-hole limpet, photographed from life. The black mantle nearly covers the shell; x½. FIG. 654. —*Diadora aspera*, the rough key-hole limpet; x3/2.

(fig. 651), has a delicate white shell with fine radiating ridges crossed by fine concentric lines. The perforation is nearly round and one-third of the length from the anterior end. A good-sized specimen is about 20 mm. long. The range is from Crescent City to Lower California.

Class—AMPHINEURA

(*The chitons*)

The chitons are the best known members of the class *Amphineura* and can easily be distinguished by the eight shingle-like plates of the shell (fig. 670). Living specimens are numerous on the rocks that are exposed at low tide and wave-washed plates of chitons are often picked up on the beach. The children call the end plates "false-teeth shells" and the central ones "butterfly-shells." To one order of this class belong worm-like, cylindrical creatures, without a shell. These however are deep water forms and out of our field.

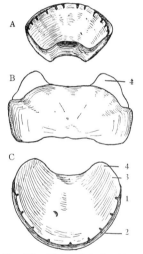

FIG. 655.—Plates of *Ischnochiton conspicuus*, upper side.
A. Anterior plate. B. 7th plate.
C. 8th or tail plate.
1. Slit.
2. Articulamentum.
3. Tegmentum.
4. Sutural laminae.

The chitons are sluggish in their movements and cling to the rocks, eating the algae which grow thereon. They are abundant on or under the rocks that are exposed at low tide while many species live at greater depths. Dr. Heath* says that the majority of the species in Monterey Bay are nocturnal and withdraw into shaded places on the approach of day. *Katharina, Tonicella,* and *Nuttallina* occupy exposed positions and do not

---

*H. Heath—The Development of *Ischnochiton;* Jena, Gustav Fischer, 1899, p. 4.

often conceal themselves, so they are apparently not so sensitive to the light as are some of the other species. *Mopalia* and *Cryptochiton*, he says, may remain out on the feeding ground if the day is dark and foggy. Of all the species, he found *Ischnochiton magdalenensis* to be the most sensitive to the light. During the day they are under boulders between tide marks, perhaps half buried in the sand. At night they come out to eat the vegetation, either emerging completely or just coming far enough to reach the algae.

When chitons are detached, they roll up into a ball, armadillo fashion. If the collector wishes to preserve them in their normal shape he binds them flat on a small piece of board or other flat surface as soon as he has removed them from the rocks.

Each of the eight plates of the chitons partially overlaps the one posterior to it. The girdle surrounds the plates, in most cases scarcely covering them at all. Each plate or valve is made up of two layers, the upper tegmentum and the deeper, porcellaneous layer, the articulamentum (fig. 655). This lower layer sometimes projects into the girdle forming insertion plates. The anterior margins of all plates except the first have two projections, sutural laminae, which fit under the hinder margin of the plate in front. The girdle is covered with tiny scales or spines or is horny.

The foot is a broad, flat, creeping surface which enables the animal to attach itself firmly to the rocks. Between the girdle and the foot, extending all around the animal is a groove, in which lie 6–80 pairs of gills (fig. 656). The mouth opens from the anterior end of this groove into the pharynx which is provided with a long radula or rasping organ, bearing numerous teeth with which the food is scraped from the rocks and torn into bits. The ducts from the liver-mass open into the stomach. The intestine is 6–7 times as long as the body, coiled, and opens through the anus into the posterior mantle cavity.

There are no tentacles and the sense organs are poorly developed. Certain species have sense buds, or definite

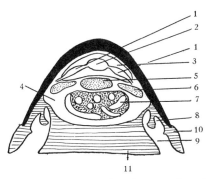

FIG. 656.—Diagram of a cross section of a chiton near the posterior end of the body; x4/3.

1. Auricles.
2. Ventricle.
3. Pericardium.
4. Body cavity.
5. Gonad.
6. Nephridium.
7. Coiled intestine and liver mass.
8. Gill.
9. Mantle cavity.
10. Girdle.
11. Foot.

groups of nerve cells in the shell, which are even modified to form "eyes" in a few species. The ganglia are small and largely scattered throughout the two pairs of longitudinal nerve cords which unite in the mouth region.

The heart lies in the pericardial cavity in the posterior third of the dorsal region of the body and consists of an elongated ventricle and two auricles. The nephridia open into the pericardial space and into the mantle cavity. Most chitons are unisexual but there are some hermaphroditic species. The genital ducts open into the mantle cavity anterior to the nephridial openings.

Dr. Heath[†] who has observed the egg laying states that a short time after the males have begun to liberate spermatozoa in a tide pool the females begin to deposit eggs. That the liberation of the spermatozoa is the necessary stimulus for egg laying seems likely, for isolated females in tide pools delayed their egg laying two months beyond their ordinary

---

[†]H. Heath—The Breeding Habits of Chitons of the California Coast; Zoologischen Anzeiger, Band 29, p. 391.

FIG. 657.—*Lepidochitona lineata*, the lined chiton, from a photograph of a living specimen; x1. FIG. 658.—*Lepidochitona hartwegii*; x3/2. FIG. 659.—*Ischnochiton mertensii*, the red chiton; x3/2. FIG. 660.—*Ischnochiton regularis*, the regular chiton; x3/2.

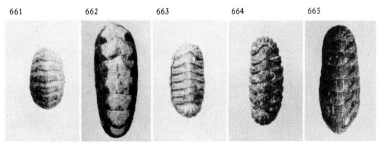

FIG. 661.—*Lepidochitona raymondi*; x3/2. FIG. 662.—*Nuttallina californica*, the California chiton; x¾. FIG. 663.—*Callistochiton palmulatus*, the palm chiton; x3/2. FIG. 664.—*Callistochiton palmulatus mirabilis*; x3/2. FIG. 665.—*Callistochiton crassicostatus*, the thick-ribbed chiton; x3/2.

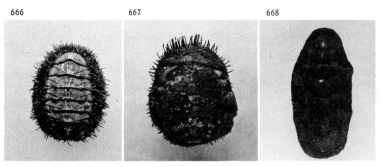

FIG. 666.—*Mopalia muscosa*, the moss chiton; x¾. FIG. 667.—*Placiphorella velata*, the veiled chiton; x⅔. FIG. 668.—*Katharina tunicata*, the black chiton; x¾.

breeding time. The eggs in some species are laid separately, with a chitinous envelope, in other species they are in gelatinous strings, or masses, more or less readily broken into fragments by the beating of the surf. Other species retain the eggs in the mantle cavity while they undergo development and in one species the ova develop in the oviduct of the mother.

FAMILY—LEPIDOCHITONIDAE

Lepidochitona (Tonicella) lineata (Wood), the lined chiton (fig. 657), has a smooth surface and is moderately arched. In living specimens, the light ground color tinged with lavender, brown, yellow, or red, is marked with wavy, dark-brown lines bordered above with light blue which often fades to white in the dry specimens. The girdle is yellowish brown, smooth and thin. The anterior and posterior valves have 8–10 slits each. The length may reach 37 mm. They are reported from Alaska to San Diego but are not common south of Monterey.

Lepidochitona (Trachydermon) hartwegii (Carpenter), is oval, low, dull olive-green outside (figs. 658 and 673). Inside the shell, the color is an intense blue-green. Sometimes it is marked with light stripes and a black blotch on each valve. The surface and the narrow girdle are finely granulated. The anterior valve has 10–11 slits and the posterior one 9–12. The length reaches 25 mm. The range is from Alaska to the Gulf of California.

Lepidochitona (Trachydermon) raymondi Pilsbry, Raymond's chiton, has a longer and narrower shell than *L. hartwegii*. The back is somewhat keeled and may be blackish, olivaceous, or dark brown, sometimes with white or dark markings. It is evenly sculptured with minute granules. The interior is light blue with some darker stains. The anterior valve has 8 slits and the tail valve 11. The girdle is narrow, leathery, and minutely papillose. The length reaches 23 mm. (Pilsbry). The range is from Alaska to San Pedro (fig. 661).

Nuttallina californica (Reeve), the California chiton (fig. 662), lives higher up on the rocks in situations that are more exposed to the light than do some of the other species. The surface is granulated, and the head valve has many radiating ribs. The color is dark brown with whitish streaks. The girdle is spiny to scaly. The foot of the animal is reddish and the interior of the valves blue-green. The length is 37 mm., breadth 15 mm. The range is from Puget Sound to the Coronado Islands (Stephens).

Family—ISCHNOCHITONIDAE

Ischnochiton magdalenensis (Hinds), the gray chiton (fig. 669), has a long, narrow shell. The light ground color is mottled with olive and the interior is bluish or pink. The sides and ends of the shell have fine radiating ribs. The front slope of the anterior valve is straight and the umbo of the posterior valve is central, projecting but little. The girdle is covered with fine close scales. The anterior valve has 10–13, and the posterior one 10–12 slits. The length is 75 mm., breadth 30 mm. The species ranges from Oregon to Lower California.

Dr. Heath* found that the eggs of this species are laid in May and June on the days when the low tides come in the early morning. The eggs are in jelly-masses which take the form of spiral strings nearly a yard in length. He estimated the number of eggs laid by an individual to be on an average about 116,000. At the end of twenty-four hours, the larvae have begun to rotate within the membrane and six days later free themselves and swim about. The free swimming stage probably lasts not longer than two hours and the larvae settle down upon the rocks or seaweed, there to

---

*H. Heath—The Development of *Ischnochiton*; Jena, Gustav Fischer, 1899, p. 5.

FIG. 669.—*Ischnochiton magdalenensis*, the gray chiton; x¾. FIG. 670.—*Ischnochiton cooperi*; x3/2. FIG. 671.—*Ischnochiton conspicuus:* x¾.

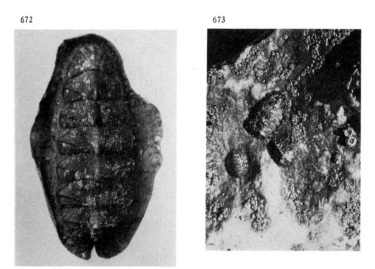

FIG. 672.—*Mopalia ciliata*, the hairy chiton; x3/2. FIG. 673.—Living chitons, *Mopalia muscosa* and *Lepidochitona hartwegii*, attached to a rock on the seashore.

undergo a metamorphosis. In 10–12 days, though small, they resemble the adult in form.

Ischnochiton conspicuus (Carpenter), the conspicuous chiton (fig. 671), also conceals itself under rocks during the day time. It resembles *S. magdalenensis* but often reaches 100 mm. in length. The girdle is thickly covered with bristles which give it a velvety appearance. The ridges of the valves are pink. The front valve is concave in contrast to the last species in which the front valve is straight. The range of the species is from Monterey to the Gulf of California.

Ischnochiton mertensii (Middendorff), the red chiton (fig. 659), is oval with angular dorsal ridges and straight sides. The color is orange to dark brown, sometimes with white blotches. The valves are sculptured, the mantle covered with fine rounded scales. The length is 25–35 mm. Its range is from Alaska to San Pedro.

Ischnochiton cooperi Carpenter, resembles *I. mertensii* but the color is olivaceous or dull brown clouded with light blue (fig. 670). The lateral areas are raised and in the adults bear 6–8 irregular rows of rounded pustules. A lens shows fine granulation of the surface between the pustules. The central areas have fine, even, radial striations crossed by raised threads parallel to the dorsal ridge. The end valves are radially ridged with long pustules. The head and tail valves each have 11 slits. The range is from Mendocino to Catalina.

Ischnochiton regularis (Carpenter), the regular chiton (fig. 660), is an elongated oval with width equal to half the length, the lateral areas have concentric, slightly granular ridges and fine radiating lines. Under the lens, the border looks like fine bead work. The outside is olive, or slaty blue and the interior light blue. The length is 35 mm. The range is from Mendocino to Monterey.

Callistochiton palmulatus Carpenter, the palm chiton (fig. 663), has small high-arched valves, marked with raised sculpturing. There are eleven ribs on the anterior valve and seven bifurcated ones on the posterior one. The color is dark brown and the length is 11 mm. It ranges from Monterey to Lower California. A variety, *mirabilis* Pilsbry, found at San Diego, has the last valve enormously thickened and the interior bluish white (fig. 664).

Callistochiton crassicostatus Pilsbry, the thick-ribbed chiton (fig. 665), has an elevated, oblong shell. The color is green or brown. The front valve has seven strong ribs. The interior is bluish white, the length is a little less than an inch. The species is found from Alaska to San Diego.

FAMILY—MOPALIIDAE

Mopalia ciliata (Sowerby), the hairy chiton (fig. 672), has an oblong shell, rather depressed. The surface is lusterless and finely sculptured. The most usual coloring is green, spotted with black or brown, the girdle yellow. There may be maroon spots or chestnut spots on the ridge, or some of the valves may be scarlet mixed with olive and white, or entirely white. Again, some specimens are light olive-buff with a brownish girdle. The central areas are sculptured with longitudinal, curving riblets, somewhat granulated. The anterior valve has 8 slits, the tail valve has a deep caudal sinus and a single slit on each side. The girdle is wide, generally notched behind, and sparsely covered with curling, strap-like, brown hairs which bear near their bases a bunch of minute, white, acute spines. The length is 7–20 mm. The range is from Vancouver Island to Lower California.

Mopalia muscosa (Gould), the mossy chiton (figs. 666 and 673), is oval with strong valves. The surface is lusterless. The color is usually dull brown, grayish, or blackish though sometimes it may be bright orange, scarlet, or vivid green. The central areas have close, fine longitudinal riblets with crenulated or latticed interstices which may diverge. The first valve has about 10 narrow, radiating, granose riblets.

The numbers of slits in the valves are the same as in *M. ciliata*. The girdle is narrow, densely covered with round, curved, or curled hairs. The length is 5–50 mm. The range is from Alaska to Lower California.

Placiphorella velata Carpenter, the veiled chiton (fig. 667), is almost circular and from its anterior end the mantle projects in such a way as to suggest a veil. This veil bears scattering hairs. The valves are low, dull reddish or greenish without and whitish within. The length is 30–50 mm. The range is from Trinidad to Lower California.

FAMILY—ACANTHOCHITONIDAE

Katharina tunicata (Wood), the black chiton (fig. 668), is easily distinguished by its black, leathery girdle which nearly covers the valves, only about a third of the width of the valve being left exposed. The valves, where they are exposed, are dark brown, but the parts that are covered are white. The anterior valve has 7–8 slits, and the tail valve 1–4 small slits on each side. The soft parts are salmon colored. Keep* says that this chiton is sometimes eaten raw by the natives of the northwest coast. The length may reach 75 mm. The range is from Alaska to the Catalina Islands and the species is abundant on the rocky beaches north of Point Conception.

FAMILY—CRYPTOCHITONIDAE

Cryptochiton stelleri (Middendorff), the giant chiton (fig. 674), is the largest chiton of the coast and can easily be recognized by its size, color and by the fact that the girdle completely covers the valves. The girdle is thick, red, and leathery, covered with minute clusters of vermilion spinelets. The first valve has 4–7 slits and the tail valve has one slit on each side of the sinus. It is found just below low tide line. Miss Rogers† says it is eaten raw by the Indians. The

---

*Josiah Keep—West Coast Shells; Whitaker and Ray-Wiggin Co., p. 261.
†Julia Ellen Rogers—The Shell Book; Doubleday, Page and Co., p. 240.

length frequently reaches 20 cm. The range is from Alaska to San Nicolas Islands.

Class—CEPHALOPODA

(*Squids, devilfish, and nautilus*)

The cephalopods (figs. 675-679) are mentioned in popular books and newspapers more often than almost any other of

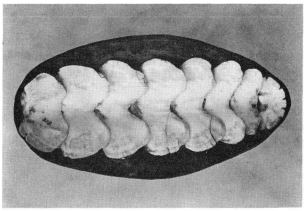

FIG. 674.—*Cryptochiton stelleri*, the giant chiton. Inside view of plates and girdle after the body of the animal has been removed; x3/8.

the marine invertebrates. To most people they are exceedingly repulsive; this fact has probably led to the exaggerated stories of their evil deeds. One California newspaper once printed a drawing, supposedly of an octopus, which represented all the arms as terminating in pincers, and the mouth at the posterior end of the body with a pair of huge jaws or beaks like those of an eagle but set with numerous teeth. Other old pictures have represented ships being crushed by octopods. As a matter of fact the octopus is rather retiring and is not so likely to attack you and tear you limb from limb as is your family cat or dog, and the squid, although it sometimes jumps out of the water for short distances much as the flying fish does, is intent upon catching fish and other animals which it finds far more palatable than wooden ships.

Dr. Paul Bartsch* of the U. S. National Museum gives under the title *Pirates of the Deep*, some interesting stories of the squid and octopus. Concerning Victor Hugo's account of Gilliatt's fight with the devil fish in *Toilers of the Sea* he says, "we regret greatly that the author's powers of observation were not on a par with his wonderful gift of dramatic diction, for a trifle more knowledge would have raised this chapter from the limbo of silly yarns to a production worthy of Victor Hugo."

Dr. Bartsch gives an account of an octopus hunt in the Philippines. He says that the natives went in a procession from the Moro village down to a sand spit which fringed a reef. Each native carried a bamboo torch in his left hand and a bolo or spear in the right. The light from the torches was sufficient to show the curious humped-up devil fishes in the shallow water. The spears descend and the devil fishes are strung on a rattan string where they squirm until they are dead. He says, "We secured enough specimens that night to enable us to spare some to the cook, for Ming assured us that they were 'vely good.' So they were—rather, I should say it was, for I chewed a single tentacle the greater part of the following forenoon and relinquished it only, and that with regret, when my jaws, aching from overexertion, refused to operate more."

On the island of Guam he found that the natives tied a holothurian to a line with a sinker and lowered it in the water. The animal is so distasteful to the octopus that when it comes near, the octopus prepares to leave the vicinity immediately and is speared by the native who watches from the bow of the canoe.

One of the methods used by the Japanese is to lower earthen pots by cords and leave them on the bottom for a

---

*Paul Bartsch—Pirates of the Deep—Stories of the Squid and Octopus; Smithsonian Report, 1916, pp. 347–375.

time. The octopodia in seeking secluded spots will often get into the jars and are secured when the pots are drawn to the surface.

Dr. Bartsch describes an experience that he had while fishing with a submarine light in the harbor of Jolo on a dark night. A 16-candle power electric light, enclosed in a glass globe, was connetted to a water tight cable. The light soon attracted a cloud of minute forms which were followed by crustaceans and worms, and these by small fish and then larger and still larger ones all preying upon the smaller forms. He says, "It was a mad dance, this whirling, circling host of creatures. Soon a new element entered; living arrows, a school of *Loligopsis* shooting across our lighted field, apparently not so much attracted by the light as by the feast before them. They were wonderful creatures, unlike anything else; they shot forward or back like a shuttle, with lightning rapidity. Not only that, but they were able to divert their course into any direction with equal speed. Shooting forward, their tentacles would seize a small fish, and instantly they would come to a full stop, only to dart backward like a flash at the least sign of danger. . . . . Never before nor since have I seen anything that appeared to me more beautifully equipped for an aquatic existence than these squids."

The old tales of sea serpents can probably be traced to the giant squids, since the regions in which the sea serpents were said to have been seen are the ones in which the giant squids exist. Professor Verrill[*] gives the measurement of the largest of these giants as follows: total length 55 feet, length of tentacular arms 35 feet, diameter of largest sucker 2.25 inches, and breadth of eye opening 7 by 9 inches. Tentacles thirty-five feet in length would furnish a good basis for a sea-serpent yarn. The largest squid (*Moroteuthis robusta*) ever reported from the west coast was 14 feet

[*] A. E. Verrill—In Annual Report Commissioner of Fish and Fisheries for 1879, p. 22.

long, and was cast up on the beach at Unalaska, Alaska.*
The sperm whale is an enemy of the squids. Remains of enormous squids are found in the whale's stomachs, and frequently circular scars from wounds inflicted by the squid's suckers may be found on the skin of the whale. One hesitates to use the term "suckers" for it leads some readers to jump to the conclusion that the animal sucks out the blood of its victim by means of them. This, of course, is as impossible as that dust should be sucked into an automobile tire through the "vacuum cups" on its surface. The "sucker" is merely an arrangement for securing a foothold. Naturally when the octopus applies the disks to one's skin and gets a firm hold, the area will appear red afterward or the skin might even be broken with some loss of blood, but blood has not been sucked from it. Furthermore, we know of no authentic accounts of bathers having been drowned because of being held under water by octopods. All the octopods we have ever seen, have tried to escape, with all possible speed, to deep water or to the shelter of overhanging rocks.

The wonderful series of color changes shown by cephalopods is brought about by contraction or enlargement of various pigment cells in the skin. These are of various colors, rosin-colored, yellow, blue-green, or brown, and have muscular walls which enable them to contract until they are almost invisible and then expand to many times their former size. The different colored cells can expand alone or in combination with the other colors and enable the animal to produce flashes of color over its whole surface. Dr. Heath† (1917) says, "As a devilfish crawls about on the sea bottom its color can be seen to change in a twinkling from deep chocolate through dull red and brown to gray. If sand or rock is encountered on the journey the skin is

---

*S. S. Berry—A Review of the Cephalopods of Western North America; Bulletin of Bureau of Fisheries, Vol. 30, p. 315.

†H. Heath—Devilfish and Squid; California Fish and Game, Vol. 3, p. 105.

usually thrown into lumps and ridges, so that under all conditions the body is practically invisible. The squids have an even greater range of color change, some species being capable of assuming almost every tint of the rainbow. If it should happen that some keen-sighted enemy, such as a wolf fish or eel, should spy out any of these chameleon-hued animals a further means of escape is provided by the so-called ink bag. In every case this is a bag concealed within the body and filled with a brownish or black fluid having the same appearance as india ink. This is squirted into the face of any formidable intruder and in the gloom thus produced the attacked is enabled to steal away to a place of safety."

Dr. Heath* has also described the movements of the devilfish: "If a devilfish at rest upon the sea bottom or in an aquarium be watched for a time its body will be seen to contract and expand every few seconds. With each expansion a slit in the neck region, and leading into a spacious cavity within the body, opens to permit the entrance of a stream of water supplying oxygen to the two feathery gills concealed within. At the outset of the contracting movement a valve prevents the outgoing stream of water from escaping as it came, and directs it through a tube or funnel, also known as a siphon. * * If the animal is disturbed the movements become more active and if sufficiently violent the water stream escaping by the funnel drives the animal backward as escaping gases drive a sky rocket. In this way all cephalopods can move, often with the rapidity of a fish of the same size, but the movement is invariably backward. When a devilfish moves forward it creeps along by means of its flexible snaky arms and suckers. The members of the squid group, on the other hand, move with equal facility backward or forward, the forward movement being brought about by the fins."

Cephalopods are of importance to man in that they live upon fish, devouring vast numbers of them, and in turn

*H. Heath—Devilfish and Squid; California Fish and Game, Vol. 3, p. 105.

furnish food for other fish. In the Cape Cod region about 3,000,000 pounds are caught annually for fish bait. The squid go in immense schools which are followed by seals, sea lions and various fishes. These schools usually come to the surface at night, since the food of the squids consisting of smaller creatures comes to the surface only after sundown. Dr. Heath* (1917) describes the squid fishing thus, "Several years ago when squid fishing was an important industry in Monterey Bay, the Chinese provided their sampans with metal baskets in which a fire was built and rowing about netted their unwary prey attracted by the light. Five thousand odd tons were taken in a single season and were dried on the ground, or in the case of higher grades, were cleaned and dried on racks. * * The choicer brands were used for soups and other dishes celestial only in name. Nevertheless, it is possible to place before the American epicurean a dish of tender squid, lightly boiled in oil, that not only is most nutritious but of most delicate flavor. The dishes of the California devilfish tested by the writer can not be described with such gusto. On the shores of the Mediterranean this article is tender and of excellent flavor, but the California product needs some softening influence to destroy its rubber-like consistency."

A report in *California Fish and Game*, for October, 1919,‡ is of interest in this connection, "Three Chinese firms have dried this season about 1,772,000 pounds (fresh weight) of squid. Three tons of wet squid furnish one ton dried. Due to high labor cost this year the squid were not cleaned, merely dried on the ground, raked up and sacked. Fishermen were paid $10 per ton for the catch and the dried product sacked ready for shipment is valued at 6 to 7 cents per pound. Practically all this sacked product is shipped to China.

"In addition, small quantities of squid have been canned

†H. Heath—Devilfish and Squid; California Fish and Game, Vol. 3, p. 107.
‡California Fish and Game, Vol. 5, p. 198.

in half pound rounds. The appreciation of fresh squid as a table delicacy is slowly growing, but people who delight in oysters and eels usually balk at squid tentacles till they have tried them once." According to the reports of the Fish and Game Commission, 196 cases of squid were canned in California in 1920, and 556 cases in 1921. W. L. Scofield,† in a report on squid, in *California Fish and Game* says, "The year 1924 was the record breaking one in the history of the industry and over 2500 tons were dried in a large field, known facetiously as Heliotrope Point, a couple of miles outside the town of Monterey."

The shore collector will frequently encounter small octopodia in the tide pools and the egg cases of the argonaut or paper nautilus are sometimes washed ashore in the southern region, particularly at Catalina. The living argonaut and the squids, being free swimmers, live farther out and are usually only brought in by fishermen.

Cephalopods are easily recognized by the large body and eyes, and the circle of long arms set with rows of sucking disks. The visceral mass is compact and covered with a thin wall. The mantle is thick and covers the mass, forming a large posterior mantle cavity in which are the gills and the openings from the intestine, kidneys, and reproductive organs. The water is taken into the mantle cavity between the free edge of the mantle and the body, and can pass out through the same opening if the animal is at rest; but when the cephalopod wishes to move, the free edge of the mantle fits down tightly so that it closes the opening and the water is driven out through the siphon, a tube developed from the mantle at the side of the head. The force of the water as it shoots out through the siphon, drives the animal backward through the water. The numerous arms drag straight behind and offer the least possible resistance to progress. A pair of sharp chitinous jaws, which remind one of a parrot's beak, lie just back of the circular lip which surrounds the

---

†W. L. Scofield—Squid at Monterey; California Fish and Game, Vol. 10, p. 179.

mouth. With these jaws the prey is torn to pieces. The pharynx contains the tongue and radula, the latter small for the animal's size. Salivary glands, liver, and pancreas contribute their digestive fluids. The intestine leads forward from the stomach to the mantle cavity. The ink sack is a glandular pocket, opening off from the rectum close to the anus. The fluid is brown or black and is ejected through the anus when the animal is excited or pursued by an enemy. The ctenidia, or gills, lie in the mantle cavity. Blood from them passes into two auricles and from there into the single ventricle which sends it by way of arteries to all parts of the body, whence it returns by veins to two branchial hearts at the bases of the gills, through which the blood is pumped by the branchial hearts. The pair of kidneys lie close to the veins that bring the blood to the gills.

The nervous system has well developed ganglia, the main ones being grouped together around the esophagus and surrounded by a protecting cartilage. This brain development gives the cephalopods first place among all the invertebrates that live in the ocean. The sense organs are well developed. The eye reminds one strongly of the vertebrate eye. It has a lens, transparent cornea, and retina. Lithocysts, organs of equilibrium are present, also a pair of pits behind the eyes which are thought to serve as olfactory organs.

The sexes are separate. The male places a capsule, or spermatophore, filled with sperm cells within the mantle cavity of the female. When the female deposits the eggs the capsule is ruptured and the eggs are fertilized by the cloud of spermatozoa thus set free. The eggs of squids are deposited in large gelatinous masses sometimes called "sea grapes." One of the arms of the male is swollen at the tip and specially adapted for the transfer of the sperm packet. It is spoken of as the hectocotylus arm or tentacle and in the case of the argonaut the tip of the arm is left within the mantle cavity of the female.

The shell of the paper nautilus is a modified egg case (fig. 675) secreted by shell glands near the tip of two of the arms. The animal always keeps the shell clasped in these two arms and can retreat into the shell or remain wholly outside of it at will. It will be seen from this that the shell of the paper nautilus is not comparable with the shells of other mollusks. Our cephalopods are without external shells. Our octopodia have no shell at all, but our squids have a slender pen within the mantle.

The remote ancestors of this group have left numerous records in the rocks. Many shells of great variety and some of considerable size are preserved to us as fossils. Some are straight cones, some loosely, and others tightly coiled, all with the inner part divided by partitions as in the chambered nautilus. Some of the cone shaped shells were as much as 14 feet long, others of the coiled sorts reached 6 feet in diameter. The chambered nautilus, an inhabitant of tropical waters, is a surviving relative of these prehistoric giants, but the more modern type of cephalopod represented by the octopods and squids has lost the shell and gained in size, brain development, and activity.

Order—DIBRANCHIATA

The cephalopods having one pair of branchiae belong to the order *Dibranchiata*. The *Octopoda*, which include the devilfish and argonauts, have eight arms on which the suckers are sessile and without a horny ring. The descriptive material that follows is taken largely from the publications of Dr. S. S. Berry.*

Family—ARGONAUTIDAE

Argonauta pacifica Dall, the paper nautilus (fig. 675), is a small species. The animal is orange with a sprinkling of

---

*S. S. Berry—A Review of the Cephalopods of Western North America; Bulletin Bureau of Fisheries, Vol. 30, pp. 267-336.
Notes on Some Cephalopods in the Collection of the University of California; University of California Publications in Zoology, Vol. 8, pp. 301-310.

fine purple dots which are larger and more crowded on the back. This species seems to be sporadic in its occurrence, great numbers of the shells sometimes being washed on the beaches at Santa Cruz Island and Catalina. Living specimens may sometimes be seen in the aquarium at Avalon.

We quote a few paragraphs selected from Chas. F. Holder's description of a living argonaut.* "Glancing into the shell we may see a yellow bunch of miniature grapes hanging from the interior wall—the eggs—and perched in front of them is the argonaut, looking very much like an octopus or devilfish.

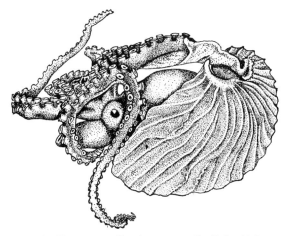

FIG. 675.—The paper nautilus, *Argonauta pacifica* Dall, with its egg case. From a preserved specimen; about natural size.

"In appearance it is one of the most beautiful of all animals as it rests in its shell, trembling with color, as waves of rose, yellow, green, violet, and all tints of brown are continually sweeping over it; now irised in the most delicate shade, of blue, now brown or green, changing to rose, vivid scarlet, or molten silver.

"The speed with which the argonaut could move backward propelled by its siphon, was remarkable when seen from the

*C. F. Holder—First Photographs ever made of a Paper Nautilus; Country Life in America, Vol. 15, p. 356.

PHYLUM—MOLLUSCA

side, but when it was observed from behind it was seen to be a perfect racing machine; the sharp keel of the shell covered by the extraordinary velamentous arms presenting a perfectly smooth surface, and the slightest current from the siphon was sufficient to send it along, the entire animal being concealed, the tentacles not trailing behind as often described. When the argonaut left the shell it crawled about in the position of an octopus, mouth down, but when in the shell its favorite position was with its mouth pointing directly upward."

Dr. Holder gives the length of living specimens that he has observed as 4–9 inches. They range from Monterey Bay southward to the Galapagos Islands.

Family—POLYPODIDAE

Polypus bimaculatus (Verrill), the two-spotted octopus (figs. 676 and 677), is the common shore devilfish of southern California. Ocular spots in front of the eyes are a dis-

Fig. 676.—*Polypus bimaculatus*, a small specimen drawn from life.

tinguishing feature of this octopus. The spots are large, round, and decidedly darker than the rest of the animal. In most animals each spot shows a dark center bounded by a bluish ring and an outer band of the center color. The surface of the body is ornamented with warty tubercles, the upper surface varying from a nearly smooth state to an extremely rugose condition where the tubercles are large and numerous. A large, conical, warted cirrus often with two smaller ones is present just over each eye. The arms are stout and 3–4 times as long as the body. The beak is strong and black. Large specimens measure 20 inches in

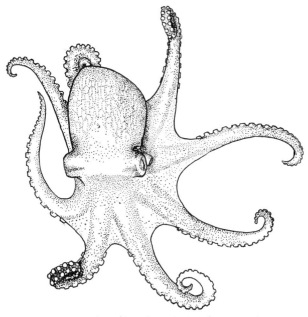

FIG. 677.—*Polypus bimaculatus*, the devil fish, dorsal view.

length (preserved). The species ranges from San Pedro south to Panama.

Polypus hongkongensis Hoyle, the Hongkong octopus, has a wide distribution, being found on the coasts of China and Japan as well as on our coast from Alaska south to San Diego. The surface is tuberculated much as in *P. bimaculatus* but the "eye spots" are absent. The hectocotylized portion of the third right arm is moderate in size, 1/9–1/20 of the total length, in contrast to *P. bimaculatus* in which the hectocotylus is minute. This species is an abundant

PHYLUM—MOLLUSCA

shore form and reaches a large size in Alaska. One was reported as 16 feet long with a radial spread of nearly 28 feet. The body of such an octopus would not be more than a foot long and six inches in diameter. Most specimens found along the coast are 1–2 feet long.

The *Decapoda* have ten arms with rows of stalked suckers each of which is usually provided with a horny ring. In the division *Myopsida*, the eyes are covered by a continuous membrane.

FAMILY—SEPIOLIDAE

Rossia pacifica Berry, the short squid (fig. 679), has a short, rounded body with ovate lateral fins; the dorsal margin of the mantle is free from the head; and both dorsal arms are hectocotylized. The arms are short, thick, and unequal, the third pair joined to the fourth by a well-developed web, functioning as a sheath for the tentacles; slight

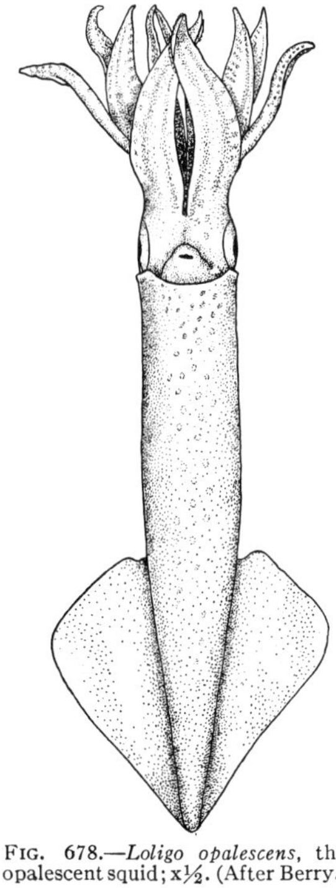

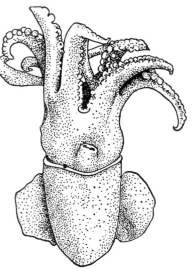

FIG. 678.—*Loligo opalescens*, the opalescent squid; x½. (After Berry.)

FIG. 679.—*Rossia pacifica*, the short squid. Ventral view; natural size. (After Berry.)

rudiments of web are also to be detected between all the other arms except the ventral pair. Width of a large specimen across the fins is 50 mm., length of mantle 42 mm., dorsal arm 29 mm., tentacle (excluding club) 45 mm.

Family—LOLIGINIDAE

Loligo opalescens Berry, the opalescent squid (fig. 678), has an elongated, pointed body, with triangular fins near the posterior end. The left ventral arm is hectocotylized. The suckers are in two rows, alternating, bowl shaped, and furnished with a toothed horny ring surrounded by a raised margin. The body is slightly swollen near the middle and tapers to a rather sharp point behind. The fins are large, about half as long as the mantle, very slightly lobed in front. The siphon is large, broad, with dorsal muscular bridles and a large terminal valve. The pen is thin, broadly lanceolate, with slender midrib. Along the sucker-bearing surface of all the arms, runs a delicate membranous swimming web. This is reduced on the ventral pair of arms and greatest on the third pair. This is the common squid of the Pacific coast and ranges from Puget Sound to San Diego. According to a report by W. L. Scofield[†] in *California Fish and Game*, the squid catch at Monterey averages from 100–150 tons per year but varies greatly from year to year. He says, "a few squid are caught throughout the year but the spring months of April, May and June are considered the regular squid season." Dr. Heath says,[‡] "The common California squid measures about ten inches in length, and there is reason to believe that it attains its full size in one year. The eggs are laid from early spring until midsummer, and looking like small shot, though whitish in color, are imbedded in translucent, milk-colored, jelly-like masses three or four inches in length and anchored to rocks, sticks and other solid objects. After a few weeks the young, a quarter of an inch

---

[†]W. L. Scofield—Squid at Monterey; California Fish and Game, Vol. 10, p. 179.
[‡]H. Heath—Devilfish and Squid; California Fish and Game, Vol. 3, p. 107.

in length, make their escape, become an inch long by the last of August, twice this length by the middle of October, and are nearly grown by May."

In the division *Oegopsida* the eye has a perforated lid.

FAMILY—OMMASTREPHIDAE

Dosidicus gigas (d'Orbigny), the large squid, is the largest of the squids to be found commonly on our coast, the length including the tentacles reaching about four feet. It has been washed ashore after storms but its large size makes preservation difficult so that specimens are not often seen.

None of the suckers are modified to form hooks; the siphon articulates with the mantle by a triangular cartilage having an ⊥-shaped groove. The suckers extend for less than half the length of the two long arms or tentacles. The arm tips are attenuate with minute and numerous suckers. On the basal half of the arms, the suckers are large, oblique, hood-shaped, having horny rings with about 19 sharp, conical teeth.

Measurements give length, 4 feet 2 inches, including the tentacles; width across fins, 1 foot 9 inches; length of dorsal arms, 17 inches; length of tentacle, 1 foot 8 inches; diameter of large sucker, ½ inch. The range extends from Monterey to San Diego.

CORRECTION FOR PAGE 428

Recent research shows that cases of "mussel poisoning" which have been reported on the Pacific Coast from Monterey to Alaska are not due to bacterial poisoning but to a preformed toxin in the mussel. This toxin is a powerful substance and the mussels (also certain clams) contain more of it during the summer months (May to October). Mussels scattered below low tide line may be even more toxic than those found higher up on shore. The poison is not destroyed in boiling nor counteracted by any known drug and in extreme cases is said to produce paralysis, coma and death. Between 1927 and 1934, 240 cases and 14 deaths were reported from this cause. In California, the sale of mussels gathered from Monterey northward between May 15th and September 30th is prohibited.

H. Sommer and K. F. Meyer: Mussel Poisoning, *California and Western Medicine*, Vol. 42, No. 6, Page 423.

CHAPTER XII.

Phulum—CHORDATA*

(Sea squirts, etc.)

Chordates include a great diversity of animals, beginning with the worm-like *Balanoglossus* and ending with man. The members of this large group differ from all the other groups, first, in the fact that they have a notochord (with the possible exception of *Balanoglossus*), an elastic rod, which lies above the digestive canal. This rod lasts through life in some of the low chordates but in the higher ones it becomes surrounded with a casing of cartilage or bone which almost obliterates it and makes it a much better support. This in the higher vertebrates becomes the bony spinal column.

A second feature of the chordates is the gill-slits which connect the pharynx, (the anterior part of the digestive canal) with the exterior. Most of the lower chordates keep their gill-slits throughout life but the higher vertebrates lose theirs after they have passed the early embryonic stages.

---

*Material used in this chapter has been taken largely from the papers by W. E. Ritter and W. E. Ritter and R. A. Forsyth listed in the bibliography, also from the following:

W. E. Ritter—The Simple Ascidians from the Northeastern Pacific in the Collection of the U. S. National Museum; Proceedings U. S. National Museum, Vol. 45, pp. 427–505.

—*Halocynthia johnsoni* n. sp. a comprehensive inquiry as to the extent of Law and Order that prevails in a single Animal Species; University of California Publications in Zoology, Vol. 6, pp. 65–114.

—The Ascidians collected by the U. S. Fisheries Bureau Steamer Albatross on the Coast of California During the Summer of 1904; University of California Publications in Zoology, Vol. 4, pp. 1–52.

W. E. Ritter and M. E. Johnson—The Growth and Differentiation of the Chain of *Cyclosalpa affinis* (Chamisso); Journal of Morphology, Vol. 22, pp. 395–454.

PHYLUM—CHORDATA

A third notable feature is the central nervous system which lies dorsal to the notochord. This nerve chord is first formed, as a longitudinal groove running along the back of the embryo. Here again *Balanoglossus* is an exception to these general statements. The groove closes and forms the nerve chord with its central canal. A relatively large brain develops from this nerve tissue in the head region, in the vertebrates.

The four subphyla of the Chordata are easily distinguished:

Enteropneusta....Body soft and worm-like.
Tunicata.........Body sac-shaped.
Leptocardia.......Body lance-shaped, firm and muscular.
Vertebrata........Backboned animals (not included in this book).

Subphylum—ENTEROPNEUSTA
Order—BALANOGLOSSIDA

A balanoglossid looks like a worm and lives in burrows on the mud flats. It has a proboscis which is separated by a narrow neck from a ring-shaped collar, back of which is the long trunk. Proboscis and collar pores lead from the outside into the body cavities, of which there are five. The mouth is in front of the collar and the anus is at the hinder end of the body. The anterior part of the digestive tract connects with the outside by rows of gill slits, and liver sacs back of them show through the outer surface of the body. The animal is unable to close the mouth, consequently, when it burrows, the sand passes through the digestive tract and leaves it in a continuous column through the terminal anus. These coiled castings left on the tide flats help in locating the creature. The proboscis and collar which can become swollen and turgid with water taken in through the pores, are the chief boring tools, being provided with strong muscles as well as a covering of cilia. Most *Enterop-*

neusta are said to have an offensive odor and one kind, at least, is reported as luminescent.

Dolichoglossus pusillus Ritter (fig. 680) has been found along the coast of southern California. The proboscis is long, with but one pore and there are no liver sacs. The length is 2–3 inches or more. It does not produce the coiled castings noted above as characteristic of most species, but at low tide the tip of the flaccid, orange-colored proboscis may often be seen lying on the sand at the mouth of the burrow. When the tide comes in, the proboscis is withdrawn into the hole. The collar and proboscis are bright orange color, the body much paler, shading into yellowish green toward the end. This species was taken in great numbers both at San Pedro and in San Diego Bay before extensive dredging operations were begun. They have also been taken in Mission Bay. Just where else they can now be found we do not know, but the mud flats of such a bay as Balboa would seem to offer a likely home for them.

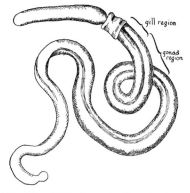

FIG. 680.—*Dolichoglossus pusillus:* about x3. (After Ritter.)

Subphylum—TUNICATA

Tunicates are so named because of their tunic or covering of cellulose tissue or cuticle. The notochord is present only during the free swimming larval stages in most cases. The nervous system, which starts out as a dorsal structure, much like that of higher vertebrates, often degenerates in the adult to a small ganglion. A special feature of tunicates is the large pharynx, also called the branchial sac, with its endostyle. The pharynx is the anterior part of the digestive tract into which the mouth opens. It is greatly enlarged,

PHYLUM—CHORDATA

perforated by slits, and has numerous blood vessels running through its walls. Hence it serves as a respiratory organ, since the blood is well oxygenated by the steady stream of water which, propelled by cilia, flows in and out of the pharynx. On leaving the branchial sac, the water goes into a large peribranchial sac and thence out through the atrial or excurrent siphon.

The water also brings in tiny plants and animals used as food. By the cilia on certain bands and grooves, the food is directed through the pharynx, going up its ventral side, around the mouth and down the dorsal side into the esophagus. The ventral and dorsal ciliated structures are termed endostyle and dorsal lamina respectively. The tubular heart has a peculiar habit of reversing the direction of the blood current periodically. This can be plainly seen in the transparent *Salpa* and nearly as well in clean specimens of *Ciona*.

Order—LARVACEA

The appendicularians are transparent, free swimming creatures which remind one of tiny tadpoles (fig. 681). They average about 10 mm. in length. They are taken in

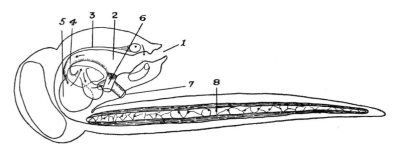

FIG. 681.—Semidiagrammatic drawing of an appendicularian; greatly enlarged. (After Ritter.)
1. Oral aperture. 3. Nerve. 5. Stomach. 7. Anus.
2. Pharynx. 4. Esophagus. 6. Atrial canal. 8. Notochord.

fine tow nets at the surface of the sea and their organization is much like that of the larval forms of some of the other tunicates, since the tail contains a notochord and dorsal nerve chord.

Order—THALIACEA

The animals of this group also swim freely in the ocean, often far out and usually near the surface. Like many others of the pelagic animals, they are transparent so that one can make out their organs without the bother of dissection. They are cylindrical or barrel-shaped with an opening at each end. They drive themselves along by taking the water in at the anterior end and forcing it out at the posterior opening. The muscles are in bands embedded in the mantle, which is just within the cellulose tunic. The body is partially divided by a partition, the respiratory organ, over which the water passes as the animal moves forward. The visceral organs usually form a small mass, often brightly colored, near the posterior end of the animal. The scheme of reproduction is peculiar in this group. An individual animal is like its brothers and sisters and grandparents, but not like its parents.

Doliolum tritonis Herdman is frequently taken with a net off southern California. It is barrel shaped and the respiratory partition is plate-like, rather than rod-like as in *Salpa*. The life history is complicated and may be graphically shown thus:

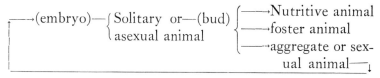

The asexual animal has a ventral stolon which produces buds, which migrate to the dorsal outgrowth of the body where they become arranged in five rows. The creature is now known as a "nurse" (fig. 682) and becomes greatly changed as the buds develop. Its muscle bands become wider and many of its internal organs disappear. It becomes hardly more than a living perambulator for its offspring.

The buds develop into three kinds of individuals. The nutritive animals (fig. 683) grow from the side rows of buds, remain attached, and furnish food and air for the colony but have very little muscle development. The foster animals

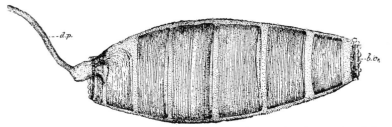

FIG. 682.—*Doliolum ehrenbergii*, the oozooid or nurse, after complete degeneration of the internal organs; approximately x35. (After Ritter.)

(fig. 684) grow from the middle row of buds on the chain. The sexual animals (fig. 685) become attached to the foster animals as very young buds. Later, when the foster animal is set free, the sexual animal remains attached to it until the latter becomes sexually mature, when the separation takes place. These two last forms resemble each other but the foster animal has no reproductive organs while the sexual animal has them and produces germ cells, which develop into tailed larvae. These grow into asexual or solitary animals and produce buds. The usual length is about 15 mm., but old "nurses," probably of this species, measure 125 mm.

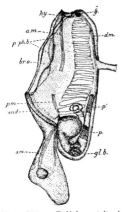

FIG. 683. — *Doliolum tritonis*, the trophozooid or nutritive animal; approximately x10. (After Ritter.)

Cyclosalpa affinis (Chamisso) has been reported from points in the Pacific, Atlantic, and Mediterranean. Although not an abundant species here it is of interest for the peculiar form of the chains (fig. 734). The animals in the chains are grouped in wheels. Each of these wheels finally separates off from the others and swims by itself. The solitary form reaches 100 mm. in length and the others are often 70 mm. long.

Salpa fusiformis-runcinata Cuvier-Chamisso (fig. 687) is the most abundant species of *Salpa* on the coast, having been

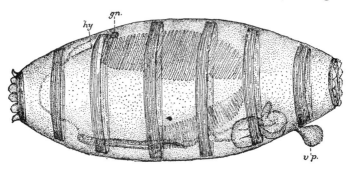

FIG. 684.—*Doliolum tritonis*, the phorozooid or foster animal; approximately x8. (After Ritter.)

reported throughout the whole region and often in large numbers. Near the base of the budding stolon, the solitary individual bears a mass of germ cells, one of which is received by each bud as it arises. The buds grow gradually. The distal ones are the largest, having been first formed and are gradually pushed outward until they extend beyond the body of the parent. Before they break off, they have begun the muscular contractions which send the water through

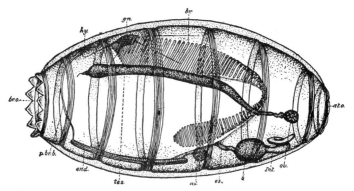

FIG. 685.—*Doliolum tritonis*, the gonozooid or sexual animal; approximately x6. (After Ritter.)

their bodies. The germ cell in each develops into a solitary animal which in turn produces a chain of aggregate individuals each supplied with a germ cell. The chain may

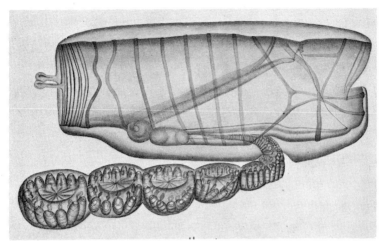

FIG. 686.—*Cyclosalpa affinis*, a solitary individual with a chain of "wheels." Each wheel contains about 13 individuals of the aggregate or sexual generation. (After Ritter and Johnson.)

drag behind the adult for some time before portions break away and swim off. A large specimen of the solitary generation is 80 mm. long; a large individual of the aggregate generation is 25 mm. long.

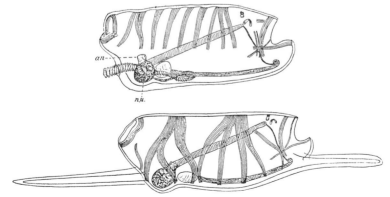

FIG. 687.—*Salpa fusiformis-runcinata.* Solitary generation (above). Aggregate generation (below). (After Ritter.)

Salpa tilesii-costata (Cuvier-Quoy et Gaim) is the largest and most conspicuous species of *Salpa* on the coast (fig. 688). The individuals of the solitary generation reach 19 cm. in length and have 18–20 muscle bands uniformly spaced and nearly parallel. Two prominent tail-like append-

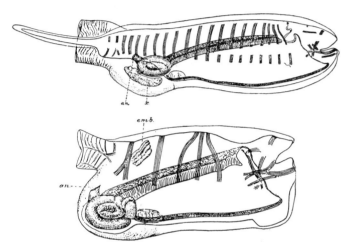

FIG. 688.—*Salpa tilesii-costata*. Solitary generation (above). Aggregate generation (below) an. anus. emb.—embryos. (After Ritter.)

ages may reach nearly 50 mm. in length. The aggregate generation may be 14 cm. long and has five muscle bands, all of which are limited to the dorsal half of the body. The first three muscles draw close together on the dorsal side and the fifth is forked at each side. This species may show a dark green coloration particularly in the mid-body region.

Order—ASCIDIACEA

Ascidians are popularly known as sea squirts because of their habit of ejecting water from their siphons. When the animal is undisturbed, a steady stream of water is flowing in at the branchial siphon and out of the atrial siphon so that the branchial sac in the body is always well filled. A disturbance causes the animal to quickly contract the muscles of the body wall, hence the jets of water leaving forcibly by the two siphon openings.

Ascidians are either fixed or free-swimming, and simple or compound. The simple ones are always fixed in the adult stage and the few that are free-swimming in the adult stage are always colonial. The sexually produced embryos are usually tailed larvae, and many of the species reproduce by budding as well as sexually.

These descriptions are necessarily brief and do not include the inner structure which, in the case of many ascidians, must be examined in order to make indentification certain. For such detailed diagnoses of Pacific coast ascidians, the reader is referred to the papers of Ritter, and Ritter and Forsyth listed in the bibliography.

Suborder—ASCIDIAE SIMPLICES

(Simple ascidians)

The simple ascidians are fixed, solitary, sac-like animals which reproduce sexually rather than by budding. Great numbers of these somewhat shapeless and unattractive creatures may be found on our shore but they are so often mistaken for seaweeds or sponges or moss-covered stones that they are practically unknown to most people, often even to those who have spent much time collecting.

A simple ascidian shows an interesting degeneracy. The larva with its notochord, nerve cord, and free-swimming habit is readily given a place among chordates. After a few hours of swimming about, the larva attaches itself to some support and degeneration begins.

When the change is complete, we have an adult animal without a notochord or means of locomotion and with only the remnants of a nerve cord. In fact, the adult characters alone would place the ascidians among the lower invertebrates along with mollusks or worms.

Halocynthia johnsoni Ritter (fig. 689) has prominent red-tipped siphons. The test is generally dark, brownish yellow, more or less covered with mud and other foreign material. The tentacles number 17–62. It is found from Point

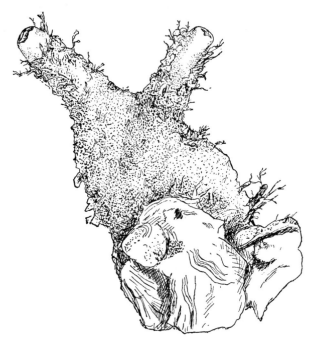

FIG. 689.—*Halocynthia johnsoni*; x1. (After Ritter.)

Conception southward being abundant at San Pedro and San Diego Bay. Large specimens reach 6–8 cm. in length.

Halocynthia haustor (Stimpson) has the characteristics mentioned above for *H. johnsoni* except that the tentacles number 12–29 and the test has many prominent tubercles. It is found on the Washington coast and is abundant at Puget Sound.

Styela montereyensis (Dall) is club-shaped and gradually merges into the peduncle which is often twice as long as the body (fig. 690). The color is a dark red which blends into lighter or yellowish on the lower part of the body. The orifices are both 4-lobed and the siphons are distinct. The branchial siphon has a downward curve as shown in the figure. The tentacles around the branchial opening number 40–130. It is found from British Columbia to San Diego. We have seen especially fine specimens attached to the wharf piles at Santa Barbara. A large specimen measures 29 cm. including the peduncle.

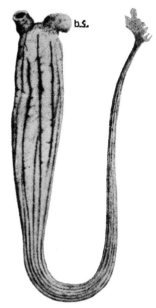

FIG. 690.—*Styela montereyensis*; x1; b.s.—branchial siphon. (After Ritter and Forsyth.)

Styela gibbsii (Stimpson) is much like the last named species but has a short peduncle. It is found from British Columbia to San Diego.

Styela stimpsoni Ritter (fig. 691) is bright orange-red with a thin, smooth test. The body is oval, the openings 4-lobed. The length is 3–4 cm. The species is found at Puget Sound, being abundant under rocks in the Friday Harbor region.

Ascidia californica Ritter and Forsyth (fig. 692), has a thick, gelatinous, translucent test, is elliptical in outline and somewhat depressed. The mantle is thin and transparent on the under side where it is attached to the rock. The

FIG. 691.—*Styella stimpsoni*, a bright orange-red species of ascidian; x1.

FIG. 692.—*Ascidia californica*; x1. b.s.—branchial siphon. (After Ritter and Forsyth.)

lobes of the incurrent opening number 8, and those of the excurrent opening number 6. The species has been found throughout the California coast. Specimens reach 3-4 cm. in length.

Chelyosoma productum (Stimpson) is flattened above (fig. 693) and the oval, upper surface is covered with plates which remind one of tortoise-shell. Both openings are 6-lobed. The outer covering is translucent and cartilage-like. It is found along the Washington and California coasts and reaches a length of 25 mm. or more.

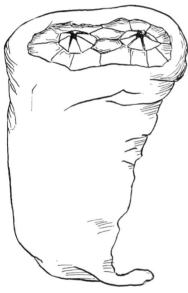

FIG. 693.—*Chelyosoma productum*, a simple ascidian; x2.3.

Ciona intestinalis (Linnaeus) is long and cylindrical (frontispiece) with a thin, soft, translucent, outer coat which is often covered with much adhering mud and debris but sometimes is so clean that the organs within are apparent. The color is usually yellowish green. The incurrent opening is 8-lobed and the excurrent opening 6-lobed. Average specimens measure 8-10 cm. long. This is a widely distributed form, being identical with the Atlantic coast and Mediterranean species. It clings to the under sides of floats and buoys and to wharf piles.

SUBORDER—ASCIDIAE COMPOSITAE

(Compound ascidians)

In compound ascidians, many individuals, called zooids, are grouped together and more or less embedded in a common mass of cellulose mantle. The whole colony or mass has been produced by budding, from a single parent. The zooids are often less than a millimeter in length or may be as much as 10 mm. long. They are frequently arranged in small groups, called systems. Some species form thin sheets over the surface of the rocks while others form thick masses or pedunculated lumps. Many times it is hard to distinguish them from sponges since both often show the same colors, red, yellow, white, or brown. Where they form large patches, they are decidedly conspicuous. Unlike the sponges, the compound ascidians contract noticeably when disturbed and force water from their openings. Moreover, the openings of the canals of the sponges are usually large and easy to see and the whole mass is stiffened with limy or glassy spicules while the ascidians are gelatinous.

Metandrocarpa dura (Ritter) forms a colony of brick-red, globular individuals, not regularly arranged in systems. They make firm encrusting masses on seaweeds and though they usually grow at considerable depths, they are often washed on shore where their bright red color makes them conspicuous. The individuals are about 2–3 mm. in diameter.

Botrylloides diegensis Ritter and Forsyth is a flat, irregular, encrusting species a few inches in extent and about 5 mm thick. The color may be pinkish yellow to purple. The zooids are arranged in somewhat elongated groups within the mass (fig. 694). The species is abundant in San Diego Bay.

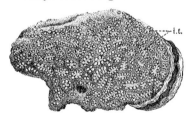

FIG. 694.—*Botrylloides diegensis*: x⅔; i.t.—island of test. (After Ritter.)

Perophora annectens Ritter (fig. 695) is an admirable species in which to study the organization of ascidians if the living material can be studied under a binocular. Dr. Ritter says that through the transparent test, one can see the pulsating heart and the blood flowing through the vessels, especially in the branchial sac, and the beating cilia on the stigmata. The course of the movement of the water, in at the branchial opening and out of the atrial siphon, can also be plainly seen. The relations of the organs can often be made out much better this way than by dissecting larger preserved specimens. The species forms irregular masses sometimes 2 inches or more long, on sticks, stones, or seaweeds. The color is pale greenish yellow. The individual zooids are about 1 mm. wide and 1¼ mm. long, short-oblong and laterally compressed, usually crowded together and wholly embedded in the test, but sometimes farther apart with only the basal part embedded. It is found from British Columbia south to San Diego.

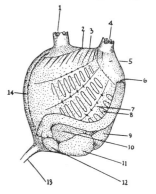

FIG. 695.—*Perophora annectens.* Individual zooid as seen through the transparent test, enlarged. (After Ritter.)
1. Branchial orifice.
2. Ganglion.
3. Radial muscle bands.
4. Atrial opening.
5. Atrial chamber.
6. Anus.
7. Transverse vessel.
8. Stigmata.
9. Esophagus.
10. Gonads.
11. Stomach.
12. Intestine.
13. Stolonic vessel.
14. Endostyle.

Distaplia occidentalis Ritter and Forsyth is variable in form, all the way from flat and encrusting to pedunculated and mushroom shaped. The groups of individuals are distinct, several being arranged around each common excurrent opening which extends above the surface like a short pipe. The color may be light green, dark brown, yellow, brick red, or dirty white. It is found from Puget Sound to San Diego and is often abundant.

Didemnum carnulentum Ritter and Forsyth (fig. 696) is conspicuous in the San Diego region and perhaps farther northward. The colonies are thin, 4 mm. or less, the color pink to white.

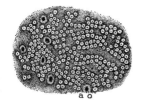

FIG. 696.—*Didemnum carnulentum*; x4/3. a.o.-atrial orifice. (After Ritter and Forsyth.)

A variety which is also abundant is always pure white. The location of the zooids in the test mass is indicated by spots caused by the accumulation of spicules. The incurrent openings of the zooids are in the center of these spicule masses.

FIG. 697.—*Glossophorum planum*; x⅔. (After Ritter and Forsyth.)

Glossophorum planum Ritter and Forsyth (fig. 697) forms large regular colonies, spherical or flattened, with thick cylindrical peduncles. The surface is smooth and free from sand or foreign substances. The systems of zooids can be plainly seen in distinct and regular groups. A large colony measured 10 cm. long, 5 cm. wide, and 1.5 cm. thick.

Amaroucium californicum Ritter and Forsyth is irregular and variable in thickness, often lobed and sometimes pedunculated. It often reaches 20 cm. in extent and is never encrusted with sand, though sand is often found in its deeper parts. The color is opalescent white to reddish brown. The zooids are often clearly visible and their grouping is more or less distinct. It is found from Alaska to San Diego and is usually abundant.

Euherdmania claviformis (Ritter) is abundant along the whole California coast. The colonies are made up of club-shaped, separate, dull green individuals (fig. 698) which arise by budding from

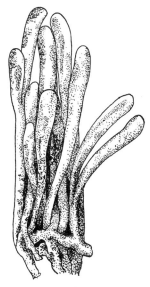

FIG. 698.—*Euherdmania claviformis*, a compound ascidian; x2.

short, branched, and interwoven stems. The colonies often cover several square inches on the surface of the rocks and individual zooids may be 3-4 cm. long. The zooids may be buried for half their length in sand and debris so that their translucent character is not apparent. The oviduct extends nearly the full length of the zooid and, during the summer months, the anterior part of the duct becomes greatly enlarged, serving as an uterus in which the embryos develop. Eighteen or twenty embryos may be present at one time. They are arranged one behind the other, the oldest nearest the siphons and the youngest near the attached end of the parent. Thus they furnish an excellent series for embryological study.

Suborder—ASCIDIAE LUCIAE

(*Luminescent ascidians*)

Free-swimming colonial tunicates are relatively rare but *Pyrosoma giganteum* Lesueur, a beautiful transparent form, is often taken in nets and is sometimes washed on shore. The colony is shaped like an elongated thimble with thickened walls (fig. 699). The individual animals are arranged at right angles to the surface with their incurrent openings

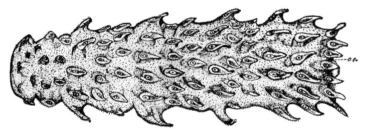

FIG. 699.—*Pyrosoma giganteum*; x2. (After Ritter.)

on the outer surface and their excurrent openings directed into the cavity of the thimble. By suddenly contracting and forcing the water out of the open end of the thimble, the colony is forced along with the closed end foremost. The largest specimens recorded were 2 feet long. The color is often pink but many specimens are colorless. Living *Pyrosoma* are beautifully luminescent. We have not had the opportunity of seeing this species alive but suppose it is like its relatives in this respect.

Subphylum—LEPTOCARDIA

Branchiostoma, commonly known as *Amphioxus* or the lancelet, is fish-shaped (fig. 700) and has a notochord or cylindrical rod extending the full length of its body. The body is narrow and pointed at both ends. There is no head, but the mouth, surrounded by long cirri, leads into a large pharynx which is perforated by gill slits. The water passes through the gill slits into a peribranchial chamber and thence out through the atriopore, somewhat anterior to the anus. The pharynx acts as a respiratory organ and also collects the food which passes on into the straight digestive tract. The nerve chord is dorsal to the notochord which in turn lies just above the digestive tube. This arrangement of organs suggests that found in the vertebrates. Although Amphioxus may be degenerate in some ways, it is probably not greatly different from the ancestral form from which developed the early chordates.

Jordan[*] says that some lancelets are found on almost every coast in semi-tropical and tropical regions. On our coast they are found as far north as San Diego Bay. The creature usually lives buried upright in the sand, in shallow waters, with the mouth projecting somewhat from the sand. The various species are distinguished by the number of muscle bands in the anterior (to the atriopore), middle (to the vent), and posterior regions of the body.

Amphioxus has always been regarded as a rarity in this country but near the island of Amoy, South China, they are said to be caught in large numbers. Four hundred

[*]David Starr Jordan—Fishes (1925); p. 164, D. Appleton and Company.

fishermen dredge for them from August to April bringing in an average daily catch of one ton.* Most of the catch is consumed locally, being eaten either fresh or dried and are esteemed a great delicacy.

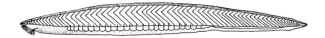

FIG. 700.—The California *Amphioxus, Branchiostoma californiense*, which lives partly buried in the mud; x1.

Branchiostoma californiense Cooper (fig. 700), reaches a length of four inches, has 31 pairs of gonads and a small head region. The number of muscle bands is 44+16+9=69. It has been dredged from the sands at the mouth of San Diego Bay and from the Gulf of California.

---

*S. F. Light—*Amphioxus* fisheries near the University of Amoy, China; Science, Vol. 58, p. 57.

APPENDIX

METHODS OF PRESERVING ANIMALS

As a usual thing, we recommend enjoying animals at the beach and leaving them there for others to enjoy. Most of the seashore animals are decidedly unattractive after they are dead and to keep them alive for a few hours after they are out of their briny home usually requires especial care.

If you must take some of your catch home alive, put back into the ocean (we are strong for conservation) everything you have collected except one or two small specimens and give these plenty of clean sea water. If, on your return from a wonderful trip to the beach, you can show some active, living specimens, you will probably find that the people who see your treasures are interested and enthusiastic too, but if you have a dish full of dead and dying animals, your most vivid descriptions of the beautiful things you have seen will usually fail to awaken an enthusiastic reponse.

For biology classes, we keep sea pansies or *Corymorpha* alive for more than a week if they are in a moderately cool room and each sea pansy by itself in a finger bowl of clean sea water. One or two *Corymorpha* will thrive in a beaker of water. We have had *Corymorpha* produce a whole forest of tiny offspring within a week after they had come into the laboratory. In such a case, we remove the adult *Corymorpha* and leave only the young ones so that there will be plenty of room for them all.

The secret of keeping the living specimens is cleanliness and ample space allowance for each individual. If much dust is likely to fall into the water, it is well to have the beakers covered, at least while sweeping is going on. If students are requested to use only clean glass rods when they feel a desire to prod the animals, the patient creatures will remain in good shape much longer than if they are poked with fingers, pencils, fountain pens, and especially dissecting instruments that have been used on formalin soaked specimens.

We have had successful exhibits of seashore animals that have been carried many miles by automobile or train, but the automobile is almost a necessity if many animals are to be carried, since the amount of clean sea water required is considerable. We always bring home a ten gallon demijohn of water to use when we transfer the animals to finger bowls and beakers in the class room. Some animals live better if they are carried home in moist seaweed which is wrapped in newspapers. Sea urchins, starfish, sand dollars, mollusks with coiled shells, and most crabs will live a number of hours in this way and can be transferred to the aquaria later. An important fact to consider is that small specimens will keep much better in the laboratory aquarium than large ones. For example, we have found it almost impossible to keep large sea urchins alive over night but often keep small ones alive for a week or two and a large sea hare will scarcely survive the trip home, while a small one, two or three inches long, will live a week or more in the laboratory.

A convenient specimen carrier is a wooden box or tray, built to fit the running board of an automobile and divided into compartments for carrying a dozen or more glass fruit jars. The animals can be transferred to these jars from the collecting pail and choice specimens can be kept by themselves so that they are not in danger of being devoured by their hungry neighbors. If it is necessary to put the covers on the jars during the ride home, they should be removed at the end of the journey. The shore forms are used to an abundant supply of air in the water and do not thrive without it. To aerate the water, shake the bottle of sea water well before pouring it into the aquarium. Other methods of aerating the water are to allow a jet of sea water to fall into the dish from a higher level, or to force air in by means of a pipette, or better yet, to put a bit of green seaweed into the aquarium.

For permanent museum exhibits, specimens are either dried or preserved in alcohol or in formalin. In drying such a specimen as the soft twenty-rayed starfish, we soak the star in formalin for a day or two and then dry it slowly. Crabs, lobsters, and their relatives can be treated this way, or better, soaked in a mixture of alcohol, glycerine, formalin, and water and then allowed to dry. By this last method the specimens do not become so brittle and the color is kept better. The quickest way to preserve specimens is to put them into alcohol (70%) or formalin (4%). The shells of mollusks and the bristles of worms are somewhat corroded by formalin, so alcohol is more commonly used. Formalin often preserves the color of a specimen better than alcohol does, but colors do not keep well at best. Bulky specimens may require a second supply of the preservative and are usually kept in good shape if the first is poured off, the specimen rinsed, and the container filled with fresh preservative.

Mollusk shells that still contain the soft body of the mollusk require special attention. Drop the mollusk into water and heat to the boiling point. The body of the mollusk can then be removed from the shell with the help of tweezers, a hook, or a bent pin. If the mollusk closes its doorway with a horny door or operculum, save this door, stuff the cleaned shell with cotton, and glue the operculum in place in the aperture of the shell.

Chitons may be kept from curling by tying them to a stick and allowing them to dry there. Some collectors prefer to let the chitons attach themselves to the container and then pour off the water. The chitons cling to the jar until they are dead and dry.

Dried specimens of any kind may be kept in the specimen

mounts so generally used for mounting insects. They are subject to attack by museum pests as are insects, so they should be treated occasionally with carbon bisulphide or some other insecticide.

BIBLIOGRAPHY

A few of the enormous number of publications dealing with shore-living invertebrates, particularly those of the Pacific coast of America north of Mexico, are included in the list below. It is divided into three parts: first, the popular or non-technical accounts; second, general textbooks; third, original papers. The literature dealing with animal life along our coast is widely scattered and our list is by no means complete. With a few exceptions, articles written in foreign languages or appearing in journals published on the continent of Europe are omitted here, as are such generally inaccessible volumes as the *Challenger* Reports. On the other hand, many of the papers cited may easily be obtained from public libraries, especially the publications of government bureaus and universities.

POPULAR WORKS

ARNOLD, AUGUSTA FOOTE
    1901. The Sea-Beach at Ebb-Tide. Century Co., New York.
CROWDER, WILLIAM
    1923. Dwellers of the Sea and Shore. Macmillan.
FLATTELY, F. W. and WALTON, C. L.
    1922. The Biology of the Sea-Shore. Macmillan.
KEEP, JOSIAH
    1911. West Coast Shells. Whitaker and Ray-Wiggin Co.
KELLOGG, JAMES L.
    1910. The Shellfish Industries. Henry Holt and Co.
MAYOR, A. G.
    1911. Sea-Shore Life. New York Zoological Society.
ROGERS, JULIA ELLEN
    1908. The Shell Book. Doubleday, Page and Co.
TEGGART, F. J. and others
    1915. Nature and Science on the Pacific Coast. A guide book for scientific travelers in the West. Paul Elder.
VERRILL, A. H.
    1917. The Ocean and Its Mysteries. Duffield and Co.

TEXT BOOKS

HARMER, S. F. and SHIPLEY, A. E., Editors
    Cambridge Natural History. Macmillan.
LANKESTER, E. RAY, Editor
    A Treatise on Zoology. A. and C. Black, London.
PARKER, T. J. and HASWELL, W. A.
    1897. A Text-book of Zoology, Vol. 1. Macmillan.
PRATT, H. S.
    1916. A Manual of the Common Invertebrate Animals. A. C. McClurg, Chicago.
SHIPLEY, A. E. and MACBRIDE, E. W.
    1901. Zoology, an Elementary Text-book. Macmillan.

ORIGINAL PUBLICATIONS

Most of the references included here occur in technical journals such as the publications of universities and research institutions or the reports of government bureaus. To secure brevity it is customary to use abbreviations for the names of the publications. The following explanatory list is included for the assistance of readers who are unfamiliar with the technical literature of the biological sciences.

Amer. Nat.—American Naturalist.
Ann. N. Y. Acad. Sci.—Annals of the New York Academy of Sciences.
Arch. f. Entwick. der Org.—Archiv für Entwicklungsmechanik der Organismen.
Biol. Bull.—Biological Bulletin.
Bull. Bur. Fish.—Bulletin of the Bureau of Fisheries.
Bull. So. Calif. Acad. Sci.—Bulletin of the Southern California Academy of Sciences, Los Angeles.
Bull. Mus. Comp. Zool. Harvard.—Bulletin of the Museum of Comparative Zoology of Harvard College.
Bull. U. S. Nat. Mus.—Bulletin of the United States National Museum.
Ill. Cat. Mus. Comp. Zool. Harvard.—Illustrated Catalogue of the Museum of Comparative Zoology of Harvard College.
Jour. Exp. Zool. Journal of Experimental Zoology.
Mem. Mus. Comp. Zool. Harvard.—Memoirs of the Museum of Comparative Zoology of Harvard College.
Mem. Nat. Acad. Sci.—Memoirs of the National Academy of Sciences.
Mem. Univ. Calif.—Memoirs of the University of California.
Pomona College Jour. Ento. and Zool.—Pomona College Journal of Entomology and Zoology.
Proc. Acad. Sci. Phil.—Proceedings of the Academy of Natural Sciences of Philadelphia.
Proc. Biol. Soc. Wash.—Proceedings of the Biological Society of Washington.
Proc. Boston Soc. Nat. Hist.—Proceedings of the Boston Society of Natural History.
Proc. Calif. Acad. Sci.—Proceedings of the California Academy of Sciences, San Francisco.

Proc. U. S. Nat. Mus.—Proceedings of the United States National Museum.
Proc. Wash. Acad. Sci.—Proceedings of the Washington Academy of Sciences.
Publ. Puget Sd. Biol. Sta.—Publications of the Puget Sound Biological Station.
Rep. U. S. Fish. Com.—Report of the United States Commissioner of Fisheries.
Smithsonian Misc. Col.—Smithsonian Miscellaneous Collection.
Trans. Amer. Mic. Soc.—Transactions of the American Microscopical Society.
Trans. Linn. Soc. London.—Transactions of the Linnean Society of London.
Trans. N. Y. Acad. Sci.—Transactions of the New York Academy of Sciences. (Since 1897 merged with the Annals).
Trans. Roy. Soc. Canada.—Transactions of the Royal Society of Canada.
Univ. Calif. Publ. Zool.—University of California Publications in Zoology.
Zool. Anz.—Zoologischer Anzeiger.
Zool. Jahr., Abt. f. Syst.—Zoologische Jahrbücher, Abteilung für Systematik.

LIFE IN THE OCEAN

JOHNSTONE, JAMES
    1908. Conditions of Life in the Sea. Cambridge University Press.
KOFOID, C. A.
    1906–1911. Publications on Dinoflagellates in Vols. 2, 3, and 8 of Univ. Calif. Publ. Zool.
KOFOID, C. A. and SWEZY, OLIVE
    1921. The Free-living Unarmored Dinoflagellata. Mem. Univ. Calif., Vol. 5, pp. 1–562.
MICHAEL, E. L.
    1919. The Problem of the Organic Fertility of the North Pacific Ocean. Bull. No. 9 of the Scripps Institution for Biological Research, pp. 51–57.
MURRAY, SIR JOHN and HJORT, J.
    1912. The Depths of the Ocean. Macmillan.
MURRAY, J., and LEE, G. V.
    1909. The Depth and Marine Deposits of the Pacific. Mem. Mus. Comp. Zool. Harvard, Vol. 38, No. 1.
PERRY, EDNA M.
    1916. Distribution of Certain Invertebrates on a Restricted Area of Sea Bottom. Publ. Puget Sd. Biol. Sta., Vol. 1, pp. 175–176.

TORREY, H. B.
    1902. An Unusual Occurrence of Dinoflagellata on the California Coast. Amer. Nat., Vol. 36, pp. 187–192.

SPONGES

LAMBE, L. M.
    1892. Sponges from the Pacific Coast of Canada and Behring Sea. Trans. Roy. Soc. Canada, Vol. 10, pp. 67–78.
    1893. Sponges from the Pacific Coast of Canada. Trans. Roy. Soc. Canada, Vol. 11, pp. 25–43.
    1894. Sponges from the Western Coast of North America. Trans. Roy. Soc. Canada, Vol. 12, pp. 113–138.
    1900. Catalogue of the Recent Marine Sponges of Canada and Alaska. Ottawa Naturalist, Vol. 14, No. 9, pp. 153–172.

MOORE, H. F.
    1910. Commercial Sponges and the Sponge Fisheries. Bull. Bur. Fish., Vol. 28, pp. 399–512.

SMITH, H. M.
    1897. The Florida Commerical Sponges. Bull. U. S. Fish. Com., Vol. 17, pp. 225–240.

WILSON, H. V.
    1904. The Sponges. Mem. Mus. Comp. Zool. Harvard, Vol. 30, No. 1. (Report of an exploration off the west coasts of Mexico and South America).

COELENTERATES

AGASSIZ, A.
    1865. North American Acalephae. Ill. Cat. Mus. Comp. Zool. Harvard, No. 2.

BIGELOW, HENRY B.
    1913. Medusae and Siphonophorae Collected by the United States Fisheries Steamer "Albatross" in the North Western Pacific, 1906. Proc. U. S. Nat. Mus., Vol. 44, pp. 1–119.
    1914. Note on the Medusan Genus *Stomolophus*, from San Diego. Univ. Calif. Publ. Zool., Vol. 13, pp. 239–241.

DAHLGREN, ULRIC
    1916–17. The Production of Light by Animals. Jour. Franklin Institute.

FRASER, C. McLEAN
    1911. The Hydroids of the West Coast of North America. Bull. from Laboratories of Natural History of State University of Iowa. Vol. 6, No. 1, pp. 3–91.
    1914. Some Hydroids of the Vancouver Island Region. Trans. Roy. Soc. Canada, 3rd Series, Vol. 8, pp. 99–216.
    1916. On the Development of *Aequorea forskalea*. Trans. Roy. Soc. Canada, Series 3, Vol. 10, pp. 97–104.

GEE, WILSON
    1913. Modifiability in the Behavior of the California Shore Anemone, *Cribrina xanthogrammica* Brandt. Jour. Animal Behavior, Vol. 3, pp. 305–328.

HAZEN, A. P.
    1903. Regeneration in the Anemone, *Sagartia luciae*. Arch. f. Entwick. der Org. 16 Band, 3 Heft., pp. 365–376.

HOWE, M. A.
    1912. The Building of "Coral" Reefs. Science, N. S., Vol. 35, pp. 837–842.

KÜKENTHAL, W.
    1913. Über die Alcyonarienfauna Californiens etc., Zool. Jahr., Abt. f. Syst., 35 Band, 2 Heft., pp. 219–270.

MAYOR, A. G.
    1910. Medusae of the World. Vol. I—The Hydromedusae; Vol. II—The Hydromedusae; Vol. III—The Scyphomedusae. Carnegie Institution Bul. 109.

MCMURRICH, J. P.
    1913. Description of a new Species of Actinian of the Genus Edwardsiella from Southern California. Proc. U. S. Nat. Mus., Vol. 44, pp. 551–553.

NUTTING, C. C.
    1900. American Hydroids. I. Plumularidae. U. S. Nat. Mus. Bul.
    1901. The Hydroids. Papers from Harriman Alaska Expedition. Proc. Wash. Acad. Sci., Vol. 3, pp. 157–216.
    1904. American Hydroids II. Sertularidae. U. S. Nat. Mus. Bul.
    1915. American Hydroids III. Campanularidae and Bonneviellidae. U. S. Nat. Mus. Bul.

PARKER, G. H.
    1900. Synopses of North American Invertebrates. XIII. The Actinaria. Amer. Nat., Vol. 34, pp. 747–758.
    1917. Actinian Behavior. Jour. Exp. Zool., Vol. 22, pp. 193–230.
    1917. The Activities of Corymorpha. Jour. Exp. Zool., Vol. 24, pp. 303–331.
    1919. The Organization of Renilla. Jour. Exp. Zool., Vol. 27, pp. 499–507.
    1920. Activities of Colonial Animals, I. Circulation of Water in Renilla. Jour. Exp. Zool., Vol. 31, pp. 343–365.
    1920. II. Neuromuscular Movements and Phosphorescence in Renilla. Jour. Exp. Zool., Vol. 31, pp. 475–515.

PETERS, A. W.
    1905. Phosphorescence in Ctenophores. Jour. Exp. Zool., Vol. 2, pp. 103–116.

TORREY, H. B.
    1902. The Hydroida of the Pacific Coast of North America. Univ. Calif. Publ. Zool., Vol. 1, pp. 1–104.

1902. Anemones, with Discussion of Variation in Metridium. Proc. Wash. Acad. Sci., Vol. 4, pp. 373–410.
1904. Biological Studies on *Corymorpha*. I. *C. palma* and Environment. Jour. Exp. Zool., Vol. 1, pp. 395–422.
1904. On the Habits and Reactions of *Sagartia davisi*. Biol. Bull., Vol. 6, pp. 203–216.
1904. The Hydroids of the San Diego Region. Univ. Calif. Publ. Zool., Vol. 2, pp. 1–43.
1904. Ctenophores of the San Diego Region. Univ. Calif. Publ. Zool., Vol. 2, pp. 45–50.
1905. The Behavior of *Corymorpha*. Univ. Calif. Publ. Zool., Vol. 2, pp. 333–340.
1906. The California Shore Anemone *Bunodactis xanthogrammica*. Univ. Calif. Publ. Zool., Vol. 3, pp. 41–45.
1907. Biological Studies on *Corymorpha*. II. The Development of *C. palma* from the egg. Univ. of Calif. Publ. Zool., Vol. 3, pp. 253–298.
1909. The Leptomedusae of the San Diego Region. Univ. Calif. Publ. Zool., Vol. 6, pp. 11–31.
1910. Biological studies on *Corymorpha*. III. Regeneration of hydranth and holdfast. Univ. Calif. Publ. Zool., Vol. 6, pp. 205–221
1910. Biological Studies on *Corymorpha*. IV. Biol. Bul., Vol. 19, pp. 280–301.

TORREY, H. B. and KLEEBERGER, F. L.
1909. Three Species of *Cerianthus* from Southern California. Univ. Calif. Publ. Zool., Vol. 6, pp. 115–125.

TORREY, H. B. and MERY, JANET RUTH
1904. Regeneration and Non-Sexual Reproduction in *Sagartia davisi*. Univ. Calif. Publ. Zool., Vol. 1, pp. 211–226.

VAUGHAN, T. WAYLAND
1919. Corals and the Formation of Coral Reefs. Smithsonian Report, 1917, pp. 189–276.

WEESE, A. O. and TOWNSEND, M. T.
1921. Some Reactions of the Jellyfish, *Aequorea*. Publ. Puget Sd. Biol. Sta., Vol. 3, pp. 117–128.

LOWER WORMS

BÜRGER, OTTO
1895. Die Nemertinen des Golfes von Neapel. Fauna und Flora des Golfes von Neapel, Vol. 22.

COE, WESLEY R.
1904. Nemerteans. Harriman Alaska Expedition, Vol. 11, pp. 1–220.
1905. Nemerteans of the West and Northwest Coasts of America. Bull. Mus. Comp. Zool. Harvard, Vol. 47.

HEATH, HAROLD and MCGREGOR, ERNEST A.
  1912. New Polyclads from Monterey Bay, California. Proc. Acad. Nat. Sci. Phil., Vol. 64, pp. 455-488.
LANG, ARNOLD
  1884. Die Polycladen des Golfes von Neapel. Fauna und Flora des Golfes von Neapel, Vol. 11.
MICHAEL, ELLIS L.
  1911. Classification and Vertical Distribution of the Chaetognatha of the San Diego Region. Univ. Calif. Publ. Zool., Vol. 8, pp. 21-186.

MOLLUSCOIDEA

BASSLER, R. S.
  1922. The Bryozoa, or Moss Animals. Smithsonian Rept. for 1920, pp. 339-380. (Publication 2633).
BROOKS, W. K. and COWLES, R. P.
  1905. *Phoronis architecta;* its Life History, Anatomy and Breeding Habits. Mem. Nat. Acad. Sci., Vol. 10, pp. 71-148.
DALL, W. H.
  1891. Notes on Some Recent Brachiopods. Proc. Acad. Nat. Sci. Phil., Vol. 43, pp. 172-175.
  1920. Annotated List of the Recent Brachiopoda in the Collection of the United States National Museum. Proc. U. S. Nat. Mus., Vol. 57, pp. 261-377.
DAVIDSON, T.
  1886-8. A Monograph of Recent Brachiopoda. Trans. Linn. Soc. London, 2nd. Ser., Vol. 4, Zoology, pp. 1-248.
MORSE, EDW. S.
  1901. Observations on Living Brachiopoda. Mem. Boston Soc. Nat. Hist., Vol. 5, pp. 313-386.
ROBERTSON, ALICE
  1900. Studies in Pacific Coast Entoprocta. Proc. Cal. Acad. Sci., Zoology, Vol. 2, pp. 323-348.
  1905. Non-Incrusting Chilostomatous Bryozoa of the West Coast of North America. Univ. Calif. Publ. Zool., Vol. 2, pp. 235-322.
  1908. The Incrusting Chilostomatous Bryozoa of the West Coast of North America. Univ. Calif. Publ. Zool., Vol. 4, pp. 253-344.
  1910. Cyclostomatous Bryozoa of the West Coast of North America. Univ. Calif. Publ. Zool., Vol. 6, pp. 225-284.
TORREY, H. B.
  1901. On *Phoronis pacifica,* sp. nov. Biol. Bull., Vol. 2, pp. 283-288.

ANNELIDS

BUSH, K. J.
    1905. Tubicolous Annelids. Reports of Harriman Alaska Expedition, Vol. 12, pp. 169-346.

CHAMBERLIN, RALPH V.
    1920. Notes on the Sipunculida of Laguna Beach. Pomona College Jour. Ento. and Zool., Vol. 12, pp. 30-31.

ESSENBERG, CHRISTINE
    1917. On Some New Species of Aphroditidae from the Coast of California. Univ. Calif. Publ. Zool., Vol. 16, pp. 401-430.
    1917. Descriptions of Some New Species of Polynoidae from the Coast of California. Ibid., Vol. 18, pp. 45-60.
    1917. New Species of Amphinomidae from the Pacific Coast. Ibid., Vol. 18, pp. 61-74.

FRASER, C. MCLEAN
    1915. The Swarming of Odontosyllis. Trans. Roy. Soc. Canada, Series III, Vol. 9, pp. 43-49.

GALLOWAY, T. W. and WELCH, PAUL S.
    1911. Studies on a Phosphorescent Bermudan Annelid, Odontosyllis enopla Verrill. Trans. Amer. Micr. Soc., Vol. 30, pp. 13-39.

HAMILTON, W. F.
    1915. Additional List of Annelids from Laguna Beach. Pomona College Jour. Ento. and Zool., Vol. 7, p. 207.
    1918. Notes on Annelids Collected during 1917 at Laguna Beach. Pomona College Jour. Ento. and Zool., Vol. 10, pp. 60-62.

JOHNSON, H. P.
    1897. A Preliminary Account of the Marine Annelids of the Pacific Coast. Proc. Calif. Acad. Sci., Third series, Vol. 1, pp. 153-198.
    1901. The Polychaeta from the Puget Sound Region. Proc. Boston Soc. Nat. Hist., Vol. 29, pp. 381-437.

MOORE, J. P.
    1904-1911. Numerous Papers on Pacific Coast Annelids. Proc. Acad. Nat. Sci. Phil., in Vols. 56, 57, 58, 59, 60, 61, 62, 63.

TREADWELL, A. L.
    1914. Polychaetous Annelids of the Pacific Coast in the Collections of the Zoological Museum of the University of California. New Syllidae from San Francisco Bay. Univ. Calif. Publ. Zool., Vol. 13, pp. 175-238.

WILSON, C. B.
    1900. North American Echiurids. Biol. Bull.,Vol. 1, pp. 163-178.

ECHINODERMS

BAKER, C. F.
    1912. Some Echinoderms Collected at Laguna. First Annual Report of Laguna Marine Laboratory, pp. 88-90.

BUSH, MILDRED
> 1921. Revised Key to the Echinoderms of Friday Harbor. Pub. Puget Sd. Biol. Sta., Vol. 3, pp. 65–77.

CLARK, H. L.
> 1901. The Holothurians of the Pacific Coast of North America. Zool. Anz., Bd. 24, pp. 162–171.
> 1911. North Pacific Ophiurans in the Collection of the United States National Museum. Bull. U. S. Nat. Mus., No. 75.
> 1908–1914. Hawaiian and Other Pacific Echini. Mem. Mus. Comp. Zool., Vol. 34, No. 4; Vol. 46, No. 1.

AGASSIZ, ALEX. and CLARK, H. L.
> 1909. Hawaiian and Other Pacific Echini. Mem. Mus. Comp. Zool. Harvard, Vol. 34, Nos. 2 and 3.

FISHER, W. K.
> 1911. Asteroidae of the North Pacific and Adjacent Waters. Part I. Phanerozonia and Spinulosa. Bull. 76, U. S. Nat. Mus.

HILTON, W. A.
> 1914. Note on the Sea Urchins of Laguna Beach. Pomona College Jour. Ento. and Zool., Vol. 6, p. 234.
> 1914. Starfish of Laguna Beach. Pomona College Jour. Ento. and Zool., Vol. 6, pp. 209–211.

JENNINGS, H. S.
> 1907. Behavior of the Starfish, *Asterias forreri* de Loriol. Univ. Calif. Publ. Zool., Vol. 4, pp. 53–185.

LYMAN, T.
> 1865. Ophiuridae and Astrophytidae. Ill. Cat. Mus. Comp. Zool. Harvard, No. 1.

McCLENDON, J. F.
> 1909. The Ophiurans of the San Diego Region. Univ. Calif. Publ. Zool., Vol. 6, pp. 33–64.

MONKS, SARAH P.
> 1903. Regeneration of the Body of a Starfish. Proc. Acad. Nat. Sci. Phil., Vol. 55, p. 351.
> 1904. Variability and Autotomy of Phataria. Proc. Acad. Nat. Sci. Phil., Vol. 56, pp. 596–600. (Note: *Phataria* is *Linkia*).

OSTERUD, H. L.
> 1918. Preliminary Observations on the Development of Leptasterias hexactis. Publ. Puget Sd. Biol. Sta., Vol. 2, pp. 1–15.

PEARSE, A. S.
> 1909. Autotomy in Holothurians. Biol. Bull., Vol. 18, pp. 42–49.

RITTER, W. E. and CROCKER, G. R.
> 1900. Multiplication of Rays and Bilateral Symmetry in the 20-rayed Starfish, Pycnopodia helianthoides. Proc. Wash. Acad. Sci., Vol. 2, pp. 247–274.

ULREY, A. B.
  1918. The Starfishes of Southern California. Bull. So. Calif. Acad. Sci., July, 1918, pp. 39-52.
VERRILL, A. E.
  1914. Monograph of the Shallow-Water Starfishes of the North Pacific Coast from the Arctic Ocean to California. Harriman Alaska Series, Smithsonian Institution, Vol. 14, parts 1 and 2. (Part two consists of plates).

ARTHROPODS

ALLEN, B. M.
  1916. Notes on the Spiny Lobster (*Panulirus interruptus*) of the California Coast. Univ. Calif. Publ. Zool., Vol. 16, No. 12, pp. 139-152.
BARROWS, A. L.
  1919. The Occurrence of a Rock-boring Isopod along the Shore of San Francisco Bay, California. Univ. Calif. Publ. Zool., Vol. 19, pp. 299-316.
BIGELOW, R. P.
  1894. Report on the Crustacea of the Order Stomatopoda. Proc. U. S. Nat. Mus., Vol. 17, pp. 489-550.
COLE, L. J.
  1901. Notes on the Habits of Pycnogonids. Biol. Bull., Vol. 2, pp. 195-207.
  1904. Pycnogonida of the West Coast of North America. Reports of Harriman Alaska Expedition, Vol. 10, pp. 247-330.
DARWIN, CHAS.
  1854. A Monograph of the Subclass Cirripedia.
ESTERLY, C. O.
  1905. Pelagic Copepoda of the San Diego Region. Univ. Calif. Publ. Zool., Vol. 2, pp. 113-233.
  1912. The Occurrence and Vertical Distribution of the Copepoda of the San Diego Region. Ibid., Vol. 9, pp. 253-340.
  1914. The Schizopoda of the San Diego Region. Ibid., Vol. 13, pp. 1-20.
  1914. The Vertical Distribution and Movements of the Schizopoda of the San Diego Region. Ibid., Vol. 13, pp. 123-145.
HALL, H. V. M.
  1912. Studies in Pycnogonida I. First Annual Report of the Laguna Marine Laboratory, pp. 91-99.
  1913. Pycnogonida from the Coast of California with Descriptions of two new Species. Univ. Calif. Publ. Zool., Vol. 11, pp. 127-142.
HANSEN, H. J.
  1913. On some Californian Schizopoda. Univ. Calif. Publ. Zool., Vol. 11, pp. 173-180.

HILTON, W. A.
  1914. The Central Nervous System of the Pycnogonid, *Lecythorhynchus*. Pomona College Jour. Ento. and Zool., Vol. 6, pp. 134-136.
  1915. Pycnogonids Collected during the Summer of 1914, at Laguna Beach. Pomona College Jour. Ento. and Zool., Vol. 7, pp. 67-70.
  1915. Pycnogonids Collected during the Summer of 1915, at Laguna Beach. Pomona College Jour. Ento. and Zool., Vol. 7, pp. 201-206.
  1916. A Remarkable Pycnogonid. Pomona College Jour. Ento. and Zool., Vol. 8, pp. 19-24.
  1916. The Life History of Anoplodactylus erectus Cole. Pomona College Jour. Ento. and Zool., Vol. 8, pp. 25-34.
  1916. Crustacea from Laguna Beach. Pomona College Jour. Ento. and Zool., Vol. 8, pp. 65-73.

HOLMES, S. J.
  1894. Notes on West American Crustacea. Proc. Calif. Acad. Sci. (2), Vol. 4, pp. 563-588.
  1900. Synopsis of the California Stalk-Eyed Crustacea. Occasional Papers of the California Academy of Sciences, No. 7.
  1904. Amphipod Crustaceans of the Expedition. Rept. Harriman Alaska Expedition. Vol. 10, pp. 231-246.
  1904. On some New and Imperfectly Known Species of West American Crustacea. Proc. Calif. Acad. Sci., Series 3, Vol. 3, pp. 307-330.

MEAD, H. T.
  1917. Notes on the Natural History and Behavior of Emerita analoga (Stimpson). Univ. Calif. Publ. Zool., Vol. 16, pp. 431-438.

NININGER, H. H.
  1918. Crabs taken at Laguna Beach in the Summer of 1916. Pomona College Jour. Ento. and Zool., Vol. 10, pp. 36-42.

PILSBRY, HENRY A.
  1907. The Barnacles. Bull. 60, U. S. Nat. Mus.
  1916. The Sessile Barnacles. Bull. 93, U. S. Nat. Mus.

RATHBUN, MARY J.
  1900. Synopses of North-American Invertebrates. VII. The Cyclometapous or Cancroid Crabs of North America. Amer. Nat., Vol. 34, pp. 131-143.
  1900. Synopses of North American Invertebrates X. The Oxyrhynchous and Oxystomatous Crabs of North America. Amer. Nat., Vol. 34, pp. 503-520.
  1900. Synopses of North American Invertebrates XI. The Catometopous or Grapsoid Crabs of North America. Ibid. Vol. 34, pp. 583-592.
  1904. Decapod Crustaceans of the Northwest Coast of North

America. Reports of the Harriman Alaska Expedition, Vol. 10, pp. 1-210.
    1918. The Grapsoid Crabs of America. Bull. 97, U. S. Nat. Mus.

RICHARDSON, HARRIETT
    1905. Monograph on the Isopods of North America. Bull. U. S. Nat. Mus., Vol. 54, pp. 1-727.

SCHMITT, WALDO L.
    1919. Early Stages of the Spiny Lobster taken by the boat "Alabacore." California Fish and Game, Vol. 5, pp. 24-25.
    1921. The Marine Decapod Crustacea of California. Univ. Calif. Publ. Zool., Vol. 23, pp. 1-470.

SCOFIELD, N. B.
    1919. Shrimp Fisheries of California. California Fish and Game, Vol. 5, pp. 1-12.

SCOFIELD, W. L.
    1924. Squid at Monterey. Ibid., Vol. 10, pp. 176-182.

STAFFORD, BLANCH E.
    1912-1913. Studies in Laguna Isopods. First Annual Report of Laguna Marine Laboratory and Pomona College Jour. Ento. and Zool., Vol. 5.

STEBBING, T. R. R.
    1906. The Amphipoda I. Gammaridea. Das Tierreich.

STOUT, VINNIE R.
    1912. Studies in Laguna Amphipoda I. First Annual Report of Laguna Marine Laboratory, pp. 134-149.
    1913. II. Zool. Jahrb., Vol. 34, Abt. f. Syst., pp. 633-659.

WAY, EVELYN
    1917. Brachyura and Crab-like Anomura of Friday Harbor, Washington. Publ. Puget Sd. Biol. Sta., Vol. 1, pp. 349-382.

WEYMOUTH, F. W.
    1910. Synopsis of the True Crabs (Brachyura) of Monterey Bay, California. Stanford Univ. Publ., Univ. Series, No. 4.
    1920. The Pacific Edible Crab and its Near Relatives. California Fish and Game, Vol. 6, pp. 7-10.

WILSON, C. B.
    1909. North American Parasitic Copepods: A List of those found upon the Fishes of the Pacific Coast. Proc. U. S. Nat. Mus., Vol. 35, pp. 431-481.

MOLLUSKS

AGERSBORG, H. P. K.
    1919. Notes on Melibe leonina (Gould). Pub. Puget Sd. Biol. Sta., Vol. 2, pp. 269-277.
    1920. The Utilization of Echinoderms and of Gasteropod Mollusks. Amer. Nat., Vol. 54, pp. 414-426.
    1921. Contribution to the Knowledge of the Nudibranchiate

Mollusk Melibe leonina (Gould). Amer. Nat., Vol. 55, pp. 222-253.

1922. Notes on the Locomotion of the Nudibranchiate Mollusk, Dendronotus giganteus O'Donoghue. Biol. Bul. Vol. 42, pp. 257-266.

ARNOLD, RALPH
1903. The Paleontology and Stratigraphy of San Pedro. Mem. Calif. Acad. Sci., Vol. 3.

BARROWS, A. L.
1917. An Unusual Extension of the Distribution of the Shipworm in San Francisco Bay, California. Univ. Calif. Publ. Zool., Vol. 18, pp. 27-43.

BARTSCH, PAUL
1916. Pirates of the Deep—Stories of the Squid and Octopus. Smithsonian Report, 1916, pp. 347-375.

BERRY, S. STILLMAN
1910. A Review of the Cephalopods of Western North America. Bull. Bur. Fish., Vol. 30, pp. 267-336.
1911. Notes on some Cephalopods in the Collection of the University of California. Univ. Calif. Publ. Zool., Vol. 8, pp. 301-310.
1917. Notes on West American Chitons I. Proc. Calif. Acad. Sci., 4th Series, Zool., Vol. 7, pp. 229-248.
1921. A Distributional Note on Haliotis. California Fish and Game, Vol. 7, pp. 254-255.

COBB, JOHN N.
1917. Neglected Pacific Fishery Resources. California Fish and Game, Vol. 3, pp. 108-114.

COCKERELL, T. D. A.
1902. Three new Species of Chromodoris. The Nautilus, Vol. 16, pp. 19-21.

DALL, WM. H.
1921. Summary of the Marine Shellbearing Mollusks of the Northwest Coast of America, etc. Bull. 112, U. S. Nat. Mus.

EDMONDSON, CHARLES H.
1922. Shellfish Resources of the Northwest Coast of the United States. Dept. of Commerce Bur. Fish. Doc., No. 920.

EDWARDS, CHAS. L.
1913. The Abalones of California. Popular Science Monthly, Vol. 82, pp. 533-550.

HEATH, HAROLD
1916. California Clams. California Fish and Game, Vol. 2, pp. 175-178.
1917. Devilfish and Squid. California Fish and Game, Vol. 3, pp. 103-108.

HOLDER, CHAS. F.
  1909. First Photographs ever made of a Paper Nautilus. Country Life in America, Vol. 15, No. 4, p. 356.
  1909. A Tame Nautilus. Scientific American, Oct. 16, 1909.
KOFOID, C. A.
  1921. The Marine Borers of the San Francisco Bay Region. Report on the San Francisco Bay Marine Piling Survey, pp. 23-61.
LLOYD, FRANCES ERNEST
  1897. On the Mechanisms in Certain Lamellibranch Boring Molluscs. Trans. N. Y. Acad. Sci., Vol. 16, pp. 307-316.
MACFARLAND, F. M.
  1906. Opisthobranchiate Mollusca from Monterey Bay, California. Bull. Bur. Fish., Vol. 25, pp. 109-151.
MILLER, R. C.
  1924. The Boring Mechanism of *Teredo*. Univ. Calif. Publ. Zool., Vol. 26, pp. 41-80.
OLIVER, J. H.
  1916. Abalone Pearl Formation. California Fish and Game, Vol. 2, pp. 182-185.
PACKARD, E. L.
  1918. Molluscan Fauna from San Francisco Bay. Univ. Calif. Publ. Zool., Vol. 14, pp. 199-452.
PARKER, G. H.
  1917. The Pedal Locomotion of the Sea-hare *Aplysia californica*. Jour. Exp. Zool., Vol. 24, pp. 139-145.
PILSBRY, H. A.
  1898. Chitons Collected by Dr. Harold Heath at Pacific Grove, near Monterey, California. Proc. Acad. Nat. Sci. Phil., Vol. 50, pp. 287-290.
RANKIN, E. P.
  1918. The Mussels of the Pacific Coast. California Fish and Game, Vol. 4, pp. 113-117.
SIGERFOOS, C. P.
  1908. Natural History, Organization, and Late Development of the Teredinidae, or Shipworms. Bull. U. S. Bur. Fish., Vol. 27, pp. 191-231.
THOMPSON, WILL F.
  1920. Abalones of Northern California. California Fish and Game, Vol. 6, pp. 45-50.
WEYMOUTH, FRANK W.
  1920. The Edible Clams, Mussels and Scallops of California. Fish Bulletin No. 4, Calif. Fish and Game Commission.
  1923. The Life-History and Growth of the Pismo Clam. Fish Bulletin No. 7, Calif. Fish and Game Commission.

CHORDATES

DAVIS, B. M.
- 1908. The Early Life-History of Dolichoglossus pusillus Ritter Univ. Calif. Publ. Zool., Vol. 4, pp. 187–226.

RITTER, W. E.
- 1900. Some Ascidians from Puget Sound, Collections of 1896. Ann. N. Y. Acad. Sci., Vol. 12, pp. 589–616.
- 1902. The Movements of the Enteropneusta and the Mechanism by which they are Accomplished. Biol. Bull., Vol. 3, pp. 255–261.
- 1905. The Pelagic Tunicata of the San Diego Region, excepting the Larvacea. Univ. Calif. Publ. Zool., Vol. 2, pp. 51–112.

RITTER, W. E. and FORSYTH, RUTH A.
- 1917. Ascidians of the Littoral Zone of Southern California. Univ. Calif. Publ. Zool., Vol. 16, pp. 439–512.

GLOSSARY

Abdomen.—The posterior part of the body of an arthropod.
Aboral.—The region or side of the body opposite the mouth.
Acontia.—In coelenterates, thread-like processes filled with nettling cells.
Actinal.—Relating to the area in which the tube feet of echinoderms are located; in *Actinaria*, to the oral region and tentacles.
Adaptation.—A structure or habit fitting an animal or plant for some environmental condition; the process whereby it becomes so fitted.
Adductor muscle.—A muscle which, in contracting, pulls parts nearer together.
Ala.—A projection that is wing-like.
Algae.—Simple plants, often unicellular; the higher forms include the seaweeds.
Alimentary canal.—The main tube of the digestive tract.
Alternation of generations.—The alternation of sexual and asexual forms in the reproductive cycle.
Ambulacral groove. In starfish, a furrow occupying a median position on the lower side of each ray. It shelters the tube feet.
Ambulatory.—Used for walking.
Amoebocytes.—Wandering cells found in the body fluids of some invertebrates. They resemble the protozoan, *Amoeba*.
Ampulla.—In echinoderms, a bulb-like structure connected with the tube foot, and acting as a reservoir for the water used to distend the latter.
Analogous.—Similar in function.
Annulation.—A ring-like segment or band.
Antenna.—One of a pair of jointed appendages attached to the head; "feelers."
Antennule.—One of the anterior pair of feelers when two pairs are present.
Anterior.—At or near the front end of the body.
Antero-lateral.—Region where the front and side join.
Anus.—The posterior opening of the alimentary tract.
Aperture.—In *Bryozoa*, the chitinous front wall of the zooecium; otherwise used to denote an opening, as in *Gastropoda*, where it refers to the opening of the shell.
Aristotle's lantern.—The teeth and their supporting framework in some echinoderms.
Articular.—Pertaining to a joint.
Articulation.—A joint. Usually a movable joint between two segments.
Auditory.—Pertaining to the ear or the sense of sound.

Auricle.—A chamber of the heart which receives blood from the veins before it enters the ventricle.

Autozooid.—One of the independent, feeding individuals of an alcyonarian colony.

Avicularium.—In *Bryozoa*, an appendage of the zooecium resembling a bird's head.

Axial.—Along mid-line of body or of appendage.

Bilateral symmetry.—Having a form which may be divided by a central axis into two similar halves.

Biramous.—Having two branches.

Bivalve.—Comprising two similar parts. A mollusk with two similar parts (or valves) to the shell.

Branchia.—A gill.

Branchial.—Pertaining to the gills or branchiae.

Brown body.—In *Bryozoa*, a brown mass resulting from the disintegration of the organs of the animal.

Buccal.—Relating to the mouth.

Bursa.—A pouch.

Byssus.—A group of threads secreted by the foot of certain mollusks and serving to attach the animal to a substratum.

Caecum.—(Plu. caeca). A pouch-like extension of some portion of the alimentary canal.

Calcareous.—Made up of carbonate of lime.

Callus.—A thickening; in gastropods a thickening of the shell near the umbilicus.

Calyx.—Cup; in crinoids, the central part of the body corresponding to the disk of the starfish.

Canal.—In sponges, a tube communicating between the cavities and the exterior; in gastropod shells, a tubular extension at the aperture to accommodate the siphon.

Capitulum.—In a stalked barnacle, the tip or portion distal to the peduncle or stalk.

Captaculum.—A ciliated filament from the side of the mouth (in *Dentalium*).

Carapace.—In crustaceans, the hard covering of the cephalothorax.

Cardinal teeth.—The teeth of the hinge, just below the umbo, in a bivalve shell.

Carina.—In barnacles (fig. 216) the median dorsal plate.

Carnivorous.—Living upon flesh.

Carpus.—Wrist, in crustaceans, the segment next the "hand."

Cartilage.—In bivalves, the internal part of the ligament.

Cellulose.—A substance related to wood, which forms the cell wall of most plant cells and of certain cells in tunicates.

Cephalic.—Pertaining to the head.

Cephalothorax.—In arthropods, the head and thorax, when united.

Cerata.—Doral projections which take the place of gills.

Cercaria.—A larva with somewhat rounded body and a long tail, the larva of a parasitic trematode.

GLOSSARY

Chela.—The pincer of a crustacean.
Chelifor (Chelophore).—The first, or claw-bearing appendage of pycnogonids.
Cheliform.—Resembling a chela.
Cheliped.—A leg bearing a chela at the end.
Chitin.—A hard, horny substance in the exoskeletons of most arthropods.
Chitinous.—Made of chitin.
Cilia.—Microscopic hair-like projections of the cells of the body, capable of vibrating.
Ciliated.—Provided with cilia.
Cirrus.—A filamentous appendage, usually sensory. In flatworms, the copulatory organ.
Class.—In classification, a group of animals or plants standing between a phylum and an order.
Clavus.—In nudibranchs, the club-shaped portion of the rhinophore.
Cloaca.—In certain chordates, a cavity near the anus which receives the discharges of intestine, genital organs, and urinary canals.
Collar cell.—A flagellate cell, having a collar-like membrane around the base of the flagellum.
Columella.—The axis or central pillar of a spiral shell.
Commensal.—Organisms living together and sharing their food; messmates.
Constriction.—A contraction at a particular point.
Corbula.—A basket; a pod-like structure on the branches of the hydroid, *Aglaophenia.*
Cornea.—The transparent covering over the eyeball; the outer covering of each unit of a compound eye.
Coxa.—In crustacea, etc., the first segment of the leg, the one next to the body.
Crenulate.—With edge minutely scalloped, having a series of notches.
Crenulations.—Marginal notches or scallops.
Ctenidium.—Type of gill characteristic of mollusks, so called from its resemblance to a comb.
Cuticle.—An outer skin; a protective covering.
Dactyl.—A finger, a term applied to the distal joint of the crustacean leg.
Deflexed.—Bent downward.
Denticle.—Literally, a little tooth, a term applied to a small, tooth-like projection.
Diatoms.—One-celled, microscopic algae with siliceous cell walls.
Distal.—Farthest from body, opposite of proximal.
Divergent.—To extend in different directions from the same point.
Dorsal.—On or in the direction of the back or dorsum.
Duct.—Any small tube in the body for the transportation of fluids.
Ectoderm.—In a many-celled animal, the outermost layer of cells.
Ectoparasite.—An animal or plant, living as a parasite upon the exterior of the host.
Elytra.—Shield-like scales of certain worms. Term also used for wing covers of beetles.

Emargination.—A notch.
Embryo.—A young animal usually undergoing development within the egg membranes.
Endoderm.—In many-celled animals, the innermost layer of cells, that line the digestive and respiratory organs and certain glands.
Endoparasite.—An animal or plant living as a parasite within the body of its host.
Endopodite.—The inner of two principal branches of a crustacean leg.
Endostyle.—In certain chordates, a ciliated groove, situated on the median ventral line in the pharyngeal sac.
Epidermis.—Outer layer of skin, horny coating of some shells.
Epimeron (Plural, epimera).—In arthropods, a part of the segment between the limb attachment and the dorsal part of the segment.
Epistome.—The transverse plate forming the forward border of the mouth cavity in *Crustacea*; in *Molluscoidea*, a projection above the mouth.
Epithelium.—The tissue which lines the cavities or covers the surface of the body.
Escutcheon.—In mollusks, an area behind the umbones enclosing the ligament.
Esophagus.—The tube connecting the mouth with the stomach.
Excurved.—Curved outward.
Exopodite.—The outer of two principal branches of a crustacean appendage.
Exoskeleton.—An external supporting structure secreted by ectoderm cells.
Exumbrella.—The upper, rounded portion of a jellyfish.
Family.—In classification, a term designating one group of an order. The name of a family of animals or plants customarily ends in *idae*, i.e., *Portunidae*.
Fascicled.—In stalks, made up of a bundle, or cluster, of stalks.
Fathom.—A nautical unit of measure, equal to six feet.
Fertilization.—The union of the sperm nucleus with that of the egg.
Finger.—The movable segment of the decapod chela or pincer; the dactyl.
Flaccid.—Relaxed, limber, lacking firmness.
Flagellated.—Possessing a flagellum, or flagella.
Flagellum.—A whip-like process from a cell, capable of vibration.
Flame cells.—Excretory cells of primitive type.
Foot.—The organ of locomotion in mollusks.
Foramen.—An opening.
Forcipiform.—Term applied to pedicellariae of starfish in which the blades are crossed so that they open and close like tweezers or pincers.
Forficiform.—Descriptive of a type of starfish pedicellaria characterized by straight jaws which open and close like forceps.
Fusiform.—Spindle-shaped.
Ganglion.—A small mass of nerve cells.
Gastric.—Pertaining to the stomach.

GLOSSARY

Gemmule.—In sponges, an asexual bud, arising internally and producing a new individual after the death of the parent.
Genital.—Pertaining to reproduction.
Genus.—In classification, a group of closely related species.
Gill.—A breathing organ which can obtain the air that is contained in water.
Girdle.—In chitons, a fold of the integument usually at the margin of the shell.
Globose.—Approaching the shape of a sphere.
Gnathopod.—In amphipods, a grasping claw; the term means "jaw-foot."
Gonad.—Organs which produce reproductive cells.
Gonangium.—An external structure or gonotheca protecting the reproductive buds of certain hydroids.
Gonotheca.—A transparent protective covering surrounding the reproductive zooid in a hydroid colony.
Granulated.—Covered with minute elevations; tuberculated minutely.
Habitat.—The locality or region where an animal or a species usually lives.
Hectocotylus arm.—Arm of male cephalopod modified to transfer sperm to the female.
Hepatic.—Pertaining to the liver.
Hermaphroditic.—Having both male and female reproductive organs, in one individual.
Homologous.—Of similar origin.
Host.—An animal supporting a parasite.
Hydranth.—A feeding polyp, or nutritive zooid, in a hydroid colony.
Hydrotheca.—In a hydroid colony, a cup-like structure, produced by the ectoderm cells and serving to protect the hydroid polyp.
Incremental.—Indicating successive increases in growth.
Integument.—The outer covering.
Internode.—The portion of a structure that occurs between two nodes or joints.
Inter-radii.—In starfish, the interspaces between the rays.
Introvert.—In sipunculids, the anterior part of the body which can be retracted or invaginated.
Invaginate.—To draw into a sheath or cavity.
Ischium.—Fifth joint of a decapod limb, counting back from the tip.
Keel.—A sharp ridge along the whorls of a gastropod shell.
Lamella.—A thin, plate-like structure.
Lamina.—A thin layer.
Lamina, dorsal.—In tunicates, a median dorsal ridge running lengthwise of the pharyngeal sac to the esophagus opening.
Lamina, sutural.—In chitons, a thin sheet extending from one valve below the adjoining one (fig. 655).
Lappets.—Little flaps or lobes.
Larva.—A young animal self-sustaining, but not yet having attained its adult form.
Lateral.—To the right or left of the mid-line of the body.

Lingual.—Referring to the tongue.
Lip.—In gastropods, the margin of the aperture.
Lithocyst.—In coelenterates, a marginal sense organ.
Liver.—In invertebrates, a digestive gland, not identical with the liver of higher animals.
Lophophore.—In the *Molluscoidea*, a horseshoe-shaped, tentacle-bearing ridge.
Lorica.—In rotifers, a protective covering or shell.
Luciferase.—An oxidizing enzyme acting on luciferine to produce luminesence.
Luciferine.—Substance produced by animals which causes luminesence when acted upon by luciferase.
Lunule.—In mollusks a depressed area in front of and close to the umbones, usually heart-shaped and limited by a ridge.
Madreporite (Madreporic plate).—A calcareous plate, perforated with minute openings and connected with the water-vascular system of echinoderms; the sieve plate.
Mandibles.—The first or anterior pair of arthropod mouth parts; they usually function as biting parts.
Mandibular.—Related in form or position to the mandibles.
Mantle.—In mollusks and brachiopods, the outer sheet of tissue which secretes the shell; in tunicates, the body wall beneath the tunic.
Mantle cavity.—In mollusks, the space enclosed by the mantle.
Manubrium.—The handle-like portion hanging down on the under surface of a jellyfish.
Maxillae.—In arthropods, paired mouth parts next behind the mandibles.
Maxillipeds.—In arthropods, appendages just posterior to the maxillae.
Median.—Middle; along the axial plane.
Medusa.—A jellyfish.
Medusoid.—Like a medusa.
Megalops.—The last larval stage of certain crustaceans, as the crabs.
Membranaceous.—Having the consistency or the structure of membrane.
Membrane.—A thin skin-like covering of cells.
Membranous.—Composed of, or resembling membrane.
Merus.—The fourth joint of a crustacean leg, counting back from the tip (fig. 302).
Mesentery.—In coelenterates, a partition extending inward from the body wall.
Mesogloea.—The noncellular tissue lying between the ectoderm and entoderm in coelenterates.
Metamorphosis.—The changes in form which an animal undergoes in developing from embryo to adult.
Microplankton.—A term which includes all microscopic plants and animals belonging to the plankton.
Mucin.—A product of mucus.
Mucus.—The slimy secretion of mucous membranes.

GLOSSARY

Mysis stage.—A larval stage of certain crustaceans resembling the schizopod, *Mysis*.
Nauplius.—In crustaceans, a larval form of certain groups.
Nematocyst.—Stinging structure within the nettling cells of coelenterates.
Nematophore.—In certain coelenterates, small polyps specialized for defense and armed with nematocysts.
Nephridium.—A tubule for the elimination of waste matter from body.
Nettle cells.—Cells containing stinging organs or nematocysts. Technically, *cnidoblasts*.
Nidamental.—Concerned with the act of laying eggs, especially the formation of the egg shell or covering.
Nodule.—A small knob-like structure.
Notochord.—An elastic cord of cellular origin between the nerve cord and the alimentary canal.
Nucleus.—A structure within the cell, usually spherical and centrally located, and essential to its life activities.
Obsolete.—Disappearing.
Ocellus.—A single eye of simple structure, found in arthropods.
Ooecium.—An ovicell.
Olfactory.—Pertaining to the sense of smell.
Operculum.—In *Bryozoa*, the lip by which the orifice is closed; in *Mollusca* and worms, a plate covering the aperture of the shell.
Oral.—Pertaining to the mouth or mouth region.
Orbits.—The eye-cavities found in the *Decapoda*.
Order.—A group of related animals or plants. The order ranks between the family and the class.
Organ.—A structure in the body with a definite function or functions.
Orifice.—In *Bryozoa*, the opening through which the polypide emerges. In general, the opening of a tubular structure.
Osculum.—In sponges, an opening by which the current of water passes out of the sponge.
Osmosis.—Diffusion between two fluids separated by a membrane.
Osphradium.—A molluscan sense organ of supposed olfactory function.
Ossicle.—A calcareous plate or rod.
Otocyst.—A primitive sense organ, supposed to be an organ of hearing.
Ovary.—The organ which produces ova or egg cells.
Ovicell.—In *Bryozoa*, the chamber in which the embryo develops.
Oviduct.—The tube by which the eggs leave the ovary.
Ovigerous.—Serving to carry eggs.
Ovotestis.—Hermaphroditic reproductive gland found in certain mollusks.
Ovum.—An egg (Plural, *ova*).
Pallial line.—Line along which shell is attached to the mantle.
Pallial sinus.—The emargination or notch in the pallial line which marks the attachment of the muscle that retracts the siphon.
Palp or palpus.—A sense organ projecting near the mouth.
Papilla.—A small elevation; in holothurians, papillae are modified tube feet not used for locomotion.

Papulae.—The dermal gills of echinoderms.
Parapodia.—In polychaets, jointless, paired processes on the segments; in certain mollusks, lateral prolongations of the foot (also called *epipodia*).
Parasite.—An animal living on or within another and using it for food. The animal upon which the parasite lives, is called the host.
Parenchyma.—The unspecialized tissue which fills the body cavity of flat worms.
Parietes.—The middle wall of each segment of the shell of barnacles.
Paxillae.—Calcareous elevations supporting a cluster of minute spines, found on the surface of some starfish.
Paxilliform.—Resembling paxillae.
Pedal.—Referring to the foot.
Pedicel.—The footstalk of a fixed organism.
Pedicellaria.—One of the minute, pincer-like structures of certain echinoderms; they may be sessile or stalked.
Peduncle.—A stalk for the attachment of an animal or an organ; in serpulid worms, a stalk for the attachment of the operculum.
Pelagic.—Referring to life in the open sea.
Penis.—The male copulatory organ.
Pericardial.—About or in the region of the heart.
Perihaemal canals.—In echinoderms, a series of canals which aid in the circulation of body fluid.
Periopods.—Ambulatory legs, thoracic appendages posterior to the maxillipeds in crustaceans.
Perisarc.—The tough protective covering of hydroids.
Periostracum.—In mollusks, the outer covering of the shell.
Periproct.—The aboral region of a sea urchin.
Peristome.—In echinoderms, the membranous area immediately around the mouth.
Pharynx.—Part of digestive tract next to the mouth.
Phyllosoma.—Larval stage of crustacean, *Panulirus.*
Phylum.—A great division of animals or plants, ranking below a subkingdom and above an order.
Pinnae.—Structure resembling a feather.
Pinnate.—Divided like a feather, having extensions on each side of a midrib.
Pinnules.—In crinoids, small branches on the arms.
Plankton.—A collective term referring to the forms of life that inhabit the surface layers of the water; the drifting species.
Planula.—The ciliated larvae of certain coelenterates.
Pleopod.—In arthropods, an abdominal appendage.
Pluteus.—Larva of brittle star or sea urchin.
Podium (Plural, *podia*)—In echinoderms, tube foot.
Polian vesicle.—In echinoderms, a sack communicating with the ring canal of the water-vascular system.

GLOSSARY

Pollex.—The immovable finger of a cheliped; the point of the propodus of a cheliped.

Polyp.—An individual coelenterate or bryozoan, attached to a substratum or one of a colony, named so because of a fancied resemblance to *Polypus*, an octopus.

Polypide.—An individual of a bryozoan colony. The designation does not include the surrounding structure or zooecium.

Posterior.—At or near the hinder end of the body.

Postero-lateral.—Referring to region located posteriorly and toward the side.

Proboscis.—In arthropods, beak-like mouth parts; in worms, an extensible organ, part of the pharynx.

Proglottides.—The body segments of the free portion of a tapeworm; the segments develop from the "neck" or attached portion.

Propodus.—The second joint from the distal end of the thoracic limb of a decapod.

Prostomium.—Anterior part of the head, preceding the mouth.

Protozoa.—Unicellular animals.

Protozoea.—In certain crustaceans, a stage preceding the zoea.

Proximal.—The opposite of distal; near.

Pseudopaxillae.—False paxillae, groups of short spinelets resembling paxillae.

Pubescent.—Covered with fine, short hairs or hair-like structures.

Puerulus.—A larval stage of the spiny lobster, *Panulirus*.

Punctate.—The surface dotted with minute holes.

Pyloric.—Referring to the lower opening of the stomach.

Pyriform.—Pear-shaped.

Rachis.—The upper part of a sea pen or sea pansy, the part which bears the polyps.

Radial canal.—A canal of the water-vascular system of echinoderms which extends along the axis of a ray.

Radial symmetry.—Organs arranged symmetrically about a central point.

Radula.—The lingual ribbon, or band of limy teeth, in the pharynx of mollusks.

Ray.—One of the divisions of an echinoderm, as a starfish arm.

Rectum.—Posterior portion of the intestine.

Regeneration.—The reproduction of lost parts.

Respiration.—The process of breathing.

Retinal.—Pertaining to the retina, the part of the eye which receives the impressions.

Rhinophores.—In nudibranchs, the posterior pair of tentacles.

Ring canal.—In echinoderms, the canal of the water-vascular system which encircles the mouth; in medusae, the circular canal at the margin of the umbrella.

Rostrum.—A projection in front of the head in crustaceans.

Rudimentary.—In an incompletely developed condition.

Sarcostyle.—The defensive polyp in the nematophore of a plume hydroid.

Scolex.—In tapeworms, the anterior part provided with organs of attachment.
Scutum.—A shield-like structure (fig. 211).
Segment.—A part cut off or marked off as separate from others.
Septum.—A partition; in corals, one of the radiating ridges of the cup.
Serrate.—Having a saw-toothed edge.
Sessile.—In animals, fixed, without power of movement; in organs, attached without a stalk.
Seta.—A slender, bristle-like structure.
Sexual reproduction.—Reproduction by means of eggs and spermatozoa.
Shell.—A hard outer covering of an animal and composed of calcareous, silicious, horny, or chitinous material produced by the animal.
Sieve plate.—See Madreporite.
Silicious.—Containing silica.
Sinus.—A depression, curve, or indentation.
Siphon.—In mollusks and ascidians, tubular structures through which water enters and leaves the animal.
Siphonoglyph.—In sea anemones, a groove at each end of the mouth.
Siphonozooid.—A small, rudimentary individual of an alcyonarian colony through which the water flows into the canals of the colony.
Somite.—A segment of the body.
Species.—A group of closely allied individuals. It ranks above the individual and below the genus, and is sometimes divided into sub-species and varieites.
Sperm cell.—See spermatozoon.
Spermatocyst.—A sac for the storage of sperm.
Spermatozoon.—The male sexual cell.
Sphaeridium.—A form of sense organ found in *Echinodermata*.
Spermatotheca.—An organ for the storing of sperm in the male animal.
Spicules.—Minute structures of silicious or limy nature which stiffen the body of echinoderms and sponges.
Spinulous.—Bearing spines.
Sporocyst.—The encysted, degenerate embryo of a parasitic trematode.
Statocyst.—An organ found in many invertebrates and supposed to aid in perceiving the position of the body.
Statolith.—A sand grain or a secreted granule within a statocyst.
Stolon.—In *Bryozoa*, a cylindrical stem from which individuals branch; in *Coelenterata*, a horizontal branch from which new hydroid stems bud; in *Tunicata*, a process which develops into a chain of buds.
Stone canal.—In *Echinodermata*, a calcareous tube, which leads from the madreporic plate to the ring canal.
Strobila.—In some coelenterates, a stage in which the embryos take the form of a pile of discs separated by constrictions; the body of a tapeworm from which proglottids are shed.
Stylet.—In nemerteans, calcareous needles in the proboscis; in general, a small, sharp, pointed projection.
Subchela.—A structure resembling a chela.

GLOSSARY

Subequal.—Almost equal.
Sulcus.—A groove.
Suture.—The line of junction of two parts immovably connected.
Swimmeret.—A pleopod or abdominal appendage of decapod crustaceans.
Tabulate.—Having a flat surface.
Tactile.—Pertaining to the sense of touch.
Tegmentum.—In chitons, the external portion of the shell.
Telson.—The terminal segment of the abdomen in the *Crustacea*.
Tentacle.—An elongated structure, tactile in function.
Tergum.—In arthropods, the dorsal part of a segment; in barnacles, the dorsal plate.
Test.—In sea urchins, the hard skeleton; in ascidians, the tunic or outer covering.
Testis.—Organ which produces sperm cells.
Thoracic.—Pertaining to the thorax.
Thorax.—The middle portion of the body.
Thumb.—A synonym for pollex in *Decapoda*.
Tooth papillae.—Hard projections from the jaw of an ophiuran.
Transverse.—Lying across or at right angles to the axis of the body.
Tribe.—A group of indefinite rank.
Trochosphere.—The spherical, ciliated larva of certain worms and mollusks.
Truncate.—With the appearance of having been cut off at the tip.
Tube foot.—The podium; the principal organ of locomotion in echinoderms.
Tubercle.—A small rounded projection.
Tunic.—Outer cuticular covering in tunicates.
Umbilicus.—In gastropod shells, the depression at the base of the axis of the shell.
Umbo. (Plural, umbones)—An elevation or knob; in mollusks, the projection above the hinge; in barnacles, see figure 214.
Unisexual.—Possessing but one sort of sex organs, either male or female; not hermaphroditic.
Univalve.—Having but one part to the shell.
Uropod.—In crustacea, modified swimmerets at the side of the telson.
Valve.—One of the pieces forming the shell of a mollusk, or a barnacle; one of the parts forming a pedicellaria of an echinoderm.
Variety.—A group within the species differing from the type in some minor respect.
Varix.—A prominent ridge across the whorl of a gastropod shell, indicating the former location of the lip.
Vas deferens.—The sperm duct.
Veliger.—The second larval stage of certain mollusks. The veliger has a swimming organ called the velum.
Velum.—In jellyfish and medusae, a membrane projecting inward from the bell; in certain mollusk larvae, the swimming organ, a circle of cilia.
Ventral.—The lower side of the body; opposite of dorsal.

Ventricle.—The chamber of the heart which pumps the blood over the body.
Verrucose.—With many small wart-like structures on the surface.
Vibraculum.—In *Bryozoa* a chitinous seta or whip-like appendage extending from a chamber on the dorsal side of the zooecium.
Viscera.—The organs in the body cavities.
Viviparous.—Bringing forth living young.
Whorls.—The spiral turns of gastropod shells.
Water-vascular system.—In echinoderms, a series of canals which conduct water through the animal and which connect with the tube feet.
Zoea.—The first larval stage in crabs.
Zooecium.—In *Bryozoa*, a chamber or sac surrounding a polypide. It may be calcareous, semicalcareous, or chitinous.
Zooid.—An individual member of a colony.

INDEX

abalone, 2, 8, 11, 475, 551-554.
  black, 552, *553*.
  corrugated, *553*, 554.
  green, *553*, *554*.
  northern green, 554.
  red, 552, *553*.
abdominalis (Phyllodurus), 327.
Abietinaria filicula, 61, *62*, *67*.
Acanthina paucilirata, *521*, 522.
  spirata, *521*, 522.
Acanthochitonidae, 566.
Acanthodoris rhodoceras, 496, Pl. XI.
Acartia tonsa, *254*, 255.
Acila, 418.
Acmaea asmi, *543*, 544.
  cassis pelta, 542, *543*.
  depicta, *543*, 545.
  insessa, *543*, 544.
  instabilis, *543*, 544.
  limatula, 542, *543*.
  mitra, 542, *543*.
  paleacea, *543*, 545.
  persona, *543*, 544.
  scabra, 542, *543*, 544.
  scutum patina, 542, *543*.
  spectrum, 544.
Acmaeidae, 542.
acontia, 96, 99, 102, 103, *104*.
acorn barnacles, 257, *260*, 264-270.
Acotylea, 116.
Actaeon punctocoelatus, *483*.
Actaeonidae, 483.
Actiniaria, 96.
Actinozoa, 43, 85.
aculeata (Crepidula), *537*, 538.
  (Ophiopholis), 221, *222*.
  (Pentidotea), *289*, 290.
acutifrons (Scyra), 374, *375*.
Adams, A., 527.
Adams, H., 527.
adhaerens (Esperella), 39.
Adula, 431.
adunca (Crepidula), *537*, 538.
Aeolidiadae, 502.
aequalis (Leptasterias), *198*, 199.
  (Mediaster), 208, *209*, *211*.
aequilibra (Caprella), 280, *281*.
aequisulcatus (Pecten) circularis, 425.

Aequorea aequorea, 68, 69, *70*, 71.
  coerulescens, 69.
  forskalea, 71.
Aequoridae, 69.
aerating aquaria, 10, 602, 603.
aestuari (Cerianthus), 105, *106*.
Aetea anguina, *141*.
Aeteidae, 141.
affinis (Callianassa), *330*.
  (Cyclosalpa), 587, 589.
Agersborg, H.P.K., 500, 501, 502, 539
Aglajidae, 485.
Aglantha digitale, *73*.
Aglaophenia inconspicua, *65*, *67*.
  struthionides, *66*, *67*.
Akeratidae, 484.
alae, *264*.
Alaska, 5.
alaskensis (Astarte), *433*, 434.
alaskensis elongata (Crago), 313.
Albacore, 318.
albida (Glottidia), 153, *154*.
Albuneidae, 347.
alcohol as a preservative, 603.
Alcyonacea, 89.
Alcyonaria, 88, 95.
Alcyonidae, 89.
Aldisa sanguinea, 492, Pl. VIII.
Alectrion fossata, *512*, 513.
  mendica, *512*, 513.
  perpinguis, *512*, 513.
  tegula, *512*, 513.
Alectrionidae, 513.
Alepas pacifica, 261.
Aletes squamigerus, *532*, 533.
Allen, B. M., 319.
Alpheidae, 308.
Alpheus, 308.
alta (Metis), *444*, *449*.
alternation of generations, 42, 77, 78.
Amaroucium californicum, 597.
ambulacral groove, 181, *183*, 185.
  ossicles, 185.
American Museum of Natural History 27, 415.
amethystina (Renilla), *92*, *93*, Pl. IV.
Amiantis callosa, *443*.
Ammotheidae, 404.

INDEX

Ammothella bi-ungulata californica, 404, *405*.
amoebocytes, 189.
Amphineura, 413, 557.
Amphinomidae, 163.
Amphiodia (Amphiura) occidentalis, *219*, 220.
amphioetus (Cancer), 376, 378.
Amphioxus, 599, *600*.
Amphipoda, 3, 20, 271.
Amphiporus bimaculatus, 125, Pl. VI.
Amphissa columbiana, 514, *515*.
  versicolor, 514, *515*.
Amphithoe humeralis, *277*, 278, *279*.
Amphithoidae, 278.
Amphiura, 220.
Amphiuridae, 220.
amplicauda (Exosphaeroma), *287*, 288.
ampulla, *182*, *183*.
Amusium, *417*.
analoga (Emerita), 325, *338*, 341-347.
Anaspidea, 486.
anatifera (Lepas), 262.
anchor from body wall of Leptosynapta, 242.
anchovies, 428.
Ancula cristata, 490.
  pacifica, 497, Pl. XI.
anemone, burrowing, 97, 98, 105, 106, Pl. III.
anguina (Aetea), *141*.
angulata (Mangilia), 508, *509*.
Anisodoris nobilis, 491, Pl. VIII.
annectens (Perophora), 595, *596*.
annelid worms, 29, 158-179, 392.
Annulata, 23, 158.
annulata (Garveia), *48*.
  (Ophionereis), *219*, 220.
annulatum (Calliostoma), *548*, 550.
Anodonta, *417*.
Anomia, 519.
  peruviana, *426*.
Anomiidae, 426.
Anomura, 21, 314, 325.
Anoplodactylus erectus, 407, *408*.
anserifera (Lepas), 262, 263.
antennae, 251.
antennarius (Cancer), 376, *377*, 378.
antennules, 251.
Anthomastus ritteri, 89, *98*.
Anthomedusae, 47.
anthonyi (Cancer), 376, 379, *380*.
Anthozoa, 43, 85, 88.
Antiopella aureocincta, 500.
antiquatus (Hipponix), *535*, 536.
aperture of shell, *476*, 477.
apex of shell, *476*, 477.
Aphrodita refulgida, 162.
Aphroditidae, 162.
Aplysia, 486.
Aplysiidae, 486.
appendicularian, *585*.
aquaria, 10.
Arachnida, 20, 403.
Arachnoidea, 20, 402.
arborescens (Dendronotus), 500.
Arca, *417*.
Archidoris montereyensis, 491, 492, Pl. VIII.
arctica (Saxicava), *464*, 466.
arcticus (Ceramaster), 208.
arenaria (Mya), 463, *464*, 465.
Arenicola claperedii, *169*.
Arenicolidae, 169.
Argonautidae, 575.
argonaut, living, 576.
Argonauta pacifica, 575, *576*, 577.
Argobuccinium, 5.
  oregonensis, *529*, 530.
Aristotle, 34, 229.
Aristotle's lantern, 229.
Armata, 178.
armatus (Astropecten), *209*, *211*, 212.
arnheimi (Macoma) inquinata, 451.
Arnold, A. F., 232.
Arnold, R., 411, 427, 438, 441, 448, 451, 454, 475.
arrow worms, 113, 128.
Arthropoda, 19, 20, 23, 249.
articulamentum, *557*, 558.
Aruga oculata, *274*.
Ascidia californica, *593*.
Ascidiacea, 590.
Ascidiae compositae, 594.
  luciae, 598.
  simplices, 591.
ascidian, frontispiece, 6, 124, 590.
ascidian colonies, 519, 594.
ascidians, luminescent, 598.
asmi (Acmaea), *543*, 544.
aspera (Diadora), *555*, *556*.
Astacura, 21, 314.
Astarte alaskensis, *433*, 434.
  esquimalti, *433*, 435.
Astartidae, 434.
Asteriidae, 194.
Asterina gibbosa, 206.
  miniata, 206.
Asterinidae, 206.
Asteroidea, 181.
Asteropidae, 208.
Astraea inaequalis, *546*.
  undosa, *546*.
Astrangia sp., *107*, *108*.
Astrangidae, 108.
Astrometis sertulifera, 183, 184, 199, *200*.

INDEX 635

Astropecten, 211.
  armatus, *209*, *211*, 212.
  californicus, *209*, 211, 212.
  ornatissimus, *209*, 212.
Astropectinidae, 211.
Astrophytidae, 224.
Atlantid, *527*.
Atlantidae, 526.
atra (Polycera), 496, Pl. X.
  (Stomotoca), *49*.
Atremata, 153.
atrial opening, *596*.
attenuata (Harenactis), *98*.
augur-shell, 504, *505*.
"Aunt Fannys," 280.
aurantiaca (Eurylepta), 118, Pl. V.
Aurelia aurita, *82*.
Aureliidae, 82.
aureocincta (Antiopella), 500.
aureotincta (Tegula), *548*, 549.
aurita (Aurelia), *82*.
autozooids, 88, 89.
Avalon, glass-bottomed boats, 26.
avicularium, 133, 142, 146, 147.

bachei (Pleurobrachia), *110*.
bacula (Leptothyra), *546*, 547.
Baker, Dr. Fred, 411, 434, 453, 461, 534, 547.
bakeri (Clytia), *53*, *55*.
  (Paguristes), *333*, *335*.
Balanidae, 264.
Balanoglossida, 583.
Balanoglossus, 582, 583.
Balanomorpha, 264.
Balanophyllia, 105.
  elegans, *108*, 109, Pl. III.
Belanus cariosus, *267*, *268*.
  crenatus, *266*, *267*.
  glandula, *266*, *267*.
  tintinnabulum californicus, *260*, *264*, *265*.
balthica (Macoma), 451.
Bankia setacea, 472.
barnacles, 6, 249, 250, 252, 257-270, 325.
Barnea pacifica, *467*, 468.
Barnhart, P. S., 175.
barrel shell, *483*.
Barrows, A. L., 288, 472.
Bartsch, Dr. Paul, 525, 568.
basal fragmentation, 104.
basket cockle, *439*, 440.
  fish, *224*.
  shells, 465, *467*, *512*, 513.
  star, *224*.
bath sponge, 34.
beach fleas, 271.
beach hoppers, 3.

bean clam, *452*, 454.
beckoning of fiddler crabs, 400, 401.
bellastriatum (Epitonium), 523, *524*.
bellimanus (Crangon), 309.
bellus (Lophopanopeus), 387, *388*.
Bering Sea, 1.
Beroe forskalii, *112*.
Beroidae, 112.
Berry, S. S., 570, 575, 579.
Betaeus harfordi, *310*.
  longidactylus, 310, *311*, 312.
Bicellariidae, 144.
bifurcatus (Septifer), 429, *430*.
Bigelow, H. B., 69, 71, 76, 84.
bigelowi (Pseudosquilla), 296, *297*.
bilateral symmetry, 17, 113.
bimaculata (Heterodonax), *456*, 457.
bimaculatus (Amphiporus), 125, Pl. VI.
  (Megatebennus), *553*, 555.
  (Polypus), *577*, *578*.
Bimeriidae, 48.
binomial nomenclature, 14.
biological station, *4*, 5, 6, 7, *9*.
biplicata (Olivella), *476*, 477, 508, *509*.
bipunctata (Sagitta), *129*.
Bittium eschrichtii, *530*, 531.
  eschrichtii montereyense, 531.
  quadrifilatum, *530*, 531.
bi-ungulata californica (Ammothella), 404, *405*.
bivalve, 412.
Blepharipoda occidentalis, *346*, 347.
blind goby, 327.
bodegensis (Tellina), *444*, 449.
body cavity, 17, 42.
Bolinopsis microptera, 70, *111*.
Borradaile, L. A., 298, 326.
Botrylloides diegensis, *595*.
Botula falcata, *430*, 431.
Bougainvilleidae, 49.
Brachiopoda, 152.
Brachygnatha, 361, *362*, *363*.
Brachyrhyncha, *362*, 363, 374.
Brachyura, 21, 314, 326, 352.
branchiae, *16*, *183*, 483.
branchial orifice, *596*.
Branchiostoma, 599, *600*.
  californiense, *600*.
brevirostris (Spirontocaris), *306*, 307.
brevispinus (Pisaster), *196*, 197.
bristle-jawed worms, 128.
brittlestar, 3, 213–225.
  long-armed, *219*, 220.
  Panama, 216, 217.
brunnea (Tegula), *548*, 549.
Bryozoa, 3, 29, 43, 130.
bubble-shell, 483, *484*, *485*.
buccal plates, *213*, *214*.

INDEX

Bugula, 133, 136.
  murrayana, *146*.
  neritina, *144*, *147*.
  pacifica, 145, *147*.
Bulla, 484.
Bullaria gouldiana, 483, *484*, *485*.
Bullariidae, 483.
burchami (Planocera), 116, Pl. IV.
burrowing of Echiurus, 178.
burrowing of Glottidia, 155.
burrowing shrimp, 251, 327, *329*.
Bursa californica, *477*, *529*.
butterfly-shells, 557.
buttoni (Tellina), *447*, 448.
button-shell, netted, 504, *507*.

Cadlina flavomaculata, 493, Pl. IX.
  marginata, 493, Pl. IX.
Cadulus californicus, 475, *476*.
caeca, pyloric, 182.
Caecidae, 532.
Caecum, 477.
  californicum, *532*.
Calcarea, 38, 39.
caliculata (Orthopyxis), *57*.
California cone, *505*, *506*.
California Fish and Game, 259, 261, 299, 315, 352, 421, 422, 428, 570, 572, 573.
  sand star, *209*, 211, 212.
  spiny lobster, 315-325.
californiana (Orchestoidea), *276*, *277*.
  (Trivia), *524*, 528.
californianus (Mytilus), 427, *430*.
  (Tagelus), *456*, 457.
californica (Ammothella) biungulata, 404, *405*.
  (Ascidia), *593*.
  (Bursa), *477*, *529*.
  (Cerithidea), *530*, 531.
  (Cryptomya), *464*, 465.
  (Diopatra), 166.
  (Donax), *452*, 454.
  (Edwardsiella), *97*, Pl. III.
  (Eurythoe), *162*, 163.
  (Halosydna), 164.
  (Hippolysmata), *304*.
  (Idmonea), 138, *139*, *150*.
  (Lyonsia), 432, *433*.
  (Mactra), 460, *461*.
  (Marginella), 510, *512*.
  (Nuttallina), *560*, 562.
  (Parapholas), *469*, 470.
  (Planocera), 116, *117*.
  (Pleurophyllidia), *493*, 498.
  (Psammobia), *456*.
  (Ranella), 529.
  (Sabellaria), 172, *173*.
  (Scrupocellaria), 144.
  (Tethys), *486*, 487, *488*, *489*.
californicum (Amaroucium), 597.
  (Caecum), *532*.
  (Eudendrium), *49*, *67*.
californicus (Astropecten), *209*, 211.
  (Balanus) tintinnabulum, *260*, *264*, *265*.
  (Cadulus), 475, *476*.
  (Conus), *505*, *506*.
  (Ensis), 458.
  (Laqueus), *155*, 157.
  (Phacoides), 438, *439*.
  (Stichopus), 246, *247*, 248.
californiense (Branchiostoma), *600*.
californiensis (Callianassa), *328*, 329, *330*.
  (Chromodoris), 494, Pl. IX.
  (Hippolyte), 303, 304.
  (Speocarcinus), *389*.
Callianassa, 178, 251.
  affinis, *330*.
  californiensis, *328*, 329, *330*.
  longimana, *328*, 329, *330*.
Callianassidae, 326.
Calliostoma annulatum, *548*, 550.
  canaliculatum, *548*, 550.
  costatum, *548*, 550.
  tricolor, *548*, 550.
Callistochiton crassicostatus, *560*, 565.
  palmulatus, *560*, 564.
  palmulatus mirabilis, *560*, 565.
callomarginata (Lucapinella), *553*, 555.
callosa (Amiantis), *443*.
Calyptoblastea, 54.
Calyptraeidae, 538.
Cambridge Natural History, 77, 101, 228, 506.
camp nut-shell, 418, *419*.
Campanularia gelatinosa, *55*.
Campanulariidae, 55.
Campanulina forskalea, 69.
canal, of shell, *476*, 477.
canaliculata (Thais), *521*.
canaliculatum (Calliostoma), *548*, 550.
canals of sponges, 35.
cancellata (Trichotropis), *530*, 531.
cancellatus (Platyodon), *464*, 465.
Cancer amphioetus, 376, 378.
  antennarius, 376, *377*, 378, 395.
  anthonyi, 376, 379, *380*.
  gibbosulus, 376, 378.
  gracilis, 376, 378, 381, *382*.
  jordani, 376, *379*.
  magister, 352, 376, 379, *380*, 381.
  oregonensis, 5, 376, *382*, 383.
  productus, *359*, *375*, 376, 378.
Cancridae, 374.

INDEX

capax (Modiolus), 429, *430*.
Cape Flattery, 1.
capillata (Cyanea), 77.
capitatus (Pisaster) giganteus, 194, *195*.
capitulum, *263*.
Caprella aequilibra, 280, *281*.
    kennerlyi, *282*.
    scaura, *281*.
Caprellidea, 272, 279.
captacula, *474*.
carapace, 21, 249, 271.
Cardiidae, 440.
cardinal teeth, *416*.
Cardita subquadrata, *433*, 435.
Carditidae, 435.
carditoides (Petricola), 446, *447*.
Cardium corbis, *439*, 440.
    elatum, *437*, *440*, 441.
    quadragenarium, *439*, 440.
    substriatum, *439*, 441.
Carides, 21, 302.
carina, 262, *263*, *264*.
carinata (Columbella), *512*, 514.
    (Spirontocaris), 305, *306*.
cariosus (Balanus), *267*, *268*.
Carmel, 2.
carnulentum (Didemnum), *596*.
carpenteri (Cerithiopsis), *530*.
    (Leptothyra), *546*, 547.
    (Tellina), *447*, 448, *449*.
    (Triopha), 495, Pl. X.
carpenterianus (Cryptoconus), *505*, 506, *507*.
carpet-shell, ribbed, 445.
carpus, *355*.
cartridge-belt, 8.
caryi (Gorgonocephalus), *224*, 225.
cassis pelta (Acmaea), 542, *543*.
castrensis (Nucula), 418, *419*.
catalinensis (Menipea) occidentalis, 142, *150*.
"cat's eye," 43, 110.
caurina (Terebratalia) transversa, 157.
caurinus (Pecten), 424.
cavicauda (Hapalogaster), 337, *338*.
cells, adhesive, 64.
cellularia (Halistaura), *68*.
Cellulariidae, 142.
cementarium (Sabellaria), 173.
Cenobitidae, 325.
centipedes, 20.
Centrechinoida, 234.
Cephalaspidea, 483.
Cephalopoda, 413, 567.
    color changes of, 570.
cephalothorax, 251.
Ceramaster arcticus, 208.
    leptoceramus, 208, *209*.

cerata, 503.
Ceratium, *30*.
cercariae, 120.
Ceriantheae, 97, 105.
Cerianthidae, 105.
Cerianthus, *97*.
    aestuari, 105, *106*.
Cerithidea californica, *530*, 531.
    hyporhysa, *530*.
Cerithiidae, 531.
Cerithiopsidae, 530.
Cerithiopsis carpenteri, *530*.
    columna, *530*.
Cestida, 110.
Cestoda, 114, 121.
Chaetognatha, 23, 114, 128.
Chaetopoda, 158.
Chalinidae, 40.
chalk beds, 28.
Challenger Expedition, 300, 397.
Chama, *435*.
    pellucida, *436*.
Chamidae, 435.
cheiragonus (Telmessus), 383, *384*.
chelifera, 284.
Chelyosoma productum, *594*.
Chilostomata, 141.
Chimaera colliei, 121.
chink-shell, one-banded, *535*, 536.
    wide, *535*.
Chione fluctifraga, 445, *447*.
    succincta, *416*, *444*, 445.
    undatella, *412*, *444*, 445.
chitin, 251.
chitinoides (Psolus), 243, *245*.
chiton, 5, 411, 413, 557.
    black, *560*, 566.
    California, *560*, 562.
    conspicuous, *557*, *563*, 564.
    egg laying of, *559*, 562.
    feeding habits of, 557.
    giant, 566, 567.
    gray, 562, *563*.
    hairy, *563*, 565.
    lined, *560*, 561.
    methods of preserving, 603.
    mossy, *560*, *563*, 565.
    palm, *560*, 564.
    red, *560*, 564.
    regular, *560*, 564.
    thick-ribbed, *560*, 565.
    veiled, *560*, 566.
Chlorhaemidae, 167.
Chloridella (Squilla) polita, 296, *297*.
Chordata, 23, 582.
chordates, 17.
Chromodoris californiensis, 494, Pl. IX.
    macfarlandi, 494, Pl. XI.

638 INDEX

porterae, 494.
chronhjelmi (Cucumaria), 244, 245.
Chrysaora, food of, 78.
　gilberti, 81, Pl. I.
Chrysaora melanaster, 80, 81.
Chrysodomidae, 511.
Chrysodomus, liratus, 511.
　tabulatus, 509, 511.
Chthamalidae, 269.
Chthamalus dalli, 269, 270.
　fissus, 266, 267, 269, 270.
cilia, 113, 115, 127, 132, 153, 185, 414, 480.
　of sea anemone, 87.
　of sea urchins, 229, 230.
ciliata (Mopalia), 563, 565.
ciliates, 29.
cinctipes (Petrolisthes), 351.
cinereus (Urosalpinx), 518, 521.
Ciona, 365, 585.
　intestinalis, 594, frontispiece.
circularis (Pecten), 423, 425.
circumtexta (Tritonalia), 515, 517.
Cirolana harfordi, 285.
Cirratulidae, 163, 166.
　cirratulids, 166.
cirri, 159.
Cirripedia, 20, 257.
Cladohepatica, 498.
clam, 2, 3, 124, 391, 395, 412, 413.
　bean, 452, 454.
　bent-nosed, 447, 449.
　butter, 443, 444.
　canneries, 455.
　flat, 450, 453.
　hard-shell, 445.
　horse, 461.
　little-neck, 445.
　long, 463, 464.
　mud, 463, 464.
　purple, 456.
　razor, 458, 459.
　rock, 445.
　"Rubber neck," 6.
　soft-shell, 463, 464.
　summer, 461, 462.
claperedii (Arenicola), 169.
Clark, H. L., 218, 225, 246, 247.
clarki (Gonothyraea), 58, 59.
class, 19.
classification, 14, 17.
　difficulties of, 273.
Clathrodrillia incisa, 505, 508.
claviformis (Euherdmania), 597.
Clione kincaidi, 481, 482.
　limacina, 481.
Clionidae, 482.
Clymenella, 392.

rubrocincta, 168.
Clypeastroida, 235.
Clytia bakeri, 53, 55.
　inconspicua, 57.
cockerelli (Laila), 494, Pl. IX.
cockle, 440, 441.
　forty-ribbed, 439, 440.
　giant, 437, 440, 441.
　hard-shell, 445.
　rock, 445, 447.
　Tomales Bay, 445.
Coe, W. R., 124, 125.
Coelenterata, 22, 42.
coerulescens (Aequorea), 69.
coffee-bean, 524, 528.
Cole, L. J., 404.
Colidotea rostrata, 291.
collar cells, 34.
collecting equipment, 10, 602.
colliei (Chimaera), 121.
color changes of cephalopods, 570.
Columbella carinata, 512, 514.
columbella (Erato), 524, 528.
Columbella gausapata, 512, 514.
Columbellidae, 514.
columbiae (Linckia), 191, 210.
columbiana (Amphissa), 514, 515.
　(Orthasterias), 199.
　(Serpula), 170, Pl. VII.
columella, 476, 477.
columna (Cerithiopsis), 530.
"comb jellies," 43, 109.
commensal, 123, 164, 392, 393, 394.
commercial sponges, 37.
commissuralis (Obelia), 59.
common names, 14.
compound ascidians, 594.
　eye of lobster, 251, 324.
compressa (Orthopyxis), 58.
compta (Phasianella), 545, 546.
conchologist's paradise, 6.
cone, California, 505, 506.
Conidae, 506.
conspicuus (Ischnochiton), 557, 563, 564.
continental shelf, 26, 30.
Conus californicus, 505, 506.
Cooke, A. H., 506.
Cooperella subdiaphana, 447, 448.
Cooperellidae, 448.
cooperi (Ischnochiton), 563, 564.
　(Turritella), 530, 533.
　(Yoldia), 419.
copepod, parasitic, 255, 486.
Copepoda, 20, 31, 252.
coral, 8, 42, 43, 85, 105-109, 131, 139, 325, 335, Pl. III.
　gorgonian, 88, 94, 95, 96, 138.

INDEX 639

islands, 107.
reefs, 107.
stony, 95.
corallines, 131, *150.*
corbis (Cardium), *439,* 440.
corbula, 65, 66.
corbula luteola, 465, *467.*
Corbulidae, 465.
cordiformis (Lovenia), *236,* 238.
corniculata (Orchestoidea), 276, *277.*
corrugata (Haliotis), *553,* 554.
Corymorpha, 2, 8, 10.
  keeping alive in aquaria, 601.
  palma, 46, *50,* 51, *53,* 105.
Corymorphidae, 50.
Corynactis, 105, Pl. III.
costatum (Calliostoma), *548,* 550.
Cotylea, 118.
cowry, nut-brown, *524,* 528.
coxa, *355.*
crab, 2, 5, 8, 11, 33, 249, 252, 314.
  big, 379, *380.*
  black-clawed, 387.
  box, 338, 340, *346.*
  burrowing, *389.*
  butterfly, 339.
  commensal, 390-395.
  dwarf, *367, 369.*
  edible, 352, 376, 379, *380,* 381.
  fishing, 352, 353, 381.
  flat, 1. 6, *350.*
  graceful decorator, *364,* 365.
  great spider, 361.
  hairy, *388,* 389.
  hairy shore, *396.*
  horse, 383, *384.*
  horseshoe, 20.
  kelp, *366,* 367.
  lumpy, *388,* 389.
  moss, *373,* 374.
  mud, *396.*
  narrow-mouthed, *362.*
  narrow-nosed, 363.
  pelagic, *397.*
  purple, 363, *364.*
  purple shore, *356,* 395, 396.
  red, *375,* 378.
  rock, *377,* 378, 395.
  sharp-nosed, 374, *375.*
  short-jawed, *362.*
  short-nosed, 363.
  striped shore, *353, 354,* 355, *357, 358, 360, 361,* 395, *398,* frontispiece.
  swimming, 384, *385.*
  thick-clawed, 349, *350.*
  true, 352-402.
  yellow shore, *396.*

crabs, preserving, 603.
cracherodii (Haliotis), 552, *553.*
Crago alaskensis elongata, 313.
  franciscorum, 16, 299, *313.*
  munita, 314.
  munitella, 314.
  nigricauda, 299, *311, 312.*
  nigromaculata, 299, 313.
Cragonidae, 312.
Crangon, 16.
  (Alpheus) dentipes, 308, *309, 311.*
  bellimanus, 309.
  equidactylus, 309.
Crangonidae, 302, 308, 312.
crassicornis (Hermissenda), 502, frontispiece.
crassicostatus (Callistochiton), *560,* 565.
crassipes (Pachygrapsus), *353, 354, 355, 357, 358, 360, 361, 362,* 395, *398,* frontispiece.
crawfish, 13, 327.
crayfish, 13, 21, 314, 327.
crayfish, Pacific coast, 13.
  salt water, 13.
crebricinctum (Micranellum), *532.*
crenatus (Balanus), *266, 267.*
crenimarginatum (Epitonium), 522, *524.*
crenulata (Megathura), 555, *556.*
  (Uca), *398,* 399-402.
Crepidula aculeata, *537,* 538.
  adunca, *537,* 538.
  excavata, *537,* 538.
  nummaria, *537,* 538.
  onyx, *537.*
Crepidulidae, 537.
Cribrina xanthogrammica, *99,* Pl. IV.
Cribrinidae, 98, 99.
Crinoidea, 181, 225.
crinoids, 225.
Crisia, 134, 140.
Crisia eburna, 137.
  edwardsiana, 136.
  geniculata, *135, 136.*
  maxima, *136,* 137, *146.*
  occidentalis, 137.
  pugeti, 136.
Crisiidae, 135.
crispatus (Loxorhynchus), *373,* 374.
cristata (Ancula), 490.
  (Spirontocaris), *306,* 308.
Crisulipora occidentalis, *139,* 140.
crocea (Tubularia), 52, *53.*
Crocker, G. R., 201.
Crossaster papposus, *204,* 205.
Crucibulum spinosum, *537,* 538.
crucifera (Fissurella) volcano, *553,* 555 .
Crustacea, 19, 20, 250.
crustaceans, 29, 250-402.

Cryptochiton, 558.
  stelleri, 164, 566, 567.
Cryptochitonidae, 566.
Cryptoconus carpenterianus, 505, 506, 507.
Cryptolithodes sitchensis, 338, 339.
  typicus, 339.
Cryptomya californica, 464, 465.
ctenidia, 479, 483.
ctenidium, 488.
Ctenobranchiata, 504.
Ctenophora, 43, 109.
ctenophores, 70, 109, 110.
Ctenostomata, 151.
Cucumaria chronhjelmi, 244, 245.
  curata, 244, Pl. VII.
  lubrica, 243, 245.
  miniata, 5, 244, 245.
Cucumariidae, 243.
Cumacea, 20.
Cumingia lamellosa, 452, 454.
cup and saucer limpet, 537, 538.
curata (Cucumaria), 244, Pl. VII.
Cuvier, G., 162.
Cyanea capillata, 77, 81.
  capillata ferruginea, 81, 82.
  capillata postelsii, 82.
Cyaneidae, 81.
Cyclosalpa affinis, 587, 589.
Cyclostomata, 135.
Cycloxanthops novemdentatus, 385, 386.
Cydippida, 110.
Cylindroleberis sp., 256.
  oblonga, 256.
Cymatiidae, 530.
Cymothoidea, 284.
Cypraea spadicea, 524, 528.
Cypraeidae, 528.
Cypridinidae, 256.
cypris larva, 257.

dactyl, 355.
Dahlgren, Ulric, 71, 79.
daily migrations of chaetognaths, 128.
Dall, W. H., 157, 411, 418, 434, 435, 472, 519, 520, 525.
dalli (Chthamalus), 269, 270.
  (Pugettia), 368, 369
damage done by shipworms, 471.
Dana, 108.
danae (Pandalus), 302, 303.
Darwin, 108.
davisii (Sagartia), 102.
dawsoni (Solaster), 198, 205.
Decapoda, 19, 20, 21, 249, 298, 579.
decisa (Semele), 450, 453.
dehiscens (Lima), 425, 426.

Demospongiae, 38, 39.
Dendraster excentricus, 235, 236.
Dendronotidae, 499.
Dendronotus arborescens, 500.
  giganteus, 499, 500.
Dentaliidae, 475.
Dentalium, diagram of organs, 474.
  neohexageonum, 475.
  pretiosum, 475.
dentipes (Crangon), 308, 309, 311.
depicta (Acmaea), 543, 545.
Dermasterias imbricata, 207, 208.
destruction of piling by Limnoria, 286.
destructor (Sphaeroma), 287.
devil fish, 11, 567, 577, 578.
Diadora aspera, 555, 556.
  murina, 553, 555.
dianthus (Metridium), 102, 103, 104.
diatoms, 28, 31.
Diaulula sandiegensis, 492, 493.
Dibranchiata, 575.
dichotoma (Obelia), 59, 60.
Didemnum carnulentum, 596.
diegensis (Botrylloides), 595.
  (Lophopanopeus), 387.
  (Scrupocellaria), 143, 146.
  (Teredo), 473.
digitale (Aglantha), 73.
dinoflagellates, 30-33.
  unarmored, 30.
Dinophysus homunculus, 30.
Diopatra, 166.
  californica, 166.
Diplodonta orbella, 436, 437, 439.
Diplodontidae, 436.
dira (Searlesia), 512, 513.
Dirona picta, 499.
Dironidae, 499.
Discophora, 80.
Distaplia occidentalis, 596.
distribution of marine animals, 1.
Dolichoglossus pusillus, 584.
Doliolum ehrenbergii, 587.
  tritonis, 586, 587, 588.
Donacidae, 454.
Donax, 55, 120, 122.
  californica, 452, 454.
  gouldii, 53, 452, 454.
Dorididae Cryptobranchiatae, 491.
  Phanerobranchiatae, 494.
Doridopsidae, 498.
Doriopsis fulva, 498, Pl. XI.
Dosidicus gigas, 581.
dove-shell, 512, 514.
  keeled, 512, 514.
draconis (Polynices), 540, 541.
dredge, 25.
drill, doleful, 508.

INDEX

incised, *505*, 508.
oyster, 518, *521*.
drobachiensis (Strongylocentrotus), 233, 234.
Dromiacea, 361, 362.
dura (Metandrocarpa), 595.

earbones of whales, 28.
ear-shell, olive, 504, *505*.
eastern oyster, *415*, 422.
eburnea (Crisia), 137.
Echinarachnius parma, 237.
Echinasteridae, 203.
echinata (Paracrangon), 314.
(Pedicellina), 134.
Echinodermata, 23, 180.
Echinoderms, 42, 180-248.
Echinoidea, 181, 227.
Echiurus, *178*.
pallassii, 178.
Ectoderm, *44*.
Ectoprocta, 134, 135.
Edmondson, C. H., 422.
edulis (Mytilus), 428, *430*.
Edwardsiae, 97.
edwardsiana (Crisia), 136.
Edwardsiella, *97*.
californica, 97, Pl. III.
egg collars of Polinices, 540, *541*, 542.
egg laying of amphipods, 272.
of bivalves, 418.
of cephalopods, 574, 575.
of chitons, 559.
of Gonionemus, 73.
of lobster, 325.
of pycnogonids, 404.
of sea urchins, 230.
of worms, 160.
egg-pits of sea anemone, 101.
egg-shell cockle, *439*, 441.
eggs of Argobuccinium, 5.
of atlantid, *527*.
of Bullaria, *484*.
of Corymorpha, 51.
of crabs, 124, 356.
of medusae, 46.
of mollusks, *480*, *481*, *485*.
of Navanax, *480*, *485*, 486.
of Planocera, 117.
of sea pansy, 93.
of squid, 580.
of Thais, 520.
ehrenbergii (Doliolum), *587*.
elatum (Cardium), *437*, *440*, 441.
elegans (Balanophyllia), *108*, *109*, Pl. III.
Ellobiidae, 503.
elongata (Crago) alaskensis, 313.

(Stylatula), *91*.
emarginata (Thais), *521*.
Emerita, 184, 250.
analoga, 325, *338*, 341-347.
talpoida, 347.
Emplectonema gracile, *124*, Pl. VI.
Endoderm, 44.
endoskeleton, 17.
endostyle, *596*.
Engraulis mordax, 428.
ensifera (Yoldia), *419*.
Ensis californicus, 458.
Enteropneusta, 583.
Entodesma saxicola, *433*, 434.
Entomostraca, 20, 252.
Entoprocta, 134.
Epiactis prolifera, *99*, 101.
Epialtus nuttallii, *366*, 367.
productus, *366*, 367.
epistome, 21, 130.
epithelial cells, 35.
Epitoniidae, 522.
Epitonium bellastriatum, 523, *524*.
crenimarginatum, 522, *524*.
fallaciosum, 523, *524*.
indianorum, 523.
wroblewskii, 522, *524*.
equidactylus (Crangon), 309.
equipment for collecting, 602.
Erato columbella, *524*, 528.
vitellina, *524*, 528.
erectus (Anoplodactylus), *407*, *408*.
eriomerus (Petrolisthes), *350*, 351.
eschrichtii (Bittium), *530*, 531.
escutcheon, 418.
esmarki (Ophioplocus), *216*, 218.
Esperella adhaerens, 39.
Esperellidae, 39.
esquimalti (Astarte), *433*, 435.
Essenberg, C., 162.
Esterly, C. O., 253, 295.
eucnemis (Gorgonocephalus), *224*, 225.
Eudendriidae, 49.
Eudendrium californicum, *49*, *67*.
rameum, 50, *67*.
Eudistylia polymorpha, 169, Pl. VII.
Euherdmania claviformis, *597*.
Eulima, 525.
Eunice, 160.
Eunicea, 94, *96*.
Euphausia pacifica, *294*.
Euphausiacea, 293.
euphausids, 294.
Euphausiidae, 294.
Euplectella, 39.
Euplexaura marki, *94*, *95*.
Eupomatus, 170, *171*.
Eupsammidae, 109.

Euryalae, 217, 224.
Eurylepta aurantiaca, 118, Pl. V.
Euryleptidae, 118.
Eurythoe californica, *162*, 163.
Evasterias troschelii, *195*, 197.
everta (Orthopyxis), *58*.
excavata (Crepidula), *537*, 538.
excentricus (Dendraster), 235, *236*.
exigua (Janthina), 523, *524*.
exogyra (Pseudochama), *436*.
exoskeleton, 17, 249.
Exosphaeroma amplicanda, *287*, 288.
  oregonensis, 288.
eyes of flatworms, 115, 117.
  of gastropods, 477, *478*.
  of jellyfish, 78.
  of pecten, 422, *423*, *425*.
  of starfish, 190.

faba (Pinnixa), 393.
Fabia lowei, 391.
  subquadrata, 391.
Fabricius, 16.
fairy palm, *50*, 51, *53*.
falcata (Botula), *430*, 431.
fallaciosum (Epitonium), 523, *524*.
False Bay, *2*.
false-teeth shells, 557.
family, 19.
fascicularis (Lepas), 262, *263*.
Fasciolariidae, 511.
feather stars, 226.
feelers, 251.
fenestrata (Iselica), *535*, 536.
ferruginea, Cyanea capillata, 81, 82.
fertilization of sea urchin eggs, 231.
festiva (Tritonia), 491, Pl. VII.
festivus (Murex), *515*, 516.
fiddler crab. 2, *398*, 399–402.
file shell, *425*, 426, *430*.
filicula (Abietinaria), 61, *62*, *67*.
filosa (Mitromorpha), 510, *512*.
  (Pandora), 432, *433*.
fimbriata (Gyrocotyle), 121.
fish, 33.
fish lice, 254.
Fisher, W. K., 7, 203, 206.
fission in sea anemone, 102, 103, 104.
Fissurella volcano, *553*, 554.
  crucifera, *553*, 555.
Fissurellidae, 554.
fissus (Chthamalus), 266, *267*, 269, *270*.
flabellaris (Tubulipora), 138.
Flabellifera, 284.
Flabellina iodinea, 503, frontispiece.
flagella, 33, 35.
flagellate cells, 35.
"flame cells," 115.

flatworm, striped, 118, Pl. V.
flat worms, 113, 114, Pls. IV, V.
flavescens (Lineus), *122*, 125.
  (Magellania), *153*.
flavomaculata (Cadlina), 493, Pl. IX.
"floating meadows," 31.
fluctifraga (Chione), 445, *447*.
flukes, 119.
foliata (Purpura), 516, *518*.
foliatus (Mimulus), *367*, 368.
foliolata (Luidia), *212*.
food of anemone, 101.
  of animals, 26, 30.
  of chitons, 557.
  of clams, 414.
  of Corymorpha, 51.
  of crabs, 360.
  of fiddler crab, 400.
  of fish, 30, 31.
  of flatworms, 115.
  of Gonionemus, 72.
  of lobster, 319, 321, 322.
  of oysters, 414.
  of pycnogonids, 403.
  of sand crabs, 343.
  of sponge, 36.
  of squid and octopus, 571.
  of starfish, 187, 188.
foraminatus (Lopholithodes), *338*, 340, *346*.
forcipiform pedicellariae, *184*, 190.
Forcipulata, 184, 190, 194.
forficiform pedicellaria, *184*, 190.
formalin, 603.
fornicatus (Modiolus), *430*, 431.
Forskal, 71.
forskalea (Campanulina), 69.
forskalii (Beroe), *112*.
Forsyth, Ruth A., 582, 591, 593, 596, 597.
Fossaridae, 536.
fossata (Alectrion), *512*, 513.
fossil bryozoans, 131.
  cephalopods, 575.
  coral reefs, 107.
  crabs, 372.
  crinoids, 226.
  Donax, 455.
  jellyfish, 42.
  scallops, 424.
franciscanus (Strongylocentrotus), *233*, 234.
franciscorum (Crago), 299, *313*.
Frazer, C. McL., 48, 55, 59, 61, 161.
fresnelii (Melita), *275*.
Friday Harbor, 4, 5, 10, 68, 69.
frog shell, *477*, *529*.
frontalis (Lophopanopeus), 387.
fry, 421.

INDEX 643

Frye, Mrs. T. C., 57.
Fucus, 147.
fulgens (Haliotis), 553, 554.
fulva (Doriopsis), 498, Pl. XI.
funebralis (Tegula), 544, 547, 548.
fusiformis-runcinata (Salpa), 588, 589.
Fusinus kobelti, 505, 511.

gabbi (Zirfaea), 467. 468.
Gadinia reticulata, 504, 507.
Gadiniidae, 504.
gallina (Tegula), 548, 549.
Galloway, T. W., 161.
Galvina olivacea, 53, 502, frontispiece.
Gammaridae, 274.
Gammaridea, 271, 272.
gaper, 6, 394, 398, 461, 462.
Garveia annulata, 48.
gastric mill, 322.
Gastropoda, 413, 475.
gastrovascular cavity, 44.
gastrozooid, 75.
gausapata (Columbella), 512, 514.
Gee, Wilson, 100.
gelatinosa (Campanularia), 55.
gem shell, 446.
Gemma gemma, 446.
gemma (Murex), 515, 516.
gemmules, 37.
generosa (Panope), 465, 467.
geniculata (Crisia), 135.
  (Obelia), 60.
genital bursae, 213, 214.
genus, 15, 19.
geoduck, 465, 467.
Gephyrea, 23, 173.
ghost shrimp, 178, 325, 326.
gibbosa (Asterina), 206.
gibbosulus (Cancer), 376, 378.
gibbsii (Styela), 593.
gigantea (Lottia), 543, 545.
giganteum (Pyrosoma), 598.
giganteus (Dendronotus), 499, 500.
  (Hinnites), 423, 425.
  (Pisaster), 194.
  (Saxidomus), 443.
gigas (Dosidicus), 581.
gilberti (Chrysaora), 81, Pl. I.
gills, 16, 322, 323, 417.
glandula (Balanus), 266, 267.
glass-bottomed boats at Avalon, 26.
glaucothoe of hermit crab, 332.
Globigerina, 26, 31.
globigerina-ooze, 26, 28.
Glossophorum planum, 597.
Glottidia albida, 153, 154.
glycerine in preserving fluids, 603.
gnathopod, 271, 273, 274, 276.

Goneplacidae, 389.
Goniasteridae, 208.
Gonionemus murbachii, 72.
  vertens, 67, 71, 72.
gonotheca, 54.
Gonothyraea clarki, 58, 59.
Gonozooid, 586-588.
Gonyaulax polyedra, 33.
goose barnacles, 257, 259-263.
gopher, 13.
Gorgonaceae, 94.
gorgonian corals, 43, 88, 94, 95, 96.
Gorgonocephalus caryi, 224.
  eucnemis, 225.
gouldiana (Bullaria), 483, 484, 485.
gouldii (Donax), 53, 452, 454.
gracile (Emplectonema), 124, Pl. VI.
  (Stylatula), 92.
gracilis (Cancer), 376, 378, 381, 382.
  (Oregonia), 364, 365.
  (Pugettia), 368.
gracillima (Tritonalia), 515, 517.
grandis (Loxorhynchus), 371, 372.
granosimanus (Pagurus), 334, 335, 336.
Grantia, 39.
Grantiidae, 39.
granulata (Scleroplax), 394.
Grapsidae, 395.
green urchin, 233, 234.
gregarium (Phialidium), 56, 57.
gribble, 285.
gulfweed, 148, 397.
gurneyi (Pandalus), 297, 303.
Gymnoblastea, 47.
Gymnolaemata, 135.
Gymnosomata, 482.
Gyrocotyle fimbriata, 121.

haemoglobin, 159.
hair worms, 126.
hairy-shell, checked, 530, 531.
Halichondria panicea, 40.
Haliotidae, 551.
Haliotis, 470.
  corrugata, 553, 554.
  cracherodii, 552, 553.
  fulgens, 553, 554.
  rufescens, 552, 553.
  wallalensis, 554.
Halistraura cellularia, 68, 70.
Hall, Ivan C., 366.
Halocynthia haustor, 592.
  johnsoni, 592.
Halosoma viridintestinalis, 406, 407.
Halosydna californica, 164.
  insignis, 164.
  lordi, 164.
  pulchra, 163.

644 INDEX

hamata (Leda), 418, *419*.
Haminoea, 486.
vesicula, 484, *485*.
virescens, 484, *485*.
Hapalogaster cavicauda, 337, *338*.
mertensii, 6, 337, *338*.
Harenactis, 97, 175.
attenuata, *98*, 105.
harfordi (Betaeus), *310*.
(Cirolana), *285*.
(Synidotea), 290, *291*.
hartwegii (Lepidochitona), *560*, 561, *563*.
hastatus (Pecten), 424.
Haswell, W. A., 22.
haustor (Halocynthia), 592.
heart shells, *433*, 435, 440.
urchin, 232, *236*, 238.
Heath, H., 116, 117, 259, 428, 557, 559, 562, 570, 571, 572.
heathii (Lophopanopeus), 386, 387, *388*.
helga (Odostomia), 525, *530*.
helianthoides (Pycnopodia), 192, *201*, *202*.
heliozoan, *32*.
Hemigrapsus, 395, 397.
nudus, *356*, 395, *396*.
oregonensis, *362*, *396*.
hemphilli (Pagurus), 334, *336*.
Henricia leviuscula, *202*, 203.
Herdman, W. A., 490.
hericius (Pecten), *423*, 424.
Hermellidae, 172.
Hermissenda crassicornis, 502, frontispiece.
hermit crab, 21, 40, 314, 325, 331.
Heterocoela, 39.
Heterocrypta occidentalis, 363, *364*.
Heterodonax bimaculata, *456*, 457.
Heteronemertea, 125.
heteropods, 479.
Hexactinellida, 38, 39.
Hexactiniae, 97, 98.
hexactis (Leptasterias), 197, *198*, 199.
hexagonal tooth shell, *475*.
Hickson, S. J., 77, 101.
hilli (Lepas), 262, *263*, 267.
Hilton, W. A., 310, 408.
hindsii navarchus (Pecten), *423*, 424.
Hinnites giganteus, *423*, 425.
Hippidae, 341.
Hippolysmata californica, *304*.
Hippolyte californiensis, 303, 304.
Hippolytidae, 303.
Hipponicidae, 536.
Hipponix antiquatus, *535*, 536.
tumens, *535*, 536.
Hippothoa hyalina, 149, *150*, *151*.

Hippothoidae, 149.
hirsutiusculus (Pagurus), 334, *335*.
Hjort, J., 482.
Holder, Chas. F., 576.
Holmes, S. J., 310, 312, 330, 384, 392, 393.
Holohepatica, 491.
Holothuriidae, 246.
Holothurioidea, 181, 238.
Homarus, 318.
homunculus, Dinophysus, *30*.
hongkongensis (Polypus), 578.
hoof-shell, ancient, *535*, 536.
sculptured, *535*, 536.
hookworm, 127.
Hopkins Marine Laboratory, 6, 7.
Hopkinsia rosacea, 497, Pl. XI.
Hoplonemertea, 124.
horn-shell, California, *530*, 531.
hornmouth, leafy, 516, *518*.
Nuttall's, *515*, 516.
Hugo, Victor, 568.
humeralis (Ampithoe), *277*, 278, *279*.
Huxley, T. H., 259.
hyalina (Hippothoa), 149, *150*, *151*.
Hyas lyratus, *370*, 372.
hydra, 46.
hydranths, 44.
hydroid, 43, 131.
ostrich-plume, *66*, *67*.
small ostrich-plume, *65*, *67*.
hydroids, plume, 63.
Hydromedusae, 43, 77.
hydrotheca, *44*, 47, 54.
Hydrozoa, 43, 77.
Hyperiidae, 271.
hyporhysa (Cerithidea) californica, *530*.

idae (Strigatella), *505*, 510.
Idmonea californica, 138, *139*, *150*.
Idothea ochotensis. 290.
rectilinea, *289*.
urotoma, *289*.
Idotheidae, 289.
Ilyanthidae, 98.
imbricata (Dermasterias), *207*, 208.
Inachidae, 365.
Inachoides tuberculatus, *362*, *364*, 365.
inaequalis (Astraea), *546*.
incisa (Clathrodrillia), *505*, 508.
inconspicua (Aglaophenia), *65*, *67*.
(Clytia), *57*.
indentata (Macoma), *447*, 451.
Indians, food of, 540, 566.
Indian money, 474.
ornaments, 508.
tooth shell, *475*.
wentletrap, 523.

INDEX 645

indianorum (Epitonium), 523.
Inermia, 175.
inermis (Navanax), 485, 486, Pl. VIII.
 (Oedignathus), 337.
Infusoria, 127.
inhaerens (Leptosynapta), 241, *242*, *245*.
inquinata (Macoma), *450*, 451.
insect, 20, 249, 250.
Insecta, 20.
insessa (Acmaea), *543*, 544.
insignis (Halosydna), 164.
instabilis (Acmaea), *543*, 544.
interfossa (Tritonalia), *515*, 517.
intermedium (Tanystylum), 405, *406*.
interruptus (Panulirus), 315-325.
intestinalis (Ciona), 594, frontispiece.
introvert, 173, 174.
iodinea (Flabellina), 503, frontispiece.
ischium, *355*.
Ischnochiton conspicuus, *557*, *563*, 564.
 cooperi, *563*, 564.
 magdalenensis, 558, 562, *563*, 564.
 mertensii, *560*, 564.
 regularis, *560*, 564.
Ischnochitonidae, 562.
Iselica fenestrata, *535*, 536.
Isopoda, 20, 282.
Isthmus of Panama, closure of, 85.

jackknife clam, *456*, 457, 458, *459*.
Janidae, 500.
Janthina, 3, 397.
 exigua, 523, *524*.
Janthinidae, 523.
jellyfish, 3, 8, 10, 31, 42, 77-85, 397, Pls. I, II.
 as food, 77.
 largest, 77, 81.
 life-history of, 77.
Jennings, H. S., 182, 184.
jewettii (Marginella), *509*, 510.
jingle, *426*, 427.
Johnson, M. E., 582, 589.
johnsoni (Halocynthia), *592*.
 (Stichopus), 246-248.
Jordan, D. S., 599.
jordani (Cancer), 376, *379*.

Katharina, 557.
 tunicata, *560*, 566.
Keep, Josiah, 411, 436, 483, 536, 566.
Kellettia kellettii, *507*, 511.
kellettii (Siphonalia), 511.
Kellia laperousi, *437*, 438.
kelp, 26, 137, 138, 141, *150*, 172.
Kelsey, F. W., 411, 453, 507, 512, 519, 535, 543, 544, 553.

kennerlyi (Caprella), *282*.
key to phyla of animals, 22.
keyhole limpets, 395, 553-557.
keys, 18-24.
Kincaid T., 327, 540.
kincaidi (Clione), *481*, 482.
king crab 326, 340.
kingdom, 18.
kobelti (Fusinus), *505*, 511.
Kofoid, C. A., 33, 286, 472, 473.

"lace-coral," *150*.
laciniata (Paphia) staminea, *444*, 446.
Lacuna porrecta, *535*.
 unifasciata, *535*, 536.
Lacunidae, 535.
ladder-shell, 522, *524*.
Laila cockerelli, 494, Pl. IX.
Lambe, L. M., 38, 39, 40.
Lamellibranchiata, 413.
lamellifera (Venerupis), 446, *447*.
lamellosa (Cumingia), *452*, *454*.
 (Thais), 519, 520, *521*.
lamp shells, 152-157.
lancelet, 599, *600*.
Languste, 13.
laperousi (Kellia), *437*, 438.
Laqueus californicus, *155*, 157.
Larvacea, 585.
lata (Velella), 75, 76, Pl. I.
latiauritus (Pecten), 424, *425*.
laticauda (Synidotea), 291.
latus (Suberites), 40.
leather star, *207*, 208.
Le Conte, J., 420.
Leda hamata, 418, *419*.
 taphria, 418, *419*.
Ledidae, 418.
Loedicidae, 165.
leonina (Melibe), *500*, 501.
Lepadidae, 262.
Lepadomorpha, 259.
Lepas anatifera, 262.
 anserifera, 262, 263.
 fascicularis, 262, *263*.
 hilli, 262, *263*, *267*.
 pectinata, 262, 263.
Lepidochitona (Tonicella) lineata, *560*, 561.
 (Trachydermon) hartwegii, *560*, 561, 563.
 (Trachydermon) raymondi, *560*, 561.
Lepidochitonidae, 561.
Lepidopa myops, *346*, *348*, 349.
Leptasterias aequalis, *198*, 199.
 hexactis, 197, *198*, 199.
Leptocardia, 583, 599.
leptoceramus (Ceramaster), 208, *209*.

Leptomedusae, 44, 47, 54.
Leptonidae, 438.
Leptosynapta inhaerens, 241, 245.
Leptothyra bacula, 546, 547.
　carpenteri, 546, 547.
lessonii (Pseudosquilla), 296.
leucomanus (Lophopanopeus), 386.
Leucosiidae, 363.
leviuscula (Henricia), 202, 203.
lewisii (Polinices), 539, 541.
Lichenopora radiata, 139, 140, 150.
Lichenoporidae, 140.
life-history, barnacle, 257.
　crab, 356.
　hermit crab, 332.
　Limnoria, 286.
　lobster, 317-319, 325.
　oyster, 420, 421.
　pycnogonid, 408.
　sand crab, 344, 345.
　shrimp, 301.
　trematode, 120.
ligament, 416.
light production in coelenterates, 70, 79.
　in jellyfish, 70, 79.
light, reaction of Cribrina to, 100.
Light, S. F., 600.
lignorum (Limnoria), 285, 286.
ligulata (Tegula), 548, 550.
Ligyda occidentalis, 292.
　pallasii, 292, 293.
Lima dehiscens, 425, 426.
lima (Thais), 520, 521.
limacina (Clione), 481.
limatula (Acmaea), 542, 543.
　(Yoldia), 419, 420.
limestone, 28.
Limidae, 426.
Limnoria, 471.
　lignorum, 285, 286.
limpet, 5, 412, 413, 475.
　black seaweed, 543, 544.
　chaffy, 543, 545.
　cup and saucer, 537, 538.
　file, 542, 543.
　giant keyhole, 555, 556.
　mask, 543, 544.
　painted, 543, 545.
　plate, 542, 543.
　ribbed, 543, 544.
　rough keyhole, 555, 556.
　seaweed, 543, 544.
　shield, 542, 543.
　two-spotted keyhole, 553, 555.
　unstable seaweed, 543, 544.
　white keyhole, 553, 555.
Linckia, 8.
　columbiae, 191, 192, 210.

lineata (Lepidochitona), 560, 561.
lingual ribbon, 478.
Lingulidae, 153.
Linnaeus, 14.
Linné, 14.
Lineus flavescens, 125.
　pictifrons, 126, Pl. VI.
liratus (Chrysodomus), 511.
lirulata (Margarites), 551.
lithocyst, 44, 54, 58, 71, 417.
Lithodidae, 326, 337.
Lithophaga plumula, 430, 431.
littoralis (Pinnixa), 393, 398.
Littorina planaxis, 534, 535.
　scutulata, 534, 535.
　sitchana, 534.
littorine, checkered, 534, 535.
　gray, 534, 535.
　Sitka, 534.
Littorinidae, 534.
lituellus (Spiroglyphus), 533.
lividus (Macron), 505, 511.
Lloyd, F. E., 431, 466.
Lobata, 110, 111.
lobster, 13, 21, 249, 252, 314.
　California spiny, 8, 13, 315-325.
　dissection of, 320, 321.
　industry, 315-320.
　preserving, 603.
　traps, 315, 316.
locusta, 13.
Loeb, Jacques, 231.
Lolinginidae, 580, 581.
Loligo opalescens, 579, 580.
longidactylus (Betaeus), 310, 311.
longimana (Callianassa), 328, 329, 330.
longipes (Pinnixa), 392, 393.
longissima (Obelia), 53, 61, 502.
looping-snail, California, 535, 536.
Lopholithodes foraminatus, 338, 340, 346.
　mandtii, 340.
Lophopanopeus bellus, 387, 388.
　diegensis, 387.
　frontalis, 387.
　heathii, 387, 388.
　leucomanus, 386.
lophophore, 130, 152, 153, 156.
lordi (Halosydna), 164.
Lottia gigantea, 543, 545.
Lovenia cordiformis, 236, 238.
lowei (Fabia), 391.
Loxorhynchus crispatus, 373, 374.
　grandis, 371, 372.
lubrica (Cucumaria), 243, 245.
Lucapinella callomarginata, 553, 555.
luciae (Sagartia), 101.
lucida (Siliqua), 458, 459, 460.

INDEX 647

luciferine, 80.
Lucinidae, 438.
"lug worm," 169.
Luidia, 189, 191.
   foliolata, *212*.
Luidiidae, 212.
luminescence in dinoflagellates, 31, 32, 33.
   in jellyfish, 69, 70, 79.
   in sea pansies, *93*, 94.
luminescent worms, 161.
lunule, 416.
lurida (Ostrea), *419*, 422.
   (Tritonalis), *515*, 517.
   munda (Tritonalia), *515*, 517.
luteola (Corbula), 465, *467*.
lutkeni (Ophiura), 218.
Lyell, C., 108.
Lyonsia californica, 432, *433*.
Lyonsiidae, 432.
lyratus (Hyas), *370*, 372.
lyre crab, *370*, 372.
Lysianassidae, 274.

MacBride, E. W., 228, 231, 252.
MacFarland, F. M., 490, 495, 497, 498.
macfarlandi (Chromodoris), 494, Pl. XI.
Macoma, 394.
   balthica, 451.
   indentata, *447*, 451.
   indentata tenuirostris, *450*, 451.
   inquinata, *450*, 451.
   inquinata arnheimi, 451.
   nasuta, *447*, *449*, 454.
   secta, *449*, 451, *452*.
Macron lividus, *505*, 511.
macroschisma (Pododesmus), *426*, 427.
Mactra californica, 460, *461*.
   nasuta, 461, *462*.
Mactridae, 460.
maculata (Triopha), 495, Pl. X.
Madreporaria, 96, 106.
   imperforata, 108.
madreporic plate, 182.
madreporite, 186, 240.
magdalensis (Ischnochiton), 558, 562, *563*, 564.
magellania flavescens, *153*.
magister (Cancer), 352, 376, 379, *380*.
Malacostraca, 19, 20, 252, 270.
Maldanidae, 168.
mandibles, 251.
mandtii (Lopholithodes), 340, 341.
mangilia angulata, 508, *509*.
mantis shrimps, 252, 295-297.
mantle, 415, 416, 417.
   cavity, 417, 479.
manubrium, *44*, 45, 47, 48.

marble, 28.
Margarites lirulata, *551*.
   pupilla, *548*, 551.
marginata (Cadlina), *493*, Pl. IX.
Marginella californica, 510, *512*.
   jewettii, *509*, 510.
Marginellidae, 510.
marina (Tubularia), 52, *54*.
marine station, University of Washington, *4*, 5.
marki (Euplexaura), *94*, *95*.
maxillae, 251.
maxillipeds, 251, 270.
maxima (Crisia), *136*, 137, *146*.
Mayor, A. G., 69, 71, 72, 83, 84, 85.
McClendon, J. F., 223.
McGregor, E. A., 116, 117.
McKnew, 421.
Mead, H. T., 344.
Mediaster aequalis, 208, 209, *211*.
medusa, 43, 46, *53*.
   aequorea, 71.
medusae, luminous, 68.
medusoids, 51, 52.
megalops, 252, 302, 356, *358*.
Megatebennus bimaculatus, *553*, 555.
Megathura cremulata, 164, 395, 555, *556*.
Melampus olivaceus, 504, *505*.
melanaster (Chrysaora), 80.
Melannella (Eulima) micans, *524*, 525.
Melanellidae, 525.
meleagris (Stomolophus), *83*, *84*.
Melibe leonina, *500*, 501.
Melita fresnelii, *275*.
membranacea (Membranipora), 132, 146, 147, *148*.
Membranipora membranacea, 132, 146, 147, *148*.
Membranipora tehuelcha, *145*, 147, *148*.
   villosa, 147.
Membraniporidae, 146.
mendica (Alectrion), *512*, 513.
Menipea occidentalis, *142*.
   occidentalis catalinensis, 142, *150*.
   ternata, 143.
meropsis (Tellina), *447*, 448.
mertensii (Hapalogaster), 337, *338*.
   (Ischnochiton), *560*, 564.
merus, *355*.
mesentery, 86.
mesogloea, 42, *44*, 89.
Metandrocarpa dura, 595.
Metazoa, 16, 18.
Metcalf, M. M., 261.
meteoric dust, 28.
Metis alta, *444*, 449.
Metridium dianthus, 5, *102*, *103*, 104.

648 INDEX

Mexico, 5.
micans (Melanella), *524*, 525.
(Prorocentrum), *31*, 33.
Michael, E. L., 128.
Micranellum crebricinctum, *532*.
microplankton, 31.
microptera (Bolinopsis), *111*.
migrations of chaetognaths, 128.
Miller, R. C., 473.
Milneria minima, 435, *437*.
Mimulus foliatus, *367*, 368.
miniata (Asterina), 206, *207*.
(Cucumaria), 244, *245*.
(Patiria), 206.
minima (Milneria), 435, *437*.
minutus (Planes), *397*.
mirabilis (Callistochiton) palmulatus, *560*, 565.
(Sarsia), *47*.
(Syncoryne), *47*, 48.
Mission Bay, *2*.
Mitella polymerus, *260*, 261, *267*.
miter-shell, *505*, 510, *512*.
mitra (Acmaea), 542, *543*.
Mitridae, 510.
Mitrocomidae, 68.
Mitromorpha filosa, 510, *512*.
modestus (Tellina), *447*, 448.
Modiolus capax, 429, *430*.
fornicatus, *430*, 431.
modiolus, 429, *430*, 431.
rectus, 429, *430*.
moesta (Pseudomelatoma), *505*, 508.
Mollusca, 23, 411.
Molluscoidea, 23, 130.
mollusks, 2, 29.
collecting, 11.
nudibranch, 8, *16*, Pls. VIII, IX, X, XI.
preserving, 603.
tectibranch, 255, 483.
Monactinellida, 39.
money shells, 443, *444*.
Monia, 427.
Monks, Sarah P., 210.
monotimeris (Pecten) latiauritus, *423*, 425.
Monterey Bay, 6, 116.
montereyense (Bittium) eschrichtii, 531.
montereyensis (Archidoris), 491, 492, Pl. VIII.
(Styela), 592, *593*.
montereyi (Tegula), *548*, 549.
moon-snail, Lewis, 539, *541*.
southern, 540, *541*.
Mopalia, 558.
ciliata, *563*, 565.
muscosa, *560*, *563*, 565.

Mopaliidae, 565.
mordax (Engraulis), 428.
Moroteuthis robusta, 569.
Morse, E. S., 154, 155.
Morse, M. W., 52.
moss animals, 130-152.
moulting, 249.
of crab, 356, *359*.
of lobster, 318.
mounts, specimen, 603.
mucous gland, 488.
mud flat, *2*, 6, 8.
munita (Crago), 314.
munitella (Crago), 314.
murbachii (Gonionemus), 72.
Murex, 477, 478.
carpenteri, 15.
festivus, *515*, 516.
gemma, *515*, 516.
Muricidae, 516.
murina (Diadora), *553*, 555.
Murray, Sir John, 482.
murrayana (Bugula), *146*.
muscosa (Mopalia), *560*, *563*, 565.
museum exhibits, 603.
pests, 604.
musica (Uca), 402.
mussels, 391, 413, 427-432.
bay, 428, *430*.
branch-ribbed, 429, *430*.
California sea, 427, *430*.
horse, 429-431.
recipes for preparing, 427.
Mya, 394, 539.
arenaria, 463, *464*.
truncata, 463, *464*.
Myacidae, 463.
myops (Lepidopa), *346*, *348*, 349.
Myopsida, 579.
Myriapoda, 20, 249, 250.
mysis, 302.
Mytilidae, 427.
Mytilimeria nuttallii, *433*, 434.
Mytilus, *417*.
californianus, 427, *430*.
edulis, 428, *430*.
Myxilla parasitica, 39, 40.

names, common, 13.
scientific, 12.
nasuta (Macoma), *447*, *449*, 454.
(Mactra), 461, *462*.
Natantia, 21, 298.
Naticidae, 539.
native oyster, *419*, 420, 422.
nauplius, 252, 301.
nautilus, 567.
paper, 575, *576*.

INDEX

navalis (Teredo), 472.
navanaci (Pseudomolgus), 255, 486.
Navanax inermis, 255, 485, Pl. VIII.
navarchus (Pecten) hindsii, *423*, 424.
Nemathelminthes, 23, 114, 126.
nematocysts, 45, 46, 64, 99, 106, 490.
nematophore, 63, 66.
Nemertea, 114.
nemertean, 122-126.
Nemertinea, 122.
neohexagonum (Dentalium), *475*.
nereid, 164-165.
Nereidae, 164.
Nereis, *163*, 164.
 vexillosa, *165*.
Nereocystis, 137.
neritina (Bugula), *144, 147*.
nervous system of crab, *361*.
nest-weaving of gammarids, 272.
nets, 25.
nettle cells, 45, 46, 74, 75, 76, 78, 82, 86, 88, 95.
nigricauda (Crago), 299, 311, 312.
nigromaculata (Crago), 299, 313.
nobilis (Anisodoris), 491, 492, Pl. VIII.
noctiluca (Pelagia), 79.
normani (Tanais), *284*.
Norrisia norrisii, *546*, 547.
novemdentatus (Cycloxanthops), *385*, 386.
Nucella, 519.
Nucleobranchiata, 525.
Nucula, *417*.
 (Acila) castrensis, 418, *419*.
Nuculidae, 418.
Nuda, 110, 112.
Nudibranchiata, 483, 489.
nudibranchs, 3, 8, *16*, 77, 483, 489-503, Pls. VII, VIII, IX, X, XI.
nudus (Hemigrapsus), *356*, 395, *396*.
 (Pinnotheres), 391.
 (Sipunculus), *174*, 175, *176*.
nummaria (Crepidula), *537*, 538.
nurse, 586, 587.
nuttallii (Epialtus), *366*, 367.
 (Mytilimeria), *433*, 434.
 (Phacoides), 438, *439*.
 (Purpura), *515*, 516.
 (Sanguinolaria), *456*.
 (Saxidomus), 443, *444*.
 (Schizothaerus), 461, *462*.
Nuttallina, 557.
 californica, *560*, 562.
Nutting, C. C., 57, 59, 60, 64.
nutrians (Thyonepsolus), 243, Pl. VII.
Nyctiphanes simplex, *294*, 295.

Obelia commissuralis, 59.

dichotoma, 59, *60*.
geniculata, *60*.
longissima, *53*, *61*, 502.
oblonga (Cylindroleberis), 256.
occidentalis (Amphiodia), *219*, 220.
 (Blepharipoda), *346*, 347.
 (Crisia), 137.
 (Crisulipora), *139*, 140.
 (Distaplia), 596.
 (Heterocrypta), *363*, *364*.
 (Ligyda), *292*.
 (Menipea), *142*, 143.
 (Terebratalia), 157.
ocelli, 47, 48, 54, 123, 124, 417, 422, *425*.
ochre star, 188.
ochotensis (Idothea), 290.
 (Pagurus), 333.
ochraceus (Pisaster), 188, *193*, 194.
ochre starfish, 194.
Octopoda, 575.
octopus, 567, 413, 577-579.
 Hongkong, 578.
 hunting, 568.
 two-spotted, *577*, *578*.
oculata (Aruga), *274*.
Ocypodidae, 399.
Odontosyllis phosphorea, 161.
Odostomia helga, 525, *530*.
Oedignathus inermis, 337.
Oegopsida, 581.
Oithona sp., *255*.
Oldroyd, Ida, 411, 434.
Oligochaeta, 173.
olivacea (Galvina), *53*, 502, frontispiece.
olivaceus (Melampus), 504, *505*.
olive, purple, *476*, 477, *478*, 508, *509*.
Olivella pedroana, *509*, 510.
Olividae, 508.
Ommastrephidae, 581.
Oniscoidea, 292.
Onychophora, 20.
onyx (Crepidula), *537*.
oozooid, 586-588.
opal-shell, northern, 522, *524*.
opalescens (Loligo), *579*, 580.
operculum, 61, 141, 151, 160, 170, 172, *477*, *478*.
Ophidiasteridae, 210.
Ophiocomidae, 221.
Ophioderma panamensis, *216*, 217.
Ophiodermatidae, 217.
Ophiolepididae, 217.
Ophionereis annulata, *219*, 220.
Ophiopholis aculeata, 15, 221, *222*.
Ophioplocus esmarki, *216*, 218.
Ophiopteris papillosa, 221, *222*.
Ophiothrix spiculata, *222*, 223.
Ophiura lutkeni, 218.

sarsii, 217.
Ophiurae, 215, 217.
ophiurans, 213-225.
Ophiuroidea, 181, 213.
Opisthobranchiata, 482.
opisthobranchs, 480, 482.
Opisthopus transversus, *394*.
orbella (Diplodonta), 436, *437*, *439*.
　(Paphia) staminea, 446.
Orchestia traskiana, 278.
Orchestoidea californiana, *276*, *277*.
　corniculata, *276*, *277*.
order, 19.
Oregon, 5.
　Triton, *529*, *530*.
oregonensis (Argobuccinum), *529*, *530*.
　(Cancer), 376, *382*, 383.
　(Exosphaeroma), 288.
　(Hemigrapsus), *362*, *396*.
Oregonia gracilis, *364*, 365.
organ of Bohadsch, 488.
ornata (Randallia), *362*, 363, *364*.
ornate sand star, *209*, 212.
ornatissimus (Astropecten), *209*, 212.
Orthasterias columbiana, 199.
Orthodonta, 504.
Orthopyxis caliculata, *57*.
　compressa, 58.
　everta, *58*.
oscula, 40.
Ostracoda, 20, 255.
Ostrea lurida, *419*, 422.
　titan, 420.
　virginica, 422.
Ostreidae, 420.
otocyst, 54, 417, 480.
otter shell, 461.
ova, of sponges, 37.
owl shell, *543*, *545*.
Oxyrhyncha, *362*, 363.
Oxystomata, 361, *362*, 363.
oyster, 413, 414, *415*, 420.
　eastern, 120, *415*, 422.
　industry, 329, 420, 421, 518, 539.
　reef, 519.
　rock, 426.

Pachycheles pubescens, 350.
　rudis, 349, *350*.
Pachygrapsus crassipes, *353*, *354*, *355*, *357*, *358*, *360*, *361*, *362*, 395, 397, *398*, frontispiece.
Pacific Grove, 6, 7.
pacifica (Alepas), 261.
　(Ancula), 497, Pl. XI.
　(Argonauta), 575, *576*.
　(Barnea), *467*, 468.
　(Bugula), 145, *147*.

(Euphausia), *294*.
(Phoronis), 151.
(Retepora), 149, *150*.
(Rossia), *579*.
(Tubulipora), 137, *150*.
Packard, E. L., 411, 427, 448, 451.
Paguridae, 331.
Paguristes bakeri, 333, *335*.
　turgidus, 332.
　ulreyi, 333.
Pagurus granosimanus, 334, *335*, *336*.
　hemphilli, 334, 336.
　hirsutiusculus, 334, *335*.
　ochotensis, 333.
　samuelis, 334, *335*, *336*.
paleacea (Acmaea), *543*, 545.
Palinura, 19, 21, 314, 315.
Palinuridae, 19, 315.
Palinurus interruptus, 15.
　vulgaris, 13.
pallassii (Echiurus), 178.
　(Ligyda), *292*, 293.
pallets, 470.
pallial line, *416*.
　sinus, *416*.
palma (Corymorpha), *50*, 51, *53*.
palmulatus (Callistochiton), *560*, 564.
palpator (Spirontocaris), *306*, 307.
paludicola (Spirontocaris), *306*.
panamensis (Ophioderma), *216*, 217.
Pandalidae, 302.
Pandalus danae, 302, *303*.
　gurneyi, *297*, *303*.
　stenolepis, *303*.
Pandora filosa, 432, *433*.
　punctata, 432, *433*.
Pandoridae, 432.
panicea (Halichondria), 40.
Panope generosa, 465, *467*.
Pantapoda, 402.
Panulirus interruptus, 13, 15, 19, 315.
　japanocum, 13.
Paphia staminea, 445, *447*.
　staminea laciniata, *444*, 446.
　staminea orbella, 446.
Paphia staminea petiti, 445.
　staminea ruderata, 446.
papillosa (Ophiopteris), 221, *222*.
papillosus (Phyllolithodes), 339.
papposus (Crossaster), *204*, 205.
papulae, *183*, 185.
Paracrangon echinata, 314.
Paragrubia uncinata, *277*, 279.
Paranemertes peregrina, *124*, P. VI.
Parapholas, 468.
　californica, *469*, 470.
parapodia, 159.
parasites, 126.

INDEX

parasitic copepod, 255, 486.
parasitica (Myxilla), 39, 40.
Parker, G. H., 92.
Parker, T. J., 22.
parma (Echinarachnius), 237.
Parthenopidae, 363.
parva (Pholadidea), *464*, 470.
parvimensis (Stichopus), *247*, 248.
patina (Acmaea) scutum, 542, *543*.
Patiria miniata, 206.
patula (Siliqua), 458, *459*, 460.
paucilirata (Acanthina), *521*, 522.
paucispinus (Pisaster), brevispinus, *196*.
pea-pod rock-borer, *430*, 431.
pearls, 414.
pectens, 413.
Pecten caurinus, 424.
  circularis, *423*, 425.
  hastatus, 424.
  hericius, *423*, 424.
  hindsii navarchus, *423*, 424.
  latiauritus, 424, *425*.
  subnodosus, 424.
pectinata (Lepas), 262, 263.
Pectinibranchiata, 528.
Pectinidae, 422.
pedal laceration, 103.
pedicellariae, 182, 185, *228*.
Pedicellina echinata, 134.
Pedipes unisulcatus, 503, *505*.
pedroana (Olivella), *509*, 510.
  philippiana (Terebra), 504, *505*.
peduncle, 152, 263.
Peirson, F. W., 93, 191, 423.
Pelagia, 71, 77, 80, Pl. II.
  noctiluca, 79.
pelagic mollusk, *527*.
Pelagiidae, 80.
Pelecypoda, 413.
Pelia tumida, *367*, *369*.
pellucida (Chama), *436*.
  (Zoobotryon), *144*, 151, *152*.
pelta (Acmaea) cassis, 542, *543*.
Peneides, 21, 302.
penicillata (Polyorchis), 66, *67*.
penita (Pholadidea), *467*, 470.
Pennatulacea, 89.
Pennatulidae, 89.
Pentidotea aculeata, *289*, 290.
  resecata, *289*, 290.
  wosnesenskii, 290, *291*.
pentodon (Sphaeroma), *287*, 288.
peregrina (Paranemertes), *124*, P. VI.
Peridinium, *31*.
Perigonimus, 49.
periopoda, 270.
periostracum, 414.
Peripatus, 20.

Periploma planiuscula, 432, *433*.
Periplomatidae, 432.
perisarc, 54.
Perkins, H. F., 72, 73.
Perophora annectens, 595, *596*.
perpinguis (Alectrion), *512*, 513.
persona (Acmaea), *543*, 544.
peruviana (Anomia), *426*.
pests, museum, 604.
Petasidae, 71.
Petricola carditoides, 446, *447*.
Petricolidae, 446.
Petrolisthes, 1, 6, 250.
Petrolisthes cinctipes, 351.
  eriomerus, *350*, 351.
Phacoides californicus, 438, *439*.
  nuttallii, 438, *439*.
Phanerozonia, 191, 208.
Phasianella compta, *545*, *546*.
Phasianellidae, 545.
pheasant-shell, *545*, *546*.
Phialidium, 68.
  gregarium, *56*, 57, 70.
Pholadidae, 466.
Pholadidea, 468.
  parva, *464*, 470.
  penita, *467*, 470.
Pholas, 394.
Phoronida, 130, 151.
Phoronis pacifica, 151.
phorozooid, 586-588.
phosphorea (Odontosyllis), 161.
phosphorescence, 32, 33.
phosphorescent crustaceans, 293.
  organs, 294.
Phoxichilidiidae, 406.
Phylactolaemata, 135.
Phyllodurus abdominalis, 327.
Phyllolithodes papillosus, 339.
Phyllopoda, 20.
phyllosoma larva of lobster, 317, *318*.
phylum, 19.
picta (Dirona), 499.
  (Spirontocaris), *306*, 308.
pictifrons (Lineus), 126, Pl. VI.
piddock, 391, 394, 470.
  common, *467*, 470.
  little, *464*, 470.
  rough, *467*, 468.
  western, *467*, 468.
pileworm, giant, 472.
  plumed, 472.
piling damaged by shipworms, *469*, 471.
pill bug, 282.
Pilsbry, Henry A., 262, 264.
Pilseneer, P., 474.
Pilumnus spinohirsutus, *388*, 389.
Pinnixa faba, 393.

INDEX

littoralis, *393, 398.*
longipes, 250, 392, *393.*
tubicola, 394.
Pinnotheres nudus, 391.
pugettensis, 390.
Pinnotheridae, 390
Pisaster brevispinus, 6, *196,* 197.
 brevispinus paucispinus, *196,* 197.
 giganteus, 194.
 giganteus capitatus, 194, *195.*
 ochraceus, 188, *193,* 194.
Pismo clam, 441, *442.*
Placiphorella velata, *560,* 566.
planaxis (Littorina), *534,* 535.
Planes minutus, *397.*
planiuscula (Periploma), 432, *433.*
plankton, 30, 31.
Planocera burchami, 116, Pl. IV.
 californica, 116, *117.*
Planoceridae, 116.
planula, 55.
planum (Glossophorum), 597.
plates from body wall of Leptosynapta, 242.
plates from body wall of Stichopus, 247.
Platyhelminthes, 23, 114.
Platyodon cancellatus, *464,* 465.
pleopoda, 271.
Pleurobrachia bachei, 70, *110.*
Pleurobrachiidae, 110.
Pleurophyllidia californica, *493,* 498.
Pleurophyllidiidae, 498.
Plexauriidae, 94.
plumula (Lithophaga), *430, 431.*
Plumularia setacea, *65.*
plumularian, 63-67.
Plumulariidae, 63.
Pododesmus macroschisma, *426,* 427.
Point Conception, 5, 8.
poison darts, 45.
poisoning from mussels, 428.
Polinices draconis, 540, *541.*
 lewisii, 6, 539, *541.*
 recluziana, 540, *541.*
polita (Chloridella), 296, *297.*
pollex, *355.*
Polycera atra, *496,* Pl. X.
Polychaeta, 158.
polyedra (Gonyaulax), 33.
Polykrikos, *30.*
polymerus (Mitella), *260,* 261, *267.*
polymorpha (Eudistylia), 169, Pl. VII.
Polynices, 540.
Polynoidae, *163.*
Polyorchis penicillata, 66, *67.*
polyp, coral, 107.
 nutritive, 45, 54.
 reproductive, 45, 54.

polypide, 132, 133.
Polypodidae, 577.
Polypus bimaculatus, *577, 578.*
 hongkongensis, 578.
Polyzoa, 130.
Pomona College, 408.
porcelain crab, 250, 351.
Porcellanidae, 325, 349.
Porifera, 22, 34.
Poromya, *417.*
porrecta (Lacuna), *535.*
porterae (Chromodoris), 494.
Portuguese Man-of-War, 74.
Portunidae, 384.
Portunus xantusii, 384, *385.*
postelsii (Cyanea) capillata, 82.
poulsonii (Tritonalia), *515,* 517.
Pratt, H. S., 38, 180, 231, 249.
prawns, 21.
preservation of animals, 601.
pretiosum (Dentalium), *475.*
Prionodesmacea, 418.
productum (Chelyosoma), *594.*
productus (Cancer), *359, 375,* 376, 378.
 (Epialtus), *366, 367.*
prolifera (Epiactis), *99,* 101.
propodium, *478.*
propodus, *355.*
Prorocentrum micans, *31, 33.*
Prostheceraeus, 119.
Prosthiostomodae, 119.
Prosthiostomum, 119, Pl. V.
Protozoa, 16, 22, 26-33.
protozoea, 301.
Przibram, 20.
Psammobia californica, *456.*
Psammobiidae, 456.
Pseudoceridae, 118.
Pseudochama exogyra, *436.*
Pseudomelatoma moesta, *505,* 508.
Pseudomolgus navanaci, 255, 486.
pseudopodia, *32.*
Pseudopythina rugifera, 327.
Pseudosquilla biglowi, 296, *297*
 lessonii, 296.
Psolus chitonoides, 243, *245.*
pteropod ooze, 482.
Pteropoda, 480.
pteropods, 26, 31, 475, 479, 480, *481.*
pterotracheid mollusk, *526.*
Pterotracheidae, 526.
Ptilosarcus quadrangularis, 89, *90.*
pubescens (Pachycheles), 350.
Puget Sound, 1, 4, 5, 10.
pugeti (Crisia), 136.
pugettensis (Pinnotheres), 390.
 (Upogebia), 327, *328, 330.*
Pugettia dalli, *368,* 369.

INDEX 653

gracilis, *368*.
richii, *368*, 369, *370*.
pulchella (Sertularia), 62, *63*.
pulchra (Halosydna), 163.
  (Rostanga), 492, Pl. VIII.
  (Semele), *450*, 454.
  (Tubulipora), *137*, 138.
pulligo (Tegula), *548*, 549.
Pulmonata, 503.
pulmonates, 480.
punctata (Pandora), 432, *433*.
punctatus (Tylos), *292*.
punctocoelatus (Actaeon), *483*.
pupilla (Margarites), *548*, 551.
purple, channeled, *521*.
  short-spined, *521*.
  wrinkled, 520.
Purpura foliata, 516, *518*.
  nuttallii, *515*, 516.
purpuratus (Strongylocentrotus), *233*, 234.
pusillus (Dolichoglossus), *584*.
Pycnogonida, 402.
Pycnogonidae, 409.
pycnogonids, 20, 402-410.
Pycnogonum stearnsi, *409*.
Pycnopodia helianthoides, 192, *201*, *202*, 540.
Pyramidellidae, 525.
Pyrosoma giganteum, *598*.

quadragenarium (Cardium), *439*, 440.
quadrangularis (Ptilosarcus), 89, *90*.
quadrifilatum (Bittium), *530*, 531.

rachis, 89, 92.
radial canal, *44*, 46, 187.
  symmetry, 17, 42, 227.
radiata (Lichenopora), *139*, 140, *150*.
radii, *264*.
radiolarian, 27, 28, 31.
radiolarian-ooze, 28.
radula, 474, 478, 487, *489*, 518.
rameum (Eudendrium), 50, *67*.
Randallia ornata, *362*, 363, *364*.
Ranella californica, 529.
Ranellidae, 529.
Rankin, E. P., 428.
ratfish, 121.
Rathbun, M. J., 330, 347, 361, 392, 393.
raymondi (Lepidochitona), *560*, 561.
razor clam, 457, 470.
recluziana (Polinices), 540, *541*.
rectilinea (Idothea), *289*.
rectus (Modiolus), *429*, *430*.
red clay, 28.
"red water," 31, 32, 33.
reef-forming corals, 107.

refulgida (Aphrodita), 162.
regeneration, 8, 43, 51.
  in brittle stars, 215, 221.
  in Corymorpha, 51.
  in Leptosynapta, 241, 242.
  in sea cucumbers, 241.
  in shrimps, 301.
  in starfish, 190, *191*, 210.
  in Tubularia, 52.
regularis (Ischnochiton), *560*, 564.
Renilla, 8.
  amethystina, *92*, *93*, Pl. IV.
Renillidae, 92.
Reptantia, 19, 21, 298, 314.
resecata (Pentidotea), *289*, 290.
Retepora pacifica, 149, *150*.
Reteporidae, 149.
reticulata (Gadinia), 504, *507*.
rhinophore, *16*, 479, 503.
Rhizostomata, 77, 83.
  food of, 78.
rhodoceras (Acanthodoris), 496, Pl. XI.
ribbon worms, 122.
rice-mollusk, California, *509*, 510.
Richardson, C. H., 343.
richii (Pugettia), *368*, 369, *370*.
ring canal, *44*, 46, 78, 187.
Ritter, W. E. 201, 582, 584, 585, 587, 588, 589, 590, 591, 592, 593, 595, 596, 597, 598.
ritteri (Anthomastus), 89, *98*.
Robertson, A., 147, 148.
robin, 13.
robusta (Moroteuthis), 569.
rock-dweller, 446, *447*.
"rock-eating" mussel, *430*, 431.
rock Venus, 446, *447*.
rockweed, 147.
Rogers, J. E., 411, 474, 484, 486, 518, 566.
rosacea (Hopkinsia), 497, P. XI.
rosaceus (Solen), *457*, 458, *459*.
rose-star, *204*, 205.
Rossia pacifica, *579*.
Rostanga pulchra, 492, Pl. VIII.
rostrata (Colidotea), *291*.
rostrum, 21, *264*.
Rotifera, 127.
rotifers, 113.
round worms, 113, 126.
rozieri (Thalamoporella), *146*, 147, 148, *149*.
rubescens (Tetraclita) squamosa, *268*, 269.
rubra (Thyone), 246.
rubrocincta (Clymenella), *168*.
rubropicta (Semele), *452*, 453.
ruderata (Paphia) staminea, 446.

INDEX

rudis (Pachycheles), 349, *350*.
rufescens (Haliotis), 552, *553*.
rugifera (Pseudopythina), 327.
rupicola (Semele), *450*, 453.

Sabellaria californica, 172, *173*.
    cementarium, 173.
Sabellidae, 169.
Saemaeostomata, 80.
Sagartia sp., 101, P. III.
    davisii, 102.
    luciae, 101.
Sagartidae, 98, 101.
Sagitta bipunctata, *129*.
Salpa, 3, 8, 31, 585.
    fusiformis-runcinata, 588, *589*.
    tilesii-costata, *590*.
samuelis (Pagurus), 334, *335*, *336*.
San Francisco Bay, shrimp industry in, 299.
San Juan Island, 4, 5.
sand collars of Polinices, 540, *541*, 542.
    crab, 3, 21, 250, 325, 326, *338*, 341-347.
    dollar, 232, 235, *236*, 237, 602.
    fleas, 272-279.
    hopper, 249, 271, 272-279.
sandiegensis (Diaulula), 492, *493*.
sanguinea (Aldisa), 492, Pl. VIII.
Sanguinolaria nuttallii, *456*.
sarcostyle, 64.
Sargassum, 148, 397.
Sarsia mirabilis, *47*.
sarsii (Ophiura), 217.
Sarsiidae, 47.
Saxicava arctica, *464*, 466.
Saxicavidae, 465.
saxicola (Entodesma), *433*, 434.
Saxidomus giganteus, 443.
    nuttallii, 443, *444*.
scabra (Acmaea), 542, *543*, 544.
scallop, 39, 417, 422.
    broad-eared, 424, *425*.
    circular, *423*, 425.
    purple-hinged, *423*, 425.
    rock, *423*, 425.
    speckled, *423*, 425.
    weather-vane, 424.
Scalpellidae, 261.
scaphopod, 412.
Scaphopoda, 413, 473.
scaura (Caprella), *281*.
Schizopoda, 20, 293.
Schizothaerus, 394.
    nuttallii, 6, 461, *462*.
Schmitt, W. L., 336.
Scleroplax granulata, 394.
Scofield, N. B., 299.

Scofield, W. L., 573, 580.
Scripps Institution, 8, 9, 128, 253, 399, 469.
Scrupocellaria, 133.
    californica, 144.
    diegensis, *143*, *146*.
    varians, 144.
Scutellidae, 235.
scutulata (Littorina), *534*, 535.
scutum, 258, 262, *263*.
scutum patina (Acmaea), 542, *543*.
Scyphomedusae, 77, 80.
Scyphozoa, 43, 77.
Scyra acutifrons, 374, *375*.
sea anemone, 2, 3, 5, 8, 11, *17*, 29, 42, 43, 85, 86, 95, 96, *99*, Pls. III, IV.
    anemone, burrowing, 2, 97, *98*, 105, *106*, P. III.
    anemone, method of taking food, 86, 87.
    anemone, striped, 101, Pl. III.
    blubber, 77, 81.
    bottle shell, 434.
    cockle, *443*.
    cucumber, 33, 238-248, *393*, Pl. VII.
    cucumber, dissection of, *239*.
    cucumber, red, 5, 244, *245*.
    eggs, 227.
    fans, 88.
    hare, 476, 483, 486-489.
    lilies, 226.
    mouse, 162.
    pansies, keeping alive in aquaria, 601.
    pansy, 8, 88, *92*, *93*, Pl. IV.
    pen, 85, 88, 89-92.
    porcupines, 227.
    serpent, 569.
    slug, 475, 482-503, frontispiece, Pls. VII, VIII, IX, X, XI.
    spiders, 402-410.
    squirt, 582, 590-599, frontispiece.
    star, 2, 180-213.
    urchin, 2, 8, 11, 180, 227-234.
    urchin, green, 5, *233*, 234.
    urchin, keeping alive in aquaria, 11, 602.
    urchin, purple, *233*, 234.
Searlesia dira, *512*, 513.
secta (Macoma), *449*, 451, *452*.
segmented body, *16*, 17.
    worms, 158-179.
Sellius, G., 472, 473.
Semele decisa, *450*, 453.
    pulchra, *450*, 454.
    rubropicta, *452*, 453.
    rupicola, *450*, 453.
Semelidae, 453.

INDEX   655

Sepiolidae, 579.
Septifer bifurcatus, 429, *430*.
serpent star, 213-224.
    daisy, 222.
Serpula columbiana, 170, Pl. VII.
serpulid worms, 5.
Serpulidae, 170.
Sertularella turgida, *62*, *67*.
Sertularia pulchella, 62, *63*.
Sertulariidae, 61.
sertulifera (Astrometis), 183, 199, *200*.
setacea (Bankia), 472.
    (Plumularia), *64*, *65*.
setae, 151, 159.
shark's teeth, 28.
sheep crab, *371*, 372.
shell building of gastropods, 476.
shell of clam, parts of, 416.
Shipley, A. E., 252.
shipworms, 285, 470-473.
shrimps, 3, 21, 249, 250, 252.
    black-tailed, *311*, *312*.
    burrowing, 251, 327, 329.
    industry, 298-300.
    snapping, 308, *309*, 311.
sicarius (Solen), 458, *459*.
sieve plate, 182, 186.
Sigerfoos, C. P., 473.
Siliqua lucida, 458, *459*, 460.
    patula, 458, *459*, 460.
silver lantern-shell, 432, *433*.
simple ascidians, 591-594.
simplex (Nyctiphanes), *294*, *295*.
    (Terebra), *504*.
siphon, 414, 417, 427.
    of gastropod, *478*.
Siphonalia kellettii, 511.
siphonoglyph, 86, 87, 95.
Siphonophora, 74.
Siphonophorae, 43, 74.
siphonozooids, 88, 89, 91, 92.
siphons of clam, *412*.
sipunculid, 173-177, *176*.
Sipunculus nudus, *174*, 175, 176.
sitchana (Littorina), *534*.
sitchensis (Cryptolithodes), *338*, 339.
Sitka crab, *338*, 339.
slipper, excavated, *537*, 538.
    hooked, *537*, 538.
    onyx, *537*.
    prickly, *537*, 538.
    white, *537*, 538.
Smith, G., 301.
snail, 6, 412, 413, 475, 482, 503.
snake's head lamp shell, *156*.
snapping shrimp, 308, *309*, *311*.
soft-shelled crabs, 357.
solandri (Trivia), *524*, 528.

Solaster, 205.
    dawsoni, *198*, 205.
    stimpsoni, *204*, 205.
Solasteridae, 205.
Solen rosaceus, *457*, 458, *459*.
    sicarius, 458, *459*.
Solenidae, 458.
somites, 249.
sounding instruments, 25.
southern kelp crab, *366*, 367.
sow bug, 282.
spadicea (Cypraea), *524*, 528.
spat, 420, 421.
Spatangidae, 238.
Spatangoida, 238.
species, 15, 19.
specimen mounts, 603.
spectrum (Acmaea), 544.
Speocarcinus californiensis, *389*.
Sphaeroma destructor, 287.
    pentodon, *287*, 288.
spiculata (Ophiothrix), *222*, 223.
spicules, *34*, *35*, 38, 88, 89.
spider, 20, 249, 250.
    crabs, 363, 365.
spindle-shell, Kobelt, *505*, 511.
spinohirsutus (Pilumnus, *388*, 389.
spinosum (Crucibulum), 537, 538.
Spinulosa, 191, 203.
spiny cockle, *439*, 440.
    lobster, *16*, 315-325.
    sand crab, *346*, 347.
Spirabranchus, *171*.
spirata (Acanthina), *521*, 522.
spire, of shell, *476*, *477*.
Spiroglyphus lituellus, *533*.
Spirontocaris brevirostris, *306*, 307.
    carinata, 305, *306*.
    cristata, *306*, 308.
    palpator, *306*, 307.
    paludicola, *306*.
    picta, *306*, 308.
    taylori, *306*, 307, *311*.
Spirorbis, 160, *171*, *172*.
sponge, 8, 29, 34, 42, 325.
    bread-crumb, 40.
    fishing, 37, 38.
sponges, California, *41*.
    glass, 39.
    simple, 39.
squamigerus (Aletes), *532*, 533.
squid, 413, 567.
    fishing at Monterey, 572, 580.
    fishing in Philippines, 569.
    large, 569, 581.
    opalescent, *579*, 580.
    short, *579*.
Squilla, 296, *297*.

squilla, polished, 296, 297.
Squillidae, 296.
staminea (Paphia), 445, 447.
Stanford University, 6, 7.
starfish, 6, 8, 11, 29, 180-213.
    keeping alive, 602.
    red, 202.
    soft, 183, 199, 200.
    twenty-rayed, 8, 201, 202.
statocyst, 54.
statolith, 54, 78.
stearnsi (Pycnogonum), 409.
Stebbing, T. R. R., 273.
stelleri (Cryptochiton), 566, 567.
stenolepis (Pandalus), 303.
Stenopides, 21, 302.
Stephens, Kate, 411.
Stichopus, 246, 394.
    californicus, 247.
    johnsoni, 246, 247.
    parvimensis, 247.
stigmata, 596.
stimpsoni (Solaster), 204, 205.
    (Styela), 593.
    (Truncatella), 535, 536.
Stimpson's sun-star, 204, 205.
sting cells, 42, 45, 103, 104, 490.
stolon, 586.
Stomatopoda, 20, 295.
Stomolophus meleagris, 83, 84.
Stomotoca atra, 49.
stone canal, 187, 229, 230.
Stout, V. R., 278.
Streptodonta, 522.
Strigatella idae, 505, 510.
striped mollusk, 485.
Strong, L. H., 57.
Strongylocentrotidae, 234.
Strongylocentrotus drobachiensis, 5, 233, 234.
    franciscanus, 233, 234.
    purpuratus, 233, 234.
struthionides (Aglaophenia), 66, 67.
stultorum (Tivela), 416, 441, 442.
Styela gibbsii, 593.
    montereyensis, 592, 593.
    stimpsoni, 6, 593.
Stylatula elongata, 91.
    gracile, 92.
Stylatulidae, 91.
stylet, 122, 124, 125.
subaperta (Tegula) funebralis, 548, 549.
subclass, 19.
subdiaphana (Cooperella), 447, 448.
Suberites latus, 40.
Suberitidae, 40.
subfamilies, 19.
subgenera, 19.

subnodosus (Pecten), 424.
suborder, 19.
subquadrata (Cardita), 433, 435.
    (Fabia), 391.
subspecies, 19.
substriatum (Cardium), 439, 441.
subteres (Tagelus), 457, 459.
succincta (Chione), 416, 444, 445.
"suckers" of octopus and squid, 570.
Suctoria, 32.
summer school, 5.
sun star, Dawson's, 198, 205.
sunflower star, twenty-rayed, 201, 202.
sunset shell, 456.
superfamilies, 19.
sutural laminae, 557, 558.
suture, 477.
swarming of Odontosyllis, 161.
Swezy, O., 33.
swimmerets, 271.
swimming habit of Planes, 397, 399.
Syllidae, 160, 161.
Synapta, 2, 241.
Synaptidae, 241.
Syncoryne mirabilis, 47, 48.
Syncorynidae, 47.
Synidotea harfordi, 290, 291.
    laticauda, 291.
synonym, 15.
Systema Naturae, 14.

tables from body wall of Stichopus, 247.
tabulatus (Chrysodomus), 509, 511.
Tagelus californianus, 456, 457.
    subteres, 457, 459.
Talitridae, 275.
talpoida (Emerita), 347.
Tanaioidea, 284.
Tanais normani, 284.
Tanystylum intermedium, 405, 406.
tapeworm, 121, 122.
taphria (Leda), 418, 419.
taylori (Spirontocaris), 306, 307, 311.
    (Xanthias), 388, 389.
tectibranch mollusks, 255, 483-489.
Tectibranchiata, 483.
tegmentum, 557, 558.
tegula (Alectrion), 512, 513.
Tegula aureotincta, 548, 549.
    brunnea, 548, 549.
    funebralis, 544, 547, 548.
    funebralis subaperta, 548, 549.
    gallina, 548, 549.
    gallina tinctum, 548, 549.
    ligulata, 548, 550.
    montereyi, 548, 549.
    pulligo, 548, 549.
tehuelcha (Membranipora), 145, 147, 148.

INDEX

Tellina bodegensis, *444*, 449.
  buttoni, *447*, 448.
  carpenteri, *447*, 448, *449*.
  meropsis, *447*, 448.
  modestus, *447*, 448.
Tellinidae, 448.
Telmessus cheiragonus, 383, *384*.
Telotremata, 156.
temperature of ocean depths, 28.
  required by Aurelia, 83.
  required by Tubularia, 52.
Tentaculata, 110.
tenuicula (Turbonilla), 525, *530*.
tenuirostris (Macoma) indentata, *450*, 451.
terebellid, 167.
Terebellidae, *167*.
Terebra pedroana philippiana, 504, *505*.
  simplex, 504.
Terebratalia occidentalis, 157.
  transversa, *155*, *156*, 157.
  transversa caurina, 157.
Terebratellidae, 157.
Terebratulidae, 156.
Terebratulina unguicula, *156*.
Terebridae, 504.
Teredidae, 470.
teredo, 413, 470.
  boring mechanism of, 471.
Teredo diegensis, 473.
  navalis, 472.
tergum, *258*, *263*, *266*.
ternata (Menipea), 143.
test, 227.
Tethymelibidae, 501.
Tethys californica, *486*, 487, *488*, *489*.
Tetraclita squamosa rubescens, *268*, 269.
Thais, 5, *476*, 519.
  canaliculata, *521*.
  emarginata, *521*.
  lamellosa, 520, *521*.
  lima, 520, *521*.
Thalamoporella rozieri, *146*, 147, *149*.
Thalamoporellidae, 147.
Thaliacea, 586.
thatched barnacle, *268*, 269.
Thaumantias cellularia, 68.
Thaumantiidae, 66.
Thecosomata, 482.
Thompson, M. T., 332.
Thompson, W. F., 453, 554.
thorax, 251.
thread worms, 126.
Thyone rubra, 246.
Thyonepsolus nutrians, 243, Pl. VII.
Thysanozoon, 118, Pl. V.
tide, low or minus, 11.
  tables, 11.

tilesii-costata (Salpa), *590*.
tinctum (Tegula) gallina, *548*, 549.
Tintinnidae, 29.
titan (Ostrea), 420.
Tivela stultorum, 416, 441, *442*.
toad crab, *370*, 372.
Tomales Bay, 6.
Tonicella, 557, 561.
tonsa (Acartia), *254*, 255.
tooth papillae, *213*, *214*.
  shells, 413, 473-475.
top-shell, blue, *548*, 550.
  channeled, *548*, 550.
  red, *546*.
  ringed, *548*, 550.
  three-colored, *548*, 550.
  wavy, *546*.
Torrey, H. B., 65, 100, 112.
tower-shell, Cooper's, *530*, 533.
Tracheata, 20.
Trachomedusae, 71.
Trachydermon, 561.
Trachynemidae, 73.
transversa (Terebratalia), *155*, *156*, 157.
transversus (Opisthopus), *394*.
traskiana (Orchestia), 278.
Trematoda, 114, 119.
trematodes, ectoparasitic, 119.
  endoparasitic, 120.
trepang, 241.
triangulatus (Trophon), *518*, 519.
tribe, 19, 21.
trichina, 127.
Trichotropidae, 531.
Trichotropis cancellata, *530*, 531.
tricolor (Calliostoma), *548*, 550.
Triopha carpenteria, 495, Pl. X.
  maculata, 495, Pl. X.
Triton, Oregon, *529*, 530.
Tritonalia circumtexta, *515*, 517.
  gracillima, *515*, 517.
  interfossa, *515*, 517.
  lurida, *515*, 517.
  lurida munda, *515*, 517.
  poulsonii, *515*, 517.
Tritonia festiva, 491, Pl. VII.
Tritoniidae, 491.
tritonis (Doliolum), 586, *587*, *588*.
Trivia californiana, *524*, 528.
  solandri, *524*, 528.
Triviidae, 528.
Trochelminthes, 23, 114, 127.
Trochidae, 547.
trochosphere, 160.
Trophon triangulatus, *518*, 519.
trophozooid, 586, *587*.
troschelii (Evasterias), *195*, 197.

truncata (Mya), 463, *464*.
Truncatella stimpsoni, *535*, 536.
Truncatellidae, 536.
tube feet, 180, 181, *183*, 186, 230, 240.
tube-mollusk, California, *532*.
  close-ringed, *532*.
tuberculatus (Inachoides), *362*, *364*, 365.
tubicola (Pinnixa), 394.
Tubularia crocea, 52, 53.
  marina, 52, *54*.
tubularian, 43, 49.
Tubularidae, 52.
Tubulipora, 140.
  flabellaris, 138.
  pacifica, *137*, *150*.
  pulchra, *137*, 138.
Tubuliporidae, 137.
tumens (Hipponix), *535*, 536.
tumida (Pelia), *367*, *369*.
Tunicata, 583, 584.
tunicata (Katharina), *560*, 566.
tunicates, 365, 390, 584.
turban, banded, *548*, 550.
  berry, *546*, 547.
  black, 547, *548*.
  brown, *548*, 549.
  dusky, *548*, 549.
  gilded, *548*, 549.
  Monterey, *548*, 549.
  red, *546*, 547.
  smooth, *546*, 547.
  speckled, *548*, 549.
Turbellaria, 114.
Turbinidae, 546.
Turbonilla tenuicula, 525, *530*.
turgida (Sertularella), *62*, *67*.
turgidus (Paguristes), 332.
turret, Carpenter's, *505*, 506.
Turritella cooperi, *530*, 533.
Turritellidae, 533.
Turritidae, 506.
turtles, 397.
twenty-rayed starfish, *201*, *202*, 540, 603.
Tylos punctatus, *292*.
typicus (Cryptolithodes), *339*.
Tyrian dye, 520.

Uca crenulata, *398*, 399-402.
  musica, 402.
ulreyi (Paguristes), 333.
umbilicus, *477*.
umbo, *263*, 416.
uncinata (Paragrubia), *277*, 279.
undatella (Chione), 412, *444*, 445.
undosa (Astraea), *546*.
unguicula (Terebratulina), *156*.
unicorn, angled, *521*, 522.
  square-spotted, *521*, 522.
unifasciata (Lacuna), *535*, 536.
unisulcatus (Pedipes), 503, *505*.
univalve, 412.
University of California, 8, 9.
Upogebia pugettensis, 327, *328*, *330*.
uropods, 273.
Urosalpinx cinereus, 518, *521*.
urotoma (Idothea), 289.

Valvifera, 289.
varians (Scrupocellaria), 144.
varieties, 19.
varix, *477*.
Vaughn, T. W., 108.
velata (Placiphorella), *560*, 566.
Velella, 3, 74, *523*.
  lata, 75, 76, Pl. I.
Velellidae, 75.
veliger larva, *480*.
velum, 43, *44*, 46, 77.
Venericardia ventricosa, *433*, 435.
Veneridae, 441.
Venerupis lamellifera, 446, *447*.
ventricosa (Venericardia), *433*, 435.
Venus' flower basket, *39*.
Vermes, 23.
Vermetidae, 533.
vernacular name, 13.
Verrill, A. E., 186, 192, 206, 237, 569.
versicolor (Amphissa), 514, *515*.
Vertebrata, *583*.
vertens (Gonionemus), *67*, 71, *72*.
vertical migrations of copepods, 253.
  of Sagitta, 129.
vesicula (Haminoea), 484, *485*.
Vesiculariidae, 151.
vexillosa, (Nereis), *165*.
vibracula, 134, *143*.
villosa (Membranipora), 147.
violet snail, 523, *524*.
virescens (Haminoea), 484, *485*.
virginica (Ostrea), 422.
viridintestinalis (Halosoma), 406, *407*.
vitality of Glottidia, 154.
vitellina (Erato), *524*, 528.
volcanic ash, 28.
volcano (Fissurella), *553*, 554.
volcano-shell, *553*, 554.
von Geldern, C. E., Pl. IV.
von Lendenfield, R., 64.

wallalensis (Haliotis), 554.
Washington clams, 443, *444*.
  University of, 4, 5.
water fleas, 249.
water-vascular system, 180, 228, 229.
Way, Evelyn, 339, 340, 341.

CATALOGUE OF DOVER BOOKS

THE MUSIC OF THE SPHERES: THE MATERIAL UNIVERSE — FROM ATOM TO QUASAR, SIMPLY EXPLAINED, *Guy Murchie*
Vast compendium of fact, modern concept and theory, observed and calculated data, historical background guides intelligent layman through the material universe. Brilliant exposition of earth's construction, explanations for moon's craters, atmospheric components of Venus and Mars (with data from recent fly-by's), sun spots, sequences of star birth and death, neighboring galaxies, contributions of Galileo, Tycho Brahe, Kepler, etc.; and (Vol. 2) construction of the atom (describing newly discovered sigma and xi subatomic particles), theories of sound, color and light, space and time, including relativity theory, quantum theory, wave theory, probability theory, work of Newton, Maxwell, Faraday, Einstein, de Broglie, etc. "Best presentation yet offered to the intelligent general reader," *Saturday Review*. Revised (1967). Index. 319 illustrations by the author. Total of xx + 644pp. 5⅜ x 8½.
T1809, T1810   Two volume set, paperbound $4.00

FOUR LECTURES ON RELATIVITY AND SPACE, *Charles Proteus Steinmetz*
Lecture series, given by great mathematician and electrical engineer, generally considered one of the best popular-level expositions of special and general relativity theories and related questions. Steinmetz translates complex mathematical reasoning into language accessible to laymen through analogy, example and comparison. Among topics covered are relativity of motion, location, time; of mass; acceleration; 4-dimensional time-space; geometry of the gravitational field; curvature and bending of space; non-Euclidean geometry. Index. 40 illustrations. x + 142pp. 5⅜ x 8½.   S1771   Paperbound $1.35

HOW TO KNOW THE WILD FLOWERS, *Mrs. William Starr Dana*
Classic nature book that has introduced thousands to wonders of American wild flowers. Color-season principle of organization is easy to use, even by those with no botanical training, and the genial, refreshing discussions of history, folklore, uses of over 1,000 native and escape flowers, foliage plants are informative as well as fun to read. Over 170 full-page plates, collected from several editions, may be colored in to make permanent records of finds. Revised to conform with 1950 edition of Gray's Manual of Botany. xlii + 438pp. 5⅜ x 8½.   T332   Paperbound $2.25

MANUAL OF THE TREES OF NORTH AMERICA, *Charles Sprague Sargent*
Still unsurpassed as most comprehensive, reliable study of North American tree characteristics, precise locations and distribution. By dean of American dendrologists. Every tree native to U.S., Canada, Alaska; 185 genera, 717 species, described in detail—leaves, flowers, fruit, winterbuds, bark, wood, growth habits, etc. plus discussion of varieties and local variants, immaturity variations. Over 100 keys, including unusual 11-page analytical key to genera, aid in identification. 783 clear illustrations of flowers, fruit, leaves. An unmatched permanent reference work for all nature lovers. Second enlarged (1926) edition. Synopsis of families. Analytical key to genera. Glossary of technical terms. Index. 783 illustrations, 1 map. Total of 982pp. 5⅜ x 8.
T277, T278   Two volume set, paperbound $6.00

CATALOGUE OF DOVER BOOKS

IT'S FUN TO MAKE THINGS FROM SCRAP MATERIALS,
Evelyn Glantz Hershoff
What use are empty spools, tin cans, bottle tops? What can be made from rubber bands, clothes pins, paper clips, and buttons? This book provides simply worded instructions and large diagrams showing you how to make cookie cutters, toy trucks, paper turkeys, Halloween masks, telephone sets, aprons, linoleum block- and spatter prints — in all 399 projects! Many are easy enough for young children to figure out for themselves; some challenging enough to entertain adults; all are remarkably ingenious ways to make things from materials that cost pennies or less! Formerly "Scrap Fun for Everyone." Index. 214 illustrations. 373pp. 5⅜ x 8½.   T1251   Paperbound $1.75

SYMBOLIC LOGIC and THE GAME OF LOGIC, *Lewis Carroll*
"Symbolic Logic" is not concerned with modern symbolic logic, but is instead a collection of over 380 problems posed with charm and imagination, using the syllogism and a fascinating diagrammatic method of drawing conclusions. In "The Game of Logic" Carroll's whimsical imagination devises a logical game played with 2 diagrams and counters (included) to manipulate hundreds of tricky syllogisms. The final section, "Hit or Miss" is a lagniappe of 101 additional puzzles in the delightful Carroll manner. Until this reprint edition, both of these books were rarities costing up to $15 each. Symbolic Logic: Index. xxxi + 199pp. The Game of Logic: 96pp. 2 vols. bound as one. 5⅜ x 8.
     T492   Paperbound $2.00

MATHEMATICAL PUZZLES OF SAM LOYD, PART I
selected and edited by M. Gardner
Choice puzzles by the greatest American puzzle creator and innovator. Selected from his famous collection, "Cyclopedia of Puzzles," they retain the unique style and historical flavor of the originals. There are posers based on arithmetic, algebra, probability, game theory, route tracing, topology, counter and sliding block, operations research, geometrical dissection. Includes the famous "14-15" puzzle which was a national craze, and his "Horse of a Different Color" which sold millions of copies. 117 of his most ingenious puzzles in all. 120 line drawings and diagrams. Solutions. Selected references. xx + 167pp. 5⅜ x 8.
     T498   Paperbound $1.25

STRING FIGURES AND HOW TO MAKE THEM, *Caroline Furness Jayne*
107 string figures plus variations selected from the best primitive and modern examples developed by Navajo, Apache, pygmies of Africa, Eskimo, in Europe, Australia, China, etc. The most readily understandable, easy-to-follow book in English on perennially popular recreation. Crystal-clear exposition; step-by-step diagrams. Everyone from kindergarten children to adults looking for unusual diversion will be endlessly amused. Index. Bibliography. Introduction by A. C. Haddon. 17 full-page plates, 960 illustrations. xxiii + 401pp. 5⅜ x 8½.
     T152   Paperbound $2.25

PAPER FOLDING FOR BEGINNERS, *W. D. Murray and F. J. Rigney*
A delightful introduction to the varied and entertaining Japanese art of origami (paper folding), with a full, crystal-clear text that anticipates every difficulty; over 275 clearly labeled diagrams of all important stages in creation. You get results at each stage, since complex figures are logically developed from simpler ones. 43 different pieces are explained: sailboats, frogs, roosters, etc. 6 photographic plates. 279 diagrams. 95pp. 5⅝ x 8⅜.
     T713   Paperbound $1.00

CATALOGUE OF DOVER BOOKS

PRINCIPLES OF ART HISTORY,
H. Wölfflin
Analyzing such terms as "baroque," "classic," "neoclassic," "primitive," "picturesque," and 164 different works by artists like Botticelli, van Cleve, Dürer, Hobbema, Holbein, Hals, Rembrandt, Titian, Brueghel, Vermeer, and many others, the author establishes the classifications of art history and style on a firm, concrete basis. This classic of art criticism shows what really occurred between the 14th-century primitives and the sophistication of the 18th century in terms of basic attitudes and philosophies. "A remarkable lesson in the art of seeing," *Sat. Rev. of Literature*. Translated from the 7th German edition. 150 illustrations. 254pp. 6⅛ x 9¼. T276 Paperbound $2.00

PRIMITIVE ART,
Franz Boas
This authoritative and exhaustive work by a great American anthropologist covers the entire gamut of primitive art. Pottery, leatherwork, metal work, stone work, wood, basketry, are treated in detail. Theories of primitive art, historical depth in art history, technical virtuosity, unconscious levels of patterning, symbolism, styles, literature, music, dance, etc. A must book for the interested layman, the anthropologist, artist, handicrafter (hundreds of unusual motifs), and the historian. Over 900 illustrations (50 ceramic vessels, 12 totem poles, etc.). 376pp. 5⅜ x 8.   T25   Paperbound $2.50

THE GENTLEMAN AND CABINET MAKER'S DIRECTOR,
Thomas Chippendale
A reprint of the 1762 catalogue of furniture designs that went on to influence generations of English and Colonial and Early Republic American furniture makers. The 200 plates, most of them full-page sized, show Chippendale's designs for French (Louis XV), Gothic, and Chinese-manner chairs, sofas, canopy and dome beds, cornices, chamber organs, cabinets, shaving tables, commodes, picture frames, frets, candle stands, chimney pieces, decorations, etc. The drawings are all elegant and highly detailed; many include construction diagrams and elevations. A supplement of 24 photographs shows surviving pieces of original and Chippendale-style pieces of furniture. Brief biography of Chippendale by N. I. Bienenstock, editor of *Furniture World*. Reproduced from the 1762 edition. 200 plates, plus 19 photographic plates. vi + 249pp. 9⅛ x 12¼.   T1601   Paperbound $3.50

AMERICAN ANTIQUE FURNITURE: A BOOK FOR AMATEURS,
Edgar G. Miller, Jr.
Standard introduction and practical guide to identification of valuable American antique furniture. 2115 illustrations, mostly photographs taken by the author in 148 private homes, are arranged in chronological order in extensive chapters on chairs, sofas, chests, desks, bedsteads, mirrors, tables, clocks, and other articles. Focus is on furniture accessible to the collector, including simpler pieces and a larger than usual coverage of Empire style. Introductory chapters identify structural elements, characteristics of various styles, how to avoid fakes, etc. "We are frequently asked to name some book on American furniture that will meet the requirements of the novice collector, the beginning dealer, and . . . the general public. . . . We believe Mr. Miller's two volumes more completely satisfy this specification than any other work," *Antiques*. Appendix. Index. Total of vi + 1106pp. 7⅞ x 10¾.
T1599, T1600   Two volume set, paperbound $7.50

CATALOGUE OF DOVER BOOKS

THE BAD CHILD'S BOOK OF BEASTS, MORE BEASTS FOR WORSE CHILDREN, and A MORAL ALPHABET, *H. Belloc*
Hardly and anthology of humorous verse has appeared in the last 50 years without at least a couple of these famous nonsense verses. But one must see the entire volumes — with all the delightful original illustrations by Sir Basil Blackwood — to appreciate fully Belloc's charming and witty verses that play so subacidly on the platitudes of life and morals that beset his day — and ours. A great humor classic. Three books in one. Total of 157pp. 5⅜ x 8.
T749   Paperbound $1.00

THE DEVIL'S DICTIONARY, *Ambrose Bierce*
Sardonic and irreverent barbs puncturing the pomposities and absurdities of American politics, business, religion, literature, and arts, by the country's greatest satirist in the classic tradition. Epigrammatic as Shaw, piercing as Swift, American as Mark Twain, Will Rogers, and Fred Allen, Bierce will always remain the favorite of a small coterie of enthusiasts, and of writers and speakers whom he supplies with "some of the most gorgeous witticisms of the English language" (H. L. Mencken). Over 1000 entries in alphabetical order. 144pp. 5⅜ x 8.   T487   Paperbound $1.00

THE COMPLETE NONSENSE OF EDWARD LEAR.
This is the only complete edition of this master of gentle madness available at a popular price. *A Book of Nonsense, Nonsense Songs, More Nonsense Songs and Stories* in their entirety with all the old favorites that have delighted children and adults for years. The Dong With A Luminous Nose, The Jumblies, The Owl and the Pussycat, and hundreds of other bits of wonderful nonsense: 214 limericks, 3 sets of Nonsense Botany, 5 Nonsense Alphabets, 546 drawings by Lear himself, and much more. 320pp. 5⅜ x 8.   T167   Paperbound $1.75

THE WIT AND HUMOR OF OSCAR WILDE, *ed. by Alvin Redman*
Wilde at his most brilliant, in 1000 epigrams exposing weaknesses and hypocrisies of "civilized" society. Divided into 49 categories—sin, wealth, women, America, etc.—to aid writers, speakers. Includes excerpts from his trials, books, plays, criticism. Formerly "The Epigrams of Oscar Wilde." Introduction by Vyvyan Holland, Wilde's only living son. Introductory essay by editor. 260pp. 5⅜ x 8.   T602   Paperbound $1.50

A CHILD'S PRIMER OF NATURAL HISTORY, *Oliver Herford*
Scarcely an anthology of whimsy and humor has appeared in the last 50 years without a contribution from Oliver Herford. Yet the works from which these examples are drawn have been almost impossible to obtain! Here at last are Herford's improbable definitions of a menagerie of familiar and weird animals, each verse illustrated by the author's own drawings. 24 drawings in 2 colors; 24 additional drawings. vii + 95pp. 6½ x 6.   T1647   Paperbound $1.00

THE BROWNIES: THEIR BOOK, *Palmer Cox*
The book that made the Brownies a household word. Generations of readers have enjoyed the antics, predicaments and adventures of these jovial sprites, who emerge from the forest at night to play or to come to the aid of a deserving human. Delightful illustrations by the author decorate nearly every page. 24 short verse tales with 266 illustrations. 155pp. 6⅝ x 9¼.
T1265   Paperbound $1.50

CATALOGUE OF DOVER BOOKS

THE PRINCIPLES OF PSYCHOLOGY,
William James
The full long-course, unabridged, of one of the great classics of Western literature and science. Wonderfully lucid descriptions of human mental activity, the stream of thought, consciousness, time perception, memory, imagination, emotions, reason, abnormal phenomena, and similar topics. Original contributions are integrated with the work of such men as Berkeley, Binet, Mills, Darwin, Hume, Kant, Royce, Schopenhauer, Spinoza, Locke, Descartes, Galton, Wundt, Lotze, Herbart, Fechner, and scores of others. All contrasting interpretations of mental phenomena are examined in detail—introspective analysis, philosophical interpretation, and experimental research. "A classic," *Journal of Consulting Psychology.* "The main lines are as valid as ever," *Psychoanalytical Quarterly.* "Standard reading...a classic of interpretation," *Psychiatric Quarterly.* 94 illustrations. 1408pp. 5⅜ x 8.
T381, T382 Two volume set, paperbound $6.00

VISUAL ILLUSIONS: THEIR CAUSES, CHARACTERISTICS AND APPLICATIONS,
M. Luckiesh
"Seeing is deceiving," asserts the author of this introduction to virtually every type of optical illusion known. The text both describes and explains the principles involved in color illusions, figure-ground, distance illusions, etc. 100 photographs, drawings and diagrams prove how easy it is to fool the sense: circles that aren't round, parallel lines that seem to bend, stationary figures that seem to move as you stare at them — illustration after illustration strains our credulity at what we see. Fascinating book from many points of view, from applications for artists, in camouflage, etc. to the psychology of vision. New introduction by William Ittleson, Dept. of Psychology, Queens College. Index. Bibliography. xxi + 252pp. 5⅜ x 8½. T1530 Paperbound $1.50

FADS AND FALLACIES IN THE NAME OF SCIENCE,
Martin Gardner
This is the standard account of various cults, quack systems, and delusions which have masqueraded as science: hollow earth fanatics. Reich and orgone sex energy, dianetics, Atlantis, multiple moons, Forteanism, flying saucers, medical fallacies like iridiagnosis, zone therapy, etc. A new chapter has been added on Bridey Murphy, psionics, and other recent manifestations in this field. This is a fair, reasoned appraisal of eccentric theory which provides excellent inoculation against cleverly masked nonsense. "Should be read by everyone, scientist and non-scientist alike," R. T. Birge, Prof. Emeritus of Physics, Univ. of California; Former President, American Physical Society. Index. x + 365pp. 5⅜ x 8. T394 Paperbound $2.00

ILLUSIONS AND DELUSIONS OF THE SUPERNATURAL AND THE OCCULT,
D. H. Rawcliffe
Holds up to rational examination hundreds of persistent delusions including crystal gazing, automatic writing, table turning, mediumistic trances, mental healing, stigmata, lycanthropy, live burial, the Indian Rope Trick, spiritualism, dowsing, telepathy, clairvoyance, ghosts, ESP, etc. The author explains and exposes the mental and physical deceptions involved, making this not only an exposé of supernatural phenomena, but a valuable exposition of characteristic types of abnormal psychology. Originally titled "The Psychology of the Occult." 14 illustrations. Index. 551pp. 5⅜ x 8. T503 Paperbound $2.75

FAIRY TALE COLLECTIONS, *edited by Andrew Lang*
Andrew Lang's fairy tale collections make up the richest shelf-full of traditional children's stories anywhere available. Lang supervised the translation of stories from all over the world—familiar European tales collected by Grimm, animal stories from Negro Africa, myths of primitive Australia, stories from Russia, Hungary, Iceland, Japan, and many other countries. Lang's selection of translations are unusually high; many authorities consider that the most familiar tales find their best versions in these volumes. All collections are richly decorated and illustrated by H. J. Ford and other artists.

THE BLUE FAIRY BOOK. 37 stories. 138 illustrations. ix + 390pp. 5⅜ x 8½.
T1437   Paperbound $1.95

THE GREEN FAIRY BOOK. 42 stories. 100 illustrations. xiii + 366pp. 5⅜ x 8½.
T1439   Paperbound $1.75

THE BROWN FAIRY BOOK. 32 stories. 50 illustrations, 8 in color. xii + 350pp. 5⅜ x 8½.
T1438   Paperbound $1.95

THE BEST TALES OF HOFFMANN, *edited by E. F. Bleiler*
10 stories by E. T. A. Hoffmann, one of the greatest of all writers of fantasy. The tales include "The Golden Flower Pot," "Automata," "A New Year's Eve Adventure," "Nutcracker and the King of Mice," "Sand-Man," and others. Vigorous characterizations of highly eccentric personalities, remarkable imaginative situations, and intensely fast pacing has made these tales popular all over the world for 150 years. Editor's introduction. 7 drawings by Hoffmann. xxxiii + 419pp. 5⅜ x 8½.   T1793   Paperbound $2.25

GHOST AND HORROR STORIES OF AMBROSE BIERCE,
edited by E. F. Bleiler
Morbid, eerie, horrifying tales of possessed poets, shabby aristocrats, revived corpses, and haunted malefactors. Widely acknowledged as the best of their kind between Poe and the moderns, reflecting their author's inner torment and bitter view of life. Includes "Damned Thing," "The Middle Toe of the Right Foot," "The Eyes of the Panther," "Visions of the Night," "Moxon's Master," and over a dozen others. Editor's introduction. xxii + 199pp. 5⅜ x 8½.   T767   Paperbound $1.50

THREE GOTHIC NOVELS, *edited by E. F. Bleiler*
Originators of the still popular Gothic novel form, influential in ushering in early 19th-century Romanticism. Horace Walpole's *Castle of Otranto*, William Beckford's *Vathek*, John Polidori's *The Vampyre*, and a *Fragment* by Lord Byron are enjoyable as exciting reading or as documents in the history of English literature. Editor's introduction. xi + 291pp. 5⅜ x 8½.
T1232   Paperbound $2.00

BEST GHOST STORIES OF LEFANU, *edited by E. F. Bleiler*
Though admired by such critics as V. S. Pritchett, Charles Dickens and Henry James, ghost stories by the Irish novelist Joseph Sheridan LeFanu have never become as widely known as his detective fiction. About half of the 16 stories in this collection have never before been available in America. Collection includes "Carmilla" (perhaps the best vampire story ever written), "The Haunted Baronet," "The Fortunes of Sir Robert Ardagh," and the classic "Green Tea." Editor's introduction. 7 contemporary illustrations. Portrait of LeFanu. xii + 467pp. 5⅜ x 8.   T415   Paperbound $2.50

CATALOGUE OF DOVER BOOKS

EASY-TO-DO ENTERTAINMENTS AND DIVERSIONS WITH COINS, CARDS, STRING, PAPER AND MATCHES, *R. M. Abraham*
Over 300 tricks, games and puzzles will provide young readers with absorbing fun. Sections on card games; paper-folding; tricks with coins, matches and pieces of string; games for the agile; toy-making from common household objects; mathematical recreations; and 50 miscellaneous pastimes. Anyone in charge of groups of youngsters, including hard-pressed parents, and in need of suggestions on how to keep children sensibly amused and quietly content will find this book indispensable. Clear, simple text, copious number of delightful line drawings and illustrative diagrams. Originally titled "Winter Nights' Entertainments." Introduction by Lord Baden Powell. 329 illustrations. v + 186pp. 5⅜ x 8½.   T921   Paperbound $1.00

AN INTRODUCTION TO CHESS MOVES AND TACTICS SIMPLY EXPLAINED, *Leonard Barden*
Beginner's introduction to the royal game. Names, possible moves of the pieces, definitions of essential terms, how games are won, etc. explained in 30-odd pages. With this background you'll be able to sit right down and play. Balance of book teaches strategy — openings, middle game, typical endgame play, and suggestions for improving your game. A sample game is fully analyzed. True middle-level introduction, teaching you all the essentials without oversimplifying or losing you in a maze of detail. 58 figures. 102pp. 5⅜ x 8½.   T1210   Paperbound $1.25

LASKER'S MANUAL OF CHESS, *Dr. Emanuel Lasker*
Probably the greatest chess player of modern times, Dr. Emanuel Lasker held the world championship 28 years, independent of passing schools or fashions. This unmatched study of the game, chiefly for intermediate to skilled players, analyzes basic methods, combinations, position play, the aesthetics of chess, dozens of different openings, etc., with constant reference to great modern games. Contains a brilliant exposition of Steinitz's important theories. Introduction by Fred Reinfeld. Tables of Lasker's tournament record. 3 indices. 308 diagrams. 1 photograph. xxx + 349pp. 5⅜ x 8.   T640   Paperbound $2.50

COMBINATIONS: THE HEART OF CHESS, *Irving Chernev*
Step-by-step from simple combinations to complex, this book, by a well-known chess writer, shows you the intricacies of pins, counter-pins, knight forks, and smothered mates. Other chapters show alternate lines of play to those taken in actual championship games; boomerang combinations; classic examples of brilliant combination play by Nimzovich, Rubinstein, Tarrasch, Botvinnik, Alekhine and Capablanca. Index. 356 diagrams. ix + 245pp. 5⅜ x 8½.   T1744   Paperbound $2.00

HOW TO SOLVE CHESS PROBLEMS, *K. S. Howard*
Full of practical suggestions for the fan or the beginner — who knows only the moves of the chessmen. Contains preliminary section and 58 two-move, 46 three-move, and 8 four-move problems composed by 27 outstanding American problem creators in the last 30 years. Explanation of all terms and exhaustive index. "Just what is wanted for the student," Brian Harley. 112 problems, solutions. vi + 171pp. 5⅜ x 8.   T748   Paperbound $1.35

CATALOGUE OF DOVER BOOKS

SOCIAL THOUGHT FROM LORE TO SCIENCE,
H. E. Barnes and H. Becker
An immense survey of sociological thought and ways of viewing, studying, planning, and reforming society from earliest times to the present. Includes thought on society of preliterate peoples, ancient non-Western cultures, and every great movement in Europe, America, and modern Japan. Analyzes hundreds of great thinkers: Plato, Augustine, Bodin, Vico, Montesquieu, Herder, Comte, Marx, etc. Weighs the contributions of utopians, sophists, fascists and communists; economists, jurists, philosophers, ecclesiastics, and every 19th and 20th century school of scientific sociology, anthropology, and social psychology throughout the world. Combines topical, chronological, and regional approaches, treating the evolution of social thought as a process rather than as a series of mere topics. "Impressive accuracy, competence, and discrimination . . . easily the best single survey," Nation. Thoroughly revised, with new material up to 1960. 2 indexes. Over 2200 bibliographical notes. Three volume set. Total of 1586pp. 5⅜ x 8.
T901, T902, T903   Three volume set, paperbound $9.00

A HISTORY OF HISTORICAL WRITING, Harry Elmer Barnes
Virtually the only adequate survey of the whole course of historical writing in a single volume. Surveys developments from the beginnings of historiography in the ancient Near East and the Classical World, up through the Cold War. Covers major historians in detail, shows interrelationship with cultural background, makes clear individual contributions, evaluates and estimates importance; also enormously rich upon minor authors and thinkers who are usually passed over. Packed with scholarship and learning, clear, easily written. Indispensable to every student of history. Revised and enlarged up to 1961. Index and bibliography. xv + 442pp. 5⅜ x 8½.
T104   Paperbound $2.50

JOHANN SEBASTIAN BACH, Philipp Spitta
The complete and unabridged text of the definitive study of Bach. Written some 70 years ago, it is still unsurpassed for its coverage of nearly all aspects of Bach's life and work. There could hardly be a finer non-technical introduction to Bach's music than the detailed, lucid analyses which Spitta provides for hundreds of individual pieces. 26 solid pages are devoted to the B minor mass, for example, and 30 pages to the glorious St. Matthew Passion. This monumental set also includes a major analysis of the music of the 18th century: Buxtehude, Pachelbel, etc. "Unchallenged as the last word on one of the supreme geniuses of music," John Barkham, Saturday Review Syndicate. Total of 1819pp. Heavy cloth binding. 5⅜ x 8.
T252   Two volume set, clothbound $15.00

BEETHOVEN AND HIS NINE SYMPHONIES, George Grove
In this modern middle-level classic of musicology Grove not only analyzes all nine of Beethoven's symphonies very thoroughly in terms of their musical structure, but also discusses the circumstances under which they were written, Beethoven's stylistic development, and much other background material. This is an extremely rich book, yet very easily followed; it is highly recommended to anyone seriously interested in music. Over 250 musical passages. Index.
viii + 407pp. 5⅜ x 8.   T334   Paperbound $2.25

CATALOGUE OF DOVER BOOKS

THREE SCIENCE FICTION NOVELS,
John Taine
Acknowledged by many as the best SF writer of the 1920's, Taine (under the name Eric Temple Bell) was also a Professor of Mathematics of considerable renown. Reprinted here are *The Time Stream*, generally considered Taine's best, *The Greatest Game*, a biological-fiction novel, and *The Purple Sapphire*, involving a supercivilization of the past. Taine's stories tie fantastic narratives to frameworks of original and logical scientific concepts. Speculation is often profound on such questions as the nature of time, concept of entropy, cyclical universes, etc. 4 contemporary illustrations. v + 532pp. 5⅜ x 8⅜.
T1180 Paperbound $2.00

SEVEN SCIENCE FICTION NOVELS,
H. G. Wells
Full unabridged texts of 7 science-fiction novels of the master. Ranging from biology, physics, chemistry, astronomy, to sociology and other studies, Mr. Wells extrapolates whole worlds of strange and intriguing character. "One will have to go far to match this for entertainment, excitement, and sheer pleasure . . ."*New York Times.* Contents: The Time Machine, The Island of Dr. Moreau, The First Men in the Moon, The Invisible Man, The War of the Worlds, The Food of the Gods, In The Days of the Comet. 1015pp. 5⅜ x 8.
T264 Clothbound $5.00

28 SCIENCE FICTION STORIES OF H. G. WELLS.
Two full, unabridged novels, *Men Like Gods* and *Star Begotten*, plus 26 short stories by the master science-fiction writer of all time! Stories of space, time, invention, exploration, futuristic adventure. Partial contents: *The Country of the Blind, In the Abyss, The Crystal Egg, The Man Who Could Work Miracles, A Story of Days to Come, The Empire of the Ants, The Magic Shop, The Valley of the Spiders, A Story of the Stone Age, Under the Knife, Sea Raiders,* etc. An indispensable collection for the library of anyone interested in science fiction adventure. 928pp. 5⅜ x 8. T265 Clothbound $5.00

THREE MARTIAN NOVELS,
Edgar Rice Burroughs
Complete, unabridged reprinting, in one volume, of Thuvia, Maid of Mars; Chessmen of Mars; The Master Mind of Mars. Hours of science-fiction adventure by a modern master storyteller. Reset in large clear type for easy reading. 16 illustrations by J. Allen St. John. vi + 490pp. 5⅜ x 8½.
T39 Paperbound $2.50

AN INTELLECTUAL AND CULTURAL HISTORY OF THE WESTERN WORLD,
Harry Elmer Barnes
Monumental 3-volume survey of intellectual development of Europe from primitive cultures to the present day. Every significant product of human intellect traced through history: art, literature, mathematics, physical sciences, medicine, music, technology, social sciences, religions, jurisprudence, education, etc. Presentation is lucid and specific, analyzing in detail specific discoveries, theories, literary works, and so on. Revised (1965) by recognized scholars in specialized fields under the direction of Prof. Barnes. Revised bibliography. Indexes. 24 illustrations. Total of xxix + 1318pp.
T1275, T1276, T1277 Three volume set, paperbound $7.50

CATALOGUE OF DOVER BOOKS

HEAR ME TALKIN' TO YA, *edited by Nat Shapiro and Nat Hentoff*
In their own words, Louis Armstrong, King Oliver, Fletcher Henderson, Bunk Johnson, Bix Beiderbecke, Billy Holiday, Fats Waller, Jelly Roll Morton, Duke Ellington, and many others comment on the origins of jazz in New Orleans and its growth in Chicago's South Side, Kansas City's jam sessions, Depression Harlem, and the modernism of the West Coast schools. Taken from taped conversations, letters, magazine articles, other first-hand sources. Editors' introduction. xvi + 429pp. 5⅜ x 8½. T1726 Paperbound $2.00

THE JOURNAL OF HENRY D. THOREAU
A 25-year record by the great American observer and critic, as complete a record of a great man's inner life as is anywhere available. Thoreau's Journals served him as raw material for his formal pieces, as a place where he could develop his ideas, as an outlet for his interests in wild life and plants, in writing as an art, in classics of literature, Walt Whitman and other contemporaries, in politics, slavery, individual's relation to the State, etc. The Journals present a portrait of a remarkable man, and are an observant social history. Unabridged republication of 1906 edition, Bradford Torrey and Francis H. Allen, editors. Illustrations. Total of 1888pp. 8⅜ x 12¼.
T312, T313 Two volume set, clothbound $25.00

A SHAKESPEARIAN GRAMMAR, *E. A. Abbott*
Basic reference to Shakespeare and his contemporaries, explaining through thousands of quotations from Shakespeare, Jonson, Beaumont and Fletcher, North's *Plutarch* and other sources the grammatical usage differing from the modern. First published in 1870 and written by a scholar who spent much of his life isolating principles of Elizabethan language, the book is unlikely ever to be superseded. Indexes. xxiv + 511pp. 5⅜ x 8½. T1582 Paperbound $2.75

FOLK-LORE OF SHAKESPEARE, *T. F. Thistelton Dyer*
Classic study, drawing from Shakespeare a large body of references to supernatural beliefs, terminology of falconry and hunting, games and sports, good luck charms, marriage customs, folk medicines, superstitions about plants, animals, birds, argot of the underworld, sexual slang of London, proverbs, drinking customs, weather lore, and much else. From full compilation comes a mirror of the 17th-century popular mind. Index. ix + 526pp. 5⅜ x 8½.
T1614 Paperbound $2.75

THE NEW VARIORUM SHAKESPEARE, *edited by H. H. Furness*
By far the richest editions of the plays ever produced in any country or language. Each volume contains complete text (usually First Folio) of the play, all variants in Quarto and other Folio texts, editorial changes by every major editor to Furness's own time (1900), footnotes to obscure references or language, extensive quotes from literature of Shakespearian criticism, essays on plot sources (often reprinting sources in full), and much more.

HAMLET, *edited by H. H. Furness*
Total of xxvi + 905pp. 5⅜ x 8½.
T1004, T1005 Two volume set, paperbound $5.25

TWELFTH NIGHT, *edited by H. H. Furness*
Index. xxii + 434pp. 5⅜ x 8½. T1189 Paperbound $2.75

CATALOGUE OF DOVER BOOKS

LA BOHEME BY GIACOMO PUCCINI,
translated and introduced by Ellen H. Bleiler
Complete handbook for the operagoer, with everything needed for full enjoyment except the musical score itself. Complete Italian libretto, with new, modern English line-by-line translation—the only libretto printing all repeats; biography of Puccini; the librettists; background to the opera, Murger's La Boheme, etc.; circumstances of composition and performances; plot summary; and pictorial section of 73 illustrations showing Puccini, famous singers and performances, etc. Large clear type for easy reading. 124pp. 5⅜ x 8½.
T404 Paperbound $1.25

ANTONIO STRADIVARI: HIS LIFE AND WORK (1644-1737),
W. Henry Hill, Arthur F. Hill, and Alfred E. Hill
Still the only book that really delves into life and art of the incomparable Italian craftsman, maker of the finest musical instruments in the world today. The authors, expert violin-makers themselves, discuss Stradivari's ancestry, his construction and finishing techniques, distinguished characteristics of many of his instruments and their locations. Included, too, is story of introduction of his instruments into France, England, first revelation of their supreme merit, and information on his labels, number of instruments made, prices, mystery of ingredients of his varnish, tone of pre-1684 Stradivari violin and changes between 1684 and 1690. An extremely interesting, informative account for all music lovers, from craftsman to concert-goer. Republication of original (1902) edition. New introduction by Sydney Beck, Head of Rare Book and Manuscript Collections, Music Division, New York Public Library. Analytical index by Rembert Wurlitzer. Appendixes. 68 illustrations. 30 full-page plates. 4 in color. xxvi + 315pp. 5⅜ x 8½.
T425 Paperbound $2.25

MUSICAL AUTOGRAPHS FROM MONTEVERDI TO HINDEMITH,
Emanuel Winternitz
For beauty, for intrinsic interest, for perspective on the composer's personality, for subtleties of phrasing, shading, emphasis indicated in the autograph but suppressed in the printed score, the mss. of musical composition are fascinating documents which repay close study in many different ways. This 2-volume work reprints facsimiles of mss. by virtually every major composer, and many minor figures—196 examples in all. A full text points out what can be learned from mss., analyzes each sample. Index. Bibliography. 18 figures. 196 plates. Total of 170pp. of text. 7⅞ x 10¾.
T1312, T1313 Two volume set, paperbound $5.00

J. S. BACH,
Albert Schweitzer
One of the few great full-length studies of Bach's life and work, and the study upon which Schweitzer's renown as a musicologist rests. On first appearance (1911), revolutionized Bach performance. The only writer on Bach to be musicologist, performing musician, and student of history, theology and philosophy, Schweitzer contributes particularly full sections on history of German Protestant church music, theories on motivic pictorial representations in vocal music, and practical suggestions for performance. Translated by Ernest Newman. Indexes. 5 illustrations. 650 musical examples. Total of xix + 928pp. 5⅜ x 8½.
T1631, T1632 Two volume set, paperbound $4.50

CATALOGUE OF DOVER BOOKS

THE METHODS OF ETHICS, *Henry Sidgwick*
Propounding no organized system of its own, study subjects every major methodological approach to ethics to rigorous, objective analysis. Study discusses and relates ethical thought of Plato, Aristotle, Bentham, Clarke, Butler, Hobbes, Hume, Mill, Spencer, Kant, and dozens of others. Sidgwick retains conclusions from each system which follow from ethical premises, rejecting the faulty. Considered by many in the field to be among the most important treatises on ethical philosophy. Appendix. Index. xlvii + 528pp. 5⅜ x 8½.
T1608 Paperbound $2.50

TEUTONIC MYTHOLOGY, *Jakob Grimm*
A milestone in Western culture; the work which established on a modern basis the study of history of religions and comparative religions. 4-volume work assembles and interprets everything available on religious and folkloristic beliefs of Germanic people (including Scandinavians, Anglo-Saxons, etc.). Assembling material from such sources as Tacitus, surviving Old Norse and Icelandic texts, archeological remains, folktales, surviving superstitions, comparative traditions, linguistic analysis, etc. Grimm explores pagan deities, heroes, folklore of nature, religious practices, and every other area of pagan German belief. To this day, the unrivaled, definitive, exhaustive study. Translated by J. S. Stallybrass from 4th (1883) German edition. Indexes. Total of lxxvii + 1887pp. 5⅜ x 8½.
T1602, T1603, T1604, T1605 Four volume set, paperbound $11.00

THE I CHING, *translated by James Legge*
Called "The Book of Changes" in English, this is one of the Five Classics edited by Confucius, basic and central to Chinese thought. Explains perhaps the most complex system of divination known, founded on the theory that all things happening at any one time have characteristic features which can be isolated and related. Significant in Oriental studies, in history of religions and philosophy, and also to Jungian psychoanalysis and other areas of modern European thought. Index. Appendixes. 6 plates. xxi + 448pp. 5⅜ x 8½.
T1062 Paperbound $2.75

HISTORY OF ANCIENT PHILOSOPHY, *W. Windelband*
One of the clearest, most accurate comprehensive surveys of Greek and Roman philosophy. Discusses ancient philosophy in general, intellectual life in Greece in the 7th and 6th centuries B.C., Thales, Anaximander, Anaximenes, Heraclitus, the Eleatics, Empedocles, Anaxagoras, Leucippus, the Pythagoreans, the Sophists, Socrates, Democritus (20 pages), Plato (50 pages), Aristotle (70 pages), the Peripatetics, Stoics, Epicureans, Sceptics, Neo-platonists, Christian Apologists, etc. 2nd German edition translated by H. E. Cushman. xv + 393pp. 5⅜ x 8.
T357 Paperbound $2.25

THE PALACE OF PLEASURE, *William Painter*
Elizabethan versions of Italian and French novels from *The Decameron*, Cinthio, Straparola, Queen Margaret of Navarre, and other continental sources — the very work that provided Shakespeare and dozens of his contemporaries with many of their plots and sub-plots and, therefore, justly considered one of the most influential books in all English literature. It is also a book that any reader will still enjoy. Total of cviii + 1,224pp.
T1691, T1692, T1693 Three volume set, paperbound $6.75

CATALOGUE OF DOVER BOOKS

THE WONDERFUL WIZARD OF OZ, *L. F. Baum*
All the original W. W. Denslow illustrations in full color—as much a part of "The Wizard" as Tenniel's drawings are of "Alice in Wonderland." "The Wizard" is still America's best-loved fairy tale, in which, as the author expresses it, "The wonderment and joy are retained and the heartaches and nightmares left out." Now today's young readers can enjoy every word and wonderful picture of the original book. New introduction by Martin Gardner. A Baum bibliography. 23 full-page color plates. viii + 268pp. 5⅜ x 8.
T691 Paperbound $1.75

THE MARVELOUS LAND OF OZ, *L. F. Baum*
This is the equally enchanting sequel to the "Wizard," continuing the adventures of the Scarecrow and the Tin Woodman. The hero this time is a little boy named Tip, and all the delightful Oz magic is still present. This is the Oz book with the Animated Saw-Horse, the Woggle-Bug, and Jack Pumpkinhead. All the original John R. Neill illustrations, 10 in full color. 287pp. 5⅜ x 8.
T692 Paperbound $1.75

ALICE'S ADVENTURES UNDER GROUND, *Lewis Carroll*
The original *Alice in Wonderland*, hand-lettered and illustrated by Carroll himself, and originally presented as a Christmas gift to a child-friend. Adults as well as children will enjoy this charming volume, reproduced faithfully in this Dover edition. While the story is essentially the same, there are slight changes, and Carroll's spritely drawings present an intriguing alternative to the famous Tenniel illustrations. One of the most popular books in Dover's catalogue. Introduction by Martin Gardner. 38 illustrations. 128pp. 5⅜ x 8½.
T1482 Paperbound $1.00

THE NURSERY "ALICE," *Lewis Carroll*
While most of us consider *Alice in Wonderland* a story for children of all ages, Carroll himself felt it was beyond younger children. He therefore provided this simplified version, illustrated with the famous Tenniel drawings enlarged and colored in delicate tints, for children aged "from Nought to Five." Dover's edition of this now rare classic is a faithful copy of the 1889 printing, including 20 illustrations by Tenniel, and front and back covers reproduced in full color. Introduction by Martin Gardner. xxiii + 67pp. 6⅛ x 9¼.
T1610 Paperbound $1.75

THE STORY OF KING ARTHUR AND HIS KNIGHTS, *Howard Pyle*
A fast-paced, exciting retelling of the best known Arthurian legends for young readers by one of America's best story tellers and illustrators. The sword Excalibur, wooing of Guinevere, Merlin and his downfall, adventures of Sir Pellias and Gawaine, and others. The pen and ink illustrations are vividly imagined and wonderfully drawn. 41 illustrations. xviii + 313pp. 6⅛ x 9¼.
T1445 Paperbound $1.75

Prices subject to change without notice.

Available at your book dealer or write for free catalogue to Dept. Adsci, Dover Publications, Inc., 180 Varick St., N.Y., N.Y. 10014. Dover publishes more than 150 books each year on science, elementary and advanced mathematics, biology, music, art, literary history, social sciences and other areas.